Universitext

*Editorial Board
(North America):*

S. Axler
K.A. Ribet

For other titles in this series, go to
www.springer.com/series/223

Corrado De Concini • Claudio Procesi

Topics in Hyperplane Arrangements, Polytopes and Box-Splines

 Springer

Corrado De Concini
Università di Roma
"La Sapienza"
Dipartimento di Matematica
Piazzale Aldo Moro 5
00185 Roma
Italy
deconcin@mat.uniroma1.it

Claudio Procesi
Università di Roma
"La Sapienza"
Dipartimento di Matematica
Piazzale Aldo Moro 5
00185 Roma
Italy
procesi@mat.uniroma1.it

ISBN 978-0-387-78962-0   e-ISBN 978-0-387-78963-7
DOI 10.1007/978-0-387-78963-7
Springer New York Dordrecht Heidelberg London

Library of Congress Control Number: 2010934230

Contents

Part II The Differentiable Case

Preface

The main purpose of this book is to bring together some areas of research that have developed independently over the last 30 years. The central problem we are going to discuss is that of the computation of the number of integral points in suitable families of variable polytopes. This problem is formulated in terms of the study of partition functions. The partition function $\mathcal{T}_X(b)$, associated to a finite set of integral vectors X, counts the number of ways in which a variable vector b can be written as a linear combination of the elements in X with positive integer coefficients. Since we want this number to be finite, we assume that the vectors X generate a *pointed cone* $C(X)$.

Special cases were studied in ancient times, and one can look at the book of Dickson [50] for historical information on this topic.

The problem goes back to Euler in the special case in which X is a list of positive integers, and in this form it was classically treated by several authors, such as Cayley, Sylvester [107] (who calls the partition function *the quotity*), Bell [15] and Ehrhart [53], [54].

Having in mind only the principal goal of studying the partition functions, we treat several topics but not in a systematic way, by trying to show and compare a variety of different approaches. In particular, we want to revisit a sequence of papers of Dahmen and Micchelli, which for our purposes, culminate in the proof of a slightly weaker form of Theorem 13.54, showing the quasipolynomial nature of partition functions [37] on suitable regions of space.

The full statement of Theorem 13.54 follows from further work of Szenes and Vergne [110].

This theory was approached in a completely different way a few years later by various authors, unaware of the work of Dahmen and Micchelli. We present an approach taken from a joint paper with M. Vergne [44] in Section 13.4.5, which has proved to be useful for further applications to the index theory.

In order to describe the regions where the partition function is a quasipolynomial, one needs to introduce a basic geometric and combinatorial object: the *zonotope $B(X)$* generated by X. This is a compact polytope defined in

2.12. The theory then consists in dividing $C(X)$ into regions Ω, called *big cells*, such that in each region $\Omega - B(X)$, the partition function is a quasipolynomial (see Definition 5.31).

The quasipolynomials appearing in the description of the partition function satisfy a natural system of difference equations of a class that in the multidimensional case we call Eulerian (generalizing the classical one-dimensional definition); see Theorem 5.32. All these results can be viewed as generalizations of the theory of the Ehrhart polynomial [53], [54].

The approach of Dahmen and Micchelli to partition functions is inspired by their study of two special classes of functions: the *multivariate spline* $T_X(x)$, supported on $C(X)$, and the *box spline* $B_X(x)$, supported on $B(X)$, originally introduced by de Boor and deVore [39]; see Section 7.1.1 for their definition.

These functions, associated to the given set of vectors X, play an important role in approximation theory. One of the goals of the theory is to give computable closed formulas for all these functions and at the same time to describe some of their qualitative behavior and applications.

These three functions can be described in a combinatorial way as a finite sum over *local pieces* (see formulas (9.5) and (14.28)). In the case of $B_X(x)$ and $T_X(x)$ the local pieces span, together with their derivatives, a finite-dimensional space $D(X)$ of polynomials. In the case of $\mathcal{T}_X(b)$ they span, together with their translates, a finite-dimensional space $DM(X)$ of quasipolynomials.

A key fact is the description of:

- $D(X)$ as solutions of a system of differential equations by formula (11.1).
- $DM(X)$ as solutions of a system of difference equations by formula (13.3).
- A strict relationship between $D(X)$ and $DM(X)$ in Section 16.1.

In particular, Dahmen and Micchelli compute the dimensions of both spaces, see Theorem 11.8 and 13.21. This dimension has a simple combinatorial interpretation in terms of X. They also decompose $DM(X)$ as a direct sum of natural spaces associated to certain special points $P(X)$ in the torus whose character group is the lattice spanned by X. In this way, $DM(X)$ can be identified with a space of distributions supported at these points. Then $D(X)$ is the subspace of $DM(X)$ of the elements supported at the identity. The papers of Dahmen and Micchelli are a development of the theory of splines, initiated by I.J. Schoenberg [95]. There is a rather large literature on these topics by several authors, such as A.A. Akopyan; A. Ben-Artzi, C.K. Chui, C. De Boor, H. Diamond, N. Dyn, K. Höllig, Rong Qing Jia, A. Ron, and A.A. Saakyan. The interested reader can find a lot of useful historical information about these matters and further references in the book [40] (and also the notes of Ron [93]).

The results about the spaces $D(X)$ and $DM(X)$, which, as we have mentioned, originate in the theory of splines, turn out to have some interest also in the theory of hyperplane arrangements and in commutative algebra in connection with the study of certain Reisner–Stanley algebras [43]. Furthermore, the space $DM(X)$ has an interpretation in the theory of the index of transversally elliptic operators (see [44]).

The fact that a relationship between this theory and hyperplane arrangements should exist is pretty clear once we consider the set of vectors X as a set of linear equations that define an arrangement of hyperplanes in the space dual to that in which X lies. In this respect we have been greatly inspired by the results of Orlik–Solomon on cohomology [84],[83] and those of Brion, Szenes, Vergne on partition functions [109], [110], [27], [22], [28], [29], [26], [108].

In fact, a lot of work in this direction originated from the seminal paper of Khovanskiĭ and Pukhlikov [90] interpreting the counting formulas for partition functions as Riemann–Roch formulas for toric varieties, and of Jeffrey–Kirwan [68] and Witten [120], on moment maps. These topics are beyond the scope of this book, which tries to remain at a fairly elementary level. For these matters the reader may refer to Vergne's survey article [116].

Due to the somewhat large distance between the two fields, people working in hyperplane arrangements do not seem to be fully aware of the results on the box spline.

On the other hand, there are some methods that have been developed for the study of arrangements that we believe shed some light on the space of functions used to build the box spline. Therefore, we feel that this presentation may be useful in making a bridge between the two theories.

For completeness and also to satisfy our personal curiosity, we have also added a short discussion on the applications of box splines to approximation theory and in particular the aspect of the *finite element method,* which comes from the *Strang–Fix conditions* (see [106]). In Section 18.1 we present a new approach to the construction of *quasi-interpolants* using in a systematic way the concept of *superfunction.*

Here is a rough description of the method we shall follow to compute the functions that are the object of study of this book.

- We interpret all the functions as *tempered distributions supported in the pointed cone $C(X)$.*
- We apply the Laplace transform and change the problem to one in algebra, essentially a problem of developing certain special rational functions into partial fractions.
- We solve the algebraic problems by module theory under the algebra of differential operators or of difference operators.
- We interpret the results by inverting the Laplace transform directly.

The book consists of five parts.

In the first part we collect basic material on convex sets, combinatorics and polytopes, the Laplace and Fourier transforms, and the language of modules over the Weyl algebra. We then recall some simple foundational facts on suitable systems of partial differential equations with constant coefficients. We develop a general approach to linear difference equations and give a method of reduction to the differentiable case (Section 5.3). We discuss in some detail the classical Tutte polynomial of a matroid and we take a detour to compute such a polynomial in the special case of root systems. The reader is advised to use these chapters mainly as a tool and reference to read the main body of the book. In particular, Chapter 6 is used only in the fourth part.

In the second part, on the *differentiable case,* we start by introducing and studying the splines. We next analyze the coordinate ring of the complement of a hyperplane arrangement using the theory of modules over the Weyl algebra. We apply this analysis to the computation of the multivariate splines. We next give a simple proof of the theorem of Dahmen–Micchelli on the dimension of $D(X)$ using elementary commutative algebra (Theorem 11.13), and discuss the similar theory of E-splines due to Amos Ron [92].

After this, we discuss the graded dimension of the space $D(X)$ (Theorem 11.13) in terms of the combinatorics of bases extracted from X. This is quite similar to the treatment of Dyn and Ron [51]. The answer happens to be directly related to the classical Tutte polynomial of a matroid introduced in the first part. We next give an algorithmic characterization in terms of differential equations of a natural basis of the top-degree part of $D(X)$ (Proposition 11.10), from which one obtains explicit local expressions for T_X (Theorem 9.7). We complete the discussion by presenting a duality between $D(X)$ and a subspace of the space of polar parts relative to the hyperplane arrangement associated to X (Theorem 11.20), a space which can also be interpreted as distributions supported on the regular points of the associated cone $C(X)$.

The third part, on the *discrete case,* contains various extensions of the results of the second part in the case in which the elements in X lie in a lattice. This leads to the study of *toric arrangements,* which we treat by module theory in a way analogous to the treatment of arrangements. Following a joint paper with M. Vergne [44], we discuss another interesting space of functions on a lattice that unifies parts of the theory and provides a conceptual proof of the quasipolynomial nature of partition functions on the sets $\Omega - B(X)$.

We next explain the approach (due mainly to Szenes–Vergne) (see Theorem 10.11 and Formula (16.3)) of computing the functions under study via residues.

We give an explicit formula relating partition functions to multivariate splines (Theorem 16.12) which is a reinterpretation, with a different proof, of the formula of Brion–Vergne [27]. We use it to discuss classical computations including Dedekind sums and generalizations.

As an application of our methods, we have included two independent chapters, one in the second and one in the third part, in which we explain how to compute the de Rham cohomology for the complement of a hyperplane or of a toric arrangement.

The fourth and fifth parts essentially contain complements. The fourth part is independent of the third. In it we present a short survey of the connections and applications to approximation theory: the role of the Strang–Fix conditions and explicit algorithms used to approximate functions by splines, such as, for example, can be found in the book *Box Splines* [40]. We also discuss briefly some other applications, such as the theory of stationary subdivision.

The fifth and final part is completely independent of the rest of the book. It requires some basic algebraic geometry, and it is included because it justifies in a geometric way the theory of Jeffrey–Kirwan residues which we have introduced as a purely computational tool, and the regularization of certain integrals. Here we give an overview of how the residues appear in the so-called *wonderful models* of subspace arrangements. These are particularly nice compactifications of complements of hyperplane arrangements that can be used to give a geometric interpretation of the theory of residues.

As the reader will notice, there is a class of examples which we investigate throughout this book. These are *root systems*. We decided not to give the foundations of root systems, since several excellent introductions are available in the literature, such as, for example, [20] or [67].

Finally, let us point out that there is an overlap between this material and other treatments, as for instance the recent book by Matthias Beck and Sinai Robins, in the Springer Undergraduate Texts in Mathematics series: *Computing, the Continuous Discretely. Integer-Point Enumeration in Polyhedra* [14].

The actual computation of partition functions or of the number of points in a given polytope is a very difficult question due to the high complexity of the problem. A significant contribution is found in the work of Barvinok [12], [10], [8], [9], [11] and in the papers of Baldoni, De Loera, Cochet, Vergne, and others (see [6], [34], [48]). We will not discuss these topics here. A useful survey paper is the one by Jésus A. De Loera [47].

Acknowledgments

We cannot finish this preface without expressing our sincere thanks to various colleagues and friends. First, Michèle Vergne, who through various lectures and discussions arouse our interest in these matters. We also thank Welleda Baldoni and Charles Cochet for help in computational matters, Carla Manni for pointing out certain aspects of quasi-interpolants, Amos Ron for sharing his notes, and Carl de Boor for useful suggestions on the literature. Finally, we thank several colleagues who pointed out some unclear points: Alberto De Sole, Mario Marietti, Enrico Rogora, Damiano Testa, and Ramadas Trivandrum.

Although we have made an effort to give proper attributions, we may have missed some important contributions due to the rather large span of the existing literature, and we apologize if we have done so.

The authors are partially supported by the Cofin 40%, MIUR.

Corrado de Concini
Claudio Procesi

Notations

When we introduce a new symbol or definition we will use the convenient
form :=, which means that the term introduced at its left is defined by the
expression at its right.

A typical example is $P := \{x \in \mathbb{N} \mid 2 \text{ divides } x\}$, which stands for P *is by
definition the set of all natural numbers x such that 2 divides x.*

The symbol $\pi : A \to B$ denotes a mapping named π from the set A to the
set B.

Given two sets A, B we set

$$A \setminus B := \{a \in A \mid a \notin B\}.$$

We use the standard notation

$$\mathbb{N}, \ \mathbb{Z}, \ \mathbb{Q}, \ \mathbb{R}, \ \mathbb{C}$$

for the natural numbers (including 0), the integers, and the rational, real, and
complex numbers.

When V is a vector space we denote by V^* its dual. The canonical pairing
between $\phi \in V^*$ and $v \in V$ will be denoted by $\langle \phi \mid v \rangle$ or sometimes by $\phi(v)$
when we want to stress ϕ as a function.

For a finite set A we denote by $|A|$ its cardinality.

Given two points a, b in a real vector space V we have the closed segment
$[a, b] := \{ta + (1-t)b, \ 0 \le t \le 1\}$ the open segment $(a, b) := \{ta + (1-t)b, \ 0 <
t < 1\}$, and the two half-open segments $[a, b) := \{ta + (1 - t)b, \ 0 < t \le
1\}, (a, b] := \{ta + (1 - t)b, \ 0 \le t < 1\}$.

The closure of a set C in a topological space is denoted by $\overline{C}$.

The interior of a set C in a topological space is denoted by $\overset{\circ}{C}$.

Sets and lists: Given a set A, we denote by χ_A its characteristic function.
The notation $A := \{a_1, \ldots, a_k\}$ denotes a set called A with elements the a_i
while the notation $A := (a_1, \ldots, a_k)$ denotes a list called A with elements the
a_i. The elements a_i appear in order and may be repeated. A *sublist* of A is a
list $(a_{i_1}, \ldots, a_{i_r})$ with $1 \le i_1 < \cdots < i_r \le k$.

WARNING: By abuse of notation, when X is a list we shall also denote a sublist Y by writing $Y \subset X$ and by $X \setminus Y$ the list obtained from X by removing the sublist Y.

Part I

Preliminaries

1

Polytopes

Our basic datum is a list $X := (a_1, \ldots, a_m)$ of nonzero elements in a real s-dimensional vector space V (we allow repetitions in the list, since this is important for the applications). Sometimes we take $V = \mathbb{R}^s$ and then think of the a_i as the columns of an $s \times m$ matrix A.

From X we shall construct several geometric, algebraic, and combinatorial objects such as the cone they generate, a hyperplane arrangement, some families of convex polytopes, certain algebras and modules, and some special functions. Our aim is to show how these constructions relate to each other, and taken together, they provide structural understanding and computational techniques.

1.1 Convex Sets

In this Section we shall discuss some foundational facts about convex sets. For a general treatment the reader can consult the book of Rockafellar [91].

1.1.1 Convex Sets

Recall that a *convex set* in a real vector space V is a subset $A \subset V$ with the property that if $a, b \in A$, the closed segment $[a, b]$ is contained in A.

On the other hand, an *affine subspace* is a subset $A \subset V$ with the property that if $a, b \in A$, the line $\ell_{a,b} := \{ta + (1-t)b, \forall t\}$ is contained in A. An affine subspace passing through a point p is of the form $p + W$, where W is a linear subspace, which one can think of as the space of translations that preserve A. This group of translations acts in a simply transitive way on A.

Clearly, the intersection of any family of convex sets is convex, and similarly, the intersection of any family of affine spaces is affine. Thus, given any set A there is a minimal convex set $C(A)$ containing A called the *convex*

C. De Concini and C. Procesi, *Topics in Hyperplane Arrangements, Polytopes and Box-Splines*, Universitext, DOI 10.1007/978-0-387-78963-7_1,
© Springer Science+Business Media, LLC 2010

envelope, or convex hull, of A, and there is a minimal affine subspace $\mathrm{Aff}(A)$ containing A called the *affine envelope or affine hull of* A.

In general, given m points $a_1, \ldots, a_m \in V$, an *affine combination* of these points is any point

$$p = t_1 a_1 + t_2 a_2 + \cdots + t_m a_m, \ t_i \in \mathbb{R}, \ \sum_{i=1}^{m} t_i = 1,$$

while a *convex combination* of these points is any point

$$p = t_1 a_1 + t_2 a_2 + \cdots + t_m a_m, \ 0 \le t_i \le 1, \ \sum_{i=1}^{m} t_i = 1.$$

It is easily verified that the convex envelope (resp. the affine envelope) of A is formed by all the convex (resp. affine) combinations of points in A.

Definition 1.1. We say that m points $a_1, \ldots, a_m \in V$ are *independent* in the affine sense if the coordinates t_i with $\sum_i t_i = 1$ expressing a point in their affine envelope are uniquely determined.

Remark 1.2. m points are independent if and only if their affine envelope has dimension $m - 1$.

Given two vector spaces V_1, V_2 and two sets $A_1 \subset V_1$, $A_2 \subset V_2$ we have that $A_1 \times A_2 \subset V_1 \times V_2$ is convex if and only if A_1, A_2 are convex, and it is affine if and only if A_1, A_2 are affine.

If $f : V_1 \to V_2$ is an affine map, i.e., the composition of a linear map with a translation, then such an f preserves affine combinations, that is, if $t_i \in \mathbb{R}$, $\sum_{i=1}^{m} t_i = 1$, we have

$$f(t_1 a_1 + t_2 a_2 + \cdots + t_m a_m) = t_1 f(a_1) + t_2 f(a_2) + \cdots + t_m f(a_m).$$

Therefore, if $A \subset V_1$ is convex, then $f(A)$ is convex, and if A is affine, then $f(A)$ is affine.

From two sets $A_1, A_2 \subset V_1$ we can construct a new set, the *Minkowski sum of* A_1, A_2, as $A_1 + A_2 := \{a_1 + a_2, \ a_1 \in A_1, \ a_2 \in A_2\}$, namely the image of $A_1 \times A_2$ under the sum, which is a linear (hence affine) map. Similarly, we may construct the difference $A_1 - A_2$. By the above observations we have that if A_1, A_2 are convex (resp. affine), then $A_1 \pm A_2$ are convex (resp. affine) sets.

Lemma 1.3. *Let* A, B *be two convex sets. The convex envelope* C *of* $A \cup B$ *is the set of convex combinations* $ta + (1 - t)b$, $0 \le t \le 1$, $a \in A$, $b \in B$.

Proof. Clearly, any such convex combination is in C. Conversely, a typical element of C is of the form $p = \sum_{i=1}^{h} t_i a_i + \sum_{j=1}^{k} s_j b_j$, with $\sum_{i=1}^{h} t_i + \sum_{j=1}^{k} s_j = 1$, $0 \le t_i$, $0 \le s_j$, $a_i \in A, b_j \in B, \forall i, j$. We may assume that

$t := \sum_i t_i > 0$, $s = 1 - t := \sum_j s_j > 0$; otherwise, the point p is in $A \cup B$. Then $p = t\sum_{i=1}^{h}(t_i/t)a_i + s\sum_{j=1}^{k}(s_j/s)b_j = ta + (1-t)b$, where $a = \sum_{i=1}^{h}(t_i/t)a_i \in A$, $b = \sum_{j=1}^{k}(s_j/s)b_j \in B$.

Definition 1.4. The *dimension* of a convex set is the dimension of its affine envelope.

The *relative interior* $\mathring{A}$ of A is the interior of A considered as a subset of its affine envelope, that is, the maximal set open in $\mathrm{Aff}(A)$ and contained in A. The *relative boundary* ∂A of A is $A \setminus \mathring{A}$.

Remark 1.5. Given m points $a_1, \ldots, a_m$, independent in the affine sense, their convex envelope is called a *simplex*. The relative interior of this simplex is the set $\{\sum_{i=1}^{m} t_i a_i \mid 0 < t_i, \sum_i t_i = 1\}$.

Proposition 1.6. *Let A be a convex set:*

(i) If $a \in A, b \in \mathring{A}$, then $(a, b) \subset \mathring{A}$.
(ii) $\mathring{A}$ is convex.
(iii) If A is a closed convex set, then A is the closure of its relative interior.

Proof. (i) Let $a \in A$. We may assume that the affine hull of A is V and reduce to the case $a = 0$. Let $s := \dim(V)$.

Take a hyperplane H passing through b and not through 0. Since $b \in \mathring{A}$, we have that $H \cap \mathring{A}$ contains a basis $a_1, \ldots, a_s$ of V, so that b lies in the interior of the simplex generated by $a_1, \ldots, a_s$. Then $(0, b)$ is contained in the interior of the simplex generated by $0, a_1, \ldots, a_k$, which also lies in the interior of A.

Clearly, (ii) is a consequence of (i). Moreover, (iii) is a consequence of (i) once we prove that $\mathring{A} \neq \emptyset$. This follows once we observe that if A spans an s-dimensional affine space V, there must be $s + 1$ points $a_0, \ldots, a_s \in A$ that are independent (in the affine sense). Thus the simplex they generate is contained in A and its interior is open in V.

Remark 1.7. In particular, we have shown that if ℓ is a line lying in the affine envelope of a closed convex set A and $\ell \cap \mathring{A} \neq \emptyset$, then $\ell \cap \mathring{A}$ is the interior of $\ell \cap A$, which is either a closed segment or a half-line or ℓ. Moreover, the relative boundary of a closed convex set A is empty if and only if A is an affine space.

1.1.2 Duality

A hyperplane $H \subset V$ divides V into two closed half-spaces which intersect in H (the closures of the two connected components of $V \setminus H$). If H is given by the linear equation $\langle \phi \mid x \rangle = a$ (where $\phi \in V^*$ is a linear form), the two half-spaces are given by the inequalities $\langle \phi \mid x \rangle \geq a$, $\langle \phi \mid x \rangle \leq a$.

From now on we assume that we have fixed an Euclidean structure on V.

Theorem 1.8. *Let A be a nonempty closed convex set and $p \notin A$ a point.*

(1) There is a unique point $q \in A$ of minimal distance from p.

(2) Let H be the hyperplane passing through q and orthogonal to the vector $q - p$. Fix a linear equation $\langle \phi \mid x \rangle = a$ for H such that $\langle \phi \mid p \rangle < a$. Then A lies entirely in the closed half-space $\langle \phi \mid x \rangle \geq a$.

(3) There is a hyperplane K such that p and A lie in the two different components of $V \setminus K$.

Proof. 1) Let R be a number so large that the closed ball $B_p(R)$ centered at p and of radius R has a nonempty intersection with A. Then $A \cap B_p(R)$ is compact and nonempty, and thus there is at least one point $q \in A \cap B_p(R)$ of minimum distance $\leq R$ from p. We cannot have two distinct points $q_1, q_2 \in A$ of minimum distance from p, since any point in the interior of the segment $[q_1, q_2]$ has strictly smaller distance from p, and by convexity, $[q_1, q_2] \subset A$.

2) If by contradiction we had a point $r \in A$ in the half-space $\langle \phi \mid x \rangle < a$, the entire half-open segment $[r, q)$ would be in A, and in this segment the points close to q have distance from p strictly less than that of q.

3) Let us now take any hyperplane K parallel to H and passing through a point in the open segment (q, p). One easily sees that p and A lie in the two different connected components of $V \setminus K$.

Definition 1.9. Given a closed convex set A and a point $p \notin A$, the hyperplane H through the point of minimal distance q and orthogonal to $p - q$ will be denoted by $H_p(A)$.

We want to deduce a duality statement from Theorem 1.8.

Consider the vector space $\tilde{V} = V^* \oplus \mathbb{R}$ of all inhomogeneous polynomials $\langle \phi \mid x \rangle - a$ of degree ≤ 1 with $\phi \in V^*$, $a \in \mathbb{R}$. The polynomials of the previous type with $\phi \neq 0$ can be thought of as linear equations of hyperplanes. Given any subset $A \subset V$, consider in $\tilde{V}$ the set

$$\tilde{A} := \{ f \in \tilde{V} \mid f(A) \leq 0 \}.$$

Clearly, $\tilde{A}$ is convex. In fact, it is even a convex cone (cf. Section 1.2.3). An immediate consequence of Theorem 1.8 is the following

Corollary 1.10. *If A is closed and convex, then*

$$A = \{ p \in V \mid f(p) \leq 0, \ \forall f \in \tilde{A} \}. \tag{1.1}$$

1.1.3 Lines in Convex Sets

Our next task is to show that, when studying closed convex sets, one can reduce the analysis to sets that do not contain any line.

Lemma 1.11. *Let A be convex and $p \in A$. Then there is a unique maximal affine subspace Z with $p \in Z \subset A$.*

Proof. If Z_1, Z_2 are two affine spaces passing through p, then the convex envelope of $Z_1 \cup Z_2$ coincides with the affine space $Z_1 + Z_2 - p$ generated by $Z_1 \cup Z_2$. The statement follows.

We denote by A_p this maximal affine space.

Proposition 1.12. *1. If A is a closed convex set and $p, q \in A$, then $A_q = A_p + (q - p)$.*

2. If a line ℓ intersects the closed convex set A in a nonempty bounded interval, then any line parallel to ℓ intersects A in a bounded interval.

3. Let A be a closed convex set and $p \in A$. Let $A_p^\perp$ be the affine subspace passing through p and orthogonal to the affine subspace A_p. We can identify V with $A_p \times A_p^\perp$. Then $A = A_p \times (A \cap A_p^\perp)$.

Proof. 1. By exchanging the roles of the two points, it is enough to show that $A_p + (q - p) \subset A_q$, and we may assume that $p = 0$. Thus we need to show that $q + A_0 \subset A$. If $a \in A_0$, we have

$$q + a = \lim_{t \uparrow 1} t(q + a) = \lim_{t \uparrow 1} tq + (1 - t)\frac{t}{1 - t} a \in A,$$

since if $t \neq 1$ we have $\frac{t}{1-t} a \in A_0 \subset A$ and A is convex and closed.

2. Assume $\ell \cap A = [a, b]$ and let $p \in A$. Let ℓ' be the line through p parallel to ℓ. Then ℓ and ℓ' can be respectively parametrized by $b + h(a - b)$ and $p + h(a - b)$, $h \in \mathbb{R}$. If $\ell' \cap A$ were unbounded, we would have either $p + h(a - b) \in A, \forall h \in \mathbb{R}^+$, or $p - h(a - b) \in A, \forall h \in \mathbb{R}^+$. Let us assume that we are in the first case:

$$b + h(a - b) = \lim_{t \uparrow 1} tb + (1 - t)\left(\frac{h}{1 - t}(a - b) + p\right) \in \ell \cap A, \ \forall h \in \mathbb{R}^+.$$

Thus if $A \cap \ell'$ contained a half-line in ℓ' originating from p, then $\ell \cap A$ would be unbounded.

3. Follows easily from 1.

In other words, there is a maximal vector space W, called the *lineal* of A, such that A is closed under translations by elements of W and for every $p \in A$, we have $A_p = p + W$. We will need also the following:

Proposition 1.13. *1. Let A be a closed convex set. If the relative boundary of A is convex (or empty), then either A coincides with its affine envelope or it is a half-space in its affine envelope.*

2. Let A be a closed convex set containing a half-space B determined by a hyperplane H. Then either $A = V$ or A is a half-space containing B with boundary a hyperplane parallel to H.

Proof. 1. After applying a translation, we can assume that the affine envelope of A is a linear subspace. Also, by replacing V with the affine envelope of A, we can assume that V is the affine envelope of A.

Notice that if a line ℓ meets $\mathring{A}$, then $\ell \cap \mathring{A}$ is either ℓ or a half-line in ℓ. Indeed, if $\ell \cap \mathring{A} = (q_1, q_2)$ is a bounded interval, then $q_1, q_2 \in \partial A$, and since ∂A is convex, $\ell \cap \mathring{A} = (q_1, q_2) \subset \partial A$, a contradiction. The same argument shows that if ℓ meets $\mathring{A}$, then $\ell \cap \partial A = \emptyset$ if and only if $\ell \subset \mathring{A}$.

If $A = V$ there is nothing to prove, so assume $A \neq V$. By Remark 1.7 the relative boundary ∂A is not empty.

Since ∂A is convex and closed, we can apply Theorem 1.8, and given a point $p \in \mathring{A}$, we have the point $q \in \partial A$ of minimal distance from p, the associated hyperplane $H_p(\partial(A))$, and an equation $\langle \phi \,|\, x \rangle = a$ for $H_p(\partial A)$ with $\langle \phi \,|\, p \rangle = b < a$ and $\langle \phi \,|\, x \rangle \geq a$ for $x \in \partial A$.

Consider the hyperplane K parallel to $H_p(\partial A)$ and passing through p. Its equation is $\langle \phi \,|\, x \rangle = b$, and since $\langle \phi \,|\, x \rangle \geq a > b$ on ∂A we have $K \cap \partial A = \emptyset$. We deduce from the previous discussion that each line $\ell \subset K$ passing through p lies entirely in $\mathring{A}$. It follows that $K \subset A$, so that $A_p = K$. Hence by Proposition 1.12, statement 3, we have that $A = K \times I$, where $I = A \cap K^\perp$ and $K^\perp$ is the line through p perpendicular to K. The above observations imply that I is a half-line, and in fact, $A = \{x \,|\, \langle \phi \,|\, x \rangle \leq a\}$.

2. By Proposition 1.12 we have that A is the product of H and a closed convex set in a line containing a half-line. Clearly, such a set is either a half-line or the whole line. In the first case, A is a half-space, while in the second $A = V$. The claim follows easily.

1.1.4 Faces

Definition 1.14. Given a convex set A we say that:

1. A subset U of A is a *face of A* if U is convex and given $a, b \in A$ such that $(a, b) \cap U \neq \emptyset$, we have $[a, b] \subset U$.

2. A point $p \in A$ is called an *extremal point* or also a *vertex*[1] if $\{p\}$ is a face.

3. A half-line ℓ that is a face is called an *extremal ray*.

4. The relative interior of a face is called an *open face*.

Proposition 1.15. *Let U be a face of a convex set A and $\langle U \rangle$ the affine space that U generates. Then $U = A \cap \langle U \rangle$. In particular, U is closed in A.*

Proof. We can immediately reduce to the case $A \subset \langle U \rangle$. By hypothesis U contains an open simplex Δ, thus any point of A is in a segment contained in A which contains in its interior a point in Δ. The claim follows.

Lemma 1.16. *Let A be a closed convex set, $p \notin A$. Let $H = H_p(A)$ be the hyperplane defined in Definition 1.9. Then:*

   1. $H \cap A$ is a face of A.

   2. If $A \not\subset H$, then $H \cap A \subset \partial A$.

---

[1] Vertex is used mostly for polytopes; see Section 1.2.

Proof. 1. Let $a, b \in A$ be such that $(a, b) \cap H \cap A \neq \emptyset$. By construction A lies entirely in one of the half-spaces determined by H, so $[a, b]$ lies in this half-space. If an interior point of $[a, b]$ lies in H we must have that the entire segment is in H, hence in $H \cap A$.

2. Let W be the affine envelope of A. By hypothesis, $H \cap W$ is a hyperplane in W. A point q of the relative interior of A has an open neighborhood contained in A, and thus q cannot lie in H, for otherwise, this neighborhood would intersect the two open half-spaces determined by H in W, contradicting the fact that A lies on one side of H.

Proposition 1.17. *Let F_1, F_2 be two faces of a closed convex set. Assume that we have a point $p \in F_2$ that is also in the relative interior of F_1. Then $F_1 \subset F_2$.*

Proof. If $F_1 = \{p\}$, there is nothing to prove. Let $q \in F_1$, $q \neq p$, and consider the line ℓ joining p and q. By the convexity of F_1, the intersection $F_1 \cap \ell$ is also convex. Since p lies in the interior of F_1, the intersection of ℓ with F_1 contains a segment $[a, q]$ with $p \in (a, q)$. Since F_2 is a face we have $[a, q] \subset F_2$, so that $q \in F_2$.

Remark 1.18. In particular, if a face F intersects the interior of A, then $F = A$. Thus all the proper faces of A are contained in the boundary ∂A.

Proposition 1.19. *Let A be a closed convex set that is not an affine space or a half-space in its affine envelope. Then A is the convex envelope of its relative boundary ∂A which is the union of its proper faces.*

Proof. By Proposition 1.13, we know that ∂A is not convex. Then we can take two points $a, b \in \partial A$ such that $[a, b] \not\subset \partial A$. By the first part of Proposition 1.6, it follows easily that $(a, b) = \ell \cap \mathring{A}$ and $[a, b] = \ell \cap A$. It follows from Proposition 1.12 that if we take any point $p \in A$, the line through p parallel to ℓ also meets A in a bounded segment $[c, d]$. It is clear that $c, d \in \partial A$, so the first statement follows.

Next, given $p \in \partial A$, we need to show that there is a face $F \subset \partial A$ with $p \in F$. For this take an affine space passing through p and maximal with the property that $W \cap A \subset \partial A$. We claim that $W \cap A$ is a face. In fact, it is convex, and if $[a, b]$ is a segment in A meeting $W \cap A$ in a point $q \in (a, b)$, we claim that $[a, b] \subset W$. Otherwise, $a, b \notin W$, and let W' be the affine space spanned by W and a. Then W is a hyperplane in W' and a, b lie in the two distinct half-spaces in which W divides W'. By maximality there is a point $r \in W' \cap \mathring{A}$, and we may assume that it is in the same half-space as a. By Proposition 1.6 (i) the segment (r, b) is contained in $\mathring{A}$, but it also intersects W, giving a contradiction.

The main result of this Section is the following:

Theorem 1.20. *If a closed convex set A does not contain any line then A is the convex envelope of its extremal points and extremal rays.*

Proof. By Proposition 1.19, A is the convex envelope of the union of its proper faces unless it is a point or a half-line. Each of the proper faces is a convex set containing no lines, so by induction on the dimension it is the convex envelope of its extremal points and extremal rays. Clearly, an extremal point (resp. extremal ray) of a face of A is also an extremal point (resp. extremal ray) of A, and the theorem follows.

1.2 Polyhedra

In this Section we shall discuss some foundational facts about polytopes. For a general treatment, the reader can consult the books of Ziegler [123] and Grünbaum [58].

1.2.1 Convex Polyhedra

We give now the main definition:

Definition 1.21. We say that a convex set A is *polyhedral* or that it is a *convex polyhedron* if there exist a finite number of homogeneous linear functions $\phi_i \in V^*$ and constants a_i such that

$$A := \{p \mid \langle \phi_i \mid p \rangle \leq a_i\}. \tag{1.2}$$

Remark 1.22. Notice that formula (1.2) implies that A is convex. Moreover, the intersection of any finite family of convex polyhedra is also a convex polyhedron.

In general, by *polyhedron* one means a finite (or sometimes just locally finite) union of convex polyhedral sets. At this point it is not clear that this is compatible with the previous definition.

Take a set A defined by the inequalities (1.2), and let us assume for simplicity that $0 \in A$. This means that all $a_i \geq 0$. Under this assumption, with the notation of Section 1.1.3 we have the following:

Proposition 1.23.
$$A_0 = \{p \mid \langle \phi_i \mid p \rangle = 0, \ \forall i\}.$$

Proof. Clearly, the set defined by these linear equations is a vector space contained in A. Conversely, if the line through 0 and p lies in A, it is clear that we must have $\langle \phi_i \mid p \rangle = 0, \ \forall i$.

We apply now Proposition 1.12 and easily verify that if A is polyhedral and W is the linear space orthogonal to A_0, then $A = A_0 \times (A \cap W)$, and $A \cap W$ is also polyhedral.

We can therefore restrict ourselves to the study of polyhedral convex sets $A := \{p \mid \langle \phi_i \mid p \rangle \leq a_i, \ 1 \leq i \leq n\}$ such that $0 \in A$ and $A_0 = \{0\}$. Given a convex polyhedral set $A := \{p \mid \langle \phi_i \mid p \rangle \leq a_i\}$, $i = \{1, \ldots, n\}$ we define a map from subsets $S \subset \{1, \ldots, n\}$ to subsets $B \subset A$ given by

$$S \mapsto A_S := \{p \in A \mid \langle \phi_i \mid p \rangle = a_i, \forall i \in S\},$$

and a map in the opposite direction given by

$$B \mapsto T_B := \{i \mid \langle \phi_i \mid p \rangle = a_i, \forall p \in B\}.$$

Lemma 1.24.   *(i) For any $S \subset \{1, \ldots, n\}$ the subset A_S is a face of A.*
(ii) For any face F of A we have $F = A_{T_F}$ and $\mathring{F} = \{p \mid T_{\{p\}} = T_F\}$.

Proof. (i)   Let $p, q \in A$ be such that there is a point $r \in (p, q) \cap A_S$. This means that for all $i \in S$, we have $\langle \phi_i \mid r \rangle = a_i$. Since by definition $\langle \phi_i \mid p \rangle \leq a_i$, $\langle \phi_i \mid q \rangle \leq a_i$, and ϕ_i is linear, we must have also $\langle \phi_i \mid x \rangle = a_i$ for all $x \in [p, q]$ and thus $[p, q] \subset A_S$.

(ii)   If a point p is such that $T_{\{p\}} = S$, we claim that p lies in $\mathring{A}_S$. For this we can restrict to the affine envelope of A_S, which is the space of equations $\langle \phi_i \mid r \rangle = a_i$ for $i \in S$. A point $p \in A$ satisfies $T_{\{p\}} = S$ if and only if it lies in the open set of this subspace defined by the inequalities $\langle \phi_i \mid r \rangle < a_i$ for $i \notin S$. This proves the claim.

Conversely, let F be a face of A. Consider a point p in the relative interior of F and let $S := T_{\{p\}} = \{i \mid 1 \leq i \leq n, \langle \phi_i \mid p \rangle = a_i\}$. Since obviously $p \in A_S$, by Proposition 1.17 we have that $F \subset A_S$. By definition, $\langle \phi_i \mid p \rangle < a_i$, $\forall i \notin S$. So p lies in the relative interior of A_S, and again by Proposition 1.17, $A_S \subset F$. From here it follows that $F = A_S$ and hence $T_{\{p\}} = T_F$. The statement follows.

Remark 1.25. (i) It is possible that $A_S = \emptyset$ for some S and also that $A_S = A_T$ when $S \neq T$.

(ii) The relative interiors of the faces decompose the convex polyhedron; they are also called *open faces*.

(iii) The faces of a convex polyhedron form a partially ordered set.

Definition 1.26. Two polyhedra are said to be combinatorially equivalent if there is an order-preserving bijection between their faces.

We can apply now Theorem 1.20 to a polyhedron containing no lines:

Theorem 1.27. *Let A be a convex polyhedron containing no lines. Then A is the convex envelope of its finitely many extremal points and extremal half-lines.*

In particular, when A is compact it contains no half-lines, and so it is the convex envelope of its finitely many extremal points.

Corollary 1.28. *Let A be a convex polyhedron. Then A is the convex envelope of finitely many points and half-lines.*

Proof. Let $p \in A$. Using Proposition 1.12, write $A = A_p \times (A \cap W)$, where W is the affine space through p orthogonal to A_p. Clearly, (by Remarks 1.22), $A \cap W$ is polyhedral and contains no lines, so by Theorem 1.27 it is the convex envelope of finitely many points and half-lines. On the other hand, an affine space is clearly the convex envelope of finitely many half-lines, and everything follows.

Remark 1.29. In what follows, a compact convex polyhedron will often be called a *convex polytope*.

Remark 1.30. It will be important in the arithmetic theory to discuss *rational convex polytopes* in $\mathbb{R}^s$. By this we mean a polytope defined by the inequalities (1.2), in which we assume that all the ϕ_i, a_i have integer coefficients.

It is then clear that for a rational polytope, the extremal points and half-lines are also rational.

We need now a converse to the above result.

Theorem 1.31. *Assume that A is the convex envelope of finitely many points $p_1, \ldots, p_k$ and half-lines $\ell_1, \ldots, \ell_h$. Then A is a polyhedron.*

Proof. As usual, we can assume that V is the affine envelope of A. We can parametrize each $\ell_i := \{a_i + tb_i, \ t \geq 0\}$, with $a_i, b_i \in V$.

Consider the vector space $\tilde{V} = V^* \oplus \mathbb{R}$ of all (inhomogeneous) polynomials of degree ≤ 1. In $\tilde{V}$ consider the set

$$\tilde{A} := \{f \in \tilde{V} \mid f(A) \leq 0\}$$

and recall that by Corollary 1.10, $A = \{p \in V \mid f(p) \leq 0, \ \forall f \in \tilde{A}\}$.

Each element in $\tilde{V}$ is of the form $\phi - a$, with $\phi \in V^*$, $a \in \mathbb{R}$. Then $\tilde{A}$ is defined by the inequalities $\langle \phi \mid p_i \rangle - a \leq 0$ and $\langle \phi \mid a_i \rangle + t\langle \phi \mid b_i \rangle - a \leq 0$, for all $t \geq 0$. These last inequalities are equivalent to saying that $\langle \phi \mid a_i \rangle - a \leq 0$ and $\langle \phi \mid b_i \rangle \leq 0$.

Thus $\tilde{A}$ is a polyhedron in the space $\tilde{V}$, and obviously $0 \in \tilde{A}$. We claim that it contains no lines (or equivalently no line through 0). In fact, a line in $\tilde{A}$ through 0 is made of linear functions vanishing on A, and therefore, since V is the affine envelope of A, on V.

We can thus apply Theorem 1.27 to $\tilde{A}$ and deduce that $\tilde{A}$ is the convex envelope of its finitely many extremal points and extremal half-lines. Reasoning as above, we get that each extremal point gives an inequality and each half-line gives two inequalities, and these inequalities define A.

These two characterizations of convex polyhedra are essentially due to Motzkin [82].

1.2.2 Simplicial Complexes

The simplest type of convex polytope is a k-dimensional *simplex*, that is the convex envelope of $k + 1$ points $p_0, \ldots, p_k$ spanning a k-dimensional affine space. In topology and combinatorics it is often important to build spaces, in particular polytopes, using simplices.

Definition 1.32. A *finite simplicial complex* in $\mathbb{R}^s$ is a finite collection of distinct simplices $A_i \subset \mathbb{R}^s$ with the property that if $i \neq j$, we have that $A_i \cap A_j$ is a face of A_i and of A_j (possibly empty).

We also say that the simplices A_i give a *triangulation* of their union $K = \cup_i A_i$.

A simple but important fact is contained in the following lemma

Lemma 1.33. *A convex polytope admits a triangulation.*

Proof. Choose, for every nonempty face F of K, a point p_F in the relative interior of F. For two faces F, G we say that $F < G$ if $F \neq G$ and $F \subset G$. If we have a sequence of faces $F_1 < F_2 < \cdots < F_k$, then by Lemma 1.24, p_{F_i} does not lie in the affine space generated by F_{i-1}, and hence the points $p_{F_1}, p_{F_2}, \ldots, p_{F_k}$ generate a simplex. The fact that these simplices define a triangulation follows by induction using the following remark.

Given a convex polytope P and a triangulation of its boundary, then a triangulation of P can be constructed by choosing a point p in the interior of P and taking for each simplex Σ in ∂P the simplex that is the convex envelope of Σ and p. We leave to the reader to verify that in this way we obtain a triangulation. By this remark the claim follows.

Remark 1.34. The assumption that K is convex is not necessary for triangulation. In fact, one can prove that the union and the intersection of subspaces of $\mathbb{R}^s$ that can be triangulated, can also be triangulated.

1.2.3 Polyhedral Cones

Let us recall that a *cone* is a set stable under multiplication by the elements of $\mathbb{R}^+$.

Let A be a closed convex cone. If $p \in A, p \neq 0$, we have that the half-line $\{tp \mid t \in \mathbb{R}^+\}$ is also in A, and thus $\{p\}$ cannot be a face. That is, the only extremal point of A is possibly 0. Now 0 is an extremal point if and only if A contains no lines. Indeed, if 0 is not a face, it is in the interior of a segment contained in A, and since A is a cone, the entire line in which this segment lies is contained in A. We deduce the following result

Proposition 1.35. *If A is a closed convex cone containing no lines and $A \neq \{0\}$, then A is the convex envelope of its extremal half-lines all of which originate from 0.*

Consider a list $X := (a_1, \ldots, a_m)$ of nonzero vectors in V.

Definition 1.36. The convex cone spanned by X is:

$$C(X) := \Big\{ \sum_{a \in X} t_a a \mid 0 \le t_a, \ \forall a \Big\}. \tag{1.3}$$

Namely, $C(X)$ is the cone of linear combinations of vectors in X with positive coefficients.

By Theorem 1.31 we deduce that $C(X)$ is a convex polyhedral cone. If we assume that 0 is not in the convex hull of the vectors in X, then by Theorem 1.8 we have that there is a linear function ϕ with $\langle \phi \mid v \rangle > 0$, $\forall v \in X$, and so also for all $v \in C(X)$, $v \neq 0$. Thus $C(X)$ does not contain lines; its points different from 0 are contained entirely in some open half-space. Thus $C(X)$ is a *pointed cone* with vertex at the origin.

Define the *dual cone* $\hat{C}(X) := \{ \phi \in V^* \mid \langle \phi \mid a \rangle \ge 0, \ \forall a \in X \}$.

Proposition 1.37. *(1)* $\hat{C}(X)$ *is a convex polyhedral cone.*
(2) $\hat{C}(X)$ *is a pointed cone if and only if X spans V.*
(3) $\hat{\hat{C}}(X) = C(X)$.

Proof. The fact that $\hat{C}(X)$ is a polyhedral cone is clear by definition.

The cone $\hat{C}(X)$ contains a line if and only if there is a nonzero $\phi \in V^*$ with both ϕ and $-\phi$ lying in $\hat{C}(X)$. But this holds if and only if ϕ vanishes on $C(X)$, that is, $C(X)$ lies in the hyperplane of equation $\langle \phi \mid v \rangle = 0$.

Finally, $\hat{\hat{C}}(X) \supset C(X)$. To see the converse, notice that if $p \notin C(X)$, then by Theorem 1.8 there exist a $\phi \in V^*$ and a constant a such that $\langle \phi \mid p \rangle a$, while $\langle \phi \mid v \rangle \le a$ for $v \in C(X)$. Since $0 \in C(X)$, we deduce $0 \le a$. If $v \in C(X)$, then also $tv \in C(X)$ for all $t \ge 0$, so that $\langle \phi \mid v \rangle \le 0$ for $v \in C(X)$, and hence $-\phi \in \hat{C}(X)$. We have $\langle -\phi \mid p \rangle < -a \le 0$, so that $p \notin \hat{\hat{C}}(X)$. This implies $\hat{\hat{C}}(X) \subset C(X)$, thus completing the proof.

Remark 1.38. The previous duality tells us in particular that a cone $C(X)$ defined as the positive linear combinations of finitely many vectors can also be described as the set defined by finitely many linear inequalities.

It may be useful to decompose cones in the same way in which we have triangulated polytopes.

Definition 1.39. A cone $C(X)$ is *simplicial* if it is of the form $C(a_1, \ldots, a_k)$ with the elements a_i linearly independent.

The analogue for cones of triangulations are *decompositions into simplicial cones*, that is, a presentation of a cone as union of finitely many simplicial cones each two of which meet in a common face.

Lemma 1.40. *A polyhedral cone admits a decomposition into simplicial cones.*

Proof. One can first decompose the cone into pointed cones, for instance by intersecting the cone with all the quadrants into which the space is subdivided. We thus reduce to this case.

Recall as first step that if the cone $C(X)$ is pointed, we can choose a linear function $\phi \in V^*$ with $\langle \phi | v \rangle > 0, \forall v \in C(X),\ v \neq 0$. Let $H := \{v \mid \langle \phi | v \rangle = 1\}$ be a corresponding hyperplane. The intersection $K := C(X) \cap H$ is a convex polytope, the envelope of the points $x/\langle \phi | x \rangle,\ x \in X$. We see immediately that $C(X)$ is obtained from K by projection from the origin, that is, every element $a \in C(X), a \neq 0$ is uniquely of the form tb, for some $t \in \mathbb{R}^+,\ b \in K$. In fact, $t = \langle \phi | a \rangle,\ b = a/\langle \phi | a \rangle$.

We can thus apply Lemma 1.33 directly to K. A triangulation of K induces, by a projection, a decomposition of $C(X)$ into simplicial cones.

In the sequel we will need to study cones generated by integral vectors, also called *rational polyhedral cones*. It is clear from the previous discussion and the proof of Lemma 1.33 that we can also take the simplicial cones decomposing $C(X)$ to be generated by integral vectors, i.e., *rational simplicial cones*.

It is useful to extend the notion of cone so that its vertex need not be at 0.

Definition 1.41. We say that a set A is a *cone* with respect to a point $p \in A$ and that p is a *vertex* if given any point $a \in A$, the half-line starting from p and passing through a lies entirely in A.

A set A may be a cone with respect to several vertices. We say that it is a *pointed cone* with vertex p if p is the only vertex of A.

Proposition 1.42. *1. If A is a convex cone, the set of points p for which A is a cone with vertex p is an affine space $V(A)$.*

2. If $V(A)$ is nonempty, we can decompose the ambient space as a product $V(A) \times W$ of affine spaces such that $A = V(A) \times C$ with C a convex pointed cone in W.

Proof. 1. If p, q are two distinct vertices, by applying the definition we see that the entire line passing through p, q lies in A. The fact that each point of this line ℓ is also vertex follows easily by convexity. In fact, take a point $p \in A$ and $p \notin \ell$, and let π be the plane generated by ℓ and p. The intersection $\pi \cap A$ is convex, and we may apply Proposition 1.13 to it to deduce that every point of ℓ is a vertex of A.

2. Choose $q \in V(A)$ and W orthogonal to $V(A)$ and passing through q. Identify V with $V(A) \times W$, so by convexity, applying Proposition 1.12 we have $A = V(A) \times C$, where $C := A \cap (p \times W)$. One easily verifies that C is a convex cone with unique vertex q.

1.2.4 A Dual Picture

Given a convex compact polyhedron Π we have a dual picture of its faces in the dual space. For this we leave to the reader to verify the following statement

Proposition 1.43. *1. Given a linear form $f \in V^*$, the set of points in which f takes its maximum on Π is a face.*

2. Every face of Π is obtained as the maxima of a linear function.

3. The set of linear functions that take their maximum on a given face form a polyhedral cone.

Therefore, a compact convex polyhedron determines a decomposition of the dual space into polyhedral cones. The mapping between faces and corresponding polyhedral cones reverses inclusion, so that the vertices of the polyhedron correspond to the cones of maximum dimension.

1.3 Variable Polytopes

1.3.1 Two Families of Polytopes

We assume from now on that V is a vector space over the real numbers, that X spans V, and that 0 is not in the convex hull of the elements of X. This is equivalent, by Theorem 1.8, to saying that there is a linear function $\phi \in V^*$ with $\langle \phi \,|\, a \rangle > 0$, $\forall a \in X$.

Choose a basis and identify V with $\mathbb{R}^s$. The usual Euclidean structure of $\mathbb{R}^s$ induces on $\mathbb{R}^s$ and all its linear subspaces canonical translation invariant Lebesgue measures, which we shall use without further ado. Think of the vectors $a_i \in X$ as the columns of a real matrix A. As in linear programming, from the system of linear equations

$$\sum_{i=1}^{m} a_i x_i = b, \quad \text{or} \quad Ax = b,$$

we deduce two families of variable convex (bounded) polytopes:

$$\Pi_X(b) := \{x \,|\, Ax = b, x_i \geq 0, \ \forall i\},$$

$$\Pi_X^1(b) := \{x \,|\, Ax = b, 1 \geq x_i \geq 0, \ \forall i\}.$$

The hypothesis made (that 0 is not in the convex hull of elements of X) is necessary only to ensure that the polytopes $\Pi_X(b)$ are all bounded. In fact, if $\sum_i x_i a_i = b$, we have $\sum_i x_i \langle \phi \,|\, a_i \rangle = \langle \phi \,|\, b \rangle$, so that $\Pi_X(b)$ lies in the bounded region $0 \leq x_i \leq \langle \phi \,|\, b \rangle \langle \phi \,|\, a_i \rangle^{-1}$. No restriction is necessary if we restrict our study to the polytopes $\Pi_X^1(b)$.

It may be convenient to translate these polytopes so that they appear as variable polytopes in a fixed space. For this, assuming that the vectors a_i span V, we can find a right inverse to A or a linear map $U : V \to \mathbb{R}^m$ with $AUb = b$, $\forall b$. Then the translated polytopes $\Pi_X(b) - U(b)$ vary in the kernel of A (similarly for $\Pi_X^1(b) - U(b)$).

Definition 1.44. We denote by $V_X(b)$, $V_X^1(b)$ the volumes of $\Pi_X(b)$, $\Pi_X^1(b)$ respectively.

Suppose, furthermore, that the vectors a_i and b lie in $\mathbb{Z}^s$.

Definition 1.45. The *partition function* on $\mathbb{Z}^s$, is given by

$$\mathcal{T}_X(b) := \#\Big\{(n_1,\dots,n_m) \mid \sum n_i a_i = b,\ n_i \in \mathbb{N}\Big\}. \qquad (1.4)$$

It may be interesting to study also the function

$$Q_X(b) := \#\Big\{(n_1,\dots,n_m) \mid \sum n_i a_i = b,\ n_i \in [0,1]\Big\}. \qquad (1.5)$$

The functions $\mathcal{T}_X(b)$, $Q_X(b)$ count the number of integer points in the polytopes $\Pi_X(b)$, $\Pi_X^1(b)$.

Alternatively, $\mathcal{T}_X(b)$ counts the number of ways in which we can decompose $b = t_1 a_1 + \cdots + t_m a_m$ with t_i nonnegative integers. From this comes the name *partition function*.

Exercise. Every convex polytope can be presented in the form $\Pi_X(b)$ for suitable X, b.

1.3.2 Faces

Clearly, the polytope $\Pi_X(b)$ is empty unless b lies in the cone $C(X)$ defined by formula (1.3). We need some simple geometric properties of this cone.

Lemma 1.46. *A point b is in the interior of $C(X)$ if and only if it can be written in the form $b = \sum_{i=1}^m t_i a_i$, where $t_i > 0$ for all i.*

Proof. Assume first that $b = \sum_{i=1}^m t_i a_i$, $t_i > \epsilon > 0$, $\forall i$. If (say) $a_1,\dots,a_s$ is a basis of V, then $b - \sum_{j=1}^s s_j a_j \in C(X)$ whenever all the $|s_i|$ are less than ϵ. This proves that b is in the interior of $C(X)$. Conversely, let b be in the interior and choose an expression $b = \sum_{i=1}^m t_i a_i$ for which the number of i with $t_i > 0$ is maximal. We claim that in fact, all the t_i are strictly greater than 0. If not, then for some i, for instance $i = 1$, we have $t_1 = 0$. Since b is in the interior of $C(X)$, we must have that for some $\epsilon > 0$, we have $b - \epsilon a_1 \in C(X)$; hence we can write $b - \epsilon a_1 = \sum_{i=1}^m t_i' a_i$, $t_i' \geq 0$, and we have $b = \frac{1}{2}[(\epsilon + t_1')a_1 + \sum_{i=2}^m (t_i + t_i')a_i]$. This expression has a larger number of positive coefficients, contradicting the maximality.

From this lemma we want to derive the following result

Proposition 1.47. *If x is in the interior of $C(X)$, the polytope $\Pi_X(x)$ has dimension $m - s$; hence its volume is strictly positive.*

Proof. In fact, assume that $a_1,\dots,a_s$ is a basis, so that we have that $a_j = \sum_{i=1}^s c_{ji} a_i, \forall j > s$. Write $x = \sum_{i=1}^s c_i a_i$. From this expression we obtain

$$\Pi_X(x) = \Big\{(t_1,\dots,t_m) \mid t_i \geq 0,\ \text{and } t_i = c_i - \sum_{j=s+1}^m t_j c_{ji},\ i = 1,\dots,s\Big\}.$$

In other words, we represent the polytope in the $(m - s)$-dimensional space of coordinates t_i, $i = s + 1, \ldots, m$, as

$$\Pi_X(x) = \left\{ (t_{s+1}, \ldots, t_m) \,|\, t_i \geq 0, \quad c_i - \sum_{j=s+1}^{m} t_j c_{ji} \geq 0, \ i = 1, \ldots, s \right\}.$$

If x lies in the interior of $C(X)$, we can choose a point $(t_{s+1}, \ldots, t_m) \in \Pi_X(x)$ at which all the functions $t_i, i = 1, \ldots, m$, are strictly larger than 0. By Lemma 1.24, this point is the interior of $\Pi_X(x)$, which therefore is of dimension $m - s$.

By definition, $\Pi_X(b)$ is defined in the space of solutions of the linear system $At = b$ by the family of linear inequalities $t_i \geq 0$. We apply Lemma 1.24 in order to describe its faces. Given a subset $S \subset X$, we define the face $\Pi_X^S(b) := \{(t_1, \ldots, t_m) \in \Pi_X(b) \,|\, t_i = 0, \ \forall a_i \in S\}$. On the other hand, if we start with a face F, we can take the subset $S_F = \{a_i \in X \,|\, t_i = 0 \text{ on } F\}$. We obtain clearly that $\Pi_X^S(b) = \Pi_{X \setminus S}(b)$ and have the following

Proposition 1.48. *1. $\Pi_X^S(b) \neq \emptyset$ if and only if $b \in C(X \setminus S)$.*

2. A coordinate t_i vanishes on $\Pi_X^S(b) = \Pi_{X \setminus S}(b)$ if and only if we have $-a_i \notin C((X \setminus S) \setminus \{a_i\}, -b)$.

Proof. The first part is clear, so let us prove the second.

We have that t_i does not vanish on $\Pi_{X \setminus S}(b)$ if and only if $a_i \in X \setminus S$ and there is an expression $\sum_{a_h \in X \setminus S} t_h a_h = b$ with $t_h \geq 0$, $t_i > 0$. This is equivalent to saying that $-a_i \in C((X \setminus S) \setminus \{a_i\}, -b)$.

1.3.3 Cells and Strongly Regular Points

Given any basis $\underline{b} = (b_1, \ldots, b_s)$ of $\mathbb{R}^s$, denote by $C(\underline{b})$ the positive quadrant spanned by $\underline{b}$. This induces a partition of $\mathbb{R}^s$ into $2^s + 1$ parts, the complement of $C(\underline{b})$ and the 2^s convex parts of the cone $C(\underline{b})$ given by the points $\sum_i t_i b_i$, $t_i \geq 0$, where the t_i vanish only on a prescribed subset of the s indices.

Theorem 1.49 (Carathéodory's theorem). *The cone $C(X)$ is the union of the quadrants $C(\underline{b})$ as $\underline{b}$ varies among the bases extracted from X.*

Proof. We proceed by induction on the cardinality of X. If X is a basis there is nothing to prove. Otherwise, we can write $X = (Y, z)$, where Y still generates V. Take now an element $u = \sum_{y \in Y} t_y y + tz \in C(X)$, $t_y, t \geq 0$. By induction, we can rewrite $u = \sum_{a \in \underline{b}} t_a a + tz$, where $\underline{b}$ is a basis and all $t_a \geq 0$. We are thus reduced to the case $X = \{v_0, \ldots, v_s\}$, spanning V, and $u = \sum_{i=0}^{s} t_i v_i$, $t_i \geq 0$. Let $\sum_{i=0}^{s} \alpha_i v_i = 0$ be a linear relation. We may assume that $\alpha_i > 0, \forall i \leq k$ for some $k \geq 0$ and $\alpha_i \leq 0, \forall i > k$. Let us consider this minimum value of the t_i / α_i, $i \leq k$. We may assume that the minimum is taken for $i = 0$. Substituting v_0 with $-\sum_{i=1}^{s} \alpha_i / \alpha_0^{-1} v_i$ we see that u is expressed as combination of $v_1, \ldots, v_s$ with nonnegative coefficients.

Definition 1.50.   (i) Let $b \in C(X)$. A basis $a_{i_1}, \ldots, a_{i_s}$ extracted from X with respect to which b has positive coordinates will be called *b-positive*.

  (ii) A point $p \in C(X)$ is said to be *strongly regular* if there is no sublist $Y \subset X$ lying in a proper vector subspace, such that $p \in C(Y)$.

  (iii) A point in $C(X)$ which is not strongly regular will be called *strongly singular*.

  (iv) A connected component of the set $C^{\mathrm{reg}}(X)$ of strongly regular points will be called *a big cell*.

Remark 1.51. The notion of strongly regular should be compared with that of *regular for hyperplane arrangements*; cf. Section 2.1.2. In this case the arrangement is formed by all the hyperplanes of V that are generated by subsets of X and a regular point is one which does not lie in any of these hyperplanes.

Example 1.52. Let us show in cross section the cone and the big cells for the positive roots of A_3.

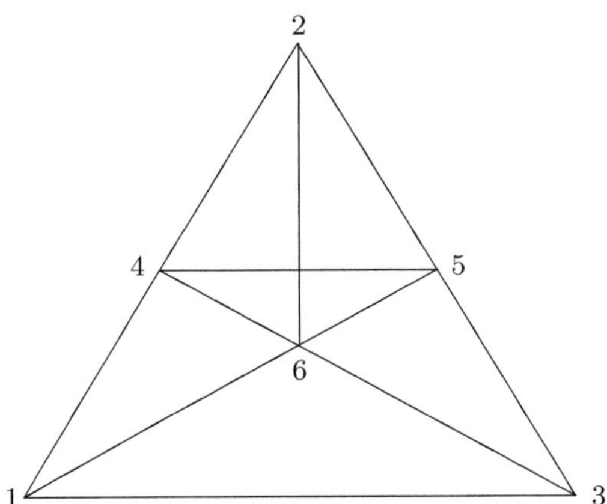

Fig. 1.1. The big cells for type A_3 in cross section

As one sees the big cell with vertices $1, 3, 6$ contains points which lie on a plane of the arrangement, so they are not regular.

With the notion of stongly singular points we obtain a stratification of $C(X)$ in which the big cells are open strata. By construction, if Ω is a big cell, its boundary $\partial(\Omega)$ is contained in the set of singular points, and unless the closure $\overline{\Omega}$ equals $C(X)$ (this happens only when X is linearly independent), we have that $\partial(\Omega)$ disconnects $C(X)$.

Let us denote by $C^{\mathrm{sing}}(X)$ the set of singular points. It is the union of finitely many cones $C(\underline{c})$, where $\underline{c} \subset X$ is a basis of a hyperplane. In particular,

$C^{\text{sing}}(X)$ is an $(s-1)$-dimensional polyhedron. Inside the singular points we have those points that are either on the boundary of one of the cones $C(\underline{c})$ or in the intersection of two of these cones lying in different hyperplanes. In the previous example we have six points of the first type and one of the second. This set we denote by $C^{\text{Sing}}(X)$, and it is even of dimension $s-2$; its complement in $C^{\text{sing}}(X)$ is formed by points q that are in the interior of some $(s-1)$-dimensional cones $C(\underline{c}_i)$ generated by bases of a fixed hyperplane K and in no other cone. They have the property that there is a ball $B(\epsilon)$ of some radius ϵ centered at q and a hyperplane K (the one defined above) passing through q such that $B(\epsilon) \cap K = B(\epsilon) \cap C^{\text{sing}}(X)$. Let us denote by $C^{\text{sing}}(X)^0 := C^{\text{sing}}(X) \setminus C^{\text{Sing}}(X)$.

Proposition 1.53. *Let p be a strongly regular point. The big cell containing p is the intersection of the interiors of all the quadrants $C(\underline{b})$ containing p in their interior.*

Proof. Let Ω denote the big cell to which p belongs. Take a basis $\underline{b} \subset X$. Clearly, the boundary of $C(\underline{b})$ disconnects the space. Thus if $\Omega \cap C(\underline{b}) \neq \emptyset$, we must have that $\Omega \subset \mathring{C}(\underline{b})$, where $\mathring{C}(\underline{b})$ denotes the interior of $C(\underline{b})$. Since $C(X)$ is the union of the cones $C(\underline{b})$ as $\underline{b} \subset X$ runs over all the bases, we must have at least one basis $\underline{b}$ with $\Omega \subset \mathring{C}(\underline{b})$.

Let now $\Omega' := \cap_{p \in \mathring{C}(\underline{b})} \mathring{C}(\underline{b})$. We have seen that $\Omega \subset \Omega'$ and need to see that they are equal. Otherwise, since Ω' is convex and thus connected, the boundary of Ω must intersect Ω' in a nonempty set. We claim that this must contain a point $q \in C^{\text{sing}}(X)^0$. In fact, since Ω is a connected component of the set of regular points and Ω' is open, we must have that the boundary of Ω disconnects Ω'; hence it cannot be contained in the $(s-2)$-dimensional space $C^{\text{Sing}}(X)$. Take now the cone $C(\underline{c})$ generated by a basis of a hyperplane K and a ball $B(\epsilon)$ centered at q as before. We must have that $\Omega \cap B(\epsilon)$ contains one of the two half-balls, while we also must have $B(\epsilon) \subset \Omega'$ and hence $B(\epsilon) \subset C(X)$. Therefore, there must be points in X contained in both the open half-spaces into which K divides the space V. Therefore, choose a point v in the same half-space in which Ω meets $B(\epsilon)$. We have that the interior of the cone generated by the basis $\underline{c}, v$ is contained in this half-space, and by our initial remarks, that Ω is contained in this interior. Thus Ω' is also contained in this interior, and we have a contradiction.

Definition 1.54. Given a lattice $\Lambda \subset V$ and a finite set $X \subset \Lambda$ generating V, the *cut locus* is the union of all hyperplanes that are translations, under elements of Λ, of the linear hyperplanes spanned by subsets of X.

We want to point out a simple relation between the notion of singular points and that of cut locus.

Proposition 1.55. *The cut locus is the union of all the translates, under Λ, of the strongly singular points in $C(X)$.*

Proof. Clearly this union is contained in the cut locus. Conversely, given a vector v in the cut locus, let us write $v = \lambda + \sum_{a \in Y} x_a a$, $\lambda \in \Lambda$, where Y is a sublist of X contained in a hyperplane. Clearly, there is integer linear combinations $\mu = \sum_{a \in Y} n_a a$, such that $n_a + x_a > 0$ for each $a \in Y$. So $v = \lambda - \mu + \sum_{a \in Y}(n_a + x_a)a$, and $\sum_{a \in Y}(n_a + x_a)a$ is strongly singular.

We have thus that the complement of the cut locus is the set of regular points for the periodic arrangement generated, under translation by Λ, by the arrangement $\mathcal{H}_X^*$ of hyperplanes generated by subsets of X.

1.3.4 Vertices of $\Pi_X(b)$.

Let us understand the vertices of $\Pi_X(b)$ using the concepts developed in the previous sections in the case in which b is a strongly regular point. Such a vertex must be of the form $\Pi_Y(b)$ for some sublist Y with $X \setminus Y = S_{\{b\}} = \{a_i \in X \mid t_i = 0 \text{ on } b\}$. We have (recall Definition 1.50) the following theorem

Theorem 1.56. *Let $b \in C(X)$ be a strongly regular point. Then*

   (i) *The vertices of $\Pi_X(b)$ are of the form $\Pi_Y(b)$ with Y a b-positive basis.*

   (ii) *The faces of $\Pi_X(b)$ are of the form $\Pi_Z(b)$ as Z runs over the subsets of X containing a b-positive basis. Such a face has dimension $|Z| - s$ and its vertices correspond to the positive bases in Z.*

  (iii) *Around each vertex the polytope is simplicial. This means that near the vertex the polytope coincides with a quadrant.*

Proof. From Proposition 1.48 we know that the faces of $\Pi_X(b)$ are of the form $\Pi_Y(b)$, where $Y \subset X$ and $b \in C(Y)$. Since b is strongly regular for X it is also strongly regular for Y, and thus for any Y for which $b \in C(Y)$ we have from Proposition 1.47 that $\Pi_Y(b)$ has dimension $|Y| - s$, and thus (1) and (2) follow immediately.

It remains to prove (3). Let Y be the b-positive basis such that our vertex is $q := \Pi_Y(b)$. For simplicity assume that $Y = (a_1, \ldots, a_s)$ and write $b = \sum_{i=1}^{s} t_i a_i$, $t_i > 0$. Since Y is a basis, there exist linear functions $x_i(u_{s+1}, \ldots, u_m)$, $i = 1, \ldots, s$, such that $\sum_{i=s+1}^{m} u_i a_i = \sum_{i=1}^{s} x_i a_i$. Therefore, the linear system $\sum_{i=1}^{m} u_i a_i = b = \sum_{i=1}^{s} t_i a_i$ can be transformed as $\sum_{i=1}^{s}(u_i + x_i(u_{s+1}, \ldots, u_m))a_i = \sum_{i=1}^{s} t_i a_i$; hence the affine space of solutions of $\sum_{i=1}^{m} u_i a_i = b$ is parametrized by $(u_{s+1}, \ldots, u_m)$, where we set $u_i = t_i - x_i(u_{s+1}, \ldots, u_m)$ for each $i = 1, \ldots, s$. In this parametrization the polytope $\Pi_X(b)$ can be identified with the polytope of $(m - s)$-tuples $u_{s+1}, \ldots, u_m$ with $u_j \geq 0$ for $j = s + 1, \ldots, m$ and $t_i \geq x_i$ for $i = 1, \ldots, s$.

At our vertex q, the coordinates u_j vanish for $j = s + 1, \ldots, m$, and thus we also get $x_i = 0$ at q for each $i = 1, \ldots, s$. Therefore for small enough $u_j \geq 0$, $j = s + 1, \ldots, m$ the conditions $t_i \geq x_i$ are automatically satisfied. It follows that in a neighborhood of q our polytope coincides with the quadrant $u_i \geq 0$, $i = s + 1, \cdots, m$.

The condition that a set Z corresponds to a face of $\Pi_X(b)$ is clearly independent of b as long as b varies in a big cell. According to Definition 1.26 we have the following corollary:

Corollary 1.57. *The polytopes $\Pi_X(b)$, as b varies in a big cell, are all combinatorially equivalent.*

When we pass from a big cell Ω to another, the set of bases Y that are positive on the cell, i.e., for which $C(Y) \supset \Omega$, clearly changes. Nevertheless, it is still possible that combinatorially the corresponding polytopes are equivalent. For instance, in the example of B_2 or the ZP element. We have 3 cells on two of which the polytope is a triangle while in the third it is a quadrangle. We can generalize this example:

Example 1.58. Consider m distinct nonzero vectors $a_1, \ldots, a_m$ in $\mathbb{R}^2$, which we order by the clockwise orientation of their arguments, that generate a pointed cone. The cone they generate is then the quadrant defined by a_1 and a_m. A strongly regular vector b in this quadrant appears uniquely in the given order, between some a_h and a_{h+1}; thus a basis $\{a_i, a_j\}$ is b-positive if and only if $i \leq h$, $j \geq h+1$. We have thus $h(m-h)$ vertices in the corresponding $m-2$ dimensional polytope. For $m = 4$ we have the previous case.

In this example one sees quite clearly how the combinatorics of the faces of the polytope $\Pi_X(b)$ changes as b passes from the big cell generated by a_h, a_{h+1} to the next cell generated by a_{h+1}, a_{h+2}. Nevertheless, there is a symmetry passing from h to $m - h$.

Example 1.59.

$$X = \begin{vmatrix} 1 & 1 & 0 & -1 \\ 1 & 0 & 1 & 1 \end{vmatrix}$$

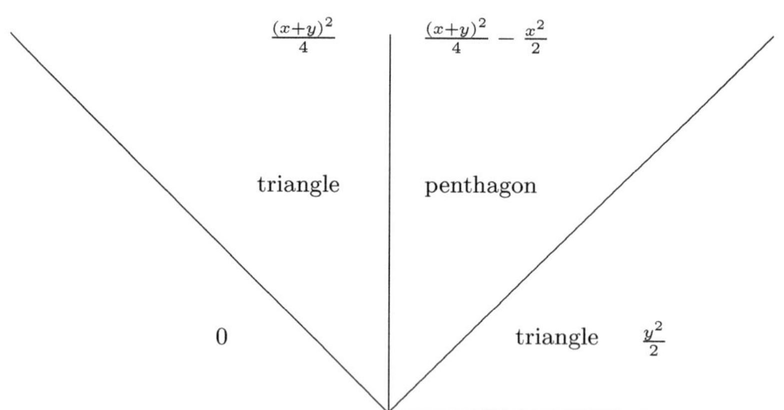

Fig. 1.2. The area and the shape as $b = (x, y)$ varies in the cone.

1.3.5 Piecewise Polynomial Functions

In the theory of the multivariate spline, which we will discuss in the second part, polyhedral cones are associated to piecewise polynomial functions such as, for instance, the area that we described in example (1.59). We shall make the following definition:

Definition 1.60. Given a polyhedron A, a family $\{A_i\}_{i \in I}$ of polyhedra is called a locally finite decomposition of A into polyhedra if

  (i) $A = \cup_{i \in I} A_i$;
  (ii) for every point $p \in A$ there is a neighborhood of p which meets only finitely many A_i's;
  (iii) for any two indices $i \neq j$, the intersection $A_i \cap A_j$ is a face of both.

Given a function f on a polyhedron A, we shall say that f is *piecewise polynomial* if A has a locally finite decomposition into polyhedra A_i such that f restricted to the (relative) interior of each A_i coincides with a polynomial.

In numerical analysis (cf. [96]), sometimes the word *spline* is used as equivalent to *piecewise polynomial function with suitable smoothness conditions*.

For example, given a positive integer m, assume that we have a function that is locally a polynomial of degree $\leq m$.

Gluing together polynomials often involves allowing discontinuity of some higher-order derivatives when one crosses a common face of codimension-one of two polyhedra, which we call a *wall*. It is easily seen that in this way, besides the functions that are globally polynomials, we may obtain functions of class at most C^{m-1}.

In particular, one can consider piecewise linear functions. A way in which these functions appear in the theory is the following.

Example 1.61. Take n polynomial functions of degree ≤ 1, $a_i(x)$, $i = 1, \ldots, n$ on V, and set $s(x) := \max a_i(x)$ (or $\min a_i(x)$). Then V decomposes in the (possibly overlapping) pieces $V_i := \{x \in V \mid s(x) = a_i(x)\}$. Of course, V_i is defined by the inequalities $a_i(x) - a_j(x) \geq 0$, $\forall j \neq i$, and it is thus a polyhedron. We have therefore decomposed V into polyhedra on each of which $s(x)$ coincides with one of the linear functions $a_i(x)$.

If the a_i are also homogeneous, then $s(x)$ is piecewise linear with respect to a decomposition into cones C_i.

2

Hyperplane Arrangements

This chapter is an introduction to the theory of hyperplane arrangements, which appear in our treatment as the hyperplanes associated to the list of linear equations given by a list X of vectors. We discuss several associated notions, including matroids, Tutte polynomials, and zonotopes. Finally, we expand the example of root systems and compute in this case all the Tutte polynomials.

2.1 Arrangements

2.1.1 Hyperplane Arrangements

We shall use several types of *arrangements*. Let us start from the simplest kind.

Definition 2.1. A *linear arrangement* $\mathcal{H}$ in an affine space U is a family of affine hyperplanes.

An arrangement in an affine space U (over $\mathbb{R}$ or $\mathbb{C}$) is *locally finite* if given any point $p \in U$, there is a neighborhood of p that meets only finitely many elements of the arrangement.

In our setting, two types of arrangements appear. We start from a finite list of vectors X in $V = U^*$ and numbers $\mu := (\mu_a)$, $a \in X$. The hyperplanes $H_a := \{u \in U \mid \langle a \mid u \rangle + \mu_a = 0, \ a \in X\}$ define an affine hyperplane arrangement, given by explicit linear equations. We shall denote this arrangement by $\mathcal{H}_{X,\mu}$. There is also another arrangement associated to X, this time in V, which we occasionally use. When X spans V one can also consider the arrangement $\mathcal{H}_X^*$ formed by all the hyperplanes of V that are generated by subsets of X.

The second type of arrangement will be a *periodic arrangement*. Fix a lattice Λ in U and a finite hyperplane arrangement $\mathcal{H}_0$ with the property that any hyperplane $H \in \mathcal{H}_0$ has an equation of the form $\langle v \mid x \rangle - b = 0$ with $v \in U^*$

C. De Concini and C. Procesi, *Topics in Hyperplane Arrangements, Polytopes and Box-Splines,* Universitext, DOI 10.1007/978-0-387-78963-7_2,
© Springer Science+Business Media, LLC 2010

such that $\langle v \mid \lambda \rangle \in \mathbb{Z}$, $\langle b \mid \lambda \rangle \in \mathbb{Z}$ for $\lambda \in \Lambda$. We associate to $\mathcal{H}_0$ the *periodic arrangement* $\mathcal{H}$ consisting of all the hyperplanes $H + \lambda$, $H \in \mathcal{H}_0$, $\lambda \in \Lambda$.

We leave it to the reader to verify that a periodic arrangement is locally finite.

Any (nonempty) affine subspace obtained as the intersection of a set of hyperplanes of the arrangement is called a *space of the arrangement* or also a *flat*. In particular, the zero-dimensional subspaces of the arrangement are called *points of the arrangement* and denoted by $P(X, \mu)$.

We shall denote the set of spaces of the arrangement $\mathcal{H}$ by $\mathcal{Z}(\mathcal{H})$. In the case of an arrangement of the form $\mathcal{H}_{X,\mu}$ for a finite list of vectors X in $V = U^*$ and numbers $\underline{\mu} := \{\mu_a, \ a \in X\}$, we shall set $\mathcal{Z}(\mathcal{H}_{X,\mu}) = \mathcal{Z}(X, \underline{\mu})$.

The set of spaces of an arrangement is a *ranked poset*.[1] We shall take as definition of ranked poset the following.

Definition 2.2. A ranked poset P is a partially ordered set P with a minimum element 0, such that given any element $a \in P$, all the maximal chains $0 < x_1 < x_2 < \cdots < x_k = a$ have the same length k.

The number k is called the *rank* of the element a, denoted by $\varrho(a)$.

In the theory of hyperplane arrangements, it is usual to choose *opposite inclusion* as the partial order. Thus the minimal element is the entire space. The fact that the poset is ranked is an obvious consequence of the fact that a proper nonempty intersection of a subspace A with a hyperplane H has dimension $\dim(A \cap H) = \dim(A) - 1$. Thus, the *rank function* $\varrho(A)$ is the codimension of the subspace A. This is the first combinatorial object associated to the arrangement.

If all the hyperplanes of the arrangement have a nonempty intersection A, then the arrangement is said to be *central* and A is its center. In this case, one can, up to translations, assume $0 \in A$, so that all the spaces of the arrangement are vector spaces. Then one can decompose $V = A \oplus B$ using some complement B to A and see that the spaces of the arrangement are all of the form $A \oplus C$, where the spaces C are the subspaces of an arrangement in B with center 0.

A subspace K of the arrangement determines two induced arrangements. The first is given by all the hyperplanes of the arrangement containing K. This is an arrangement with center K.

To define the second arrangement, take the hyperplanes H of the arrangement that do not contain K. They cut on K a hyperplane arrangement given by the elements $K \cap H$. We denote this arrangement by $\mathcal{H}_K$. Notice that if $\mathcal{H}$ is finite, so is $\mathcal{H}_K$, and if $\mathcal{H}$ is periodic with respect to the lattice Λ, then $\mathcal{H}_K$ is periodic with respect to the lattice $\Lambda \cap K$.

Example 2.3. The arrangement $\mathcal{H}_{X,\underline{\mu}}$.

---

[1] Poset $=$ **partially ordered set**.

Given a subspace W of the arrangement associated to a list of vectors X and parameters μ, we set

$$X_W := \{a \in X \mid a + \mu_a = 0, \text{ on } W\}. \tag{2.1}$$

When $p \in P(X, \mu)$ we set $X_p := X_{\{p\}}$. Clearly, W is of codimension k if and only if X_W spans a space of dimension k. Any basis of this space gives a (minimal) set of equations for W.

In particular, to a basis $\underline{b} := (b_1, \ldots, b_s)$ extracted from X we associate the unique point $p_{\underline{b}} \in U$ such that $\langle b_i \mid p_{\underline{b}} \rangle = -\mu_{b_i}$ for each $i = 1, \ldots, s$.

The set $P(X, \mu)$ consists of the points $p_{\underline{b}}$ as $\underline{b}$ varies among the bases extracted from X.

For generic μ all these points are distinct (cf. Section 8.1.4), while for $\mu = 0$ they all coincide with 0. In the other cases we may have various ways in which the points $p_{\underline{b}}$ will coincide, depending on the compatibility relations among the parameters μ.

It is clear by definition that if we restrict to the subset X_p, the points of this restricted arrangement reduce to p. Moreover, by a translation we can center the arrangement at 0.

The arrangement $\mathcal{H}_{X, \mu}$ depends strongly on the parameters μ. We will see how the combinatorics and the formulas change from the generic values of μ to the degenerate case in which all the μ_i's equal 0.

2.1.2 Real Arrangements

In the case of a vector space over the reals, the complement of a hyperplane is formed by two disjoint open half-spaces.

Let us fix a locally finite real arrangement $\mathcal{H} := \{H_i\}_{i \in I}$ and set $\mathcal{A}_\mathcal{H}$ to be the complement of the union of all the hyperplanes in $\mathcal{H}$. A point $p \in \mathcal{A}_\mathcal{H}$ is called a *regular point of the arrangement*.

Definition 2.4. The connected components of the complement $\mathcal{A}_\mathcal{H}$ of the arrangement are called *chambers*.

More generally, we make the following definition

Definition 2.5. Let K be a space of the arrangement $\mathcal{H}$. A chamber of the arrangement $\mathcal{H}_K$ cut by $\mathcal{H}$ on K, is called a *face* of $\mathcal{H}$.

Let us fix an equation $f_H(x) = 0$ for each hyperplane H of the arrangement $\mathcal{H}$. A point p is in $\mathcal{A}_\mathcal{H}$ if and only if $f_H(p) \neq 0,\ \forall\, H \in \mathcal{H}$. Define a (sign) function $s : U \times \mathcal{H} \to \{+1, -1, 0\}$ by

$$s(p, H) := \begin{cases} 0 & \text{if } f_H(p) = 0, \\ f_H(p)/|f_H(p)| \in \{+1, -1\} & \text{otherwise.} \end{cases}$$

These functions define an equivalence relation by setting

$$p \equiv q \iff s(p, H) = s(q, H),\ \forall H \in \mathcal{H}.$$

Proposition 2.6. *1. Each equivalence class is a convex set.*

2. $\mathcal{A}_{\mathcal{H}}$ is a union of equivalence classes that are its connected components.

3. The equivalence classes coincide with the faces of the arrangement.

Proof. 1. The first statement is clear.

2. A point p lies in $\mathcal{A}_{\mathcal{H}}$ if and only if $s(p, H) \neq 0$, $\forall H \in \mathcal{H}$. If p, q lie in two different equivalence classes in $\mathcal{A}_{\mathcal{H}}$, there must be an H such that $s(-, H)$ takes different signs on the two points. Hence p, q lie in opposite open half-spaces with respect to H.

3. Given a point p, consider $\mathcal{H}_p := \{H \in \mathcal{H} \mid f_H(p) = 0\}$. Then p is a regular point of the arrangement induced on the subspace of equations $f_H(x) = 0$ for $H \in \mathcal{H}_p$. Thus by statement 2, we have the required equality.

Theorem 2.7. *(1) If $\mathcal{H}$ is a finite arrangement, each of its faces is the relative interior of a polyhedron.*

(2) If $\mathcal{H}$ is finite and has center 0 each chamber is the interior of a pointed polyhedral cone.

Assume that $\mathcal{H}$ is periodic with respect to a lattice Λ and there is a space of the arrangement consisting of a point:

(3) The chambers of $\mathcal{H}$ are bounded polytopes.

(4) There are only finitely many distinct chambers up to translation by elements of Λ.

Proof. (1) We can reduce immediately to the case of regular points and chambers. As we have seen, the component of a regular point $p \in \mathcal{A}_{\mathcal{H}}$ is given by the inequalities $f_i(p) > 0$ if $s_i(p) > 0$, $f_i(p) < 0$ if $s_i(p) < 0$. Since I is finite these define the interior of a polyhedron.

(2) Since our arrangement is centered at zero, it is clear that its chambers are polyhedral cones. Now by assumption there are hyperplanes $H_1, \ldots, H_s$ in $\mathcal{H}$ whose intersection is 0. Thus each chamber is contained in a quadrant relative to these hyperplanes, and hence it is pointed.

(3) By our assumptions we must have s linearly independent hyperplanes in the arrangement meeting in a point p such that for some $h \in \mathbb{N}$, $hp \in \Lambda$. Applying translations and choosing coordinates x_i, we can assume that the point is 0 and Λ is of finite index in $\mathbb{Z}^s$. The periodic arrangement generated by the equations $x_i = 0$ and by Λ has as chambers parallelepipeds that are convex polytopes. It follows that the chambers of $\mathcal{H}$ are contained in these parallelepipeds and hence are bounded. Each parallelepiped P intersects only finitely many hyperplanes of the arrangement; hence we have only finitely many chambers contained in P. Each of these chambers is defined by a finite number of linear inequalities, and hence it is the interior of a polyhedron.

(4) From the proof of statement 3, we have that the chambers are obtained by decomposing the parallelepipeds associated to the periodic arrangement generated by a basis. These parallelepipeds form a unique orbit under translation, so each chamber up to translation can be brought into a fixed parallelepiped.

Proposition 2.8. *The closure of a face is a union of faces.*

Proof. We claim that a point p is in the closure of a face F if and only if, for each of the linear functions f_i we have

(i) If f_i vanishes on F then it vanishes on p.
(ii) If f_i is positive on F we have $f_i(p) \geq 0$.
(iii) If f_i is negative on F we have $f_i(p) \leq 0$.

In fact, if p is in the closure of a face F, it clearly satisfies these properties. Conversely, take a point p that satisfies these properties and choose any point $q \in F$. We see immediately that the segment $(p, q]$ lies entirely in F. This implies our claim.

Remark 2.9. One easily sees that the notion of face for hyperplane arrangements coincides, for the closure of a chamber, with that given for polytopes or cones.

Also, given two chambers, the intersection of their closures is either empty, or a common face.

2.1.3 Graph Arrangements

Given a finite set V let us denote by $\binom{V}{2}$ the set consisting of subsets of V with two elements.

Definition 2.10. A graph Γ with labeled edges is a pair $\Gamma := (V, m)$ where V is a finite set called the set of *vertices* and m is a function $m : \binom{V}{2} \to \mathbb{N}$.

The subset $L \subset \binom{V}{2}$ consisting of elements $\{u, v\} \in \binom{V}{2}$ for which $m(\{u, v\}) > 0$ is called the set of *edges*. Given an edge $a \in L$, $m(a)$ is called the multiplicity of a. We can realize such a combinatorial object geometrically as a topological space $|\Gamma|$ by joining two vertices u, v that form an edge with a segment that we denote by $[u, v]$ labeled by the integer $n := m([u, v])$. We have thus a 1-dimensional simplicial complex. One can realize it concretely as subset of the $|V|$-dimensional vector space with basis the elements of V by taking the union of all the segments joining two basis elements which form an edge each with its appropriate label. We then have the obvious notion of a path (sequence of edges) connecting two vertices and of connected components.

When we want to stress the fact that the vertices V are the vertices of the graph Γ we denote them by V_Γ.

Definition 2.11. A *cycle* of a graph Γ is a sequence of distinct vertices $v_1, v_2, \ldots, v_{k-1}$ such that, setting $v_k = v_1$, the sets $\{v_i, v_{i+1}\}$, $i = 1, \ldots, k-1$ are all edges. A cycle is *simple* if no proper sublist is also a cycle.

A graph is called a *tree* if it is connected and has no cycles.

Often, when taking a subset of the set of edges we will speak of the graph they generate, meaning that the vertices are exactly all the vertices of the given set of edges.

It is convenient to orient the graph Γ. This means that each set in L is represented by a chosen ordered pair (u, v). Once we do this, each $a \in L$ has an initial vertex $i(a)$ and a final vertex $f(a)$.

From homology theory one has in this case a simple construction.

Consider the two free abelian groups C_1, C_0 with basis the edges and the vertices and the *boundary map* δ defined on an edge a by $\delta(a) := f(a) - i(a)$. Then the kernel and cokernel of δ are the homology groups $H_1(|\Gamma|), H_0(|\Gamma|)$ of the topological space $|\Gamma|$ associated to the graph.[2]

Denote by b_1, b_0 the two Betti numbers of the graph, that is, the dimension of the two homology groups. One easily sees that b_0 is the number of connected components of Γ and if $p := |V|$, $\ell = |L|$ (the number of vertices and edges), we have that $p - \ell = b_0 - b_1$ is the Euler characteristic.[3]

Elementary topology shows that a graph is a tree if and only if $b_0 = 1$, $b_1 = 0$, that is, the graph is connected and $p = \ell + 1$. Moreover, it is easy to interpret the number b_1 as the *number of independent cycles*.

Consider the space V_Γ^* of functions on the set V_Γ of all vertices with the restriction that on each connected component of $|\Gamma|$, the sum equals 0.

If for each edge a we take the hyperplane H_a of all functions taking the same value on the two vertices of a, we get a hyperplane arrangement $\mathcal{H}_\Gamma$ in V_Γ^*, which we shall call a graph arrangement.

In order to get equations of the hyperplanes H_a let us again orient Γ. Given a vertex $v \in V_\Gamma$, let e_v denote the characteristic function of the vertex v. Consider the set of vectors $\Delta_\Gamma := \{x_a := e_{f(a)} - e_{i(a)}\} \subset V_\Gamma^*$ as $a \in L$. Then x_a is a linear equation of the hyperplane H_a.

Notice that since the edges are labeled, once we fix a total ordering on L, we can also associate to Γ the list X_Γ consisting of the vectors x_a, $a \in L$ each repeated $m(a)$ times.

Lemma 2.12. *The vectors x_a span V_Γ^*.*

Proof. Clearly, the spaces spanned by vectors in different connected components form a direct sum, so it suffices to prove the formula when Γ is connected. These vectors span a space U contained in the subspace generated by the vectors e_v where the sum of the coordinates is 0. We claim that U coincides with this subspace; in fact, choose a vector e_v and add it to U then by connectedness, each e_w is in $U + Re_v$, whence the claim.

Proposition 2.13. *If Γ is connected, $v - 1$ edges a_i are such that the vectors x_{a_i} form a basis of V_Γ^* if and only if these edges span a maximal tree.*

---

[2] Of course, this is a trivial example of a chain complex.

[3] Our use of homology is extremely limited, and our statements can be easily justified.

Proof. Edges in a cycle add to zero, thus are linearly dependent, so the graph generated by a basis cannot contain cycles. Let Γ' be the graph generated by the edges a_i. By construction, Γ' has $\ell' = v - 1$ edges and no cycles. Suppose that it has k connected components and v' vertices. We thus have that Γ' is the union of k trees and $v' = \ell' + k = v - 1 + k$. Since $v' \leq v$ and $k \geq 1$, this implies $v' = v, k = 1$.

Conversely, a tree contained in Γ has at most v vertices and hence at most $v - 1$ edges. In any case, these edges are linearly independent and thus can be completed to a basis, so if the tree is maximal, it is associated to a basis.

We deduce the following

Proposition 2.14. *A sublist of the list X_Γ is linearly dependent if and only if either the corresponding edges contain a cycle or the list contains the same vector twice (we may think to this as the case in which we have a cycle consisting of two copies of the same edge).*

A minimal dependent list thus coincides with a cycle and it is called a *circuit*.[4]

2.1.4 Graphs Are Unimodular

The previous proposition has a simple but important consequence. In order to explain it, let us observe that the vectors x_a for $a \in L$ span a lattice Λ in V_Γ^*.

The property to which we are referring, that of being *unimodular,* can in fact be defined, in the case of a list of vectors X in a lattice Λ of covolume 1 in a vector space V with a fixed Lebesgue measure, by either of the following equivalent conditions.

Proposition 2.15. *Given a list of vectors X in Λ, the following are equivalent:*
   1. Any sublist Y of X spans a direct summand in Λ.
   2. Any basis $\underline{b}$ of V extracted from X has $|\det(\underline{b})| = 1$.

The proof is left to the reader.

Corollary 2.16. *A graph arrangement is unimodular.*

Proof. We can reduce to the case that Γ is connected. Thus a basis corresponds to a maximal tree; clearly it is enough to see that each such basis $\underline{b}$ is also a basis of the lattice Λ. To see this choose an edge a not in $\underline{b}$, complete a to a circuit $a = a_0, a_1, \ldots, a_k$ with $a_i \in \underline{b}$, $\forall i \geq 1$. Since we have a cycle the sum of these elements with coefficients ± 1 is 0. This implies that a lies in the lattice spanned by $\underline{b}$, which is then equal to Λ giving the claim.

---

[4] In matroid theory a circuit is a minimal dependent set, the axioms of matroids can also be formulated in term of circuits.

There is a dual form in which unimodular lists are associated to graphs. Take the spaces $C_1(\Gamma)$ with basis the edges of the graph, which we assume oriented, and $C_0(\Gamma)$ with basis the vertices of the graph. We have the *boundary map* from $C_1(\Gamma)$ to $C_0(\Gamma)$, mapping an edge e with vertices a, b to $\delta(e) = a - b$ (if a is the initial vertex and b the final). The kernel of this map is called the *homology* $H_1(\Gamma)$. We have a surjective dual map $C_1(\Gamma)^* \to H_1(\Gamma)^*$, so the edges induce a list of vectors in $H_1(\Gamma)^*$. This list is unimodular in the lattice they span. In fact, a deep Theorem of Seymour implies that any unimodular list of vectors in a lattice is constructed through a simple rule from the the two lists associated to a graph in one of the two previous ways, and from a unique exceptional list denoted by R_{10} [100].

If the given orientation of our graph is such that the vectors x_a span an acute cone, i.e., they are all on the same side of a hyperplane which does not contain any of them, we will say that the graph with its orientation is a *network*. Let us explain how to obtain networks.

Proposition 2.17. *1. Fix a total order on the set of vertices and orient the edges according to the given order. Then the corresponding oriented graph is a network.*

2. An oriented graph is a network if and only if it does not contain oriented cycles.

3. In every network we can totally order the vertices in a way compatible with the orientation of the edges.

Proof. 1. Suppose we have a total order on the vertices. Consider any linear form α such that $\langle \alpha \,|\, e_v \rangle > \langle \alpha \,|\, e_u \rangle$ if $v > u$. It is the clear that $\langle \alpha \,|\, x_a \rangle > 0$ for each edge $a \in L_\Gamma$.

Let us now prove 2 and 3.

An oriented cycle $a_1, \ldots, a_k$ gives vectors $x_{a_1}, \ldots, x_{a_k}$ with $x_{a_1} + \cdots + x_{a_k} = 0$. This is impossible in a network.

Conversely, assume that there are no oriented cycles. Take a maximal oriented chain $a_1, \ldots, a_k$. This chain is not a cycle, and necessarily $i(a_1)$ is a source. Otherwise, we could increase the oriented chain. We take this source to be the smallest vertex, remove it and all the edges coming from it, and then start again on the remaining graph by recursion. We then obtain a total order on the vertices compatible with the orientation of the edges. By 1 our oriented graph is a network.

Since oriented graphs produce unimodular lists (Corollary 2.16), we may apply formula (2.16) to them. In this case we need to interpret the result. We write first the list X of vectors x_a in the basis $e_1, \ldots, e_v$ given by the v vertices. The entry $a_{i,j}$ of the matrix $A := XX^t$ has the following meaning:

1. When $i = j$ then $a_{i,i}$ equals the number of edges that have e_i as a vertex.
2. When $i \neq j$ then $-a_{i,j}$ equals the number of edges connecting e_i and e_j.

We call A the *incidence matrix* of the graph.

We cannot apply formula (2.16) directly, since the columns of X span a proper subspace. Thus in order to apply it, we take the image of these vectors in the quotient space, where we set one of the vectors e_i equal to 0. In the quotient space the lattice generated by the classes of the elements x_a coincides with the lattice with basis the classes of the elements e_j, $j \neq i$. In matrix notation this amounts to replacing X with the same matrix with the i-th row removed and hence replace A with the same matrix with both the i-th row and i-th column removed, and we obtain the following result

Proposition 2.18. *Let A be the incidence matrix of a graph Γ. Given any index i, let A_i be obtained from A by removing the i-th row and the i-th column. Then $\det(A_i)$ equals the number of maximal trees contained in Γ.*

A special important example is the *complete graph* of vertices $1, 2, \ldots, n$, which we can orient so that each edge goes from a higher-numbered vertex to a lower-numbered one. Its associated set of vectors $e_i - e_j$, $i < j$ is the root system of type A_{n-1}. For this example we have the following result known as Cayley's theorem

Example 2.19 (*Cayley's theorem*).
  The number of marked trees on n vertices is n^{n-2}.

Proof. This set equals the set of maximal trees contained in the complete graph on n vertices. For this graph, the matrix A is an $n \times n$ matrix with $n - 1$ on the main diagonal and -1 elsewhere. When we delete one row and the corresponding column we obtain the $(n - 1) \times (n - 1)$ matrix with $n - 1$ on the main diagonal and -1 elsewhere. Let us call this matrix $A(n - 1)$. We need to show that $\det(A(k)) = (k + 1)^{k-1}$. We compute as follows: let E_k denote the $k \times k$ matrix with all entries equal to 1, and I_k the identity matrix. E_k has rank 1, so it has 0 as an eigenvalue with multiplicity $k - 1$. Moreover, the vector with all coordinates 1 is an eigenvector with eigenvalue k. It follows that $\det(xI_k - E_k) = (x - k)x^{k-1}$. Substituting x with $k + 1$, one obtains the result.

2.2 Matroids

The next subsections deal with fundamental definitions that belong to the theory of matroids. Given the list X, we shall study its *matroid structure*, in other words, the combinatorics of the linearly dependent subsets of X. Let us recall, for the convenience of the reader, the notion of *matroid*. This is a concept introduced by Whitney in order to axiomatize the ideas of linear dependence [117], [118]; see also the book of Tutte [115]. We restrict our attention to the finite case and present one of the several possible axiomatic approaches.

Definition 2.20. A matroid M on a ground set E is a pair $(E, \mathcal{I})$, where $\mathcal{I}$ is a collection of subsets of E (called the independent sets) with the following properties:

(a) The empty set is independent.
(b) Every subset of an independent set is independent (hereditary property).
(c) Let A and B be two independent sets and assume that A has more elements than B. Then there exists an element $a \in A \setminus B$ such that $B \cup \{a\}$ is still independent (augmentation or exchange property).

A maximal independent set is called a *basis*. It follows immediately from axiom (c) that all bases have the same cardinality, called the *rank* of the matroid.

If M is a matroid on the ground set E, we can define its *dual matroid* M^*, which has the same ground set and whose bases are given by the set complements $E \setminus B$ of bases B of M. We have $\mathrm{rank}(M^*) = |E| - \mathrm{rank}(M)$.

We shall mainly be concerned with the matroid structure of a list of vectors. In this case, independence is the usual linear independence. One can in general define a rank function $\rho(A)$ on the subsets of E; in our case the dimension of the linear span $\langle A \rangle$ of A. Observe that given a list of vectors, their matroid structure depends only on the vectors up to nonzero scaling. We should keep in mind the special example of graphs with the associated arrangements and polytopes, which motivates some of our more general definitions.

2.2.1 Cocircuits

As we have seen in the case of oriented graphs, a set of vectors corresponding to edges is linearly independent if and only if the graph that they span is a tree or a disjoint union of trees. Thus a minimal dependent set is formed by edges that generate a simple *cycle*, or in the words of matroids, a *circuit*.

Definition 2.21. Let M be a matroid.
  A minimal dependent set in M is called a *circuit*.
  A circuit in the dual matroid M^* is called a *cocircuit*.

If B is a basis of M and $p \notin B$, then there is a unique circuit $cir(B, p)$ contained in $B \cup \{p\}$. Dually, if $q \in B$, then there is a unique cocircuit $cocir(B, q)$ contained in $(E \setminus B) \cup \{q\}$. These *basic circuits and cocircuits* are related in the following way:

$$q \in cir(B, p) \iff (B \setminus q) \cup p \text{ is a basis } \iff p \in cocir(B, q).$$

Let us now specialize to the case that is most interesting for us, in which the ground set X is a list of vectors and the collection $\mathcal{I}$ consists of linearly independent sublists.

In this case, if $\underline{b} \subset X$ is a basis and $a \notin \underline{b}$, then the unique circuit $cir(\underline{b}, a)$ is obtained by writing a as a linear combination of $\underline{b}$ and taking the list formed by a and the elements of $\underline{b}$ that appear with nonzero coordinates. Also, a cocircuit is a sublist $Y \subset X$ such that $X \setminus Y$ does not span $\langle X \rangle$ and is minimal with this property.

Let us assume that X spans V. The cocircuits can thus be obtained as follows: fix a hyperplane $H \subset V$ spanned by elements in X and consider $Y := \{x \in X \mid x \notin H\}$; it is immediately verified that *this is a cocircuit and every cocircuit is of this type.* In fact, we leave to the reader to verify that if $\underline{b} \subset X$ is a basis and $q \in \underline{b}$, the unique cocircuit $cocir(\underline{b}, q)$ contained in $(X \setminus \underline{b}) \cup \{q\}$ is the complement in X of the hyperplane generated by $\underline{b} \setminus \{q\}$.

If a sublist $Z \subset X$ consists of all the vectors in X lying in a given subspace, we say that Z is *complete.*

Thus a cocircuit is obtained by removing from X a complete set spanning a hyperplane. The set of all cocircuits will be denoted by $\mathcal{E}(X)$.

Example 2.22. In the case of oriented graphs, a cocircuit is a minimal subgraph whose complement does not contain any maximal tree.

Definition 2.23. The minimal cardinality of a cocircuit is an important combinatorial invariant of X; it will be denoted by $m(X)$.

We shall say that X is *nondegenerate* if $m(X) \geq 2$.

Proposition 2.24. *We can uniquely decompose X as $X = \{Y, z_1, z_2, \ldots, z_r\}$ where $V = \langle Y \rangle \oplus \oplus_{i=1}^{r} \mathbb{R}z_i$ and Y is nondegenerate or empty.*

Proof. This is easily proved by induction. The elements z_i are just the ones which form cocircuits of length 1.

Remark 2.25. This is a special case of the decomposition into irreducibles that will be developed in Section 20.1.

2.2.2 Unbroken Bases

Our next basic combinatorial notion has been used extensively in the theory of hyperplane arrangements (cf. [30], [117], [118], [119]). It is the notion of broken circuit and unbroken basis. These are again notions that belong to matroid theory and are related to the notion of *external activity*, which for our purposes plays a more important role than its dual notion of *internal activity.*

So let us first start for completeness in an abstract way and introduce these notions for a matroid $(E, \mathcal{I})$. From now on, let us assume that the ground set E is linearly ordered.

Definition 2.26. Given a basis B and an element $p \notin B$, we say that p is *externally active* in B if p is the least element of $cir(B, p)$. Dually, an element $q \in B$ is said to be *internally active* in B if q is the least element of $cocir(B, p)$.

Note that these concepts are dual: p is externally active in the basis B of M if and only if p is internally active in the basis $E \setminus B$ of M^*.

Let us now reinterpret the same concepts with slightly different notation for a list of vectors X. Take $\underline{c} := (a_{i_1}, \ldots, a_{i_k})$, $i_1 < i_2 < \cdots < i_k$, a sublist of linearly independent elements in X.

We say that $\underline{c}$ is a *broken circuit*[5] if there are an index $1 \le e \le k$ and an element $a_i \in X \setminus \underline{c}$ such that:

- $i < i_e$.
- The list $(a_i, a_{i_e}, \ldots, a_{i_k})$ is linearly dependent.

We say that a_i *breaks* $\underline{c}$.

In other words, a broken circuit is obtained from a circuit by removing its minimal element.

In particular, for a basis $\underline{b} := (a_{i_1}, \ldots, a_{i_s})$ extracted from X, an element $a \in X$ is breaking if and only if it is externally active, and we set

$$E(\underline{b}) := \{a \in X \mid a \text{ externally active or breaking for } \underline{b}\}. \tag{2.2}$$

Similarly, given a basis $\underline{b}$ extracted from X, an element $b \in \underline{b}$ is *internally active* with respect to $\underline{b}$ if there is no element a in the list X preceding b such that $\{a\} \cup (\underline{b} \setminus \{b\})$ is a basis of V.

Definition 2.27. The number $e(\underline{b})$ of externally active elements is called the *external activity* of $\underline{b}$. The number $i(\underline{b})$ of internally active elements is called the *internal activity* of $\underline{b}$.

If $E(\underline{b}) = \emptyset$, i.e., $e(\underline{b}) = 0$, the basis $\underline{b}$ is called *unbroken*. Equivalently, a basis is unbroken if it does not contain any broken circuit.

Example 2.28. The list of positive roots for type A_3.

$$X = \{\alpha_1, \alpha_2, \alpha_3, \alpha_1 + \alpha_2, \alpha_2 + \alpha_3, \alpha_1 + \alpha_2 + \alpha_3\}.$$

We have 16 bases and 6 unbroken bases; all necessarily contain α_1:

$$\alpha_1, \ \alpha_2, \ \alpha_3, \ \alpha_1 + \alpha_2, \ \alpha_2 + \alpha_3, \ \alpha_1 + \alpha_2 + \alpha_3.$$
$$\alpha_1, \ \alpha_2, \ \alpha_3, \ \alpha_1 + \alpha_2, \ \alpha_2 + \alpha_3, \ \alpha_1 + \alpha_2 + \alpha_3.$$
$$\alpha_1, \ \alpha_2, \ \alpha_3, \ \alpha_1 + \alpha_2, \ \alpha_2 + \alpha_3, \ \alpha_1 + \alpha_2 + \alpha_3.$$
$$\alpha_1, \ \alpha_2, \ \alpha_3, \ \alpha_1 + \alpha_2, \ \alpha_2 + \alpha_3, \ \alpha_1 + \alpha_2 + \alpha_3.$$
$$\alpha_1, \ \alpha_2, \ \alpha_3, \ \alpha_1 + \alpha_2, \ \alpha_2 + \alpha_3, \ \alpha_1 + \alpha_2 + \alpha_3.$$
$$\alpha_1, \ \alpha_2, \ \alpha_3, \ \alpha_1 + \alpha_2, \ \alpha_2 + \alpha_3, \ \alpha_1 + \alpha_2 + \alpha_3.$$

In what follows we shall need a relative version of unbroken basis. Let us start from a list X of vectors and parameters $\underline{\mu}$ that defines an affine arrangement.

Given a subspace W of this arrangement, recall that X_W is the list consisting of the elements a in X such that $a + \mu_a = 0$ on W.

---

[5] Notice that a broken circuit is **not** a circuit.

Definition 2.29. Let W be a space of the arrangement of codimension k. A subset of X_W that is a basis of the span $\langle X_W \rangle$ will be called a *basis on* W (or a basis in X_W).

Since X_W is a list, we also have the notion of unbroken basis on W.

For interesting information on these topics in the context of matroids, see Björner [16].

The combinatorics of unbroken bases is usually very complicated. We will see later how to unravel it for the list of positive roots of type A_n.

There is only one case in which the theory is trivial, the case when X is a list of (nonzero) vectors in 2-dimensional space. In this case, any unbroken basis contains the first vector, and for the second vector we can take any other vector in the list, provided a multiple of it does not appear earlier in the list. In particular, if the m vectors are mutually nonproportional, we have $m - 1$ unbroken bases.

2.2.3 Tutte Polynomial

A basic combinatorial invariant of matroids is the *Tutte polynomial*, introduced by W. T. Tutte in [114] as a generalization of the *chromatic polynomial*, which counts colorings of graphs.

Let $M = (X, I)$ be a matroid, with rank function ρ. The Tutte polynomial $T(M; x, y)$ is the polynomial in x and y (with integer coefficients) given by

$$T(M; x, y) := \sum_{A \subset X} (x - 1)^{\rho(X) - \rho(A)} (y - 1)^{|A| - \rho(A)}. \tag{2.3}$$

It is immediate that $T(M, 1, 1)$ equals the number of bases that one extracts from X. Although it is not apparent from the formula, this is a polynomial with positive integer coefficients. In combinatorics, polynomials with positive integer coefficients usually appear associated to *statistics*. For instance, a polynomial in one variable may appear when we have a function f on a finite set S with values in $\mathbb{N}$. In this situation, the polynomial $\sum_{s \in S} q^{f(s)}$ has the property that the coefficient of q^i counts the number of elements $s \in S$ for which $f(s) = i$. Indeed, we shall see that the Tutte polynomial encodes two statistics on the set of bases (formula (2.6)).

Let us discuss the Tutte polynomial for a list X of vectors, where for $A \subset X$ we have $\rho(A) = \dim\langle A \rangle$. In this case we denote it by $T(X; x, y)$.

Example 2.30. Assume that the list is *generic*; that is, if $s = \dim\langle X \rangle$, any subset with $k \leq s$ elements is independent. Then, letting $m := |X|$,

$$T(M; x, y) := \sum_{i=0}^{s} \sum_{A \subset X,\ |A|=i} (x-1)^{s-i} + \sum_{i=s+1}^{m} \sum_{A \subset X,\ |A|=i} (y-1)^{i-s}$$

$$= \sum_{i=0}^{s} \binom{m}{i} (x-1)^{s-i} + \sum_{i=s+1}^{m} \binom{m}{i} (y-1)^{i-s}$$

$$= \sum_{h=1}^{s} \binom{m-h-1}{s-h} x^{h} + \sum_{h=1}^{m-s} \binom{m-1-h}{s-1} y^{h}.$$

In particular, let us note two trivial cases:

Case 1. X is the list of m vectors equal to 0. Then we see that

$$T(X; x, y) := \sum_{A \subset X} (y-1)^{|A|} = y^{m}. \tag{2.4}$$

Case 2. X is a list of m linearly independent vectors. Then we see that

$$T(X; x, y) := \sum_{A \subset X} (x-1)^{m-|A|} = x^{m}. \tag{2.5}$$

In general, we shall give a simple recursive way to compute $T(X; x, y)$ that will also explain its meaning in terms of a statistic on the set of bases extracted from X. This method is called *method of deletion and restriction* and goes back to Zaslavsky [122].

Remark 2.31. Assume that the list X is the union of two lists $X = X_1 \cup X_2$ such that $\langle X \rangle = \langle X_1 \rangle \oplus \langle X_2 \rangle$ (in the notation of Sections 7.1.4, and 20.1 this is called a *decomposition*).

Then we have

$$T(X; x, y) = T(X_1; x, y) T(X_2; x, y).$$

Let us now see the basic step of a recursion.

Given a list of vectors X and a nonzero vector v in X, let X_1 be the list obtained by deleting v. If $\langle X \rangle = \langle v \rangle \oplus \langle X_1 \rangle$, we have by the previous formulas

$$T(X; x, y) = T(X_1; x, y)\, x.$$

Otherwise, we have $\langle X \rangle = \langle X_1 \rangle$ and we let X_2 be the list obtained from X_1 by reducing modulo v.

In this case the following identity holds:

$$T(X; x, y) = T(X_1; x, y) + T(X_2; x, y).$$

Indeed, the sum expressing $T(M; x, y)$ splits into two parts, the first over the sets $A \subset X_1$:

$$\sum_{A \subset X_1} (x-1)^{\rho(X)-\rho(A)}(y-1)^{|A|-\rho(A)}$$

$$= \sum_{A \subset X_1} (x-1)^{\rho(X_1)-\rho(A)}(y-1)^{|A|-\rho(A)} = T(X_1; x, y),$$

since clearly $\rho(X_1) = \rho(X)$. The second part is over the sets A with $v \in A$. For these sets, set $\overline{A}$ to be the image of $A \setminus v$ modulo v. We have that

$$|\overline{A}| = |A| - 1, \ \rho(\overline{A}) = \rho(A) - 1, \ \rho(X_2) = \rho(X) - 1,$$

and we get

$$\sum_{A \subset X, \ v \in A} (x-1)^{\rho(X)-\rho(A)}(y-1)^{|A|-\rho(A)}$$

$$= \sum_{\overline{A} \subset X_2} (x-1)^{\rho(X_2)-\rho(\overline{A})}(y-1)^{|\overline{A}|-\rho(\overline{A})} = T(X_2; x, y),$$

which proves our claim.

This allows us to compute the Tutte polynomial recursively starting from the case in which X consists of m vectors all equal to 0, the case for which we have formula (2.4). The final formula, due to Crapo [36], involves in a clear way the notions of external and internal activity (Tutte [114]). Notice that from this recursion it follows that the Tutte polynomial has positive integer coefficients.

Theorem 2.32 (Crapo). *The Tutte polynomial $T(X, x, y)$ equals*

$$\sum_{\underline{b} \text{ basis in } X} x^{i(\underline{b})} y^{e(\underline{b})}. \tag{2.6}$$

Proof. Set

$$\tilde{T}(X, x, y) = \sum_{\underline{b} \text{ basis in } X} x^{i(\underline{b})} y^{e(\underline{b})}.$$

If X is the list of m vectors equal to 0, then there is only the empty basis for which the external activity is m and the internal activity is 0. Thus $\tilde{T}(X, x, y) = y^m = T(X, x, y)$.

Assume now that X contains at least one nonzero vector and let v be the last such vector. Set $X_1 = X \setminus \{v\}$. We have two cases: $V = \langle X_1 \rangle \oplus \langle v \rangle$ and $V = \langle X_1 \rangle$.

If $V = \langle X_1 \rangle \oplus \langle v \rangle$, then every basis extracted from X contains v, v is internally active with respect to every basis, and we clearly see that $\tilde{T}(X, x, y) = x\tilde{T}(X_1, x, y)$.

Let $\mathcal{B}_X$ denote the sublists extracted from X that give bases of V. If $V = \langle X_1 \rangle$, denote by $\mathcal{B}_1$ the set of bases in $\mathcal{B}_X$ not containing v and $\mathcal{B}_2 = \mathcal{B}_X \setminus \mathcal{B}_1$ those containing v. Since v is the last nonzero element in X, we immediately get that

$$\tilde{T}(X_1, x, y) = \sum_{\underline{b} \in \mathcal{B}_1} x^{i(\underline{b})} y^{e(\underline{b})}.$$

On the other hand, let us consider the list X_2 consisting of the images of the elements of X_1 in $V/\langle v \rangle$. If $v \in \underline{b}$, then the image $\underline{\tilde{b}}$ of $\underline{b} \setminus \{v\}$ modulo v is a basis extracted from X_2, and we have that $e(\underline{b}) = e(\underline{\tilde{b}})$ and $i(\underline{b}) = i(\underline{\tilde{b}})$, as the reader will easily verify. Thus

$$\tilde{T}(X_2, x, y) = \sum_{\underline{b} \in \mathcal{B}_2} x^{i(\underline{b})} y^{e(\underline{b})}.$$

It follows that

$$\tilde{T}(X, x, y) = \tilde{T}(X_1, x, y) + \tilde{T}(X_2, x, y).$$

Since both $T(X, x, y)$ and $\tilde{T}(X, x, y)$ satisfy the same recursive relation, the claim follows.

We have normalized the previous algorithm, that at each step produces lists of vectors, by choosing as vector v the last (or the first) in the list that is nonzero. Once we do this, we can visualize the full algorithm by a rooted tree. The root is given by the full list X, the nodes are the lists of vectors that we obtain recursively, the internal nodes are the lists that contain a nonzero vector, and the leaves are the lists of vectors that are all zero.

The steps of the first type produce a new node, that we mark by the value x by which we multiply, and we join to the previous node by an edge marked by the vector v that we have removed. The second type of step produces a branching into two nodes, one relative to the list X_1 and the other relative to X_2. We mark the edge joining X and X_2 with the vector v that we have used to construct $V/\langle v \rangle$. Observe that in passing from a node to the next by an edge, we reduce the rank by 1 if and only if the edge is marked.

At each step, every vector of the list of vectors we obtain remembers the position of the vector from which it came (by successive quotients). Consider now a maximal chain C of this tree. To C associate the subset of vertices v marking the edges appearing on C. It is clear that this subset is a basis $\underline{b}$ extracted from X. In this maximal chain there are k vertices marked by x for some k, and the final leaf, which is the list of 0 with some multiplicity h. Clearly, this chain contributes $x^k y^h$ to the Tutte polynomial. A simple interpretation shows that $k = i(\underline{b})$ and $h = e(\underline{b})$. In other words:

1. The maximal chains are in one-to-one correspondence with the bases.
2. Each produces a single monomial $x^{i(\underline{b})} y^{e(\underline{b})}$.
3. The sum of all these monomials is the Tutte polynomial.

In particular, setting $x = 1$, we obtain the polynomial

$$T(X; 1, y) = \sum_{A \subset X, \, | \, \rho(X) = \rho(A)} (y - 1)^{|A| - \rho(A)}, \tag{2.7}$$

giving the statistic of the external activity.

The same recursive algorithm computes all the unbroken bases that one can extract from a list $X := \{a_1, \ldots, a_m\}$. To begin, observe that each such basis contains a_1.

The starting point is given by two observations:

1. if $m = s$ and $a_1, \ldots, a_m$ is a basis, then it is the only unbroken basis one can extract from X.

2. If the elements of X do not span $\mathbb{R}^s$, no unbroken basis can be extracted.

Given this, we may divide the set of unbroken bases into two subsets. The first consists of the bases extracted from $\{a_1, \ldots, a_{m-1}\}$. The second is made of unbroken bases containing a_m. These are in one-to-one correspondence with the unbroken bases of the image modulo a_m of the list $\{a_1, \ldots, a_{m-1}\}$.

Exercise. Let X be indecomposable and assume that there is only one basis $\underline{b}$ in X with $e(\underline{b}) = 0$; then $s = 1$.

2.2.4 Characteristic Polynomial

An interesting combinatorial invariant of a hyperplane arrangement is its *characteristic polynomial*. Such a polynomial can be associated to any (finite) poset $\mathcal{P}$ with a rank function $r(x)$ and a minimum element $\hat{0}$ using the *Möbius function*. The Möbius function of a ranked poset is a generalization of the classical function defined on the poset of positive integers ordered by divisibility. It is introduced by Rota in [94]. It may be defined for a pair of elements $a, b \in \mathcal{P}$ by

$$
\mu_{\mathcal{P}}(a, b) := \begin{cases} 0 & \text{if} \quad a \not\leq b, \\ 1 & \text{if} \quad a = b, \\ -\sum_{a \leq x < b} \mu_P(a, x) & \text{if } a < b. \end{cases}
$$

The Möbius function arises in the following context. Define a transform T on functions on $\mathcal{P}$ by $(Tf)(x) := \sum_{y \leq x} f(y)$. It is clear (we are assuming that the poset is finite only for simplicity, but one can work with much weaker hypotheses) that the matrix of T is triangular with 1 on the diagonal and at all positions (y, x) with $y \leq x$ and zero in every other position. Then we have the *Möbius inversion formula* for T^{-1}:

$$
(T^{-1}g)(x) = \sum_{y \leq x} \mu_P(x, y) g(y).
$$

We now define the characteristic polynomial of $\mathcal{P}$ by

$$
\chi_{\mathcal{P}}(q) := \sum_{x \in \mathcal{P}} \mu(\hat{0}, x) q^{s - r(x)}, \tag{2.8}
$$

where s is the maximum of the rank function on $\mathcal{P}$.

Take a list X of vectors in V, and consider the partially ordered set $\mathcal{P}_X$ whose elements are the subspaces $\langle A \rangle$ spanned by sublists A of X. The order is given by inclusion and the rank function given by the dimension of $\langle A \rangle$. The poset $\mathcal{P}_X$ contains a unique minimal element $\hat{0} = \{0\}$, that is the origin of V. We can thus consider the characteristic polynomial of $\mathcal{P}_X$ that we shall denote by $\chi_X(q)$.

The following Theorem tells us that the characteristic polynomial can be computed as a specialization of the Tutte polynomial.

Theorem 2.33. *Assume X only contains nonzero vectors. We have then* $\chi_X(q) = (-1)^s T(X, 1 - q, 0)$.

Proof. For any $W \in \mathcal{P}_X$ set $X_W := X \cap W$ and $\mathcal{A}_W := \{A \subset X_W \mid \langle A \rangle = W\}$. We claim that for each $W \in \mathcal{P}_X$,

$$\tilde{\mu}(W) := \sum_{A \in \mathcal{A}_W} (-1)^{|A|} = \mu(\hat{0}, W).$$

We proceed by induction on dimension. If $W = \{0\} = \hat{0}$ then $X_W = \emptyset$ and $\tilde{\mu}(\hat{0}) = 1 = \mu(\hat{0}, \hat{0})$. Take $W \in \mathcal{P}_X \setminus \{\hat{0}\}$. We have

$$0 = \sum_{A \subset X_W} (-1)^{|A|} = \sum_{U \in \mathcal{P}_{X_W}} \tilde{\mu}(U) = \sum_{U \in \mathcal{P}_X, U \subset W} \tilde{\mu}(U).$$

If $U \subsetneq W$ the inductive hypothesis implies that $\tilde{\mu}(U) = \mu(\hat{0}, U)$. Thus

$$\tilde{\mu}(W) = - \sum_{U \in \mathcal{P}_X, U \subsetneq W} \mu(\hat{0}, U) = \mu(\hat{0}, W)$$

proves the first claim. Computing, we get,

$$(-1)^s T(X, 1 - q, 0) = (-1)^s \sum_{A \subset X} (-q)^{s - \rho(A)} (-1)^{|A| - \rho(A)}$$

$$= \sum_{A \subset X} q^{s - \rho(A)} (-1)^{|A|} = \sum_{W \in \mathcal{P}_X} \Big(\sum_{A \in \mathcal{A}_W} (-1)^{|A|} \Big) q^{s - \dim(W)}$$

$$= \sum_{W \in \mathcal{P}_X} \mu(\hat{0}, W) q^{s - \dim(W)} = \chi_X(q).$$

We shall see later that in the complex case, by a theorem of Orlik–Solomon (cf. [83]), the characteristic polynomial computes the Betti numbers of the complement $\mathcal{A}$ of the hyperplane arrangement corresponding to X. Indeed,

$$\sum_{i \geq 0} \dim H^i(\mathcal{A}, \mathbb{R}) q^i = (-q)^s \chi_{\mathcal{A}}(-1/q).$$

In the real case one has the following result due to Zaslavsky [122].

Theorem 2.34 (Zaslavsky). *For a real hyperplane arrangement the number of open chambers is equal to $(-1)^s \chi_{\mathcal{A}}(-1)$. In the affine case the number of bounded regions is $(-1)^s \chi_{\mathcal{A}}(1)$.*

2.2.5 Identities

Although in principle one can compute the Tutte polynomial by the recursive formula, in practice this may be very cumbersome. For instance, for exceptional root systems and in particular E_8 (see [67]), it seems to take an enormous amount of computer time. There are other simple formulas that we want to discuss and that allow us to compute the Tutte polynomial for root systems. Indeed, we are going to use these formulas together with results contained in [84] to compute the Tutte polynomials for root systems.

Definition 2.35. Given a matroid on a set A, we say that $B \subset A$ is complete in A or a *flat* if, once $a \in A$ is dependent on B, then $a \in B$.

Given $B \subset A$, we set $\overline{B}$ to be the set of elements $a \in A$ dependent on B.

Clearly, for every B we have that $\overline{B}$ is complete and B is complete if and only if $B = \overline{B}$.

Assume that the rank of A is s. Let L_A be the set of complete subsets of A. Set $E_A(y) := T(A, 1, y)$, the polynomial expressing external activity.

Proposition 2.36. *The following identities hold:*

$$T(A; x, y) = \sum_{B \in L_A} (x - 1)^{s - \rho(B)} E_B(y). \tag{2.9}$$

$$y^{|A|} = \sum_{B \in L_A} (y - 1)^{\rho(B)} E_B(y). \tag{2.10}$$

Proof. We can clearly reformulate (2.7) as

$$E_A(y) = \sum_{B \subset A, \, | \, \overline{B} = A} (y - 1)^{|B| - s}$$

Thus

$$T(A; x, y) := \sum_{B \in L_A} \sum_{C \, | \, \overline{C} = B} (x - 1)^{s - \rho(B)} (y - 1)^{|C| - \rho(B)}$$

$$= \sum_{B \in L_A} (x - 1)^{s - \rho(B)} E_B(y).$$

As for our second formula.

$$\sum_{B \in L_A} (y - 1)^{\rho(B)} E_B(y) = \sum_{B \in L_A} (y - 1)^{\rho(B)} \sum_{C \subset A, \, | \, \overline{C} = B} (y - 1)^{|B| - \rho(B)}$$

$$= \sum_{B \subset A} (y - 1)^{|B|} = (1 + (y - 1))^{|A|} = y^{|A|}.$$

One can use the Möbius inversion formula to get

$$E_A(y) = \sum_{X \in L_A} y^{|X|} \mu(X, A).$$

We can use formula (2.10) for computing $E_A(y)$ once we know $E_B(y)$ for all proper complete subsets of A. It follows that using this and formula (2.9), we obtain an inductive procedure to compute the Tutte polynomial. Of course, in order to achieve this goal we need to be able to enumerate the proper complete subsets of A.

Notice that if A has a group of symmetries W, as in the case in which A is a root system and W is the Weyl group, we can first classify the orbits of complete subsets of A and then compute the number of elements $n(O)$ of each orbit O. Denoting by $\mathcal{O}_k$ the set of orbits of rank-k complete subsets of A, we then have the formula

$$y^{|A|} = \sum_{k=0}^{s} (y-1)^k \sum_{O \in \mathcal{O}_k} n(O) E_O(y). \qquad (2.11)$$

In the case of root systems, the computation of all orbits, their cardinality, and the corresponding root systems has been carried out by a lengthy case-by-case analysis by Orlik and Solomon, in the paper [84].[6] Using induction and the tables in [84], we are going to compute the Tutte polynomials for root systems in Section 2.4.

2.3 Zonotopes

For a systematic treatment see Ziegler [123].

2.3.1 Zonotopes

We have already seen that the Minkowski sum of two convex sets is convex. We shall be interested in a very special case, the polytope $B(X)$ associated to a list of vectors X, that is characterized as the set of points $x \in V$ for which the variable polytope $\Pi_X^1(x)$ defined in Section 1.3.1 is nonempty. Let us start with an observation. Let A, B be two convex polytopes, the first being the convex envelope of its extremal points $a_1, \ldots, a_h$ and the second the convex envelope of its extremal points $b_1, \ldots, b_k$.

Proposition 2.37. *$A + B$ is a convex polytope and its extremal points are contained in the set $a_i + b_j$.*

Proof. Take any point $p \in A + B$. It is of the form

$$p = \sum_{i=1}^{h} s_i a_i + \sum_{j=1}^{k} t_j b_j, \qquad \sum_i s_i = \sum_j t_j = 1.$$

---

[6]We thank John Stembridge for calling our attention to this paper.

Thus

$$p = \sum_{i=1}^{h} s_i \left(\sum_{j=1}^{k} t_j a_i \right) + \sum_{j=1}^{k} t_j \left(\sum_{i=1}^{h} s_i b_j \right) = \sum_{i=1}^{h} \sum_{j=1}^{k} s_i t_j (a_i + b_j), \quad \sum_{i=1}^{h} \sum_{j=1}^{k} s_i t_j = 1.$$

Thus $A + B$ is the convex envelope of the points $a_i + b_j$. Simple examples show that in general, not all of them are extremal.

A particularly interesting example is obtained by the sum of segments originating from the origin. In other words, when we take a list of vectors $X = (a_1, \ldots, a_m)$, we can construct the Minkowski sum of the corresponding segments $[0, a]$, $a \in X$:

$$B(X) := \left\{ \sum_{a \in X} t_a a \,\middle|\, 0 \leq t_a \leq 1 \right\}. \tag{2.12}$$

This convex polytope will be called the *box* or the *zonotope* associated to X. The polytope $B(X)$ can be also visualized as the image or *shadow* of the cube $[0, 1]^m$ under the projection $(t_1, \ldots, t_m) \mapsto \sum_{i=1}^{m} t_i a_i$:

$$X = \begin{vmatrix} 1 & 0 & 1 & -1 \\ 0 & 1 & 1 & 1 \end{vmatrix}$$

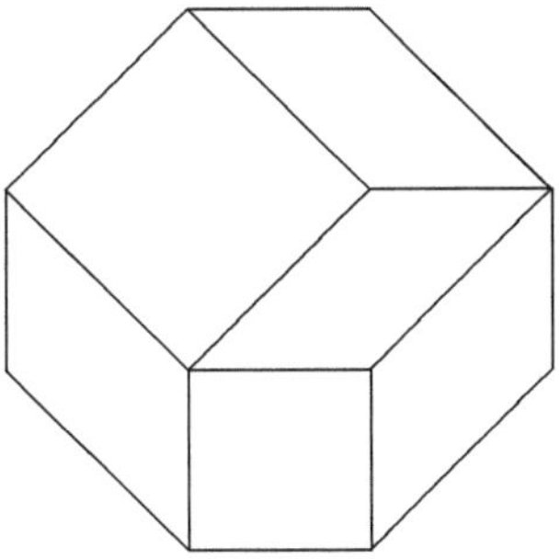

Fig. 2.1. The zonotope $B(X)$.

Remark 2.38. By the previous proposition it follows that $B(X)$ is the convex envelope of the points $\sum_{a \in S} a$ as S runs over the sublists of X.

The importance of zonotopes in this book comes from the fact that they appear as the support of an interesting class of special functions, the *box splines*. This will be discussed in Chapter 7. Furthermore, they play an important role in the theory of the partition function (cf. Section 13.2.2).

It is often convenient to shift $B(X)$ by the element $\rho_X := \frac{1}{2} \sum_{a \in X} a$. The resulting polytope $B(X) - \rho_X$ is then symmetric with respect to the origin. Moreover, given a subset $S \subset X$, let us define $P_S := \frac{1}{2} (\sum_{a \in S} a - \sum_{a \in X \setminus S} a) =$

$\sum_{a \in S} a - \rho_X$. The shifted box $B(X) - \rho_X$ is, by Remark 2.38, the convex envelope of the points P_S.

Two zonotopes.[7]

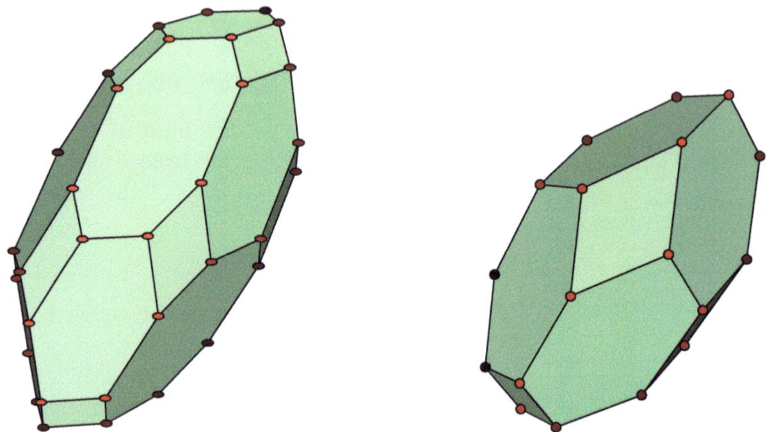

Fig. 2.2. The zonotope associated to the root system B_3 and A_3.

From now on, we shall assume that X spans V, so that $B(X)$ has a nonempty interior. We are now going to show that each face of a zonotope is itself a zonotope (shifted by a vector). We begin with a lemma

Lemma 2.39. *If a point* $v := \sum_{a \in X} v_a a$, $0 \le v_a \le 1$ *is in the boundary of* $B(X)$, *then the set* $A_v := \{a \in X \,|\, 0 < v_a < 1\}$ *does not span* V.

Proof. If A_v spans V, extract a basis $\underline{b} := \{b_1, \ldots, b_s\}$ from A_v.

Let $\epsilon := \min_{a \in A_v} [\min[v_a, 1 - v_a]] > 0$. The set of points $v + \sum_{i=1}^{s} t_i b_i$, $|t_i| < \epsilon$, is an open neighborhood of v contained in $B(X)$; hence v is in the interior of $B(X)$.

This lemma can be interpreted as follows. Let A, B be two disjoint sublists of X such that A does not span V, and $B(A)$ its associated zonotope.

Let $B(X)_{A,B} := \sum_{a_i \in A} t_i a_i + \sum_{b \in B} b, 0 \le t_i \le 1 = B(A) + \lambda_B$ with $\lambda_B := \sum_{b \in B} b$. Then the boundary $\partial B(X)$ is contained in the nonconvex polytope $Q = \cup_{A,B} B(X)_{A,B}$.

We now want to describe the faces of $B(X)$. We start from codimension-one faces. Consider a sublist A of X spanning a hyperplane H_0 and such that $A = H_0 \cap X$. Take a linear equation ϕ_{H_0} for H_0 and set

$$B = \{a \in X | \langle \phi_{H_0} | a \rangle > 0\}, \quad C = \{a \in X | \langle \phi_{H_0} | a \rangle < 0\}.$$

---

[7]The two pictures were produced with *polymake*, http://www.math.tu-berlin.de/polymake/.

Proposition 2.40. *$B(X)_{A,B}$ and $B(X)_{A,C}$ are exactly the two codimension-one faces in $B(X)$ parallel to H_0.*

Proof. Let F be a codimension-one face spanning a hyperplane H parallel to H_0. By Proposition 1.15, $F = H \cap B(X)$. Since $B(X)$ lies entirely in one of the two closed half-spaces with boundary H, we may assume that ϕ_{H_0} is positive on H with constant value a and that it takes values less than a in the open half-space meeting $B(X)$. Thus F coincides with the set of points in $B(X)$ where ϕ_{H_0} takes its maximum value. Now let $A := \{x \in X \mid \phi_{H_0}(x) = 0\}$ and $B := \{x \in X \mid \phi_{H_0}(x) > 0\}$. Clearly, the maximum taken by ϕ_{H_0} on $B(X)$ is $\sum_{x \in B} \phi_{H_0}(x)$, and it is taken exactly on the points of $B(X)_{A,B}$. Thus $F = B(X)_{A,B}$, similarly for the minimum value.

Let $\mathcal{H}_X$ be the real hyperplane arrangement defined, in dual space, by the vectors of X thought of as linear equations. As a consequence, one has the following theorem

Theorem 2.41. *There is an order-reversing one-to-one correspondence between faces of $B(X)$ and faces of $\mathcal{H}_X$.*

To a face G of the hyperplane arrangement we associate the set of points in $B(X)$ on which some, and hence every, element $f \in G$ takes its maximum value.

This set is the face $B(X)_{A,B}$ where $A := \{x \in X \mid \langle x \mid f \rangle = 0\}$ and $B := \{x \in X \mid \langle x \mid f \rangle > 0\}$.

Proof. The previous discussion implies that there is a one-to-one correspondence between codimension-one faces of $B(X)$ and dimension-one faces of the arrangement $\mathcal{H}_X$. We can finish our proof by induction. Let ℓ be a half-line of the arrangement $\mathcal{H}_X$ and let $A \subset X$ be the set of vectors vanishing on ℓ. To ℓ is also associated the face $B(A) + \lambda_B$ of $B(X)$. The faces of $B(A) + \lambda_B$ are of the form $F + \lambda_B$, with F a face of $B(A)$. The vectors A define the hyperplane arrangement $\mathcal{H}_A$ generated by the hyperplanes on $\mathcal{H}_X$ containing ℓ. By induction, there is an order-reversing one-to-one correspondence between faces of $B(A)$ and faces of $\mathcal{H}_A$. So the theorem follows from the following lemma

Lemma 2.42. *Every face of $\mathcal{H}_X$ that contains ℓ in its closure is contained in a unique face of the arrangement $\mathcal{H}_A$.*

In fact, the faces of the arrangement $\mathcal{H}_A$ are unions of faces of the arrangement $\mathcal{H}_X$, and each of them has the line spanned by ℓ in its closure. If we have two distinct faces F_1 and F_2 of $\mathcal{H}_X$, they must lie in opposite half-spaces for some hyperplane H of the arrangement. If they both have ℓ in their closure, we must have that H contains ℓ, and this proves the claim.

We can deduce now from the previous theorem a dual description of $B(X)$ as the polytope defined by linear inequalities. Each 1-dimensional face F of $\mathcal{H}_X$

is a half-line. Choose a generator $u_F \in F$. Then $u_F = 0$ is the linear equation of a rational hyperplane in V. Denote by $\mu_F = \sum_{b \in X, \langle u_F, b \rangle > 0} \langle u_F, b \rangle$ the maximum value of u_F on $B(X)$. We then have that the set $\{v \in V \mid \langle u_F, v \rangle \le \mu_F\}$ is the half-space containing $B(X)$ and bounded by the hyperplane spanned by the face $B(X)_{A,B}$, where $A := \{a \in X \mid \langle u_F, a \rangle = 0\}$, $B := \{b \in X \mid \langle u_F, b \rangle > 0\}$. Since every convex polytope is the intersection of such subspaces, we get the following dual description of $B(X)$:

Proposition 2.43.

$$B(X) = \{v \in V \mid \langle u_F \mid v \rangle \le \mu_F\}, \tag{2.13}$$

as F runs over the 1-dimensional faces of the arrangement $\mathcal{H}_X$.
The interior $\mathring{B}(X)$ is given by strict inequalities:

$$\mathring{B}(X) = \{v \in V \mid \langle u_F \mid v \rangle < \mu_F, \ \forall F\}.$$

Let us see how we can construct $B(X)$ inductively.

Let $y \in X$, $y \ne 0$, and consider the projection $p : V \to V/\mathbb{R}y$. The 1-parameter group of translations $v \mapsto v + ty$ has as orbits the fibers of p. Set $Z = X \setminus \{y\}$, and $\tilde{Z} := p(Z)$, a subset of $V/\mathbb{R}y$.

Clearly, $p(B(X)) = p(B(Z)) = B(\tilde{Z})$. Moreover, for each $w \in B(\tilde{Z})$, the set $p^{-1}(w) \cap B(X)$ is an oriented segment $[v_0, v_0 + t_1 y]$, $t_1 \ge 0$, from which we can choose the point $s(w) := v_0 + t(w)y$ where the translation group *exits* the zonotope $B(Z)$.

Lemma 2.44. *The map $s : B(\tilde{Z}) \to B(Z)$ is a piecewise linear homeomorphism of the zonotope $B(\tilde{Z})$ with the piece $B(Z)_y$ of the boundary of $B(Z)$ where the translation group* exits *the zonotope $B(Z)$:*

$$B(Z)_y := \{p \in B(Z) \mid p + ty \notin B(Z), \ \forall t > 0\}.$$

Proof. If Z spans a hyperplane, we clearly get that $B(Z)_y = B(Z)$, and the linear map p gives a homeomorphism between $B(Z)$ and $B(\tilde{Z})$ so there is nothing to prove.

Assume that Z spans V. Choose first a complement C to $\mathbb{R}y$ and let $f : V/\mathbb{R}y \to C$ be the corresponding linear isomorphism with $pf(w) = w, \forall w \in V/\mathbb{R}y$. If $w \in B(\tilde{Z})$, we have that $s(w) = f(w) + t(w)y$, where $t(w)$ is the maximum number t for which $f(w) + ty \in B(Z)$. In other words, $t(w)$ is the maximum t for which $\langle u_F \mid f(w) \rangle + t \langle u_F \mid y \rangle \le \mu_F$, for every codimension-one face of $B(Z)$. Thus

$$t(w) = \min(\langle u_F \mid y \rangle^{-1}(\mu_F - \langle u_F \mid f(w) \rangle)), \ \forall F \mid \langle u_F \mid y \rangle > 0.$$

By Example (1.61), the functions $t(w)$ and hence $s(w)$ are piecewise linear. Clearly, $p(s(w)) = w$.

Remark 2.45. Geometrically, $B(Z)_y$ is what one *sees* when looking at the polytope $B(Z)$ from infinity in the direction of y. One sees the projected zonotope $B(\tilde{Z})$, tiled by zonotopes (the faces that are visible).

We have a map $\pi : B(Z)_y \times [0,1] \to B(X)$, $\pi(p,t) = p + ty$. We easily see that the following result holds

Proposition 2.46. *If Z does not span V, we have $B(Z)_y = B(Z)$ and π is a homeomorphism. Assume that Z spans V:*

(a) $B(Z)_y$ is the union of the faces corresponding to faces of $\mathcal{H}_Z$ that are positive on y.
(b) π is injective, and hence it is a homeomorphism onto its image.
(c) $B(Z) \cap \pi(B(Z)_y \times [0,1]) = \pi(B(Z)_y \times [0]) = B(Z)_y$.
(d) $B(X) = B(Z) \cup \pi(B(Z)_y \times [0,1])$.

Proof. (a) Consider a k-dimensional face $F = B(Z)_{A,B}$ of $B(Z)$ associated to a face ℓ of $\mathcal{H}_Z$. If y vanishes on ℓ, then given $x \in F$ an interior point, we have $x + ty \in F$ for all sufficiently small t, and so $x \notin B(Z)_y$. Moreover, under the map $p : V \to V/\mathbb{R}$ we have that F maps to a $(k-1)$-dimensional polyhedron.

Otherwise, y is not orthogonal to ℓ, the map p projects F homeomorphically to its image in $B(\tilde{Z})$, and since by Theorem 2.41 any element $u \in \ell$ takes its maximum on F, we see that $F \subset B(Z)_y$ if and only if y is positive on ℓ.

(b) Take $p, q \in B(Z)_y$. If $p + t_1 y = q + t_2 y$, $t_1 \leq t_2$, we have $p = q + (t_2 - t_1)y$. So by the definition of $B(Z)_y$, $t_2 - t_1 = 0$ and $p = q$.

(c) Follows by the definition of $B(Z)_y$.

(d) Any element b of $B(X)$ is of the form $b = c + ty$, $c \in B(Z)$, $0 \leq t \leq 1$. If $b \notin B(Z)$, we need to show that $b \in \pi(B(Z)_y \times [0,1])$. Since $b \notin B(Z)$, there is a maximal $s \geq 0$, $s < t$, so that $c + sy \in B(Z)$. Then $c + sy \in B(Z)_y$ and $b = (c + sy) + (t - s)y$.

Given linearly independent vectors $\underline{b} := \{b_1, \ldots, b_h\}$ and $\lambda \in V$, define

$$\Pi_\lambda(\underline{b}) := \lambda + B(\underline{b}) = \{\lambda + \sum_{i=1}^{h} t_i b_i\}, \ 0 \leq t_i \leq 1.$$

We call such a set a *parallelepiped*.[8] Its faces are easily described. Given a point $p = \lambda + \sum_{i=1}^{h} t_i b_i$ in $\Pi_\lambda(\underline{b})$ consider the pair of disjoint sets of indices $A(p) := \{i \,|\, t_i = 0\}$, $B(p) = \{i \,|\, t_i = 1\}$. For A, B given disjoint subsets of $\{1, 2, \ldots, h\}$, set $F_{A,B} := \{p \in \Pi_\lambda(\underline{b}) \,|\, A(p) = A, \ B(p) = B\}$. One easily verifies that these sets decompose $\Pi_\lambda(\underline{b})$ into its *open* faces. Furthermore, the interior of $\Pi_\lambda(\underline{b})$ equals $F_{\emptyset,\emptyset}$, while the closure of the open face $F_{A,B}$ is the face

$$\overline{F}_{A,B} = \cup_{A \subset C, \ B \subset D} F_{C,D}.$$

---

[8] This is the simplest type of zonotope.

Definition 2.47. We say that a set $B \subset V$ is *paved* by a finite family of parallelepipeds Π_i if $B = \cup_i \Pi_i$ and any two distinct parallelepipeds Π_i, Π_j in the given list intersect in a common, possibly empty, face.

The following facts are easy to see:

(i) If B is paved by parallelepipeds any two faces of two parallelepipeds of the paving intersect in a common face.

(ii) A closed subset $C \subset B$ that is a union of faces of the parallelepipeds of the paving has an induced paving.

(iii) Assume that B is convex. If F is a face of B and a parallelepiped P of the paving of B meets F in an interior point of P, then $P \subset F$. In particular, F is a union of faces of the parallelepipeds of the paving and has an induced paving.

We are going to construct in a recursive way a paving $B(X)$ by parallelepipeds indexed by all the bases extracted from X. Notice that when $X = \{Z, a\}$, the bases of V extracted from X decompose into two sets: those contained in Z index the parallelepipeds decomposing $B(Z)$ and those containing a are in one-to-one correspondence with the bases extracted from $\tilde{Z}$ that index the remaining parallelepipeds decomposing $B(Z)_a + [0, a]$. Let us see the details.

For any basis $\underline{b} = \{a_{i_1}, \ldots, a_{i_s}\}$ extracted from X, we set, for any $1 \le t \le s$, V_t equal to the linear span of $\{a_{i_1}, \ldots, a_{i_t}\}$. We have that V_t is a hyperplane in V_{t+1}, and for each $1 \le t \le s - 1$, we let ϕ_t be the equation for V_t in V_{t+1}, normalized by $\langle \phi_t \,|\, a_{i_{t+1}} \rangle = 1$. Define

$$B_t := \{a_i \in X \,|\, i < i_{t+1}, \ a_i \in V_{t+1} \ \text{ and } \ \langle \phi_t \,|\, a_i \rangle > 0\}$$

and set

$$\lambda_{\underline{b}} = \sum_{t=1}^{s-1} \left(\sum_{a_i \in B_t} a_i \right). \tag{2.14}$$

We can then consider the parallelepiped $\Pi_{\lambda_{\underline{b}}}(\underline{b})$, and deduce a paving of $B(X)$ due to Shephard [101].

Theorem 2.48. *[Shephard] The collection of parallelepipeds $\Pi_{\lambda_{\underline{b}}}(\underline{b})$, as $\underline{b}$ varies among the bases extracted from X, is a paving of $B(X)$.*

Proof. If X consists of a single vector, there is nothing to prove. So we proceed by induction and as before we write $X = (Z, y)$.

If Z spans a hyperplane H in V, then by Proposition 2.46, we have that $B(X) = B(Z) + [0, y]$. By induction, $B(Z)$ is paved by the parallelepipeds $\Pi_{\lambda_{\underline{c}}}(\underline{c})$ as $\underline{c}$ varies among the bases of H extracted from Z. A basis of V extracted from X is of the form $\underline{b} = \{\underline{c}, y\}$, where $\underline{c}$ is a basis of H contained in Z. Furthermore, $\lambda_{\underline{b}} = \lambda_{\underline{c}}$, so that $\Pi_{\lambda_{\underline{b}}}(\underline{b}) = \Pi_{\lambda_{\underline{c}}}(\underline{c}) + [0, y]$, and our claim follows.

If, on the other hand, Z spans V, then by induction the parallelepipeds $\Pi_{\lambda_{\underline{b}}}(\underline{b})$ as $\underline{b}$ varies among the bases of V extracted from Z are a paving of

$B(Z)$. We know that $B(Z)_y$ is the union of the faces $B(A) + \sum_{b \in B} b$, where A spans a hyperplane H with $A = X \cap H$ and $y \notin H$. Given a linear equation ϕ for H equal to 1 on y, we have that B is the set of elements of Z on which ϕ is positive. By induction, $B(A)$ is paved by the parallelepipeds $\Pi_{\lambda_{\underline{c}}}(\underline{c})$ as $\underline{c}$ varies among the bases of H extracted from A, so that $B(A) + \sum_{b \in B} b$ is paved by $\Pi_{\lambda_{\underline{c}}}(\underline{c}) + \sum_{b \in B} b$.

We have by the definitions that $\lambda_{\underline{c}} + \sum_{b \in B} b = \lambda_{\{\underline{c}, y\}}$, so that $B(Z)_y + [0, y]$ is paved by the parallelepipeds $\Pi_{\lambda_{\{\underline{c},y\}}}(\{\underline{c}, y\})$. In order to finish, we need to verify that on $B(Z)_y$ the given paving is equal to the one induced by $B(Z)$. In other words, we claim that the parallelepipeds $\Pi_{\lambda_{\underline{c}}}(\underline{c}) + \sum_{b \in B} b$ are faces of those paving $B(Z)$. If B is nonempty, let a be the largest element of B; in particular, $\underline{b} = \{\underline{c}, a\}$ is a basis of V. We claim that $\Pi_{\lambda_{\underline{c}}}(\underline{c}) + \sum_{b \in B} b$ is a face of $\Pi_{\lambda_{\underline{b}}}(\underline{b}) = \lambda_{\underline{b}} + B(\underline{b})$.

In fact, with the notation of (2.14) we have $B_{s-1} = B \setminus \{a\}$. Thus $\lambda_{\underline{b}}$ equals $\lambda_{\underline{c}} + \sum_{b \in B} b - a$, so $\Pi_{\lambda_{\underline{c}}}(\underline{c}) + \sum_{b \in B} b = \lambda_{\underline{b}} + B(\underline{c}) + a$. The claim follows from the fact that $B(\underline{c}) + a$ is a face of $B(\underline{b})$.

If B is empty, let a be the minimum element of $Z \setminus A$. Then again, $\underline{b} = \{\underline{c}, a\}$ is a basis of V and $\Pi_{\lambda_{\underline{c}}}(\underline{c})$ is a face of $\Pi_{\lambda_{\underline{b}}}(\underline{b})$, since we see that $\lambda_{\underline{c}} = \lambda_{\underline{b}}$.

2.3.2 $B(X)$ in the Case of Lattices

A lattice Λ in V is by definition the abelian subgroup of all integral linear combinations of some basis $v_1, \ldots, v_s$ of V. Of course, when $V = \mathbb{R}^s$, we have the *standard lattice* $\mathbb{Z}^s$. Sometimes we shall speak of a lattice Λ without specifying the vector space V, which will be understood as the space spanned by Λ. A region A of V with the property that V is the disjoint union of the translates $\lambda + A$, $\lambda \in \Lambda$, is called a *fundamental domain*. The volume of A with respect to some chosen translation-invariant measure in V is also called the *covolume* of the lattice. If $V = \mathbb{R}^s$ with the standard Lebesgue measure and Λ is generated by a basis $b_1, \ldots, b_s$, a fundamental domain is $\{\sum_{i=1}^{s} t_i b_i, \ 0 \le t_i < 1\}$ of volume $|\det(b_1, \ldots, b_s)|$.

Assume that the list $X \subset \Lambda$ generates V.

Definition 2.49. We let $\delta(X)$ denote the volume of the box $B(X)$.

As a simple application of Theorem 2.48 we obtain the following

Proposition 2.50. *1. Under the previous hypotheses*

$$\delta(X) = \sum_{\underline{b}} |\det(\underline{b})| \tag{2.15}$$

as $\underline{b}$ runs over all bases that one can extract from X.

2. Assume that the lattice $\Lambda \subset \mathbb{R}^s$ has covolume 1, that is, an integral basis of Λ has determinant ± 1. Let x_0 be a point outside the cut locus (cf. Definition 1.54). Then $(B(X) - x_0) \cap \Lambda$ consists of $\delta(X)$ points.

Proof. 1. The volume of $\Pi_\lambda(\underline{b})$ equals $|\det(\underline{b})|$. Thus statement 1 follows from Theorem 2.48.

Observe that, in the case of a lattice Λ of covolume 1 and $\underline{b} \subset \Lambda$, the number $|\det(\underline{b})|$ is a positive integer equal to the index in Λ of the sublattice generated by $\underline{b}$.

2. By definition, $\mu + x_0$ does not lie in the cut locus for every $\mu \in \Lambda$. On the other hand the boundary of each parallelepiped $\Pi_\lambda(\underline{b})$ lies in the cut locus; thus $(\Pi_\lambda(\underline{b}) - x_0) \cap \Lambda$ is contained in the interior of $\Pi_\lambda(\underline{b})$. It is therefore enough to see that $(\Pi_\lambda(\underline{b}) - x_0) \cap \Lambda$ consists of $|\det(\underline{b})|$ points, and summing over all parallelepipeds, we obtain our claim.

For a given parallelepiped, let $\overset{\circ}{\Pi}$ denote the interior of $\Pi_\lambda(\underline{b}) - x_0$. Since x_0 does not lie in the cut locus, the intersection $\{a_1, \ldots, a_k\} := \Lambda \cap (\Pi_\lambda(\underline{b}) - x_0)$ is contained in $\overset{\circ}{\Pi}$. Let M be the lattice generated by $\underline{b}$. We have that the union $\cup_{\alpha \in M}(\alpha + \overset{\circ}{\Pi})$ is a disjoint union and contains Λ. Thus Λ is the disjoint union of the cosets $a_i + M$. This means that the a_i form a full set of coset representatives of M in Λ; in particular, $k = [\Lambda : M] = |\det(\underline{b})|$, which is the volume of $\Pi_\lambda(\underline{b})$.

In general, the computation of $\delta(X)$ is rather cumbersome, but there is a special case in which this can be expressed by a simple determinantal formula.

Definition 2.51. The list X is called *unimodular* if $|\det(\underline{b})| = 1$ for all bases extracted from X.

In the unimodular case case $\delta(X)$ equals the number of bases that one can extract from X. This number, which will always be denoted by $d(X)$, also plays a role in the theory.

Proposition 2.52. *Let X be unimodular and $A := XX^t$. Then*

$$\delta(X) = d(X) = \det(A). \tag{2.16}$$

Proof. By Binet's formula we have

$$\det(A) = \sum_{\underline{b}} |\det(\underline{b})|^2. \tag{2.17}$$

So if $|\det(\underline{b})| = 1$ for all bases, the claim follows.

We have seen that interesting unimodular arrangements exist associated to graphs (cf. Corollary 2.16).

In general, for X in a lattice of covolume 1 we have $d(X) \leq \delta(X)$ and $d(X) = \delta(X)$ if and only if X is unimodular.

Example 2.53. In the next example,

$$X = \begin{vmatrix} 1 & 0 & 1 & -1 & 2 & 1 \\ 0 & 1 & 1 & 1 & 1 & 2 \end{vmatrix},$$

we have 15 bases and 15 parallelograms.

We pave the box inductively:

Start with
$$\begin{vmatrix} 1 & 0 \\ 0 & 1 \end{vmatrix}$$

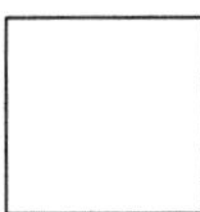

Then
$$\begin{vmatrix} 1 & 0 & 1 \\ 0 & 1 & 1 \end{vmatrix}$$

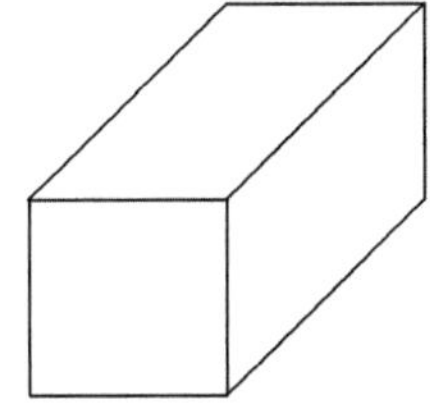

$$\begin{vmatrix} 1 & 0 & 1 & -1 \\ 0 & 1 & 1 & 1 \end{vmatrix}$$

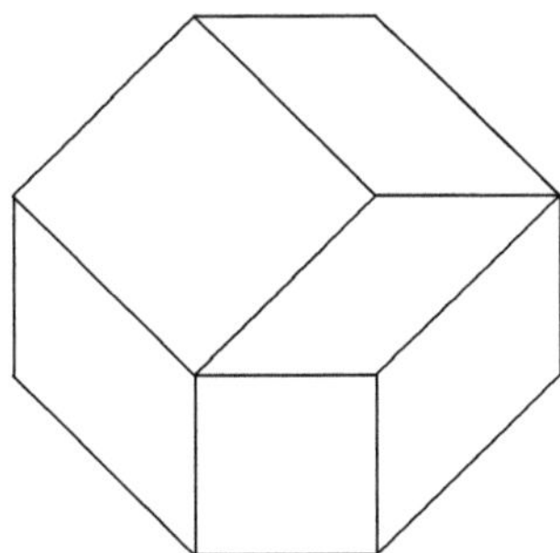

$$\begin{vmatrix} 1 & 0 & 1 & -1 & 2 \\ 0 & 1 & 1 & 1 & 1 \end{vmatrix}$$

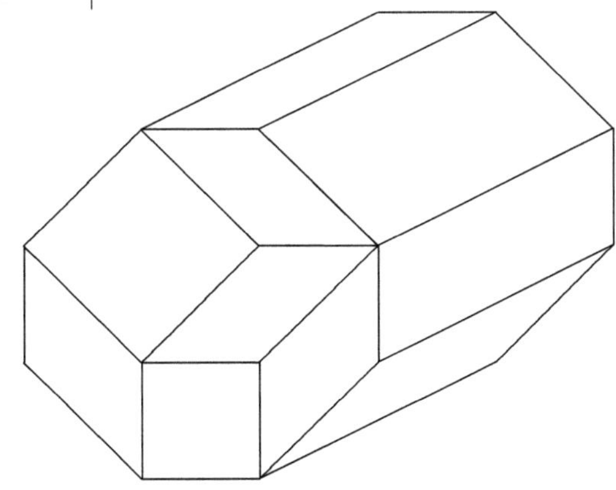

$$\textbf{Finally,} \qquad X = \begin{vmatrix} 1 & 0 & 1 & -1 & 2 & 1 \\ 0 & 1 & 1 & 1 & 1 & 2 \end{vmatrix}$$

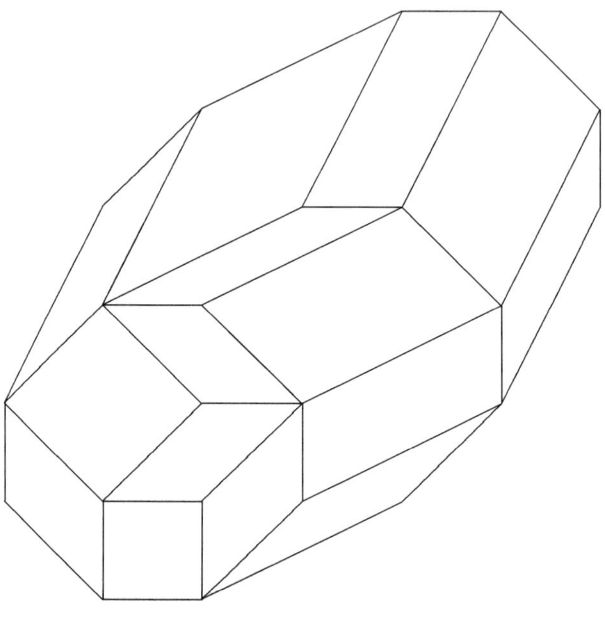

Fig. 2.3. The final paving

Remark 2.54. (1) The parallelepipeds involved in the paving have in general different volumes, given by the absolute value of the determinant of the corresponding basis (in our example we get areas 1, 2, 3).

(2) The paving of the box that we have presented is given by an algorithm that depends on the ordering of the vectors in X.

(3) Different orderings may give rise to different pavings.

For instance, for the three vectors $e_1, e_2, e_1 + e_2$ one can easily verify that we may obtain two different pavings from six different orderings.

$$A = \begin{vmatrix} 1 & 0 & 1 \\ 0 & 1 & 1 \end{vmatrix} \qquad\qquad\qquad A = \begin{vmatrix} 1 & 1 & 0 \\ 0 & 1 & 1 \end{vmatrix}$$

2.4 Root Systems

2.4.1 The Shifted Box

We want to make explicit many of the previous ideas in the particularly interesting case in which the list X coincides with the positive roots of a root system. For details about root systems one can consult [67].

Let R be a root system spanning a Euclidean space E of dimension r with Weyl group W acting by isometries. We choose a set R^+ of positive roots. Let $\Delta = \{\alpha_1, \ldots, \alpha_r\}$ be the set of simple roots and $P = \{p_1, \ldots, p_r\}$ the basis dual to Δ. We denote by C the fundamental Weyl chamber

$$C = \Big\{ v \in E \,|\, v = \sum_{i=1}^{r} a_i p_i, a_i \geq 0 \Big\}.$$

We want to discuss some properties of the zonotope $B(X)$ for $X = R^+$. In order to do so it is convenient to shift it by the element $\rho_X = \frac{1}{2} \sum_{\alpha \in R^+} \alpha$ and study $B(X) - \rho_X$. It is well known that ρ_X is an object that plays a fundamental role in the theory.[9] We need some preliminaries.

For a given $w \in W$ set $X_w := \{a \in X \,|\, w(a) \in -X\}$, so that with the notation of Section 2.3.1, $w(\rho_X) = P_{X_w} = \frac{1}{2}(\sum_{\alpha \in R^+ \setminus X_w} \alpha - \sum_{\alpha \in X_w} \alpha)$.

There is a characterization of the subsets X_w. Define the *convexity property* for a subset $A \subset X$ to be that for all $\alpha, \beta \in A$, the condition $\alpha + \beta \in X$ implies $\alpha + \beta \in A$. We then have the following result (see [85], or [72], Proposition 5.10).

Proposition 2.55. *A set $S \subset X$ is of the form X_w if and only if S and $X \setminus S$ satisfy the convexity property.*

Theorem 2.56. *If X is the set of positive roots of a root system, then $B(X) - \rho_X$ coincides with the convex envelope of the W-orbit of ρ_X.*

Proof. We know by Remark 2.38 that $B(X) - \rho_X$ coincides with the convex envelope of the elements P_S, $S \subset X$. Thus it suffices to show that if S is not of the form X_w, then P_S is not an extremal point of $B(X) - \rho_X$. Take S such that either S or $X \setminus S$ does not satisfy the convexity property. Assume, for instance, that S does not satisfy it, the argument being similar for $X \setminus S$. There are $\alpha, \beta \in S$ with $\alpha + \beta \in X \setminus S$. Consider now $S' = S \cup \{\alpha + \beta\}$ and $S'' = S \setminus \{\alpha, \beta\}$. Then $P_{S'} = P_S + \alpha + \beta$ and $P_{S''} = P_S - \alpha - \beta$. Thus $P_S = \frac{1}{2} P_{S'} + \frac{1}{2} P_{S''}$ is not extremal.

Remark 2.57. The convex envelope of the W-orbit of ρ_X is a remarkable polytope, known in the literature as a *generalized permutohedron*.[10]

---

[9] ρ_X is a very fundamental and ubiquitous object in Lie theory.

[10] The usual permutohedron is obtained in the case A_n.

We are now going to illustrate various computations for root systems. By the general theory it is enough to discuss the case of an irreducible root system R of rank s.

The following table gives information on the numbers $m(X)$, $d(X)$, $\delta(X)$:

Type	mrs	$m(X)$	$d(X)$	$\delta(X)$
A_s	A_{s-1}	s	$(s+1)^{s-1}$	$(s+1)^{s-1}$
D_s	D_{s-1}	$2(s-1)$	$d(s)$	$\delta(s)$
B_s	B_{s-1}	$2s-1$	$b(s)$	$\beta(s)$
C_s	C_{s-1}	$2s-1$	$b(s)$	$\gamma(s)$
G_2	A_1	5	15	24
F_4	B_3	15	7560	12462
E_6	D_5	16	846720	895536
E_7	E_6	27	221714415	248454360
E_8	E_7	57	348607121625	438191214480

We start with the computation of $m(X)$. A cocircuit is the complement in $X = R^+$ of a hyperplane spanned by some of is elements. In particular, a cocircuit is the set of positive roots which do not lie in a root system of rank $s-1$ contained in R.

Thus in order to compute $m(X)$ we need to find a proper root system contained in R of maximal cardinality. This is achieved by a simple case-by-case analysis. The column mrs lists the type of a sub-root system of rank $s-1$ having maximal cardinality.

We now pass to the computation of $d(X)$ and $\delta(X)$.

A possible way to compute the volume $\delta(X)$ is to use formula (2.15). This is not very efficient, since one needs to enumerate all bases $\underline{b}$ that can be extracted from X and for each such basis determine $|\det(\underline{b})|$. Of course, with this method one also computes $d(X)$.

Let us do this in the case of classical groups.

1. Type A_{s-1}. In this case the positive roots $e_i - e_j$, $1 \leq i < j \leq s$, correspond to the edges of the complete graph on s elements. Thus the bases extracted from the list correspond to maximal trees, that is trees with s labeled vertices (Proposition 2.13), and by Cauchy's formula and unimodularity, $d(X) = \delta(X) = (s+1)^{s-1}$.

2. Type D_s. The positive roots now are the vectors $e_i + e_j, e_i - e_j$, $i < j$. They can be interpreted as the edges of the complete graph on s elements marked by either $+$ or $-$.

In this case one reasons as in Proposition 2.13 and verifies that a subset of the roots, thought of as marked graph, is a basis if and only if:

(a) It contains all vertices.
(b) Each connected component contains a unique cycle in which the number of occurrences of $+$ is odd.

3. Type B_s, C_s are dual types, the positive roots in the B_s case are the vectors $e_i + e_j, e_i - e_j$, $i < j, e_i$, while in the C_s case, they are the vectors $e_i + e_j, e_i - e_j$,

$i < j, 2e_i$. Besides the marked edges the element e_i (resp. $2e_i$) can be thought of as a cycle at the vertex i that we mark by $+$.

In this case one easily verifies that a subset of the roots, thought of as a marked graph, is a basis if and only if:

(a) It contains all vertices.
(b) Each connected component contains a unique cycle in which the number of occurrences of $+$ is odd.

A recursive formula for $d(X)$ and $\delta(X)$ can be deduced as follows.

There is a general method to pass from connected graphs to any type of graph.

For any type of admissible graph let c_s count the number of connected admissible graphs on s vertices. The number b_s of marked graphs on s vertices all of whose connected components are admissible is

$$b_s = \sum_{k_1+2k_2+3k_3+\cdots=s} \frac{s!}{k_1!1!^{k_1}\,k_2!2!^{k_2}\cdots k_i!i!^{k_i}\cdots} c_1^{k_1} c_2^{k_2} \cdots .$$

Therefore

$$\log\left(\sum_{n=0}^{\infty} \frac{b_n}{n!} t^n\right) = \sum_{i=0}^{\infty} \frac{c_i}{i!} t^i. \tag{2.18}$$

It is easy to develop recursive formulas for $d(X), \delta(X)$ in the classical cases by this method, but we are going to explain, in the following section, a different method that easily allows us to compute both $d(X)$ and $\delta(X)$ in all cases in terms of the same invariants for maximal parabolic subsystems.

2.4.2 The Volume of $B(X)$

As in the previous section, let X be the set of positive roots in a root system with simple roots $\Delta = \{\alpha_1, \ldots, \alpha_s\}$. Given a subset $I \subset \Delta$ and a simple root $\alpha \in I$, we let $X_{I,\alpha}$ equal the set of positive roots in the root system spanned by I whose support contains α. We set $m_{I,\alpha} = |X_{I,\alpha}|$ and $z_{I,\alpha}$ equal to the sum over all the roots $\beta \in X_{I,\alpha}$ of the coefficient of α in the expression of β in the basis I. Given a permutation σ of Δ, we define, for $1 \leq k \leq s$, the subset $I_k = \{\sigma(\alpha_k), \ldots, \sigma(\alpha_s)\}$ and set

$$m_\sigma := \prod_{k=1}^{s-1} m_{I_k,\sigma(\alpha_k)}, \qquad z_\sigma := \prod_{k=1}^{s-1} z_{I_k,\sigma(\alpha_k)}. \tag{2.19}$$

Theorem 2.58. *We have the following expressions for $d(X)$ and $\delta(X)$*

$$d(X) = \frac{|W|}{2^s s!} \sum_{\sigma \in S_\Delta} m_\sigma, \qquad \delta(X) = \frac{|W|}{2^s s!} \sum_{\sigma \in S_\Delta} z_\sigma.$$

Proof. The proofs of the two formulas are very similar. Let us start with the case of $d(X)$.

Consider the set

$$\mathcal{S} := (\{b_1, \ldots, b_{s-1}\}, b_s, \phi),$$

where $\{b_1, \ldots, b_{s-1}, b_s\}$ is a basis extracted from R out of which we have chosen an element b_s, and ϕ is a vector of norm 1 orthogonal to $\{b_1, \ldots, b_{s-1}\}$. Clearly, since $R = X \cup -X$, this set is formed by $2^{s+1} s d(X)$ elements. We are now going to count this number in a different way. We use the fact that the fundamental Weyl chamber C is a fundamental domain for the action of W, so each such element can be carried into one in which $\phi \in C$. Since the set of roots orthogonal to ϕ spans a hyperplane, we must have that ϕ lies in one of the s half-lines ℓ_i generated by the fundamental weights ω_i and the intersection of $s - 1$ walls. This determines $\phi = \omega_i/|\omega_i|$. Thus $\mathcal{S} = \cup_i \mathcal{S}_i$ where $\mathcal{S}_i$ is the subset where $\phi = \omega_i/|\omega_i|$. The stabilizer of $\omega_i/|\omega_i|$ is the Weyl subgroup W_i of the root system R_i obtained by removing the i-th node of the Dynkin diagram. On the other hand, if X_i denote the positive roots of R_i, clearly the set $\mathcal{S}_i$ is formed by $2^{s-1}d(X_i)|R - R_i| = 2^s d(X_i)|X - X_i|$ elements. We deduce

$$d(X) = \frac{1}{2s} \sum_{i=1}^{s} \frac{|W|}{|W_i|} |X - X_i| d(X_i). \qquad (2.20)$$

A simple induction gives the required expression for $d(X)$.

We now pass to the expression for $\delta(X)$.

We use the fact that the boundary of the polytope $B(X) - \rho$ (convex envelope of the W orbit of ρ) is decomposed into s orbits under W, where s is the rank of the root system. For each node i of the Dynkin diagram, the convex envelope of $W_i\rho$ is a face F_i stabilized exactly by W_i. Its orbit consists of $|W/W_i|$ distinct faces. As i varies over all nodes we obtain all the codimension-one faces exactly once. Thus the volume of $B(X) - \rho$ is the sum of the volumes of the pyramids P_i with bases the faces F_i and their transforms. In order to compute it we make several remarks.

First we are computing volumes assuming that the root lattice has covolume 1. Suppose we compute volumes using any W-invariant scalar product. Then we have a different normalization and we need to multiply the previous volume by the square root $c^{\frac{1}{2}}$ of the determinant $c = \det(M)$ of the matrix $M = ((\alpha_i, \alpha_j))$.

We now compute using the metric structure. If v_i is the volume (relative) of the face H_i, and h_i the height of the corresponding pyramid, we have the formula

$$\delta(X) = c^{-\frac{1}{2}} \frac{|W|}{s} \sum_i \frac{h_i v_i}{|W_i|}.$$

We need to compute h_i, v_i. The barycenter of F_i is a vector fixed under W_i. Now by general facts such a vector is necessarily a multiple of the fundamental weight ω_i. The orthogonal projection of F_i to the subspace orthogonal to ω_i is an isometry, and one easily sees that F_i maps to the corresponding polytope $B(X_i) - \rho_i$ for the root system R_i. Thus we may assume that this has been computed by induction, and we have $v_i = c_i^{\frac{1}{2}} \delta(X_i)$. As for the height, we have to project ρ on the line generated by ω_i, and we obtain $(\rho, \omega_i)/(\omega_i, \omega_i)\omega_i$. Thus $h_i = (\rho, \omega_i)/(\omega_i, \omega_i)^{\frac{1}{2}}$ and

$$\delta(X) = c^{-\frac{1}{2}} \frac{|W|}{s} \sum_i \frac{c_i^{\frac{1}{2}}(\rho, \omega_i)\delta(X_i)}{(\omega_i, \omega_i)^{\frac{1}{2}}|W_i|}.$$

Let us compute h_i. By homogeneity we can substitute ω_i with its positive multiple p_i having the property that $(p_i, \alpha_j) = \delta_{i,j}$. We claim that $c_i = c(p_i, p_i)$. Indeed, the elements (p_i, p_j) are the entries of the M^{-1}. Thus the identity $c_i/c = (p_i, p_i)$ is the usual expression of the entries of the inverse as cofactors divided by the determinant. To finish the analysis of our formula, we have $\rho = \frac{1}{2}\sum_{\alpha \in X} \alpha$, and for any root α, (α, p_i) equals the coefficient of α_i in the expansion of α in the basis of simple roots. Thus if P_i is the positive integer sum of all these coordinates, we finally have the following

$$\delta(X) = \frac{1}{2s} \sum_i \frac{|W|}{|W_i|} P_i \, \delta(X_i). \tag{2.21}$$

Again a simple induction gives the required expression for $\delta(X)$.

It is clear that the previous formula can be expanded and generalized to give a formula for the volume of the convex hull of the W-orbit of any given vector v in Euclidean space.

Fix a vector $v \in E$. We define for any subset $I \subset \Delta$ the vector v_I as the orthogonal projection of v in the subspace E_I spanned by the set of simple roots in I. Given a complete flag of subsets

$$f = (I_1 \subset I_2 \subset \cdots \subset I_{s-1} \subset I_s = \Delta),$$

with $|I_h| = h$, we define the sequence of vectors $(u_1^f, \ldots, u_{s-1}^f)$ by setting $u_h^f = v_{I_h} - v_{I_{h-1}}$ $(u_1^f = v_{I_1})$. We then set

$$z_f := \prod_{j=1}^r (u_j^f, u_j^f)^{\frac{1}{2}}.$$

Now take v in the interior of C and denote by P_v the convex hull of the orbit of v under W.

Proposition 2.59. *Let $\mathcal{F}$ be the set of all complete flags in $\{1, 2, \ldots, s\}$. The volume of P_v equals*

$$\frac{|W|}{s!} \sum_{f \in \mathcal{F}} z_f. \tag{2.22}$$

Proof. We proceed by induction on s. If $s = 1$ then P_v is the segment $[-v, v]$, there is a unique flag, and $u_1^f = v$. Thus formula (2.22) gives $2(v, v)^{\frac{1}{2}}$ that is indeed the length of the segment $[-v, v]$.

Passing to the general case, let us start by remarking that the volume $\mathrm{Vol}(P_v)$ of P_v is the sum of the volumes of the pyramids with bases the faces of codimension-one and vertex the origin. If we take such a face F, one knows, by the theory of root systems, that there are a unique $i \in \{1, 2, \ldots, s\}$ and a unique coset $\Sigma_F \subset W/W_i$ with W_i equal to the subgroup of W generated by the simple reflections s_j, $j \neq i$, such that for any $w \in \Sigma_F$, wF is the face F_i given by the convex hull of the W_i orbit of v.

Thus, if we denote by Π_i the pyramid with basis F_i and vertex at the origin, we get

$$\mathrm{Vol}(P_v) = \sum_{i=1}^{r} \mathrm{Vol}(\Pi_i) \frac{|W|}{|W_i|}.$$

Now notice that the face F_i is parallel to the hyperplane spanned by the roots $\Delta_i := \Delta \setminus \{\alpha_i\}$. Thus its volume equals that of its orthogonal projection on this hyperplane, that is, the volume of $P_{v_{\Delta_i}}$. Also, the height of Π_i is clearly the segment $[0, v - v_{\Delta_i}]$; thus

$$\mathrm{Vol}(\Pi_i) = \frac{1}{s} \mathrm{Vol}(P_{v_{\Delta_i}})(v - v_{\Delta_i}, v - v_{\Delta_i})^{\frac{1}{2}}.$$

Substituting, we get

$$\mathrm{Vol}(P_v) = \sum_{i=1}^{s} \mathrm{Vol}(P_{v_{\Delta_i}})(v - v_{\Delta_i}, v - v_{\Delta_i})^{\frac{1}{2}} \frac{|W|}{r|W_i|}. \tag{2.23}$$

We now use the inductive hypothesis. Take for a fixed $i = 1, \ldots, s$ the set $\mathcal{F}_i$ of flags of subsets of Δ_i. We can think of this set as the set of flags $f = (I_1 \subset I_2 \subset \cdots \subset I_{s-1} \subset I_s)$ in Δ with $I_{s-1} = \Delta_i$. For such a flag, $u_s^f = v - v_{\Delta_i}$, and by the inductive hypothesis,

$$\mathrm{Vol}(P_{v_{\Delta_i}}) = \frac{|W_i|}{(s-1)!} \sum_{f \in \mathcal{F}_i} \frac{z_f}{(v - v_{\Delta_i}, v - v_{\Delta_i})^{\frac{1}{2}}}.$$

This and formula (2.23) clearly give our claim.

A simple computation gives the following

Example 2.60. We have the recursive formulas:

$$\beta(s) = \sum_{h=1}^{s} \beta(s-h)(2h)^{h-1}\binom{s}{h}\frac{2s-h}{s},$$

$$\gamma(s) = \sum_{h=1}^{s-1} \gamma(s-h)(2h)^{h-1}\binom{s}{h}\frac{2s-h+1}{s} + (2s)^{s-2}(s+1),$$

$$\delta(s) = \sum_{h=1}^{s-2} \delta(s-h)(2h)^{h-1}\binom{s}{h}\frac{2s-h-1}{s} + (2s)^{s-2}(s-1),$$

$$d(s) = \frac{1}{2s}\left[\sum_{i=1}^{s-2}(2i)^{i-1}\binom{s}{i}(4s-3i-1)d(s-i) + (2s)^{s-1}(s-1)\right],$$

$$b(s) = \frac{1}{2s}\left[\sum_{i=1}^{s-2}2^i\binom{s}{i}\left(s^2 - \left[\binom{i}{2} + (s-i)^2\right]\right)i^{i-2}b(s-i)\right]$$
$$+ 2^{s-2}\left(\frac{s(s+3)}{2} - 2\right)(s-1)^{s-3} + 2^{s-2}(s+1)s^{s-2}.$$

For type A_s one gets the identity

$$(s+1)^{s-1} = \frac{1}{2s}\sum_{h=1}^{s}\binom{s+1}{h}h^{h-1}(s-h+1)^{s-h}.$$

2.4.3 The External Activity and Tutte Polynomials

In this section we are going to give the generating functions for both the Tutte and the external activity polynomials, defined in formulas (2.3) and (2.7), for classical root systems, and in exceptional cases, their explicit expression. This computations for type A_n were done by Tutte in [114] and for other classical groups by Ardila [4] by a different method.

We follow the method explained in Section 2.2.5, so we first express the external activity polynomial in terms of the external activity polynomial of complete subsystems using formula (2.10). After this. we use formula (2.9) to get the Tutte polynomials. As we have already mentioned, in order to use this formula, we need know all proper complete subsystems, and for this we use the results and tables contained in [84].

1. Type A_{s-1}. We are in the subspace of $\mathbb{R}^s$ where the sum of coordinates is 0. Here a space of the arrangement is determined by a partition of the set $[1,\cdots,s]$. To such a partition we associate the subspace in which the coordinates are equal on each part. If the partition is of type $1^{h_1}2^{h_2}\cdots n^{h_s}$ with $\sum_{i=1}^{s} h_i i = s$, we have that the corresponding root system is $A_0^{h_1}A_1^{h_2}\cdots A_{s-1}^{h_s}$ with rank $\sum_{i=0}^{s}(i-1)h_i$ (here and later A_0 will have external activity equal to 1). There are exactly

$$\frac{s!}{h_1!1!^{h_1}h_2!2!^{h_2}\cdots h_s!s!^{h_s}}$$

such subspaces, so that if we denote by $e_s(y)$ the external activity of A_{s-1}, we have, by formula (2.10),

$$\frac{y^{\binom{s}{2}}}{s!} = \sum_{h_1 h_2 \cdots h_s \ | \ \sum_{i=1}^{s} h_i i = s} (y-1)^{\sum_{i=1}^{s}(i-1)h_i} \frac{e_1(y)^{h_1} e_2(y)^{h_2} \cdots e_s(y)^{h_s}}{h_1! 1!^{h_1} h_2! 2!^{h_2} \cdots h_s! s!^{h_s}}.$$

Writing the generating function, we get

$$\sum_{n=0}^{\infty} t^n \frac{y^{\binom{n}{2}}}{n!}$$

$$= \sum_{n=0}^{\infty} t^n \sum_{h_1 h_2 \cdots h_n \ | \ \sum_{i=1}^{n} h_i i = n} \frac{[e_1(y)]^{h_1} [(y-1)e_2(y)]^{h_2} \cdots [(y-1)^{n-1} e_n(y)]^{h_n}}{h_1! 1!^{h_1} h_2! 2!^{h_2} \cdots h_n! n!^{h_n}}$$

$$= \exp\left(\sum_{i=1}^{\infty} t^i \frac{(y-1)^{i-1} e_i(y)}{i!} \right)$$

so that

$$\sum_{i=1}^{\infty} t^i \frac{(y-1)^{i-1} e_i(y)}{i!} = \log\left(\sum_{n=0}^{\infty} t^n \frac{y^{\binom{n}{2}}}{n!} \right). \tag{2.24}$$

We can also rewrite this identity, if for t we substitute $t/(y-1)$, as

$$(y-1)^{-1} \sum_{i=1}^{\infty} t^i \frac{e_i(y)}{i!} = \log\left(\sum_{n=0}^{\infty} \left(\frac{t}{y-1} \right)^n \frac{y^{\binom{n}{2}}}{n!} \right).$$

For the Tutte polynomial $T_n(x,y)$ of A_{n-1} we have

$$\sum_{i=1}^{\infty} t^n \frac{T_n(x,y)}{n!}$$

$$= (x-1)^{-1} \sum_{n=1}^{\infty} t^n \sum_{h_1 h_2 \cdots h_n \ | \ \sum_{i=1}^{n} h_i i = n} \frac{(x-1)^{\sum_i h_i} e_1(y)^{h_1} e_2(y)^{h_2} \cdots e_n(y)^{h_n}}{h_1! 1!^{h_1} h_2! 2!^{h_2} \cdots h_n! n!^{h_n}}$$

$$= \frac{\exp\left(\sum_{i=1}^{\infty} t^i \frac{(x-1)e_i(y)}{i!} \right) - 1}{x-1} = \frac{\left(\sum_{n=0}^{\infty} (\frac{t}{y-1})^n \frac{y^{\binom{n}{2}}}{n!} \right)^{(y-1)(x-1)} - 1}{x-1}.$$

2. Types B_s, D_s. We are in $\mathbb{R}^s$. In type D_s, the hyperplanes are those of equations $x_i \pm x_j = 0$, $i \neq j$, while in B_s we have also the equations $x_i = 0$. In both cases we may take permutations of coordinates and sign changes as group of symmetries of the arrangement (this group coincides with the Weyl group in type B, while in type D, it contains the Weyl group as a subgroup of index 2).

Here a space of the arrangement is determined, up to sign changes, by a subset A of $[1, \cdots, s]$ where the coordinates are 0, that in case D_s if nonempty must contain at least two elements, and a partition of the set $[1, \cdots, s] \setminus A$, where the coordinates are equal on each part. If $k = |A|$ and $1^{h_1} 2^{h_2} \cdots (s-k)^{h_{s-k}}$ with $\sum_{i=1}^{s-k} h_i i = s - k$ are the sizes of the parts of the partition, we have that the corresponding root system is $B_k A_0^{h_1} A_1^{h_2} \cdots A_{s-k-1}^{h_{s-k}}$ or $D_k A_0^{h_1} A_1^{h_2} \cdots A_{s-k-1}^{h_{s-k}}$. There are exactly

$$\frac{2^{s - (k + h_1 + h_2 + \cdots + h_{s-k})} s!}{k! h_1! 1!^{h_1} h_2! 2!^{h_2} \cdots h_s! (s-k)!^{h_{s-k}}}$$

such subspaces, so that if we denote by $b_s(y), d_s(y)$ the external activity polynomials of B_s, D_s, we have, by formula (2.10),

$$B_s : \quad \frac{y^{s^2}}{2^s s!} = \sum_{k, \sum_{i=1}^{s-k} h_i i = s-k} (y-1)^{k + \sum_{i=1}^{s}(i-1)h_i} \frac{b_k(y)}{2^k k!} \prod_{j=1}^{s-k} \frac{e_j(y)^{h_j}}{2(h_j! j!^{h_j})},$$

$$D_s : \quad \frac{y^{s(s-1)}}{2^s s!} = \sum_{k \neq 1, \sum_{i=1}^{s-k} h_i i = s-k} (y-1)^{k + \sum_{i=1}^{s}(i-1)h_i} \frac{d_k(y)}{2^k k!} \prod_{j=1}^{s-k} \frac{e_j(y)^{h_j}}{2(h_j! j!^{h_j})},$$

In term of generating functions we have

$$\sum_n \frac{y^{n^2}}{2^n n!} t^n = \left[\sum_k (t(y-1))^k \frac{b_k(y)}{2^k k!} \right] \exp \sum_{i=1}^{\infty} t^i \frac{(y-1)^{i-1} e_i(y)}{2(i!)}$$

$$\sum_n \frac{y^{n(n-1)}}{2^n n!} t^n = \left[\sum_{k \neq 1} (t(y-1))^k \frac{d_k(y)}{2^k k!} \right] \exp \sum_{i=1}^{\infty} t^i \frac{(y-1)^{i-1} e_i(y)}{2(i!)}$$

whence, applying formula (2.24) we deduce

$$\sum_k (t(y-1))^k \frac{b_k(y)}{2^k k!} = \sum_n \left(\frac{y^{n^2}}{2^n n!} t^n \right) \left[\sum_{n=0}^{\infty} t^n \frac{y^{\binom{n}{2}}}{n!} \right]^{-\frac{1}{2}}$$

$$\sum_{k \neq 1} (t(y-1))^k \frac{d_k(y)}{2^k k!} = \sum_n \left(\frac{y^{n(n-1)}}{2^n n!} t^n \right) \left[\sum_{n=0}^{\infty} t^n \frac{y^{\binom{n}{2}}}{n!} \right]^{-\frac{1}{2}}.$$

Now for the Tutte polynomials let us denote by $T_s^B(x, y), T_s^D(x, y)$ the two series.

We have for type B_s,

$$\frac{T_s^B(x, y)}{s!} = \sum_{k, \sum_{i=1}^{s-k} h_i i = s-k} 2^s (x-1)^{\sum_i h_i} \frac{b_k(y)}{2^k k!} \prod_{j=1}^{s-k} \frac{e_j(y)^{h_j}}{2(h_j! j!^{h_j})}.$$

For type D_s, we have for $s \neq 1$:

$$\frac{T_s^D(x,y)}{s!} = \sum_{k \neq 1, \sum_{i=1}^{s-k} h_i i = s-k} 2^s (x-1)^{\sum_i h_i} \frac{d_k(y)}{2^k k!} \prod_{j=1}^{s-k} \frac{e_j(y)^{h_j}}{2(h_j! j!^{h_j})}.$$

For $s = 1$ we still use the formula and reset by abuse of notation $T_1^D = x - 1$. In terms of generating functions we get

$$\sum_n \frac{T_n^B(x,y)}{2^n n!} t^n = \Big(1 + \sum_{k \geq 1} t^k \frac{b_k(y)}{2^k k!}\Big) \exp\Big(\sum_i t^i \frac{(x-1)e_i(y)}{2i!}\Big)$$

$$= \Big(\sum_n \frac{y^{n^2}}{2^n n!}\Big(\frac{t}{y-1}\Big)^n\Big)\Big(\sum_{n=0}^\infty \Big(\frac{t}{y-1}\Big)^n \frac{y^{\binom{n}{2}}}{n!}\Big)^{\frac{1}{2}}\Big(\sum_{n=0}^\infty \Big(\frac{t}{y-1}\Big)^n \frac{y^{\binom{n}{2}}}{n!}\Big)^{\frac{(y-1)(x-1)}{2}}$$

$$= \Big(\sum_n \frac{y^{n^2}}{2^n n!}\Big(\frac{t}{y-1}\Big)^n\Big)\Big(\sum_{n=0}^\infty \Big(\frac{t}{y-1}\Big)^n \frac{y^{\binom{n}{2}}}{n!}\Big)^{\frac{(y-1)(x-1)+1}{2}}$$

$$\sum_n \frac{T_n^D(x,y)}{2^n n!} t^n = \Big(1 + \sum_{k \geq 2} t^k \frac{d_k(y)}{2^k k!}\Big) \exp\Big(\sum_i t^i \frac{(x-1)e_i(y)}{2i!}\Big)$$

$$= \Big(\sum_n \frac{y^{n(n-1)}}{2^n n!}\Big(\frac{t}{y-1}\Big)^n\Big)\Big(\sum_{n=0}^\infty \Big(\frac{t}{y-1}\Big)^n \frac{y^{\binom{n}{2}}}{n!}\Big)^{\frac{1}{2}}\Big(\sum_{n=0}^\infty \Big(\frac{t}{y-1}\Big)^n \frac{y^{\binom{n}{2}}}{n!}\Big)^{\frac{(y-1)(x-1)}{2}}$$

$$= \Big(\sum_n \frac{y^{n(n-1)}}{2^n n!}\Big(\frac{t}{y-1}\Big)^n\Big)\Big(\sum_{n=0}^\infty \Big(\frac{t}{y-1}\Big)^n \frac{y^{\binom{n}{2}}}{n!}\Big)^{\frac{(y-1)(x-1)-1}{2}}.$$

Summarizing,

$$\sum_{i=1}^\infty (y-1)^{i-1} t^i \frac{e_i(y)}{i!} = \log\Big(\sum_{n=0}^\infty t^n \frac{y^{\binom{n}{2}}}{n!}\Big)$$

$$\sum_{n=1}^\infty ((y-1)t)^n \frac{T_n(x,y)}{n!} = \frac{\Big(\sum_{n=0}^\infty t^n \frac{y^{\binom{n}{2}}}{n!}\Big)^{(y-1)(x-1)} - 1}{x-1}$$

$$\sum_k \Big((y-1)t\Big)^k \frac{b_k(y)}{k!} = \Big(\sum_n \frac{y^{n^2}}{n!} t^n\Big)\Big(\sum_{n=0}^\infty (2t)^n \frac{y^{\binom{n}{2}}}{n!}\Big)^{-\frac{1}{2}}$$

$$\sum_n \frac{T_n^B(x,y)}{n!} t^n = \Big(\sum_n \frac{y^{n^2}}{n!} t^n\Big)\Big(\sum_{n=0}^\infty (2t)^n \frac{y^{\binom{n}{2}}}{n!}\Big)^{\frac{(y-1)(x-1)+1}{2}}$$

$$\sum_{k \neq 1} \Big((y-1)t\Big)^k \frac{d_k(y)}{k!} = \Big(\sum_n \frac{y^{n(n-1)}}{n!} t^n\Big)\Big(\sum_{n=0}^\infty (2t)^n \frac{y^{\binom{n}{2}}}{n!}\Big)^{-\frac{1}{2}}$$

$$\sum_n \frac{T_n^D(x,y)}{n!}((y-1)t)^n = \left(\sum_n \frac{y^{n(n-1)}}{n!}t^n\right)\left(\sum_{n=0}^\infty (2t)^n \frac{y^{\binom{n}{2}}}{n!}\right)^{\frac{(y-1)(x-1)-1}{2}}$$

Notice that when we set $x = 1$ in the generating series of Tutte polynomials, we get the generating series of external activity.

Under the change of variable $q := (x-1)(y-1), y$ with inverse $x = \frac{q+y-1}{y-1}, y$, one recovers the formulas of Ardila [4], that are expressed in terms of the *coboundary polynomial* introduced by Crapo [36], for a matroid $M = (X, I)$. If X is of rank r, the coboundary polynomial is

$$\overline{\chi}(x,y) := \sum_{A\subset X} (x-1)^{r-\rho(A)}(y-1)^{|A|}.$$

The coboundary polynomial is related to the Tutte polynomial by the identity

$$\overline{\chi}(x,y) := (y-1)^r T\left(M, \frac{x+y-1}{y-1}, y\right).$$

Since the external activity and Tutte polynomials depend on the list of vectors only up to scale, type C_n (where the elements x_i are replaced by $2x_i$) is identical to type B_n for these computations.

For the exceptional groups one has to compute case by case using the method explained in Section 2.2.5 and the tables in [84]. The results one obtains are listed below.

2.4.4 Exceptional Types

We present the Tutte polynomials for exceptional root systems by a computer assisted computation

For G_2 the external activity is

$$5 + 4y + 3y^2 + 2y^3 + y^4.$$

The Tutte polynomial is

$$x^2 + 4x + 4y + 3y^2 + 2y^3 + y^4$$

and $\delta(X) = 24$, $d(X) = 15$, and $m(X) = 5$.

For F_4 the external activity is

$$385 + 700y + 850y^2 + 900y^3 + 873y^4 + 792y^5 + 680y^6 + 560y^7 + 455y^8 +$$
$$364y^9 + 286y^{10} + 220y^{11} + 165y^{12} + 120y^{13} + 84y^{14} + 56y^{15} + 35y^{16} + 20y^{17} +$$
$$10y^{18} + 4y^{19} + y^{20}.$$

The Tutte polynomial is

$$240x + 124x^2 + 20x^3 + x^4 + 240y + 392xy + 68x^2y + 508y^2 + 324xy^2 +$$
$$18x^2y^2 + 660y^3 + 240xy^3 + 729y^4 + 144xy^4 + 720y^5 + 72xy^5 + 656y^6 +$$

$24\,x\,y^6 + 560\,y^7 + 455\,y^8 + 364\,y^9 + 286\,y^{10} + 220\,y^{11} + 165\,y^{12} + 120\,y^{13} + 84\,y^{14} + 56\,y^{15} + 35\,y^{16} + 20\,y^{17} + 10\,y^{18} + 4\,y^{19} + y^{20},$
and $\delta(X) = 12462$, $d(X) = 7560$. In this case $m(X) = 15$.

For E_6 the external activity is

$12320 + 33600\,y + 54600\,y^2 + 71200\,y^3 + 81000\,y^4 + 84240\,y^5 + 82080\,y^6 + 76140\,y^7 + 68040\,y^8 + 58940\,y^9 + 49728\,y^{10} + 40992\,y^{11} + 33082\,y^{12} + 26172\,y^{13} + 20322\,y^{14} + 15504\,y^{15} + 11628\,y^{16} + 8568\,y^{17} + 6188\,y^{18} + 4368\,y^{19} + 3003\,y^{20} + 2002\,y^{21} + 1287\,y^{22} + 792\,y^{23} + 462\,y^{24} + 252\,y^{25} + 126\,y^{26} + 56\,y^{27} + 21\,y^{28} + 6\,y^{29} + y^{30}.$

The Tutte polynomial is

$5040\,x + 5004\,x^2 + 1900\,x^3 + 345\,x^4 + 30\,x^5 + x^6 + 5040\,y + 17064\,x\,y + 9516\,x^2\,y + 1860\,x^3\,y + 120\,x^4\,y + 17100\,y^2 + 27660\,x\,y^2 + 9030\,x^2\,y^2 + 810\,x^3\,y^2 + 32140\,y^3 + 32220\,x\,y^3 + 6570\,x^2\,y^3 + 270\,x^3\,y^3 + 46125\,y^4 + 31140\,x\,y^4 + 3735\,x^2\,y^4 + 56250\,y^5 + 26226\,x\,y^5 + 1764\,x^2\,y^5 + 61380\,y^6 + 20034\,x\,y^6 + 666\,x^2\,y^6 + 61830\,y^7 + 14130\,x\,y^7 + 180\,x^2\,y^7 + 58635\,y^8 + 9360\,x\,y^8 + 45\,x^2\,y^8 + 53090\,y^9 + 5850\,x\,y^9 + 46290\,y^{10} + 3438\,x\,y^{10} + 39102\,y^{11} + 1890\,x\,y^{11} + 32137\,y^{12} + 945\,x\,y^{12} + 25767\,y^{13} + 405\,x\,y^{13} + 20187\,y^{14} + 135\,x\,y^{14} + 15477\,y^{15} + 27\,x\,y^{15} + 11628\,y^{16} + 8568\,y^{17} + 6188\,y^{18} + 4368\,y^{19} + 3003\,y^{20} + 2002\,y^{21} + 1287\,y^{22} + 792\,y^{23} + 462\,y^{24} + 252\,y^{25} + 126\,y^{26} + 56\,y^{27} + 21\,y^{28} + 6\,y^{29} + y^{30},$
and $\delta(X) = 895536$, $d(X) = 846720$. In this case $m(X) = 16$.

For E_7 the external activity is

$765765 + 2522520\,y + 4851000\,y^2 + 7320600\,y^3 + 9580725\,y^4 + 11439540\,y^5 + 12802020\,y^6 + 13650000\,y^7 + 14026425\,y^8 + 13998600\,y^9 + 13645800\,y^{10} + 13047048\,y^{11} + 12273786\,y^{12} + 11386620\,y^{13} + 10436139\,y^{14} + 9463104\,y^{15} + 8499015\,y^{16} + 7567056\,y^{17} + 6683418\,y^{18} + 5859000\,y^{19} + 5100354\,y^{20} + 4410540\,y^{21} + 3789891\,y^{22} + 3236688\,y^{23} + 2747745\,y^{24} + 2318904\,y^{25} + 1945440\,y^{26} + 1622376\,y^{27} + 1344708\,y^{28} + 1107540\,y^{29} + 906192\,y^{30} + 736281\,y^{31} + 593775\,y^{32} + 475020\,y^{33} + 376740\,y^{34} + 296010\,y^{35} + 230230\,y^{36} + 177100\,y^{37} + 134596\,y^{38} + 100947\,y^{39} + 74613\,y^{40} + 54264\,y^{41} + 38760\,y^{42} + 27132\,y^{43} + 18564\,y^{44} + 12376\,y^{45} + 8008\,y^{46} + 5005\,y^{47} + 3003\,y^{48} + 1716\,y^{49} + 924\,y^{50} + 462\,y^{51} + 210\,y^{52} + 84\,y^{53} + 28\,y^{54} + 7\,y^{55} + y^{56}.$

The Tutte polynomial is

$368640\,x + 290304\,x^2 + 90944\,x^3 + 14560\,x^4 + 1260\,x^5 + 56\,x^6 + x^7 + 368640\,y + 1340928\,x\,y + 672000\,x^2\,y + 129696\,x^3\,y + 10920\,x^4\,y + 336\,x^5\,y + 1419264\,y^2 + 2498496\,xy^2 + 831600\,x^2y^2 + 97860\,x^3y^2 + 3780\,x^4y^2 + 3046400\,y^3 + 3409840\,xy^3 + 801360\,x^2y^3 + 61740\,x^3y^3 + 1260\,x^4y^3 + 4959360\,y^4 + 3923640\,xy^4 + 666540\,x^2y^4 + 31185\,x^3y^4 + 6864480\,y^5 + 4055436\,xy^5 + 505260\,x^2y^5 + 14364\,x^3y^5 + 8549856\,y^6 + 3892098\,xy^6 + 354900\,x^2y^6 + 5166\,x^3y^6 + 9886584\,y^7 + 3529896\,xy^7 + 232260\,x^2y^7 + 1260\,x^3y^7 + 10815720\,y^8 + 3065490\,xy^8 + 144900\,x^2y^8 + 315\,x^3y^8 + 11339440\,y^9 + 2573060\,x\,y^9 + 86100\,x^2\,y^9 + 11496240\,y^{10} + 2100588\,x\,y^{10} + 48972\,x^2\,y^{10} + 11344368\,y^{11} + 1676220\,x\,y^{11} + 26460\,x^2\,y^{11} + 10949456\,y^{12} + 1311100\,x\,y^{12} +$

$13230\,x^2\,y^{12} + 10374840\,y^{13} + 1006110x\,y^{13} + 5670\,x^2\,y^{13} + 9676584\,y^{14} + 757665xy^{14} + 1890x^2y^{14} + 8902578y^{15} + 560148xy^{15} + 378x^2y^{15} + 8092350y^{16} + 406665xy^{16} + 7277256y^{17} + 289800xy^{17} + 6481048y^{18} + 202370xy^{18} + 5720820y^{19} + 138180xy^{19} + 5008332y^{20} + 92022xy^{20} + 4350956y^{21} + 59584xy^{21} + 3752532y^{22} + 37359xy^{22} + 3214134y^{23} + 22554xy^{23} + 2734746y^{24} + 12999xy^{24} + 2311848y^{25} + 7056\,x\,y^{25} + 1941912\,y^{26} + 3528\,x\,y^{26} + 1620808\,y^{27} + 1568\,x\,y^{27} + 1344120\,y^{28} + 588\,x\,y^{28} + 1107372\,y^{29} + 168\,x\,y^{29} + 906164\,y^{30} + 28\,x\,y^{30} + 736281\,y^{31} + 593775\,y^{32} + 475020\,y^{33} + 376740\,y^{34} + 296010\,y^{35} + 230230\,y^{36} + 177100\,y^{37} + 134596\,y^{38} + 100947\,y^{39} + 74613\,y^{40} + 54264\,y^{41} + 38760\,y^{42} + 27132\,y^{43} + 18564\,y^{44} + 12376\,y^{45} + 8008\,y^{46} + 5005\,y^{47} + 3003\,y^{48} + 1716\,y^{49} + 924\,y^{50} + 462\,y^{51} + 210\,y^{52} + 84\,y^{53} + 28\,y^{54} + 7\,y^{55} + y^{56},$

and $\delta(X) = 248454360, d(X) = 221714415$. In this case $m(X) = 27$.

For E_8 the external activity is

$215656441 + 832880048\,y + 1846820976\,y^2 + 3154271120\,y^3 + 4616982370\,y^4 + 6116101992y^5 + 7560167076y^6 + 8882835192y^7 + 10042185195y^8 + 11014370920y^9 + 11790836876\,y^{10} + 12374757528\,y^{11} + 12777173066\,y^{12} + 13013665000\,y^{13} + 13102173000\,y^{14} + 13061395200\,y^{15} + 12909698025\,y^{16} + 12664461600\,y^{17} + 12341786800\,y^{18} + 11956490000\,y^{19} + 11522046984\,y^{20} + 11050547072\,y^{21} + 10552683964\,y^{22} + 10037779560\,y^{23} + 9513837015\,y^{24} + 8987619288\,y^{25} + 8464749444\,y^{26} + 7949828968\,y^{27} + 7446561710\,y^{28} + 6957871080y^{29} + 6486009452\,y^{30} + 6032658736\,y^{31} + 5599021077\,y^{32} + 5185898640\,y^{33} + 4793761440\,y^{34} + 4422802176\,y^{35} + 4072982628\,y^{36} + 3744072816\,y^{37} + 3435684120\,y^{38} + 3147297560\,y^{39} + 2878288435\,y^{40} + 2627948520\,y^{41} + 2395505940\,y^{42} + 2180142840\,y^{43} + 1981010970\,y^{44} + 1797245304\,y^{45} + 1627975812\,y^{46} + 1472337504\,y^{47} + 1329478865\,y^{48} + 1198568800\,y^{49} + 1078802208y^{50} + 969404304y^{51} + 869633808y^{52} + 778785120y^{53} + 696189600y^{54} + 621216072y^{55} + 553270671y^{56} + 491796152y^{57} + 436270780y^{58} + 386206920y^{59} + 341149446y^{60} + 300674088y^{61} + 264385836y^{62} + 231917400y^{63} + 202927725y^{64} + 177100560\,y^{65} + 154143080\,y^{66} + 133784560\,y^{67} + 115775100\,y^{68} + 99884400\,y^{69} + 85900584\,y^{70} + 73629072\,y^{71} + 62891499\,y^{72} + 53524680\,y^{73} + 45379620\,y^{74} + 38320568\,y^{75} + 32224114\,y^{76} + 26978328\,y^{77} + 22481940\,y^{78} + 18643560\,y^{79} + 15380937\,y^{80} + 12620256\,y^{81} + 10295472\,y^{82} + 8347680\,y^{83} + 6724520\,y^{84} + 5379616y^{85} + 4272048y^{86} + 3365856y^{87} + 2629575y^{88} + 2035800y^{89} + 1560780y^{90} + 1184040\,y^{91} + 888030\,y^{92} + 657800\,y^{93} + 480700\,y^{94} + 346104\,y^{95} + 245157\,y^{96} + 170544\,y^{97} + 116280\,y^{98} + 77520\,y^{99} + 50388\,y^{100} + 31824\,y^{101} + 19448\,y^{102} + 11440\,y^{103} + 6435\,y^{104} + 3432\,y^{105} + 1716\,y^{106} + 792\,y^{107} + 330\,y^{108} + 120\,y^{109} + 36\,y^{110} + 8\,y^{111} + y^{112}.$

The Tutte polynomial is

$127733760\,x + 70154496\,x^2 + 15748992\,x^3 + 1883728\,x^4 + 130144\,x^5 + 5208\,x^6 + 112\,x^7 + x^8 + 127733760\,y + 487014912\,x\,y + 187062912\,x^2\,y + 28823872\,x^3\,y + 2164512\,x^4\,y + 78960\,x^5\,y + 1120\,x^6\,y + 544594176\,y^2 + 995541120\,x\,y^2 + 275866080\,x^2\,y^2 + 29436960\,x^3\,y^2 + 1359960\,x^4\,y^2 + 22680\,x^5\,y^2 + 1298666880\,y^3 + 1513341760\,x\,y^3 + 316755040\,x^2\,y^3 + 24726240\,x^3\,y^3 + 773640\,x^4\,y^3 + 7560\,x^5\,y^3 +$

$2337363280y^4+1944391680xy^4+316590120x^2y^4+18285120x^3y^4+352170x^4y^4+$
$3558562560y^5+2252074608xy^5+292502448x^2y^5+12802608x^3y^5+159768x^4y^5+$
$4858714728y^6+2436564312xy^6+256396392x^2y^6+8435952x^3y^6+55692x^4y^6+$
$6151371912y^7+2510748360xy^7+215485920x^2y^7+5216400x^3y^7+12600x^4y^7+$
$7369379325y^8+2494145520xy^8+175532400x^2y^8+3124800x^3y^8+3150x^4y^8+$
$8465594760y^9+2407667360xy^9+139319600x^2y^9+1789200x^3y^9+9409765316y^{10}+$
$2271720360\,x\,y^{10}+108358320\,x^2\,y^{10}+992880\,x^3\,y^{10}+10186403808\,y^{11}+$
$2104854360\,x\,y^{11}+82970160\,x^2\,y^{11}+529200\,x^3\,y^{11}+10792181106\,y^{12}+$
$1922032000\,x\,y^{12}+62695360\,x^2\,y^{12}+264600\,x^3\,y^{12}+11232554200\,y^{13}+$
$1734243840\,x\,y^{13}+46753560\,x^2\,y^{13}+113400\,x^3\,y^{13}+11518581060\,y^{14}+$
$1549144080\,x\,y^{14}+34410060\,x^2\,y^{14}+37800\,x^3\,y^{14}+11664558360\,y^{15}+$
$1371840000\,x\,y^{15}+24989280\,x^2\,y^{15}+7560\,x^3\,y^{15}+11686317405\,y^{16}+$
$1205492400\,x\,y^{16}+17888220\,x^2\,y^{16}+11600023680\,y^{17}+1051848000\,x\,y^{17}+$
$12589920\,x^2\,y^{17}+11421406080\,y^{18}+911703800\,x\,y^{18}+8676920\,x^2\,y^{18}+$
$11165341880\,y^{19}+785303400\,x\,y^{19}+5844720\,x^2\,y^{19}+10845723504\,y^{20}+$
$672483840\,x\,y^{20}+3839640\,x^2\,y^{20}+10475351712\,y^{21}+572741440\,x\,y^{21}+$
$2453920\,x^2\,y^{21}+10065843904\,y^{22}+485319240\,x\,y^{22}+1520820\,x^2\,y^{22}+$
$9627584040\,y^{23}+409285800\,x\,y^{23}+909720\,x^2\,y^{23}+9169710435\,y^{24}+$
$343605360xy^{24}+521220x^2y^{24}+8700137688y^{25}+287199360\,x\,y^{25}+282240x^2y^{25}+$
$8225609004\,y^{26}+238999320\,x\,y^{26}+141120\,x^2\,y^{26}+7751775168\,y^{27}+$
$197991080xy^{27}+62720x^2y^{27}+7283296430y^{28}+163241760xy^{28}+23520x^2y^{28}+$
$6823954920y^{29}+133909440xy^{29}+6720x^2y^{29}+6376765212y^{30}+109243120xy^{30}+$
$1120x^2y^{30}+5944078216y^{31}+88580520xy^{31}+5527677357y^{32}+71343720xy^{32}+$
$5128866000\,y^{33}+57032640\,x\,y^{33}+4748545080\,y^{34}+45216360\,x\,y^{34}+$
$4387279896\,y^{35}+35522280\,x\,y^{35}+4045355028\,y^{36}+27627600\,x\,y^{36}+$
$3722820816\,y^{37}+21252000\,x\,y^{37}+3419532600\,y^{38}+16151520\,x\,y^{38}+$
$3135183920y^{39}+12113640xy^{39}+2869334875y^{40}+8953560xy^{40}+2621436840y^{41}+$
$6511680xy^{41}+2390854740y^{42}+4651200xy^{42}+2176887000y^{43}+3255840xy^{43}+$
$1978783290y^{44}+2227680xy^{44}+1795760184y^{45}+1485120xy^{45}+1627014852y^{46}+$
$960960xy^{46}+1471736904\,y^{47}+600600\,x\,y^{47}+1329118505\,y^{48}+360360\,x\,y^{48}+$
$1198362880\,y^{49}+205920\,x\,y^{49}+1078691328\,y^{50}+110880\,x\,y^{50}+969348864\,y^{51}+$
$55440\,x\,y^{51}+869608608\,y^{52}+25200\,x\,y^{52}+778775040\,y^{53}+10080\,x\,y^{53}+$
$696186240\,y^{54}+3360\,x\,y^{54}+621215232\,y^{55}+840\,x\,y^{55}+553270551\,y^{56}+$
$120\,x\,y^{56}+491796152\,y^{57}+436270780\,y^{58}+386206920\,y^{59}+341149446\,y^{60}+$
$300674088y^{61}+264385836y^{62}+231917400y^{63}+202927725y^{64}+177100560y^{65}+$
$154143080\,y^{66}+133784560\,y^{67}+115775100\,y^{68}+99884400\,y^{69}+85900584\,y^{70}+$
$73629072\,y^{71}+62891499\,y^{72}+53524680\,y^{73}+45379620\,y^{74}+38320568\,y^{75}+$
$32224114\,y^{76}+26978328\,y^{77}+22481940\,y^{78}+18643560\,y^{79}+15380937\,y^{80}+$
$12620256\,y^{81}+10295472\,y^{82}+8347680\,y^{83}+6724520\,y^{84}+5379616\,y^{85}+$
$4272048y^{86}+3365856y^{87}+2629575y^{88}+2035800y^{89}+1560780y^{90}+1184040y^{91}+$
$888030\,y^{92}+657800\,y^{93}+480700\,y^{94}+346104\,y^{95}+245157\,y^{96}+170544\,y^{97}+$
$116280\,y^{98}+77520\,y^{99}+50388\,y^{100}+31824\,y^{101}+19448\,y^{102}+11440\,y^{103}+$
$6435\,y^{104}+3432\,y^{105}+1716\,y^{106}+792\,y^{107}+330\,y^{108}+120\,y^{109}+36\,y^{110}+$
$8\,y^{111}+y^{112},$

and $\delta(X) = 438191214480$, $d(X) = 348607121625$. In this case $m(X) = 57$.

3

Fourier and Laplace Transforms

This short chapter collects a few basic facts of analysis needed for the topics discussed in this book.

3.1 First Definitions

3.1.1 Algebraic Fourier Transform

In this chapter, by vector space we shall mean a finite-dimensional vector space over the field of real numbers $\mathbb{R}$. It is convenient to take an intrinsic and basis-free approach.

Let us fix an s-dimensional vector space U, and denote by $V := U^*$ its dual. As usual, we also think of U as the dual of V. We identify the symmetric algebra $S[V]$ with the ring of polynomial functions on U.

This algebra can also be viewed as the algebra of polynomial differential operators on V with constant coefficients. Indeed, given a vector $v \in V$, we denote by D_v the corresponding directional derivative defined by

$$D_v f(x) := \left. \frac{df(x + tv)}{dt} \right|_{t=0}.$$

Observe that if $\phi \in U$ is a linear function (on V), we have $\langle \phi \mid x + tv \rangle = \langle \phi \mid x \rangle + t\langle \phi \mid v \rangle$. Thus D_v is algebraically characterized, on $S[U]$, as the derivation that on each element $\phi \in U$ takes the value $\langle \phi \mid v \rangle$. It follows that $S[V]$ can be thus identified with the ring of differential operators that can be expressed as polynomials in the D_v.

Similarly, $S[U]$ is the ring of polynomial functions on V, or polynomial differential operators on U with constant coefficients.

It is often very convenient to work over the complex numbers and use the complexified spaces

C. De Concini and C. Procesi, *Topics in Hyperplane Arrangements, Polytopes and Box-Splines*, Universitext, DOI 10.1007/978-0-387-78963-7_3,
© Springer Science+Business Media, LLC 2010

$$V_{\mathbb{C}} := V \otimes_{\mathbb{R}} \mathbb{C}, \quad U_{\mathbb{C}} := U \otimes_{\mathbb{R}} \mathbb{C} = \hom(V, \mathbb{C}).$$

One can organize all these facts in the algebraic language of the Fourier transform. Let us introduce the *Weyl algebras* $W(V), W(U)$ of differential operators with complex polynomial coefficients on V and U respectively. Choose explicit coordinates $x_1, \ldots, x_s$ for U, that is, a basis x_i for V, the dual basis in U is given by $\frac{\partial}{\partial x_1}, \ldots, \frac{\partial}{\partial x_s}$ and:

$$W(U) = W(s) := \mathbb{R}\left[x_1, \ldots, x_s; \frac{\partial}{\partial x_1}, \ldots, \frac{\partial}{\partial x_s} \right].$$

Notice that from a purely algebraic point of view they are both generated by $V \oplus U$. In the coordinates $x_1, \ldots, x_s$, for a multi-index $\alpha := \{h_1, \ldots, h_s\}$, $h_i \in \mathbb{N}$, one sets

$$x^{\alpha} := x_1^{h_1} \cdots x_s^{h_s}, \quad \partial^{\alpha} := \frac{\partial^{h_1}}{\partial x_1} \cdots \frac{\partial^{h_s}}{\partial x_s}.$$

One easily proves that the elements $x^{\alpha} \partial^{\beta}$ form a basis of $W(s)$, in other words, as a vector space we have that $W(U)$ and $W(V)$ equal the tensor product $S[U] \otimes S[V]$.

In the first case, V is thought of as the space of directional derivatives and then we write D_v instead of v, and U is thought of as the linear functions. In $W(U)$ the two roles are exchanged.

The relevant commutation relations are

$$[D_v, \phi] = \langle \phi \,|\, v \rangle, \qquad [D_\phi, v] = \langle v | \phi \rangle = \langle \phi \,|\, v \rangle, \tag{3.1}$$

and we have a canonical isomorphism of algebras

$$\mathcal{F} : W(V) \to W(U), \quad D_v \mapsto -v, \quad \phi \mapsto D_\phi.$$

One usually writes $\hat{a}$ instead of $\mathcal{F}(a)$.

This allows us, given a module M over $W(V)$ (resp. over $W(U)$) to consider its *Fourier transform* $\hat{M}$. This is the same vector space M, whose elements we shall denote by $\hat{m}$, and it is considered as a module over $W(U)$ (resp. over $W(V)$) by the formula $a.\hat{m} := \hat{a}m$.

In coordinates the automorphism $\mathcal{F}$ is then

$$\mathcal{F} : \ x_i \mapsto \frac{\partial}{\partial x_i}, \quad \frac{\partial}{\partial x_i} \mapsto -x_i.$$

3.1.2 Laplace Transform

We use Fourier transforms as an essentially algebraic tool. The fact is that for the purpose of this theory, the Fourier transform is essentially a duality between polynomials and differential operators with constant coefficients. As long as one has to perform only algebraic manipulations, one can avoid difficult issues of convergence and the use of various forms of algebraic or geometric duality (given by residues) is sufficient.

Nevertheless, we shall use some elementary analytic facts that we shall present in the language of *Laplace transforms*. This avoids cluttering the notation with unnecessary i's.

Our conventions for Laplace transforms are the following. With the notation of the previous section, fix a Euclidean structure on V that induces Lebesgue measures dv, du on V, U and all their linear subspaces. We set

$$Lf(u) := \int_V e^{-\langle u \,|\, v\rangle} f(v)dv. \tag{3.2}$$

Then L maps suitable functions on V to functions on U; a precise definition is given below. To start, if f is a C^∞ function with compact support, (3.2) is well-defined. We have the following basic properties when $p \in U, w \in V$, writing p for the linear function $\langle p \,|\, v\rangle$, and D_w for the derivative on V in the direction of w, (and dually on U):

$$L(D_w f)(u) = wLf(u), \qquad L(pf)(u) = -D_p Lf(u), \tag{3.3}$$
$$L(e^p f)(u) = Lf(u - p), \qquad L(f(v + w))(u) = e^w Lf(u). \tag{3.4}$$

3.1.3 Tempered Distributions

In order for the formula (3.2) to have a correct analytic meaning, it is necessary to impose restrictions on the functions $f(v)$, but the restriction that f be C^∞ with compact support is too strong for the applications we have in mind. One very convenient setting, which we shall use systematically, is given by the analytic language of *tempered distributions* (cf. [97], [98], [99], [121]).

Definition 3.1. The *Schwartz space* $\mathcal{S}$ is the space of smooth C^∞ and *rapidly decreasing* functions f on $\mathbb{R}^s$. That is, such that for every pair of multi-indices α, β one has that $x^\alpha \partial^\beta f$ is a bounded function.

It follows easily that $|x^\alpha \partial^\beta f|$ takes a maximal value, which will be denoted by $|f|_{\alpha,\beta}$. This is a family of seminorms on $\mathcal{S}$ that induces a topology on $\mathcal{S}$, and

Definition 3.2. A *tempered distribution* is a linear functional on $\mathcal{S}$ continuous under all these seminorms.

Remark 3.3. We have used coordinates to fix explicitly the seminorms, but one could have given a coordinate-free definition using monomials in linear functions ϕ and derivatives D_v.

By definition, the Schwartz space is a module over the Weyl algebra $W(s)$ of differential operators. By duality, so is the space of tempered distributions via the formulas

$$\langle x_i T \,|\, f \rangle = \langle T \,|\, x_i f \rangle, \quad \left\langle \frac{\partial}{\partial x_i} T \,\Big|\, f \right\rangle = -\left\langle T \,\Big|\, \frac{\partial}{\partial x_i} f \right\rangle. \tag{3.5}$$

The usual Fourier transform is

$$\hat{f}(u) := (2\pi)^{-s/2} \int_V e^{i\langle u \,|\, v \rangle} f(v) dv. \tag{3.6}$$

A basic fact of the Fourier analysis (cf. [121]), is that the Fourier transform is an isomorphism of $\mathcal{S}$ to itself (as a topological vector space) and thus it induces, by duality, a Fourier transform on the space of tempered distributions. We shall use this fact in order to describe certain special functions g that are implicitly given by suitable formulas for the associated distribution $f \mapsto \int_V g(v)f(v)dv$.

We leave to the reader to make explicit the analogues of formulas (3.3) that make explicit the compatibility of the Fourier transform with the $W(s)$ module structure on $\mathcal{S}$.

We shall also use, in a superficial way, distributions on periodic functions, that is the dual of C^∞ functions on a compact torus with a similar topology as that of Schwartz space. In this case Fourier transform is a function on the character group, a lattice Λ. The Fourier coefficients of C^∞ functions and of distributions can be characterized by their decay or their growth at infinity; see, for instance, [71].

3.1.4 Convolution

In our treatment, functions and distributions will be very special, and most of their analytic properties can be verified directly. Nevertheless, we should recall some basic general constructions, referring to [64] for the proofs.

The *convolution* of two L^1 functions on $\mathbb{R}^s$ is defined by the formula

$$(f * g)(x) := \int_{\mathbb{R}^s} f(y)g(x - y)dy.$$

For two distributions T_1 and T_2, where T_2 has compact support, the convolution can be defined using the following fact. Given a function $f \in \mathcal{S}$, we consider for each x the function $f[x](y) := f(x - y)$ as a function of y and apply to it the distribution T_2. We set $(T_2 * f)(x) := \langle T_2 \,|\, f[x] \rangle$. Then $T_2 * f$ belongs to $\mathcal{S}$, so we can set

$$\langle T_1 * T_2 \,|\, f \rangle := \langle T_1 \,|\, T_2 * f \rangle.$$

Indeed, this definition is compatible with the definition of $T_2 * f$ once we think of f as a distribution.

One has the following formula for Fourier transform

$$\widehat{T_1 * T_2} = \hat{T}_1 \hat{T}_2.$$

Although the product of two distributions is in general not well-defined, this formula is justified by the fact that in this case $\hat{T}_2$ is a function. In fact, if one restricts to distributions with compact support, one has unconditionally that convolution of such distributions is an associative and commutative law and Fourier transform is a homomorphism to holomorphic functions with respect to usual multiplication.

Example 3.4. Let $p \in \mathbb{R}^s$ and let δ_p be the usual Dirac distribution at p: $\langle \delta_p \mid f \rangle := f(p)$. We then have $\delta_p * f(x) = f(x - p)$ for any function f. Thus $\delta_0 * T = T$ for all distributions T, and by associativity $\delta_p * \delta_q = \delta_{p+q}$.

3.1.5 Laplace Versus Fourier Transform

If we take a C^∞ function $g(x)$ with compact support, its Fourier transform extends to a global holomorphic function. Then the Fourier transform is connected with the Laplace transform by the formula

$$Lg(u) = (2\pi)^{s/2}\hat{g}(iu). \tag{3.7}$$

We need to generalize this relation. Take a list $X := (a_1, \ldots, a_m)$ of nonzero vectors in V. We assume that 0 is not in the convex hull of the vectors in X so that $C(X)$ does not contain lines.

Define the *dual cone* $\widehat{C(X)}$ of $C(X)$:

$$\widehat{C(X)} := \{u \in U \mid \langle u \mid v \rangle \geq 0, \ \forall v \in C(X)\}.$$

This cone thus consists of the linear forms that are nonnegative on $C(X)$. Its interior in not empty, since $C(X)$ contains no lines.

Take a function $T(v)$ of polynomial growth supported in $C(X)$. Such a function determines by integration a tempered distribution. Its Fourier transform is defined, for $f \in \mathcal{S}$, by

$$\langle \hat{T} \mid f \rangle = \int_{C(X)} T(v)\hat{f}(v)dv = (2\pi)^{-s/2} \int_{C(X)} T(v)\left(\int_U e^{i\langle u \mid v \rangle} f(u)du \right)dv.$$

We would like to argue that $\hat{T}(u) = (2\pi)^{-s/2} \int_{C(X)} e^{i\langle u \mid v \rangle} T(v)dv$ is a function, so that

$$\langle \hat{T} \mid f \rangle = (2\pi)^{-s/2} \int_U \left(\int_{C(X)} e^{i\langle u \mid v \rangle} T(v)dv \right) f(u)du = \int_U \hat{T}(u)f(u)du.$$

In fact, the integral $\int_{C(X)} e^{i\langle u \mid v \rangle} T(v)dv$ may diverge, in which case the exchange of integrals is not justified. We then proceed as follows:

Proposition 3.5. *(1) The formula*

$$\hat{T}(z) := (2\pi)^{-s/2} \int_{C(X)} e^{i\langle x+iy\,|\,v\rangle} T(v)\,dv = (2\pi)^{-s/2} \int_{C(X)} e^{-\langle y-ix\,|\,v\rangle} T(v)\,dv$$

$$(3.8)$$

defines an analytic function of the complex variable $z = x + iy$, in the domain where y lies in the interior of $\widehat{C(X)}$ and $x \in U$.

(2) $\hat{T}$ is the weak limit (in the sense of distributions) of the family of functions $\hat{T}(x+iy)$ of x, parametrized by y, as $y \to 0$.

Proof. (1) The fact that $\hat{T}(z)$ is holomorphic in the described region follows from the fact that the function $e^{-\langle y-ix\,|\,v\rangle} T(v)$ is rapidly decreasing at infinity when y lies in the interior of $\widehat{C(X)}$.

(2) We claim now that $\hat{T}$ is the weak limit of the functions $\hat{T}(x+iy)$ (each for fixed y) in the space of tempered distributions. This means that if f is any C^∞ function in the Schwartz space $\mathcal{S}$, we have

$$\langle \hat{T} \,|\, f \rangle = \lim_{y\to 0} \int_U \hat{T}(x+iy) f(x)\,dx.$$

Notice that formula (3.8) shows that given y, the function $\hat{T}(x+iy)$ is the Fourier transform of the function $T_y(v) := e^{-\langle y\,|\,v\rangle} T(v)$, which is easily seen to be an L^2 function.

Since the Fourier transform is an isomorphism of the Schwartz space, $\hat{T}$ is the weak limit of the functions $\hat{T}(x+iy)$ if and only if $\lim_{y\to 0} T_y = T$ in the weak topology of tempered distributions. Thus we have to show that given a Schwartz function f,

$$\lim_{y\to 0} \int_V T_y(v) f(v)\,dv = \int_V T(v) f(v)\,dv.$$

We have that $\lim_{y\to 0} T_y(v) = T(v)$ pointwise and uniformly on compact sets. For y in the interior of $\widehat{C(X)}$ we have $|T_y(v)| \leq |T(v)|$, $\forall v$. By hypothesis, there is an integer k and two positive numbers R, C such that, for points outside the ball of radius R we have an estimate $|T_y(v)| < C|v|^k$. Given $f \in \mathcal{S}$, we can also find a radius, that we may assume to be equal to R, such that if $|v| \geq R$ we have $|f(v)| < R^{-2k}$. Now given $\epsilon > 0$ we can choose R large enough that $|\int_{|v|\geq R} T_y(v) f(v)\,dv| < \epsilon$, $\forall y$. On the other hand, by the uniform convergence of T_y to T on the ball $|v| \leq R$, we can find η such that when $|y| < \eta$, y in the interior of $\widehat{C(X)}$, we have $|\int_{|v|\leq R}(T(v) - T_y(v)) f(v)\,dv| < \epsilon$. This proves the weak convergence.

Thus $LT(z) := (2\pi)^{s/2}\hat{T}(iz)$ is holomorphic for $z = x + iy$ as long as x lies in the interior of $\widehat{C(X)}$. This is the sense in which we understand formula (3.7).

We shall apply formula (3.2), defining the Laplace transform, to functions (or distributions) supported in $C(X)$ and with polynomial growth there. Thus by definition the Laplace transform will be a holomorphic function in the region of complex vectors where the real part x lies in the interior of $\widehat{C(X)}$.

In most examples these functions extend to a meromorphic functions with poles on suitable hypersurfaces, as hyperplanes or tori. We shall also considerthese functions as Laplace transforms.

4

Modules over the Weyl Algebra

All the modules over Weyl algebras that will appear are built out of some basic irreducible modules, in the sense that they have finite composition series in which only these modules appear. It is thus useful to give a quick description of these modules. Denote by F the base field (of characteristic 0) over which $V, U := V^*$ are finite-dimensional vector spaces of dimension s. We can take either $F = \mathbb{R}$ or $F = \mathbb{C}$.

4.1 Basic Modules

4.1.1 The Polynomials

With the notation of the previous chapter we are going to study certain modules over the Weyl algebra $W(U)$ of differential operators with polynomial coefficients on U. The most basic module on $W(U)$ is the polynomial ring $S[V] = F[x_1, \ldots, x_s]$ (using coordinates). The following facts are immediate, but we want to stress them, since they will be generalized soon.

Proposition 4.1. *(a) $F[x_1, \ldots, x_s]$ is a cyclic module generated by 1.*
(b) $F[x_1, \ldots, x_s]$ is an irreducible module.
(c) The annihilator ideal of 1 is generated by all the derivatives D_u, $u \in U$, equivalently by all the derivatives $\frac{\partial}{\partial x_i}$.
(d) An element $a \in F[x_1, \ldots, x_s]$ is annihilated by all the derivatives if and only if it is a constant multiple of 1.

We shall use a simple corollary of this fact.

Corollary 4.2. *Let M be any $W(U)$ module.*
If $m \in M$ is a nonzero element satisfying $D_u m = 0$, $\forall u \in U$, then the $W(U)$-module generated by m is isomorphic to $S[V]$ under a unique isomorphism mapping 1 to m.

C. De Concini and C. Procesi, *Topics in Hyperplane Arrangements, Polytopes and Box-Splines*, Universitext, DOI 10.1007/978-0-387-78963-7_4,
© Springer Science+Business Media, LLC 2010

Let $m_1, \ldots, m_k$ be linearly independent elements in M satisfying the equations $D_u m = 0$, $\forall u \in U$. Then the modules $W(U)m_i$ form a direct sum.

Proof. If I denotes the left ideal generated by the elements D_u, we have $W(U)/I = S[V]$. We have a nonzero surjective morphism $W(U)/I \to W(U)m$ mapping 1 to m. Since $S[V]$ is irreducible, this map is an isomorphism.

Let us prove the second part. Any element of $\sum_i W(U)m_i$ can clearly be written (by part 1) as a sum $\sum_i f_i m_i$, $f_i \in S[V]$ polynomials. We have to show that if $\sum_i f_i m_i = 0$ then all the f_i's are equal to 0.

Assume this is not the case. Let $r \geq 0$ be the highest degree of the polynomials f_i. Choose a monomial x^α of degree r that appears in at least one of the polynomials f_i and apply ∂^α to $\sum_i f_i m_i$. Since $D_u m_i = 0$ for all u, we have that $\partial^\alpha(f_i m_i) = \partial^\alpha(f_i)m_i$. For all i we have that $\partial^\alpha(f_i)$ is a constant, and for at least one of the f_i's it is nonzero. Thus $0 = \sum_i \partial^\alpha(f_i)m_i$ is a linear combination with nonzero constant coefficients of the m_i, a contradiction.

4.1.2 Automorphisms

From the commutation relations (3.1) it is clear that the vector subspace $V \oplus U \oplus F \subset W(U)$ is closed under commutators. Thus it is a Lie subalgebra (called the *Heisenberg algebra*). The Lie product takes values in F and induces the natural symplectic form on $V \oplus U$.

The group $G = (V \oplus U) \rtimes \mathrm{Sp}(V \oplus U)$ (where $\mathrm{Sp}(V \oplus U)$ denotes the symplectic group of linear transformations preserving the given form) acts as a group of automorphisms of the algebra $W(U)$ preserving the space $V \oplus U \oplus F$. Let us identify some automorphisms in G.[1]

- Translations. Given $u \in U$ we get the automorphism τ_u defined by $\tau_u(v) = v + \langle v, u \rangle$ for $v \in V$ and $\tau_u(u') = u'$ if $u' \in U$. Similarly, we get the automorphism τ_v associated to an element $v \in V$.
- Linear changes. Given an element $g \in Gl(V)$, we can consider the symplectic transformation
$$\begin{pmatrix} g & 0 \\ 0 & g^{*-1} \end{pmatrix},$$
where $g^* : U \to U$ is the adjoint to g.
- Partial Fourier transform. If we choose a basis x_j, $1 \leq j \leq s$ of U and denote by $\partial/\partial x_i$ the dual basis of V, we have for a given $0 \leq k \leq s$ the symplectic linear transformation ψ_k defined by
$$x_i \mapsto \frac{\partial}{\partial x_i}, \quad \frac{\partial}{\partial x_i} \mapsto -x_i, \quad \forall i \leq k,$$
$$x_i \mapsto x_i, \quad \frac{\partial}{\partial x_i} \mapsto \frac{\partial}{\partial x_i}, \quad \forall i > k.$$

---

[1] It is not hard to see that these automorphisms generate G.

We can use automorphisms in the following way.

Definition 4.3. Given a module M over a ring R and an automorphism ϕ of R, one defines $^\phi M$, the *module twisted by* ϕ, as the same abelian group (or vector space for algebras) M with the new operation $r \circ_\phi m := \phi^{-1}(r)m$.

The following facts are trivial to verify:

- $^\phi(M \oplus N) =^\phi M \oplus^\phi N$.
- If M is cyclic generated by m and with annihilator I, then $^\phi M$ is cyclic generated by m and with annihilator $\phi(I)$.
- M is irreducible if and only if $^\phi M$ is irreducible.
- If ϕ, ψ are two automorphisms, then $^{\phi\psi} M =^\phi (^\psi M)$.

What will really appear in our work are some twists, under automorphisms of the previous type, of the basic module of polynomials.

More precisely, for each affine subspace S of U we have an irreducible module N_S, which is defined as follows.

Let $S = W + p$, where W is a linear subspace of U and $p \in U$. Choose coordinates $x_1, \ldots, x_s$ such that

$$W = \{x_1 = x_2 = \cdots = x_k = 0\}$$

and $\langle x_i, p \rangle = 0$ for $i > k$. We define the module N_S as follows: N_S is generated by an element δ_S satisfying

$$x_i \delta_S = \langle x_i, p \rangle \delta_S, \; i \le k, \quad \frac{\partial}{\partial x_i} \delta_S = 0, \; i > k. \tag{4.1}$$

To our basis we associate the linear transformation ψ_k and to p the translation τ_p. It is clear that the two transformations commute and their composition gives a well-defined element $\phi \in G$ and hence a well-defined automorphism of $W(U)$, which we denote by the same letter.

Proposition 4.4. *1. N_S is canonically isomorphic to the twist $^\phi S[V]$ of the polynomial ring under the automorphism ϕ defined above. This isomorphism maps δ_S to 1. In particular N_S is irreducible.*
2. N_S is freely generated by the element δ_S, the transform of 1, over the ring

$$\phi(F[x_1, \ldots, x_s]) = F\left[\frac{\partial}{\partial x_1}, \ldots, \frac{\partial}{\partial x_k}, x_{k+1}, x_{k+2}, \ldots, x_s\right].$$

3. N_S depends only on S and not on the coordinates chosen.

Proof. The first two statements follow from the definitions.

As for the last statement, notice that the elements $x_i - \langle x_i, p \rangle$ that vanish on S span the intrinsically defined space $S^\perp$ of all (inhomogeneous) linear polynomials vanishing on S. On the other hand, the derivatives $\frac{\partial}{\partial x_i}$, $i > k$, thought of as elements of U, span W. These two spaces generate the left ideal annihilator of the element δ_S.

Remark 4.5. We can also interpret N_S in the language of distributions, leaving to the reader to verify that in this language, δ_S should be thought of as the δ–function of S, given by $\langle \delta_S \,|\, f \rangle = \int_S f dx$.

4.1.3 The Characteristic Variety

In our work, we shall need to be able to distinguish some modules as nonisomorphic. In particular, we need to show that if $S_1 \neq S_2$ are two affine spaces, the corresponding modules N_{S_1}, N_{S_2} are not isomorphic.

This is best done with the use of the *characteristic variety*. This is an important notion and deserves a more detailed treatment (cf. [35]). For our purposes only very little information is necessary, and we give it here.

We start with some simple facts on *filtered algebras*. We shall use only increasing filtrations.

Definition 4.6. A *filtration* of an algebra R is a sequence of subspaces R_k, $k = -1, \ldots, \infty$, such that

$$R = \cup_{k=0}^{\infty} R_k, \quad R_h R_k \subset R_{h+k}, \quad R_{-1} = 0. \tag{4.2}$$

The concept of a filtered algebra compares with the much more restrictive notion of a graded algebra.

Definition 4.7. A *graded algebra* is an algebra R with a sequence of subspaces $R_k \subset R$, $k = 0, \ldots, \infty$ such that

$$R = \oplus_{k=0}^{\infty} R_k, \quad R_h R_k \subset R_{h+k}. \tag{4.3}$$

An element of R_k is said to be *homogeneous of degree k*. The product of two homogeneous elements of degree h, k is homogeneous of degree $h + k$.

A graded algebra is also filtered in a natural way (we take $\oplus_{i \leq k} R_i$ as the subspace of filtration degree k), but in general filtrations do not arise from gradings.

While the algebra of polynomials can be graded in the usual way, the algebra of differential operators can only be filtered.

In the case of the algebra $W(s)$ let us consider for each $k \geq 0$ the subspace $W(s)_k$ consisting of those operators that are of degree $\leq k$ in the derivatives. This is clearly a filtration, that is known as the *Bernstein filtration* [35].

From a filtered algebra $R = \cup R_k$ we can construct its *associated graded algebra*.

Intuitively, the idea is that given an element of R_h, we wish to concentrate on the degree h and neglect elements in R_{h-1}. More concretely, we replace R with the direct sum

$$\mathrm{gr}(R) := \oplus_{h=0}^{\infty} R_h / R_{h-1}, \quad \mathrm{gr}(R)_h := R_h / R_{h-1}, \text{ for } h \geq 0.$$

Given elements $a \in \mathrm{gr}(R)_h$, $b \in \mathrm{gr}(R)_k$, in order to define their product we lift them to elements $\tilde{a} \in R_h$, $\tilde{b} \in R_k$. We then define $ab \in \mathrm{gr}(R)_{h+k}$ as the image modulo R_{h+k-1} of $\tilde{a}\tilde{b} \in R_{h+k}$. It easy to see that this gives a well-defined product

$$R_h/R_{h-1} \times R_k/R_{k-1} \to R_{h+k}/R_{h+k-1}.$$

Thus $\mathrm{gr}(R)$ with this product is a graded algebra, which we call the associated graded algebra to R.

The class in R_h/R_{h-1} of an element $a \in R_h$ is called the *symbol of* a. This is consistent with the terminology used in the theory of differential operators.

The graded algebra associated to a filtration of an algebra R can be viewed as a *simplification* or *degeneration* of R. Often it is possible to recover many properties of R from those of the (simpler) associated graded algebra. A particularly important case arises when R is noncommutative but $\mathrm{gr}(R)$ is commutative, this happens if and only if for any two elements a, b of degree (in the filtration) h, k we have $[a, b] \in R_{h+k-1}$.

This occurs in particular in the case of the Bernstein filtration. From the definitions one easily sees the following

Lemma 4.8. *The associated graded algebra of $W(s)$ relative to the Bernstein filtration is a ring of polynomials in the variables x_i and the classes ξ_i of $\frac{\partial}{\partial x_i}$.*

Proof. The elements x_j have degree 0 and the $\frac{\partial}{\partial x_i}$ have degree 1, both $x_j \frac{\partial}{\partial x_i}$ and $\frac{\partial}{\partial x_i} x_j$ have degree 1. Their commutators $[\frac{\partial}{\partial x_i}, x_j] = \delta_{i,j}$ have degree 0, and therefore their classes commute in the graded algebra.

The fact that the classes x_i and ξ_i polynomially generate the graded algebra is a simple exercise, that follows from the canonical form of the operators in $W(s)$.

One can take a more intrinsic point of view and then see that this graded polynomial algebra is just the coordinate ring of $U \oplus V$, that can be thought of as the *cotangent bundle* to U.

Consider now a module M over a filtered algebra R.

Definition 4.9. A *filtration* on M, compatible with the filtration of R, is an increasing sequence of subspaces $0 = M_{-1} \subset M_0 \subset M_1 \subset M_k \subset \cdots$ that satisfies

$$\cup_k M_k = M, \quad R_h M_k \subset M_{h+k}, \quad \forall h, k.$$

Lemma 4.10. *The graded vector space $\mathrm{gr}(M) := \oplus_{h=0}^{\infty} M_h/M_{h-1}$ is in a natural way a module over the graded algebra.*

Proof. We leave this as an exercise.

From now on, we assume that M is a finitely generated module over R. We can define on M a compatible filtration by choosing generators $m_1, \ldots, m_d$ and setting $M_k := \sum_{i=1}^{d} R_k m_i$. This filtration depends on the generators m_i, but we have the following lemma:

Lemma 4.11. *Let M_k^1, M_k^2 be the filtrations corresponding to two sets of generators $m_1, \ldots, m_d$; $n_1, \ldots, n_e$. There exist two positive integers a, b such that for each k, we have*

$$M_k^1 \subset M_{k+b}^2, \quad M_k^2 \subset M_{k+a}^1.$$

Proof. Let a be an integer such that $n_i \in M_a^1$, $i = 1, \ldots, e$ and b an integer such that $m_i \in M_b^2, i = 1, \ldots, d$. The assertion follows immediately from the definition of the filtrations, since $M_k^2 = \sum_i R_k n_i \subset \sum_i R_k M_a^1 \subset M_{k+a}^1$ (similarly for the other case).

When the graded algebra is commutative, it is useful to consider the ideal $J_{\mathrm{gr}(M)}$ in $\mathrm{gr}(R)$ that annihilates $\mathrm{gr}(M)$. Notice that $J_{\mathrm{gr}(M)}$ is a graded ideal.

A priori, this ideal depends on the generators of M we have chosen but we have the following important theorem:

Theorem 4.12. *The radical $\sqrt{J_{\mathrm{gr}(M)}}$ is independent of the choice of the generators.*

Proof. Let us consider two such filtrations and let a, b be defined as above. Let us show, for instance, that $J_{\mathrm{gr}_1(M)} \subset \sqrt{J_{\mathrm{gr}_2(M)}}$. Take $x \in J_{\mathrm{gr}_1(M)}$ homogeneous of degree h. This means that for all k we have $x M_k^1 \subset M_{h+k-1}^1$. Thus $x^t M_k^1 \subset M_{th+k-t}^1, \forall t > 0$, and thus $x^t M_k^2 \subset x^t M_{k+b}^1 \subset M_{th+k+b-t}^1 \subset M_{th+k+b+a-t}^2$. Thus if $t > a + b$, we have $x^t \in J_{\mathrm{gr}_2(M)}$, as desired.

In the case of a finitely generated module M over the Weyl algebra $W(s)$ the ideal $\sqrt{J_{\mathrm{gr}(M)}}$ is a well-defined ideal in a polynomial ring in $2s$ variables, the functions on $U \oplus V$.

Definition 4.13. The variety of the zeros of $\sqrt{J_{\mathrm{gr}(M)}}$ is the *characteristic variety* of M.

We have just seen that the characteristic variety of a finitely generated module M is a geometric invariant of the isomorphism class of the module M.

Consider now the module N_S, where as before, S is an affine subspace of the form $W + p$, with W a linear subspace. It is clear that, the filtration on N_S induced by the Bernstein filtration, using the generator δ_S, equals the filtration by degree in the variables $\frac{\partial}{\partial x_j}$ for

$$N_S = F\left[\frac{\partial}{\partial x_1}, \ldots, \frac{\partial}{\partial x_k}, x_{k+1}, x_{k+2}, \ldots, x_s\right]\delta_S.$$

The ideal $J_{\mathrm{gr}(N_S)}$ is the ideal generated by the elements $x_i - \langle x_i \,|\, p \rangle$, ξ_j with $i \leq k$, $j > k$. The associated characteristic variety is a subspace of dimension n of the $2n$-dimensional space that intrinsically can be identified to the conormal space to S, that is the set of pairs $(v, u) \in V \oplus U$ such that $u \in S$ and $v \in W^{\perp}$.

In particular, we deduce the following result

Corollary 4.14. *If S_1, S_2 are two distinct subspaces. Then the two modules N_{S_1}, N_{S_2} are not isomorphic.*

Proof. The two characteristic varieties are different.

Remark 4.15. In the theory one has a deep property of the characteristic variety, that of being *involutive* for the symplectic structure on the cotangent bundle. This implies that the characteristic variety has always dimension $\geq s$. When it has dimension exactly s it is a *Lagrangian variety* and the module is called *holonomic*. The modules N_S are all (simple examples of) holonomic modules.

5

Differential and Difference Equations

The purpose of this chapter is to recall standard facts about certain special systems of differential equations that admit, as solutions, a finite-dimensional space of exponential polynomials. The theory is also extended to difference equations and quasipolynomials.

5.1 Solutions of Differential Equations

5.1.1 Differential Equations with Constant Coefficients.

In this section we recall the algebraic approach to special systems of differential equations with constant coefficients.

We use the same notation as in Chapter 4, but for reasons coming from duality, we usually think of V as directional derivatives and U as linear functions on V.

In coordinates, we fix a basis $e_1, \ldots, e_s$ of V. We denote by y_i the coordinates of a vector $y \in V$ in this basis, so $y_1, \ldots, y_s$ form a basis of U. We then can identify $e_1, \ldots, e_s$ with $\frac{\partial}{\partial y_1}, \ldots, \frac{\partial}{\partial y_s}$, so that

$$S[U] = F[y_1, \ldots, y_s], \quad S[V] = F\left[\frac{\partial}{\partial y_1}, \ldots, \frac{\partial}{\partial y_s}\right].$$

The duality between U and V extends as follows:

Proposition 5.1. *The ring $S[U]$ is the graded dual of the polynomial ring $S[V]$.*[1]

Proof. Using coordinates, the duality pairing can be explicitly described as follows. Given a polynomial $p(\frac{\partial}{\partial y_1}, \ldots, \frac{\partial}{\partial y_s})$ in the derivatives and another $q(y_1, \ldots, y_s)$ in the variables y_i, the pairing

---

[1] By definition, the graded dual of a graded vector space $\oplus_i V_i$ is the space $\oplus_i V_i^*$, clearly contained in the dual $(\oplus_i V_i)^* = \prod_i (V_i)^*$.

C. De Concini and C. Procesi, *Topics in Hyperplane Arrangements, Polytopes and Box-Splines*, Universitext, DOI 10.1007/978-0-387-78963-7_5,
© Springer Science+Business Media, LLC 2010

$$\left\langle p\left(\frac{\partial}{\partial y_1}, \ldots, \frac{\partial}{\partial y_s}\right) \mid q(y_1, \ldots, y_s)\right\rangle = p\left(\frac{\partial}{\partial \underline{y}}\right)(q(\underline{y}))_{\underline{y}=0} \tag{5.1}$$

is obtained by applying p to q and then evaluating at 0.

Even more explicitly, the dual basis of the monomials $y_1^{h_1} \cdots y_s^{h_s}$ is given by the monomials $(1/\prod_i h_i!)\frac{\partial^{h_1}}{\partial y_1} \cdots \frac{\partial^{h_s}}{\partial y_s}$.

The algebraic theory of differential equations starts with the following simple lemma

Lemma 5.2. *A polynomial $f \in \mathbb{C}[y_1, \ldots, y_s]$ satisfies the differential equations of an ideal $I \subset \mathbb{C}[\frac{\partial}{\partial y_1}, \ldots, \frac{\partial}{\partial y_s}]$ if and only if it is orthogonal to I.*

Proof. One direction is clear: if f satisfies the differential equations of I then it is orthogonal to I. Conversely, assume that f is orthogonal to I, take $g(\frac{\partial}{\partial y_1}, \ldots, \frac{\partial}{\partial y_s}) \in I$, and consider $q := g(\frac{\partial}{\partial y_1}, \ldots, \frac{\partial}{\partial y_s})(f)$. We need to show that $q = 0$. By assumption, for every $h \in \mathbb{C}[\frac{\partial}{\partial y_1}, \ldots, \frac{\partial}{\partial y_s}]$ we have that $h(q)$ evaluated at 0 equals 0. In particular, this implies that q with all of its derivatives is 0 at 0, and this clearly implies that $q = 0$.

We shall use the algebraic dual $(S[V])^*$ of $S[V]$, which is a rather enormous space of a rather formal nature. It can be best described as a space of *formal power series* by associating to an element $f \in (S[V])^*$ the formal expression

$$\langle f \mid e^y \rangle = \sum_{k=0}^{\infty} \frac{\langle f \mid y^k \rangle}{k!}, \quad y = \sum_{i=1}^{s} y_i e_i \in V.$$

Write $y = \sum_{i=1}^{s} y_i e_i$ in the basis e_i, thus

$$y^k = \sum_{h_1+h_2+\cdots+h_s=k} \binom{k}{h_1, h_2, \ldots, h_s} y_1^{h_1} \cdots y_s^{h_s} e_1^{h_1} \cdots e_s^{h_s},$$

$$\frac{\langle f \mid y^k \rangle}{k!} = \sum_{h_1+h_2+\cdots+h_s=k} \frac{\langle f \mid e_1^{h_1} \cdots e_s^{h_s} \rangle}{h_1! h_2! \cdots h_s!} y_1^{h_1} \cdots y_s^{h_s},$$

is a genuine homogeneous polynomial of degree k on V.

We have that $S[U]$ is the subspace of $S[V]^*$ formed by those f with $\langle f \mid y^k \rangle = 0$ for k large enough. The following facts are easy to see:

Proposition 5.3. *1. Given a vector $v \in V$, the transpose of multiplication by v acting on $S[V]$ is the directional derivative D_v (acting on $(S[V])^*$ and preserving $S[U]$).*

2. Given $\phi \in U$, denote by τ_ϕ the automorphism of $S[V]$ induced by translation $y \mapsto y - \langle \phi \mid y \rangle, \forall y \in V$, and by τ_ϕ^* its transpose. We have $\tau_\phi^* f = e^{-\langle \phi \mid y \rangle} f$.

Proof. 1 follows from the way we have defined the duality and the formula $D_v(\phi) = \langle \phi \,|\, v \rangle$.

As for 2,

$$\langle \tau_\phi^* f \,|\, e^y \rangle = \langle f \,|\, \tau_\phi(e^y) \rangle = \langle f \,|\, e^{y - \langle \phi \,|\, y \rangle} \rangle = e^{-\langle \phi \,|\, y \rangle} \langle f \,|\, e^y \rangle.$$

Observe that if J is an ideal of $S[V]$ defining a subvariety $Z \subset U$, we have that $\tau_\phi(J)$ defines the subvariety $Z + \phi$.

Starting from a quotient $S[V]/I$ by an ideal I we deduce an injection $i : (S[V]/I)^* \to (S[V])^*$. The image of i is, at least formally, the space of solutions of the differential equations given by I, that is the space of formal power series solutions of the system I of linear differential equations. As we shall see, as soon as I is of finite codimension, all the formal solutions are in fact holomorphic. We denote by $\mathrm{Sol}(I)$ the space of C^∞ solutions of the system of differential equations given by I.

Assume now that $S[V]/I$ is finite-dimensional. Denote by $\{\phi_1, \ldots, \phi_k\} \subset U$ the finite set of points that are the support of I. Take the decomposition

$$S[V]/I = \oplus_{i=1}^k S[V]/I(\phi_i), \tag{5.2}$$

where $S[V]/I(\phi_i)$ is local and supported at ϕ_i. Under these assumptions we get the following result:

Theorem 5.4. *(1) If $S[V]/I$ is finite-dimensional and supported at 0 the image of i lies in $S[U]$ and coincides with $\mathrm{Sol}(I)$.*

(2) If $S[V]/I$ is finite-dimensional and supported at a point $\phi \in U$, then $S[V]/\tau_{-\phi}I$ is supported at 0

$$\mathrm{Sol}(I) = e^{\langle \phi \,|\, y \rangle}\, \mathrm{Sol}(\tau_{-\phi}I).$$

(3) For a general finite-dimensional $S[V]/I = \oplus_{i=1}^k S[V]/I(\phi_i)$, as in 5.2:

$$\mathrm{Sol}(I) = \oplus_{i=1}^k \mathrm{Sol}(I(\phi_i)).$$

Proof. (1) Consider $S[V] = \mathbb{C}[x_1, \ldots, x_s]$ as a polynomial algebra. To say that I is of finite codimension and supported at 0 means that I contains all homogeneous polynomials of high enough degree. Dually a function satisfying the differential equations defined by substituting the x_i with $\frac{\partial}{\partial y_i}$ in the elements of I have all the high enough derivatives equal to 0. Thus they are polynomials. Now we can apply Lemma 5.2 and see that a function $f \in S[U]$ satisfies the differential equations given by an ideal I if and only if it is in the subspace of $S[U]$ orthogonal to I.

As for (2) observe that for $v \in V$ and f a differentiable function we have $D_v e^{\langle \phi \,|\, x \rangle} f = \langle \phi \,|\, v \rangle e^{\langle \phi \,|\, x \rangle} f + e^{\langle \phi \,|\, x \rangle} D_v f$. Therefore, f satisfies a differential equation $gf = 0, g \in S[V]$, if and only if $e^{\langle \phi \,|\, x \rangle} f$ satisfies the differential equation $\tau_\phi g$.

For part (3) let us choose elements $g_i \in S[V]$ giving the unit elements of the various $S[V]/I(\phi_i)$. If f satisfies the differential equations of I we see that $y_i(f)$ satisfies the differential equations of $I(\phi_i)$. Moreover, $\sum_i g_i - 1 \in I$, so that $\sum_i g_i(f) = f$ and the claim follows.

We see that the space of solutions of the equations given by an ideal I of finite codimension m is an m-dimensional vector space formed by functions of type $e^\phi p$, with ϕ linear and p a polynomial. Some authors refer to these functions as *exponential polynomials*. If the variety associated to I is supported at 0, then all the solutions are polynomials. On the other hand if I defines m distinct points (the reduced case), the space of solutions has a basis given by the exponential functions associated to these distinct points. In both cases the space of solutions is closed under the operation of taking derivatives.

Proposition 5.5. *Let M be an m-dimensional vector space of C^∞ functions on some open set $A \subset V$ that is closed under the operation of taking derivatives. Then:*

1. *All the elements of M are linear combinations of functions of type $e^\phi p$, with ϕ linear and p a polynomial.*
2. *M is the space of solutions of the differential equations given by an ideal $I \subset S[V]$ of codimension m.*

Proof. Since the derivatives are commuting operators, M has a Fitting decomposition $M = \oplus_\alpha M_\alpha$, where M_α is the subspace corresponding to the generalized eigenvalue $\underline{\alpha} := \{\alpha_1, \ldots, \alpha_s\}$. This means that the elements f of $M_{\underline{\alpha}}$ satisfy, for some $k > 0$, the differential equations

$$\left(\frac{\partial}{\partial y_i} - \alpha_i\right)^k f = 0, \; \forall i. \tag{5.3}$$

By the theory already developed, the space of solutions of 5.3 is a space of functions $e^{\sum_i \alpha_i y_i} p(y)$ where $p(y)$ is a polynomial of degree $< k$.

This proves (i). As far as (ii) is concerned, the Fitting decomposition allows us to reduce to the case in which M consists only of polynomials times a fixed exponential $e^{\sum_i \alpha_i y_i}$. Applying a translation, we can reduce to the case in which $\underline{\alpha} = 0$, and the space M is a space of polynomials.

Having made this reduction, consider M as a module over $S[V]$. Its annihilator ideal I is thus the set of differential equations (with constant coefficients) satisfied by all the elements of M. We claim that I has finite codimension and $M = \mathrm{Sol}(I)$. Since $M \subset \mathrm{Sol}(I)$, by duality it is enough to see that I has finite codimension equal to the dimension of M.

By applying derivatives, we see that every subspace of M invariant under derivatives, contains the element 1.

There is a duality under which M^* is also a module over the algebra $S[V]$, the graded dual of $S[U]$, using the formula $\langle u\phi \,|\, m \rangle := \langle \phi \,|\, u(m) \rangle$. Consider the element $\epsilon \in M^*$ given by the map $\epsilon : p \mapsto p(0)$. We claim that ϵ generates M^* as an $S[V]$ module.

In fact, if this were not true, the subspace of M, orthogonal to $S[V]\epsilon$, would be a nontrivial $S[V]$ submodule of M consisting of elements vanishing at 0, contradicting the property that every submodule contains 1.

We finally claim that I is the annihilator of ϵ, so $S[V]/I \cong M$, and this completes the claim. In fact, it is clear that $I\epsilon = 0$. Conversely, assume that $u\epsilon = 0$, $u \in S[V]$, then ϵ vanishes on the submodule uM. Since every nonzero submodule of M contains 1 and $\epsilon(1) = 1$, this means that $uM = 0$.

Remark 5.6. When we take an ideal of finite codimension the polynomial solutions we have found exhaust all solutions of the system of differential equations, even in the space of tempered distributions.

This follows easily by the following lemma

Lemma 5.7. *A tempered distribution T that is annihilated by all derivatives of order k is necessarily a polynomial of degree $\leq k - 1$.*

We offer two proofs, the first uses cohomology with compact support and the second elliptic regularity.

First proof.   We proceed by induction on the order of the derivatives that annihilate the distribution.

First, if all the first derivatives are 0, we claim that the distribution is a constant. This is equivalent to showing that given any function $f \in \mathcal{S}$ in the Schwartz space, with the property that $\int_V f dv = 0$, we have $\langle T \,|\, f \rangle = 0$.

This requires several steps. Assume first that f has compact support. We use a basic fact for which we refer to [18], namely that the top de Rham cohomology with compact support of $\mathbb{R}^n$ is $\mathbb{R}$. This implies thatv if $\int_V f dv = 0$ then f is a sum of derivatives of functions with compact support. Thus if a distribution T has all the derivatives equal to zero it must vanish on such a function.

In order to pass from functions with compact support to all functions in the Schwartz space we can use theorem XV on page 75 of [99], stating that functions with compact support are dense in the Schwartz space (for the topology of seminorms).

Using induction, we can find polynomials p_i with,

$$\frac{\partial T}{\partial y_i} = p_i \ \ \forall \ i = 1, \ldots, m.$$

We have

$$\frac{\partial p_i}{\partial y_j} = \frac{\partial}{\partial y_j}\frac{\partial}{\partial y_i}T = \frac{\partial}{\partial y_i}\frac{\partial}{\partial y_j}T = \frac{\partial p_j}{\partial y_i},$$

and hence there is a polynomial q with

$$\frac{\partial q}{\partial y_i} = p_i, \ \ \forall \ i = 1, \ldots, m.$$

Thus $T - q$ is a constant, as required.

Second proof. If T satisfies the hypotheses, then T also satisfies the differential equation $\Delta^k T = 0$ where $\Delta = \sum_i \frac{\partial^2}{\partial y_i}$ is the Laplace operator. Since the operator Δ^k is elliptic, it follows (see [64]) that T is a C^∞ function. The hypothesis that its high enough derivatives vanish immediately implies that it is a polynomial.

5.1.2 Families

In many interesting examples, a finite subscheme of $\mathbb{C}^n$ appears as the limit of a family of sets of points. The classical example is for roots of a polynomial of degree n, where we can think of the identity

$$\prod_{i=1}^{n}(t - y_i) = t^n + \sum_{i=1}^{n} t^{n-i} a_i$$

as a system of n-equations in the $2n$ variables $y_1, \ldots, y_n; a_1, \ldots, a_n$. For generic values of the a_i, these equations define $n!$ distinct points. On the other hand, if $a_i = 0$, $\forall i$, we have an interesting scheme, supported at 0, whose dual is the space of *harmonic polynomials*. This is the space of polynomials f satisfying the differential equations $p(\frac{\partial}{\partial y_1}, \ldots, \frac{\partial}{\partial y_s})f = 0$ for all the symmetric polynomials p without constant coefficients.

Since a similar phenomenon will occur when we study E-splines versus multivariate splines it is useful to collect some general facts.

Assume that we have an ideal $J \subset \mathbb{C}[y_1, \ldots, y_s, x_1, \ldots, x_m]$ of a polynomial algebra defining some variety $V \subset \mathbb{C}^{s+m}$. If $J \cap \mathbb{C}[x_1, \ldots, x_m] = 0$, we have that the mapping $\pi : V \to \mathbb{C}^m$ induced by $(y_1, \ldots, y_s, x_1, \ldots, x_m) \mapsto (x_1, \ldots, x_m)$ is generically surjective. Let us specialize the variables x_i to explicit values $\underline{\mu} := (\mu_1, \ldots, \mu_m)$, so that

$$\mathbb{C}[y_1, \ldots, y_s] = \mathbb{C}[y_1, \ldots, y_s, x_1, \ldots, x_m]/(x_1 - \mu_1, \ldots, x_i - \mu_i, \ldots, x_m - \mu_m).$$

The corresponding ideal $J + (x_1 - \mu_1, \ldots, x_i - \mu_i, \ldots, x_m - \mu_m)$ gives rise to the ideal $J(\underline{\mu}) := J + (x_1 - \mu_1, \ldots, x_m - \mu_m)/(x_1 - \mu_1, \ldots, x_m - \mu_m) \subset \mathbb{C}[y_1, \ldots, y_s]$. The scheme that this ideal defines in $\mathbb{C}^s$ is by definition the *scheme-theoretic fiber* $\pi^{-1}(\underline{\mu})$ of the map π at the point $\underline{\mu}$.

There are rather general theorems of semicontinuity of the fibers (cf. [55]). The case that we will discuss is that in which $\mathbb{C}[y_1, \ldots, y_s, x_1, \ldots, x_m]/J$ is a domain that is a finite module over its subalgebra $\mathbb{C}[x_1, \ldots, x_m]$. In particular, the corresponding extension of quotient fields is of some finite degree k. In this case it is well known (cf. [61]), that the generic fiber of π consists of k distinct and reduced points (i.e., $\mathbb{C}[y_1, \ldots, y_s]/J(\underline{\mu})$ is the algebra $\mathbb{C}^k$ of functions on k-distinct points), and for every point $\underline{\mu}$ we have

$\dim \mathbb{C}[y_1, \ldots, y_s]/J(\underline{\mu}) \geq k$. Moreover, the set of points on which, for some h we have $\dim \mathbb{C}[y_1, \ldots, y_s]/J(\underline{\mu}) \geq h$ is closed.

The case that we will encounter and that often appears is that in which the following *equivariance* condition holds:

The ideal J is homogeneous for some choice of positive weights h_i for the variables x_i. In other words, if $f(y_1, \ldots, y_s, x_1, \ldots, x_m) \in J$ and $\lambda \in \mathbb{C}$ we have that $f(\lambda y_1, \ldots, \lambda y_s, \lambda^{h_1} x_1, \ldots, \lambda^{h_m} x_m) \in J$. This occurs for instance in the previous example if we give to a_i the weight i.

In this case consider the action of the multiplicative group $\mathbb{C}^*$ on $\mathbb{C}^{s+m}$ by

$$(y_1, \ldots, y_s, x_1, \ldots, x_m) \mapsto (\lambda y_1, \ldots, \lambda y_s, \lambda^{h_1} x_1, \ldots, \lambda^{h_m} x_m).$$

This is a group of symmetries of J and the set of points of $\mathbb{C}^m$ where $\dim \mathbb{C}[y_1, \ldots, y_s]/J(\underline{\mu}) \geq h$ is stable under $\mathbb{C}^*$. Thus since it is closed, if it is not empty it contains zero. In particular, one has the following consequence:

Proposition 5.8. *If the hypothesis that* $\dim \mathbb{C}[y_1, \ldots, y_s]/J(\underline{\mu}) = k$ *holds generically, we have* $\dim \mathbb{C}[y_1, \ldots, y_s]/J(\underline{\mu}) = k$ *everywhere if and only if* $\dim \mathbb{C}[y_1, \ldots, y_s]/J(0) \leq k$.

There are in this case some interesting algebraic constructions, which we mention without proofs. The first relates the algebra $\mathbb{C}[y_1, \ldots, y_s]/J(\underline{\mu})$ for $\underline{\mu}$ generic to $\mathbb{C}[y_1, \ldots, y_s]/J(0)$. One defines an increasing filtration on the algebras $\mathbb{C}[y_1, \ldots, y_s]/J(\underline{\mu})$ by giving degree $\leq h$ to the image of the polynomials of degree $\leq h$ and easily sees that $\mathbb{C}[y_1, \ldots, y_s]/J(0)$ can be obtained as the graded algebra associated to this filtration.

There is a similar description in the dual picture of functions satisfying the corresponding polynomial differential equations. As we have seen before, we can think of $(\mathbb{C}[y_1, \ldots, y_s]/J(\underline{\mu}))^*$ as a space of exponential polynomials (or just exponential functions in the generic case). The space of polynomials $(\mathbb{C}[y_1, \ldots, y_s]/J(0))^*$ coincides with the space spanned by the initial terms of the Taylor series of the elements of $(\mathbb{C}[y_1, \ldots, y_s]/J(\underline{\mu}))^*$. This remark is sometimes used as a device to compute the solutions of $J(0)$.

5.2 Tori

We start by discussing some very basic facts on tori that belong to the first chapters of the theory of linear algebraic groups.

5.2.1 Characters

Let us start recalling the main definitions in the special case of complex numbers.

Definition 5.9. The standard s-dimensional algebraic torus (over $\mathbb{C}$) is the set $(\mathbb{C}^*)^s$ of s-tuples of nonzero complex numbers. This is a group under multiplication of the coordinates.

The group $(\mathbb{C}^*)^s$ is an affine algebraic variety with coordinate ring the ring of *Laurent polynomials* $\mathbb{C}[x_1^{\pm 1}, \ldots, x_s^{\pm 1}]$. In fact, $(\mathbb{C}^*)^s$ is an affine algebraic group. We do not want to use the theory of algebraic groups, of which we need only a very small portion (see [66], [103], [17]).

Definition 5.10. A (multiplicative) character of a group G is a homomorphism $f : G \to \mathbb{C}^*$.

For tori one usually drops the term *multiplicative* and simply speaks of a character. Moreover, we look only at characters that are algebraic, that is lie in $\mathbb{C}[x_1^{\pm 1}, \ldots, x_s^{\pm 1}]$.

Proposition 5.11. *The functions in* $\mathbb{C}[x_1^{\pm 1}, \ldots, x_s^{\pm 1}]$ *that are characters of* $(\mathbb{C}^*)^s$ *are the monomials* $\prod_{i=1}^s x_i^{h_i}$, $h_i \in \mathbb{Z}$.

Proof. It is immediate to verify that any monomial is a character. Conversely, to say that a function f is a character means that $f(xy) = f(x)f(y)$ for all $x, y \in (\mathbb{C}^*)^s$.

Given any element $y \in (\mathbb{C}^*)^s$, consider the *multiplication operator* m_y mapping a function $f(x)$ into the function $f(xy)$. In coordinates $y = (y_1, \ldots, y_s)$, m_y maps $x_i \mapsto x_i y_i$. If f is a character, then f is an eigenvector for the operators m_y of eigenvalue $f(y)$.

Clearly, the monomials $\prod_{i=1}^s x_i^{h_i}$ are a basis of the ring of Laurent polynomials consisting of eigenvectors for these operators with distinct eigenvalues. Thus f must be a multiple $c \prod_{i=1}^s x_i^{h_i}$ of one of these monomials. Since $f(1) = 1$ we have $c = 1$ and f is a monomial.

Corollary 5.12. *(1) The set Λ of characters of $(\mathbb{C}^*)^s$ is an abelian group under multiplication, naturally isomorphic to $\mathbb{Z}^s$.*
(2) The ring of Laurent polynomials is naturally identified with the group algebra $C[\Lambda]$.

Proof. (1) The fact that the multiplicative characters of any group G also form a group under multiplication is clear from the definition.

In the case of the torus $(\mathbb{C}^*)^s$ we have seen that these characters are of the form $\prod_{i=1}^s x_i^{h_i}$, $(h_1, \ldots, h_s) \in \mathbb{Z}^s$.

Clearly, the mapping $(h_1, \ldots, h_s) \mapsto \prod_{i=1}^s x_i^{h_i}$ is an isomorphism between the additive group $\mathbb{Z}^s$ and the multiplicative group Λ.

(2) Follows immediately from the definitions.

We need to stress now the duality between tori and lattices, which is an important tool for computations. It is the algebraic-geometric analogue of the much more general theory of Pontryagin duality between compact and discrete abelian groups.

Take a free abelian group Λ of rank s. By definition, Λ is isomorphic (but not in a unique way) to the group $\mathbb{Z}^s$. Its group algebra $\mathbb{C}[\Lambda]$ is isomorphic to the algebra of Laurent polynomials. Like every commutative algebra, finitely generated over $\mathbb{C}$ and without zero divisors, $\mathbb{C}[\Lambda]$ is the coordinate ring of an irreducible affine variety whose points are intrinsically defined as given by the homomorphisms $\phi : \mathbb{C}[\Lambda] \to \mathbb{C}$. By a general elementary argument, to give such a homomorphism is equivalent to giving a homomorphism of Λ into the multiplicative group $\mathbb{C}^*$, that is, a *character* of Λ. We shall denote this variety by $T_{\mathbb{C}}^{\Lambda} := \hom(\Lambda, \mathbb{C}^*)$ to stress the fact that it is the torus associated to Λ.

Often when the context is clear we shall drop the supscript Λ and denote our torus by $T_{\mathbb{C}}$. We can perform the same construction when Λ is a finitely generated abelian group, so $\Lambda \simeq \mathbb{Z}^s \times G$ with G finite. In this case, the algebraic group $T_{\mathbb{C}}^{\Lambda} = \hom(\Lambda, \mathbb{C}^*)$ is isomorphic to $(\mathbb{C}^*)^s \times \hat{G}$, where $\hat{G}$, the character group of G, is noncanonically isomorphic to G.

Since $\mathbb{C}^* = S^1 \times \mathbb{R}^+$ the torus $T_{\mathbb{C}}^{\Lambda}$ decomposes as the product of a compact torus $T^{\Lambda} := \hom(\Lambda, S^1)$ and a noncompact part $T := \hom(\Lambda, \mathbb{R}^+)$ isomorphic under the logarithm to the space $U := \hom(\Lambda, \mathbb{R})$. Let $V := \Lambda \otimes \mathbb{R}$. We then have that U is the dual space to V.

In other words, start from the vector space $U_{\mathbb{C}} = \hom(\Lambda, \mathbb{C})$, the complex dual of Λ. From a function $\phi \in U_{\mathbb{C}}$ construct the function $a \mapsto e^{\langle \phi \,|\, a \rangle}$ on Λ.

One immediately verifies the following statements

Proposition 5.13. *1. For each $\phi \in U_{\mathbb{C}}$ the function $a \mapsto e^{\langle \phi \,|\, a \rangle}$ is a character.*

2. The map $\phi \mapsto e^{\langle \phi \,|\, a \rangle}$ is a homomorphism between the additive group $U_{\mathbb{C}}$ and the multiplicative group $T_{\mathbb{C}}$.

3. The elements of $\Lambda^* := \hom(\Lambda, 2\pi i \mathbb{Z})$ give rise to the trivial character.

4. $T_{\mathbb{C}} = U_{\mathbb{C}}/\Lambda^*$ is an algebraic group isomorphic to $(\mathbb{C}^*)^s$ whose group of algebraic characters is Λ.

The expression $e^{\langle \phi \,|\, a \rangle}$ has to be understood as a pairing between $T_{\mathbb{C}}$ and Λ. I.e., a map $U_{\mathbb{C}} \times \Lambda \to \mathbb{C}^*$ factoring as

$$e^{\langle \phi \,|\, a \rangle} : T_{\mathbb{C}} \times \Lambda \to \mathbb{C}^*.$$

This duality expresses the fact that Λ is the group of algebraic characters of $T_{\mathbb{C}}$, denoted by e^a for $a \in \Lambda$, and $T_{\mathbb{C}}$ the analogous group of *complex characters* of Λ.

The class of a vector $\phi \in U_{\mathbb{C}}$ in $T_{\mathbb{C}}$ will be denoted by e^{ϕ} so that we may also write $\langle e^{\phi} \,|\, e^a \rangle = e^{\langle \phi \,|\, a \rangle}$.

5.2.2 Elementary Divisors

To proceed further, we are going to use the following standard fact, known as the *theorem on elementary divisors* (for a proof we refer, for example, to [19]).

Theorem 5.14. *Let $M \subset \Lambda$ be a subgroup of a lattice Λ. There exists a basis of Λ such that M is the free abelian group with basis given by k elements of the type*

$$(d_1, 0, \ldots, 0), \ (0, d_2, 0, \ldots, 0), \ldots, (0, 0, \ldots, d_k, 0, \ldots, 0), \ d_i \in \mathbb{N}^+.$$

Furthermore, one can assume that $d_1 \mid d_2 \mid \cdots \mid d_k$ and this determines uniquely the elements d_i.

Notice that in particular, M is itself a free abelian group of some rank k, so that the algebra $\mathbb{C}[M]$ is thus also the coordinate algebra of some torus $S_{\mathbb{C}}$. By standard algebraic geometry, the inclusion $\mathbb{C}[M] \subset \mathbb{C}[\Lambda]$ corresponds to a homomorphism $\pi : T_{\mathbb{C}} \to S_{\mathbb{C}}$. In order to understand this homomorphism we can use the explicit change of basis given by Theorem 5.14. We then see that we identify $S_{\mathbb{C}} = (\mathbb{C}^*)^k$, and π is given by:

$$\pi : (c_1, \ldots, c_s) \mapsto (c_1^{d_1}, \ldots, c_k^{d_k}).$$

We thus have that the image of π is S, while the kernel $\ker(\pi)$ is the subgroup of $(\mathbb{C}^*)^s$ given by the equations $c_i^{d_i} = 1$, $i = 1, \ldots, k$.

Let us understand this kernel.

The equation $x^d = 1$ defines the d roots of unity $e^{\frac{2\pi i k}{d}}$, $0 \le k < d$. Therefore, $\ker(\pi)$ has $d_1 d_2 \cdots d_k$ connected components, given by

$$(\zeta_1, \ldots, \zeta_k, a_{k+1}, \ldots, a_s) \mid \zeta_i^{d_i} = 1.$$

The component K_0 containing the element 1 is the *subtorus* of dimension $s - k$, with the first k coordinates equal to 1.

It is now not difficult to convince ourselves that

- The characters that take the value 1 on K_0 are those that have the last $s - k$ coordinates equal to 0.
- These characters form a subgroup $\overline{M}$ with $M \subset \overline{M}$ and $\overline{M}/M = \oplus_{i=1}^k \mathbb{Z}/(d_i)$.
- $\overline{M}$ can also be characterized as the set of elements $a \in \Lambda$ for which there exists a nonzero integer k with $ka \in M$ (the torsion elements modulo M).
- Moreover, $\Lambda/\overline{M}$ is a lattice whose corresponding torus of complex characters is K_0, while M (resp. $\overline{M}$) is the character group of the torus $T_{\mathbb{C}}/\ker M$ (resp. $T_{\mathbb{C}}/K_0$).

Remark 5.15. One can in fact prove (see for instance [103]) that every Zariski closed subgroup (i.e., also a subvariety) of a torus $T_{\mathbb{C}}$ is of the previous type, obtaining a correspondence between subgroups of $T_{\mathbb{C}}$ and those of Λ.

In this context it is useful to think of a torus as a periodic analogue of a vector space, and a subtorus as an analogue of a subspace.

In general, one may prove that Λ/M is the group of characters of the, not necessarily connected, group $T_{\mathbb{C}}^{\Lambda/M} = \ker(\pi)$, thus obtaining dual exact sequences of algebraic groups and characters respectively:

$$0 \to T_{\mathbb{C}}^{\Lambda/M} \to T_{\mathbb{C}}^{\Lambda} \to T_{\mathbb{C}}^{M} \to 0, \qquad 0 \to M \to \Lambda \to \Lambda/M \to 0.$$

We immediately note a point of difference: the intersection of two subtori is not necessarily connected.

An important special case of the previous theorem arises when we have s linearly independent vectors, i.e., a basis $\underline{b} = (\chi_1, \dots, \chi_s)$, $\chi_i \in \Lambda$, of the vector space $\Lambda \otimes \mathbb{Q}$. Consider the lattice $\Lambda_{\underline{b}} \subset \Lambda$ that it generates in Λ. Let us use a multiplicative notation and identify Λ with the Laurent monomials.

Proposition 5.16. *There are a basis $x_1, \dots, x_s$ of Λ and positive integers $d_1, \dots, d_s$ such that $x_1^{d_1}, \dots, x_s^{d_s}$ is a basis of the lattice $\Lambda_{\underline{b}}$.*

The kernel $T(\underline{b}) := \{p \in T \,|\, \chi_i(p) = 1\}$ is a finite subgroup with $d_1 \cdots d_s$ elements.

$\Lambda/\Lambda_{\underline{b}}$ is the character group of $T(\underline{b})$.

Proof. We apply Theorem 5.14, observing that now $k = s$, since $\Lambda_{\underline{b}}$ is s-dimensional and we note that the restriction of a character in Λ to $T(\underline{b})$ gives the duality between $T(\underline{b})$ and $\Lambda/\Lambda_{\underline{b}}$.

Notice that if we think of $\underline{b}$ as the columns of an integer $s \times s$-matrix A, we have that $|\det(\underline{b})| := |\det(A)|$ is the order of the finite group $\Lambda/\Lambda_{\underline{b}}$.

To complete the picture, we prove the following lemma

Lemma 5.17. *Let $M \subset \Lambda$ be a subgroup generated by some elements $\mathfrak{m}_i$, $i = 1, \dots t$.*

(1) The ideal I_M of $\mathbb{C}[\Lambda]$ generated by the elements $\mathfrak{m}_i - 1$ is the ideal of definition of $T_{\mathbb{C}}^{\Lambda/M}$.
(2) Each coset in Λ/M maps to a single element in $\mathbb{C}[\Lambda]/I_M$.
(3) The quotient $\mathbb{C}[\Lambda]/I_M$ has as possible basis the elements images of the cosets Λ/M, the character group of $T_{\mathbb{C}}^{\Lambda/M}$.

Proof. (1) We may choose our coordinates so that $\mathbb{C}[\Lambda] = \mathbb{C}[x_1^{\pm 1}, \dots, x_s^{\pm 1}]$ and M is generated by the elements $x_i^{d_i}$.

Let us observe that the set of monomials $\mathfrak{m} \in \Lambda$ such that $\mathfrak{m} - 1 \in I_M$ is a subgroup of Λ. Indeed, we have that if $\mathfrak{m} - 1, \mathfrak{n} - 1 \in I$:

$$\mathfrak{m}\mathfrak{n} - 1 = \mathfrak{m}(\mathfrak{n} - 1) + \mathfrak{m} - 1 \in I, \quad \mathfrak{m}^{-1} - 1 = -\mathfrak{m}^{-1}(\mathfrak{m} - 1) \in I.$$

It follows that the ideal I_M coincides with the ideal generated by the elements $x_i^{d_i} - 1$. By the standard correspondence between ideals and subvarieties, in order to prove that this is the entire defining ideal of $T_{\mathbb{C}}^{\Lambda/M}$ it is then enough to show that $\mathbb{C}[x_1^{\pm 1}, \dots, x_s^{\pm 1}]/(x_1^{d_i} - 1, \dots, x_k^{d_k} - 1)$ is an algebra without nilpotent elements. We have a separation of variable, and hence, always by standard facts, it is enough to see that an algebra $\mathbb{C}[x^{\pm 1}]/(x^d - 1)$ has this property. One easily sees that $\mathbb{C}[x^{\pm 1}]/(x^d - 1)$ is isomorphic to $\mathbb{C}^d$ under the map $x \mapsto (1, \zeta_d, \zeta_d^2, \ldots, \zeta_d^{d-1})$, where $\zeta_d := e^{\frac{2\pi i}{d}}$.

(2) This follows since each element of M maps to 1.

(3) Since the image of a monomial $\mathfrak{m}$ in $\mathbb{C}[\Lambda]/I_M$ depends only on its coset modulo the subgroup M, it is enough to prove the statement for a given choice of coset representatives. In the coordinates x_i a set of representatives is formed by the elements $x_1^{h_1} \ldots x_s^{h_s}$, $0 \le h_i < d_i, \forall i \le k$. By the previous discussion, the algebra $\mathbb{C}[\Lambda]/I_M$ is isomorphic to $\otimes_{i=1}^{k} \mathbb{C}^{d_i} \otimes \mathbb{C}[x_{k+1}^{\pm 1}, \ldots, x_s^{\pm 1}]$. We are reduced to analyzing a single factor $\mathbb{C}[x^{\pm 1}]/(x^d - 1)$. The image of x^i, $i = 0, \ldots, d-1$ in $\mathbb{C}^{d_i}$ equals $(1, \zeta_d^i, \zeta_d^{2i}, \ldots, \zeta_d^{(d-1)i})$, where $\zeta_d = e^{\frac{2\pi i}{d}}$. These elements form a basis of $\mathbb{C}^{d_i}$ by the usual Vandermonde determinant argument.

We shall in particular apply this discussion to the problem of counting integral points for the family of polytopes associated to a list X of integer vectors $a_i \in \mathbb{Z}^s$. In this case, each (rational) basis $\underline{b}$ extracted from X gives rise to an integral matrix. This matrix has a determinant with absolute value some positive integer $m(\underline{b})$. The number $m(\underline{b})$ is the index in $\mathbb{Z}^s$ of the subgroup $M_{\underline{b}}$ generated by $\underline{b}$. This basis determines also the subgroup $T(\underline{b})$ of $(C^*)^s$, formed by the $m(\underline{b})$ points on which the elements $a_i \in \underline{b}$ take the value 1.

All the points of such subgroups will contribute to the formulas that we wish to derive and will play a special role in the theory. They will be called *points of the arrangement* (14.3).

5.3 Difference Equations

5.3.1 Difference Operators

We now study the discrete analogue of the theory of differential equations discussed in the previous sections. We start with a lattice $\Lambda \subset V$, from which we can construct the *group algebra* $\mathbb{C}[\Lambda]$ and we use the multiplicative notation e^a for the element in $\mathbb{C}[\Lambda]$ corresponding to $a \in \Lambda$.

Definition 5.18. Given any subset A of V, stable under translation by Λ, for $v \in \Lambda$ we define the translation and difference operators τ_v, ∇_v, acting on the space of functions on A, as

$$\tau_v f(u) := f(u - v), \quad \nabla_v f(u) := f(u) - f(u - v), \quad \nabla_v = 1 - \tau_v. \quad (5.4)$$

Given two elements $a, b \in \Lambda$, we have that, $\tau_a \tau_b = \tau_{a+b}$. Thus the space of complex-valued functions on A is a module over the algebra $\mathbb{C}[\Lambda]$.

Given an element $p \in \mathbb{C}[\Lambda]$ and a function f on A, we say that f satisfies the *difference equation* p if $pf = 0$.

Remark 5.19. Sometimes one uses the *forward difference operator* defined by $\nabla_{-v}f(u) = -\nabla_v f(u+v)$.

In particular, we are interested in the space $\mathcal{C}[\Lambda]$ of complex-valued functions f on the lattice Λ. We identify this last space with the *algebraic dual* $\mathbb{C}[\Lambda]^*$ of $\mathbb{C}[\Lambda]$ by the formula

$$\langle f \,|\, e^a \rangle := f(a).$$

In a similar way, by duality we act on tempered distributions by

$$\langle \tau_v D \,|\, f \rangle := \langle D \,|\, \tau_{-v} f \rangle, \quad \langle \nabla_v D \,|\, f \rangle := \langle D \,|\, \nabla_{-v} f \rangle, \tag{5.5}$$

and we have $\tau_v(\delta_a) = \delta_{a+v}$, $\nabla_v(\delta_a) = \delta_a - \delta_{a+v}$.

Definition 5.20. Given a function f on Λ, we define the distribution

$$\sum_{\lambda \in \Lambda} f(\lambda)\delta_\lambda, \tag{5.6}$$

where δ_v is the Dirac distribution supported at v.

Remark 5.21. Formula (5.6) implies that for any function f, we have the identity $\widetilde{\nabla_v f} = \nabla_v \tilde{f}$.

We want to reformulate the fact that a function is a solution of a system of difference equations as the property for such a function to vanish on an appropriate ideal J_X of $\mathbb{C}[\Lambda]$. As we have seen in the previous section $\mathbb{C}[\Lambda]$ is the algebra of regular algebraic functions on the algebraic torus $T_{\mathbb{C}} := \hom(\Lambda, \mathbb{C}^*)$. We use the notation of the previous section.

Notice that the ideal I_1 of functions in $\mathbb{C}[\Lambda]$ vanishing at $1 \in T_{\mathbb{C}}$ has as linear basis the elements $1 - e^{-a}$, $a \in \Lambda$, $a \neq 0$. If one takes another point e^ϕ, the ideal I_ϕ of functions in $\mathbb{C}[\Lambda]$ vanishing at e^ϕ has as linear basis the elements $1 - e^{-a + \langle \phi \,|\, a \rangle}$, $a \in \Lambda$, $a \neq 0$.

For every $a, x \in \Lambda$, we have by definition

$$\langle \tau_a f \,|\, e^x \rangle = \langle f \,|\, e^{x-a} \rangle, \quad \langle \nabla_a f \,|\, e^x \rangle = \langle f \,|\, (1 - e^{-a})e^x \rangle.$$

Thus the translation operator τ_a and the difference operator ∇_a are the duals of the multiplication operator by $e^{-a}, 1 - e^{-a}$. In this setting, we get a statement analogous to that of Theorem 5.4 on differential equations:

Theorem 5.22. *Given a polynomial p, a function f on Λ satisfies the difference equation*

$$p(\nabla_{a_1}, \ldots, \nabla_{a_k})f = 0$$

if and only if, thought of as element of the dual of $\mathbb{C}[\Lambda]$, f vanishes on the ideal of $\mathbb{C}[\Lambda]$ generated by the element $p(1 - e^{-a_1}, \ldots, 1 - e^{-a_k})$.

We have also the *twisted difference operators* ∇_a^ϕ, $e^\phi \in T_{\mathbb{C}}$ defined by

$$(\nabla_a^\phi f)(x) := f(x) - e^{\langle \phi \mid a \rangle} f(x - a), \tag{5.7}$$

dual to multiplication by $1 - e^{-a + \langle \phi \mid a \rangle}$.

We need some simple algebraic considerations. We use the word *scheme* in an intuitive way as meaning some sort of variety defined by a system of polynomial equations. We treat only schemes that reduce to a finite set of points. Consider a subscheme of $T_{\mathbb{C}}$ supported at a point $e^\phi \in T_{\mathbb{C}}$. This really means that we take an ideal J of $\mathbb{C}[\Lambda]$ vanishing exactly at e^ϕ. As a consequence, the algebra $\mathbb{C}[\Lambda]/J$ (the coordinate ring of such a scheme) is finite-dimensional. We want to identify this scheme with a subscheme of $U_{\mathbb{C}}$ supported at ϕ.

We can repeat the discussion made in Section 5.1.

Proposition 5.23. *There exists a linear map* $i : V \to \mathbb{C}[\Lambda]/J$ *such that:*

1. *The elements* $i(v) - \langle \phi \mid v \rangle$, $v \in V$ *are all nilpotent.*
2. *Given an element* $a \in \Lambda$, *we have that the class of* e^a *in* $\mathbb{C}[\Lambda]/J$ *equals to* $e^{i(a)} = \sum_{k=0}^{\infty} \frac{i(a)^k}{k!}$.
3. *The map* i *extends to a surjective homomorphism* $i : S[V] \to \mathbb{C}[\Lambda]/J$.
4. *Let* I *be the kernel of* i. *So that* $i : S[V]/I \cong \mathbb{C}[\Lambda]/J$. *The variety defined by the ideal* I *coincides with the point* ϕ.

Proof. 1.   Given an element $a \in \Lambda$, we have that $e^{-\langle \phi \mid a \rangle} e^a - 1$ vanishes at e^ϕ. Thus, in the ring $\mathbb{C}[\Lambda]/J$, the class t_a of $e^{-\langle \phi \mid a \rangle} e^a - 1$ is nilpotent. The power series of $u_a := \log(1 + t_a) = t_a - t_a^2/2 + t_a^3/3 - \cdots$ reduces to a finite sum which is still a nilpotent element. We set

$$i(a) := \log(1 + t_a) + \langle \phi \mid a \rangle$$

(notice that this definition depends on the choice of a representative for e^ϕ).

We have $1 + t_{a+b} = (1 + t_a)(1 + t_b)$ hence the map $i : \Lambda \to \mathbb{C}[\Lambda]/J$ is additive and thus extends to a linear map $i : V \to \mathbb{C}[\Lambda]/J$.

2.   By definition, $e^{i(a)} = (1 + t_a) e^{\langle \phi \mid a \rangle} \in \mathbb{C}[\Lambda]/J$ equals the class of e^a.

3.   The linear map i extends to a homomorphism $i : S[V] \to \mathbb{C}[\Lambda]/J$. In order to prove that i is surjective recall that, for each $a \in \Lambda$, the elements t_a and $u_a = \log(1 + t_a)$ are nilpotent. Thus the series $\sum_{k \geq 0} u_a^k/k!$ reduces to a finite sum and equals $1 + t_a$, the class $e^{-\langle \phi \mid a \rangle + a}$. Thus, for each $a \in \Lambda$, the class of $e^{-\langle \phi \mid a \rangle + a}$ lies in the image of the homomorphism i. Since these classes generate $\mathbb{C}[\Lambda]/J$, i is surjective.

4. Since, for every $a \in V$, $i(a) - \langle \phi \mid a \rangle$ is nilpotent. The element $a - \langle \phi \mid a \rangle$, vanishes on the variety defined by I. But the elements $a - \langle \phi \mid a \rangle$ vanish exactly at ϕ, hence I defines the point ϕ.

We have thus identified $\mathbb{C}[\Lambda]/J$ with a quotient $S[V]/I$ where I is an ideal of finite codimension in $S[V]$. Moreover, the class of the element e^a that at the beginning was just a symbol is the actual exponential $e^{i(a)}$ of the element $i(a)$ class of $a \in V$ modulo I.

Definition 5.24. The isomorphism $i : S[V]/I \to \mathbb{C}[\Lambda]/J$ is called the *logarithmic isomorphism*.

We have thus identified, using this algebraic logarithm, the given scheme as a subscheme in the tangent space. This may be summarized in the following proposition:

Proposition 5.25. *Let $J \subset \mathbb{C}[\Lambda]$ be an ideal such that $\mathbb{C}[\Lambda]/J$ defines (set theoretically) the point e^ϕ. Under the logarithm isomorphism $\mathbb{C}[\Lambda]/J$ becomes isomorphic to a ring $S[V]/I$ defining the point ϕ.*

When we dualize this isomorphism we get a canonical isomorphism i^* between the space of solutions of the difference equations given by J, that is a space of functions on Λ, and the space of solutions of the differential equations given by I, that instead is a space of functions on V. We want to understand this isomorphism and in particular prove the following result

Proposition 5.26. *The inverse of i^* is the restriction of a function on V to Λ.*

Proof. We have identified in Proposition 5.23, the class of e^a modulo J with $e^{u_a} e^{\langle \phi \,|\, a \rangle}$. The explicit formula is, given $f \in (\mathbb{C}[\Lambda]/J)^*$:

$$f(a) := \langle f \,|\, e^a \rangle = \langle f \,|\, e^{u_a} e^{\langle \phi \,|\, a \rangle} \rangle = e^{\langle \phi \,|\, a \rangle} \langle f \,|\, e^{u_a} \rangle = e^{\langle \phi \,|\, a \rangle} \langle f \,|\, 1 + t_a \rangle. \quad (5.8)$$

The element f corresponds to the element of $(S[V]/I)^*$ given by

$$e^{\langle \phi \,|\, a \rangle} \left\langle f \,\big|\, e^{i(a) - \langle \phi \,|\, a \rangle} \right\rangle.$$

Since $i(a) - \langle \phi \,|\, a \rangle$ is nilpotent $\langle f \,|\, e^{i(a) - \langle \phi \,|\, a \rangle} \rangle$ is a polynomial. Thus we have extended the function f on Λ, to an exponential-polynomial on V. $\qquad \square$

We can view this isomorphism in an equivalent way that will be used later. This is done by making explicit the relationship between the difference and differential equations satisfied by a given space of polynomials (invariant under such operators).

Consider the space $P^m(V)$ of polynomials on V of degree $\leq m$. We have that given $v \in V$, both $\frac{\partial}{\partial v}$ and ∇_v map $P^m(V)$ to $P^{m-1}(V)$. Therefore, these operators are nilpotent on $P^m(V)$ of degree $m+1$. Moreover, the Taylor series expansion gives:

$$\tau_v = e^{-\frac{\partial}{\partial v}}, \quad \nabla_v = 1 - e^{-\frac{\partial}{\partial v}}, \quad \frac{\partial}{\partial v} = -\log(1 - \nabla_v).$$

On $P^m(V)$ we have thus the identities

$$\frac{\partial}{\partial v} = \sum_{i=1}^{m} \frac{\nabla_v^i}{i}, \quad \nabla_v = -\sum_{i=1}^{m} \frac{(-1)^i}{i!} \frac{\partial^i}{\partial v}. \tag{5.9}$$

Therefore, we deduce from these formulas the following facts

Proposition 5.27. *1. On $P^m(V)$ the algebra of operators generated by derivatives coincides with that generated by difference operators.*
2. Using formula (5.9) we can transform (on $P^m(V)$) a differential into a difference equation and conversely.

Example 5.28. Let $a_1, \ldots, a_s$ be an integral basis of Λ. The ideal J of $\mathbb{C}[\Lambda]$ generated by $1 - e^{-a_1}, \ldots, 1 - e^{-a_s}$ defines the point 1 and $\mathbb{C}[\Lambda]/J = \mathbb{C}$. Let $I \subset \mathbb{C}[x_1, \ldots, x_s]$ be the ideal generated by $x_1, \ldots, x_s$. Consider $\mathbb{C}[\Lambda]/J^k$. Take the logarithm isomorphism mapping $x_i \mapsto -\log(e^{-a_i})$ (modulo J^k). If u_i denotes the class of $1 - e^{-a_i}$ modulo J^k we use the finite polynomial expression $-\log(e^{-a_i}) = -\log(1 - u_i) = \sum_{j=1}^{k-1} \frac{u_i^j}{j}$.

   This establishes an isomorphism between $\mathbb{C}[\Lambda]/J^k$ and $\mathbb{C}[x_1, \ldots, x_s]/I^k$. *We have shown that the restriction to Λ of the space of polynomials of degree less than or equal to $k-1$ coincides with the space of solutions of the difference operators in J^k.*

In general, assume that J is an ideal of $\mathbb{C}[\Lambda]$ of finite codimension that defines points $p_1, \ldots, p_k \in T_{\mathbb{C}}$. Thus $\mathbb{C}[\Lambda]/J = \oplus_{i=1}^k \mathbb{C}[\Lambda]/J_i$ decomposes as a sum of local algebras. Choose representatives $\phi_i \in U_{\mathbb{C}}$ for the points p_i. At each ϕ_i, under the logarithm isomorphism $\mathbb{C}[\Lambda]/J_i = S[V]/I_i$ and I_i defines a space of polynomials D_i on V.

Theorem 5.29. *The space of functions $\oplus_{i=1}^k e^{\langle \phi_i \mid v \rangle} D_i$ on V restricts isomorphically to the space of functions on Λ that are solutions of the system of difference equations given by J.*

In order to complete this theorem, we have still to understand explicitly the equations defining D_p for a point p defined by J. Thus take a representative ϕ for p. We have, for a difference operator ∇_a and a polynomial $f(v)$, that

$$\nabla_a(e^{\langle \phi \mid v \rangle} f(v)) = e^{\langle \phi \mid v \rangle} f(v) - e^{\langle \phi \mid v-a \rangle} f(v-a)$$
$$= e^{\langle \phi \mid v \rangle}(f(v) - e^{-\langle \phi \mid a \rangle} f(v-a)) = e^{\langle \phi \mid v \rangle} \nabla_a^{-\phi} f(v).$$

Thus we have proved the following corollary

Corollary 5.30. *Let $q(t_1, \ldots, t_k)$ be a polynomial. The difference equation*

$$q(\nabla_{a_1}, \ldots, \nabla_{a_k})(e^{\langle \phi \mid v \rangle} f(v)) = 0$$

is equivalent to the twisted difference equation

$$q(\nabla_{a_1}^{-\phi}, \ldots, \nabla_{a_k}^{-\phi}) f(v) = 0.$$

5.4 Recursion

5.4.1 Generalized Euler Recursion

A special role is played by the points of finite order m, i.e. characters e^ϕ on Λ whose kernel is a sublattice Λ_ϕ of index m (of course this implies that we have $m\phi \in \Lambda^* = \hom(\Lambda, 2\pi i\mathbb{Z})$).

Let $J \subset \mathbb{C}[\Lambda]$ be an ideal that defines (set-theoretically) the finite order point e^ϕ. As we have seen, a function $f(x)$ appearing in the dual of $\mathbb{C}[\Lambda]/J$, is of the form $f(x) := e^{\langle \phi \,|\, x \rangle} g(x)$ with $g(x)$ a polynomial.

Since e^ϕ is of finite order, $e^{\langle \phi \,|\, x \rangle}$ takes constant values (m-th roots of 1) on the m cosets of the sublattice Λ_ϕ. Thus $f(x)$ is a polynomial only on each such coset. This is a typical example of what is called a *periodic polynomial* or *quasipolynomial*.

Formally, we make the following definition

Definition 5.31. A function f on a lattice Λ is a quasipolynomial, or periodic polynomial, if there exists a sublattice Λ^0 of finite index in Λ such that f restricted to each coset of Λ^0 in Λ is (the restriction of) a polynomial.

As a consequence, of this definition and of Proposition 5.25 we deduce the following resut

Theorem 5.32. *Let $J \subset \mathbb{C}[\Lambda]$ be an ideal such that $\mathbb{C}[\Lambda]/J$ is finite-dimensional and defines (set- theoretically) the points $e^{\phi_1}, \ldots, e^{\phi_k}$, all of finite order.*

Then the space $(\mathbb{C}[\Lambda]/J)^*$ of solutions of the difference equations associated to J is a direct sum of spaces of periodic polynomials $e^{\langle \phi_i \,|\, a \rangle} p(a)$ for the points ϕ_i, each invariant under translations under Λ.

We shall call such an ideal J an ideal of multidimensional Euler recursions.

Remark 5.33. A solution of the difference equations associated to J is determined by the values that it takes on a subset $A \subset \Lambda$ consisting of elements that modulo the ideal J give a basis of $\mathbb{C}[\Lambda]/J$. Once such a set is chosen, in order to perform the *recursion,* one needs to compute the s- matrices giving the multiplication of the generators e^{a_i} of the lattice, with respect to this basis. The transpose matrices give the action of the lattice on the dual space, from which the recursive formulas are easily written.

We shall see that, for the partition functions, we have specially useful choices for such a subset A (see Theorem 13.54).

As in the case of differential equations, the previous theorem has a converse. Let us thus take a finite-dimensional vector space Q spanned by periodic polynomials $e^{\langle \phi_i \,|\, a \rangle} p(a)$ (for some points e^{ϕ_i} of finite order) invariant under translations by elements of Λ. We have the following

Lemma 5.34. *Any nonzero subspace M of Q, invariant under translations by elements of Λ, contains one of the functions $e^{\langle \phi_i \,|\, a \rangle}$.*

Proof. Indeed, a nonzero subspace invariant under translations under Λ contains a nonzero common Λ-eigenvector. This eigenvector must necessarily be a multiple of one of the $e^{\langle \phi_i \mid a \rangle}$.

Let J be the ideal of the difference equations satisfied by Q.

Theorem 5.35. *The ideal J is of finite codimension equal to* $\dim(Q)$, *and Q is the space of all solutions of J.*

Proof. The proof is analogous to that of Proposition 5.5. We can easily reduce to the case of a unique point ϕ. In fact, the commuting set of linear operators Λ on Q has a canonical Fitting decomposition relative to the generalized eigenvalues that are characters of Λ. Our hypotheses imply that these points are of finite order.

Applying translation in the torus, one reduces to the case of the unique eigenvalue 1 and of polynomials. From the previous lemma, we have also for polynomials that every subspace stable under difference operators contains 1. We take thus Q^* and the map ϵ, evaluation at 0.

At this point we can continue as in the proof of Proposition 5.5 or invoke directly Proposition 5.27 in order to transform the statement on difference equations into one on differential equations.

6

Approximation Theory I

In this chapter we discuss an approximation scheme as in [33] and [51], that gives some insight into the interest in box splines, which we will discuss presently.

6.1 Approximation Theory

6.1.1 A Local Approximation Scheme

Let us start with an ideal $I \subset S[V]$ of finite codimension h. According to the theory developed in Section 5.1, this set of differential operators determines a finite-dimensional space of solutions H whose elements are exponential polynomials.

The first approximation scheme starts with a C^∞ function f on some domain Ω and a point α, and then consists in choosing, in the space H, an element that shares as much as possible of the Taylor series of f in α.

In order to do this, we describe the Taylor series in α intrinsically, by acting with the differential operators in $S[V]$ on C^∞ complex-valued functions defined in Ω. Consider the pairing

$$S[V] \times C^\infty(\Omega) \to \mathbb{C}, \quad \langle p \,|\, f \rangle := p(f)(\alpha).$$

This pairing induces a mapping $\pi_\alpha : C^\infty(\Omega) \to S[V]^*$ that we can interpret as associating to a function f its Taylor series at α. Assume now for simplicity that $\alpha = 0$ and that I defines only the point 0 (this means that $I \supset V^k$ for some k).

We now fix a complement A to I such that $S[V] = I \oplus A$. We compose the map π_0 with the projection $S[V]^* \to A^*$, and we obtain a linear map $\pi_A : C^\infty(\Omega) \to A^*$. By what we have seen in Section 5.1, $H \subset C^\infty(\Omega)$ is a space of polynomials mapping isomorphically under π_A to A^*. Therefore, given $f \in C^\infty(\Omega)$, there is a unique $g \in H$ with $\pi_A(f) = \pi_A(g)$. In other words,

C. De Concini and C. Procesi, *Topics in Hyperplane Arrangements, Polytopes and Box-Splines*, Universitext, DOI 10.1007/978-0-387-78963-7_6,
© Springer Science+Business Media, LLC 2010

$$p(f - g)(0) = 0, \ \forall p \in A.$$

The function g can be understood as a *local approximant* to f chosen in H.

An explicit formula to determine g is the following: fix dual bases $p_i \in A, h_i \in H$, $i = 1, \ldots, m$ then: $g = \sum_{i=1}^{m} p_i(f)(0)h_i$.

The way in which g approximates f (at least around 0) depends on properties of I. For instance, assume that I is such that H contains all the polynomials of degree $\leq d$. Then we may choose A such that it contains all the partial derivative operators of degree $\leq d$, and thus we deduce that the Taylor series of $f - g$ starts with terms of order $> d$. In this case, an estimate of the approximation of g to f can be obtained by the standard theorem evaluating the remainder in the Taylor series, which we recall for the convenience of the reader.

Let G be a disk of radius R centered at 0 in $\mathbb{R}^s$, f a C^r function on G, $r \geq 1$. Let q_r be the polynomial expressing the Taylor expansion of f at 0 up to order r.

Theorem 6.1. *For every $1 \leq p \leq \infty$ there is a constant C, dependent on p, r, s but independent of f, such that we have*

$$\|f - q_{r-1}\|_{L^p(G)} \leq C \sum_{|\alpha|=r} \|D^\alpha f\|_{L^p(G)} R^r. \tag{6.1}$$

Proof. Assume first $p < \infty$. We work in polar coordinates $(y, \underline{u})$, setting the variable x equal to $y\underline{u}$, where $y \in \mathbb{R}^+$, $\underline{u} \in S^{s-1}$ is a unit vector, and the Lebesgue measure is $dx = dy\, d\underline{u}$. Thus set $g_{\underline{u}}(y) := f(y\underline{u}) = f(x)$. Integrating by parts, and applyig induction we have

$$\int_0^y \frac{(y-t)^{r-1}}{(r-1)!} g_{\underline{u}}^{(r)}(t)dt = f(x) - q_{r-1}(x). \tag{6.2}$$

We expand

$$g_{\underline{u}}^{(r)}(t) = \sum_\alpha \binom{r}{\alpha} \underline{u}^\alpha \partial^\alpha f(t\underline{u}) = \sum_\alpha c_\alpha \partial^\alpha f(t\underline{u}),$$

where all the c_α can be bounded by $C_0 := \max_{\alpha \in \mathbb{N}^s, \ |\alpha|=r} \binom{r}{\alpha} \leq s^r$,

$$\|f - q_{r-1}\|_{L^p(G)} \leq C_0 \sum_{|\alpha|=r} \left\| \int_0^y \frac{(y-t)^{r-1}}{(r-1)!} \partial^\alpha f(t\underline{u})dt \right\|_{L^p(G)}.$$

We now start making some estimates

$$\left\| \int_0^y \frac{(y-t)^{r-1}}{(r-1)!} \partial^\alpha f(t\underline{u})dt \right\|_{L^p(G)}$$

$$= \left(\int_{S^{s-1}} \int_0^R \left| \int_0^y \frac{(y-t)^{r-1}}{(r-1)!} \partial^\alpha f(t\underline{u})dt \right|^p dy d\underline{u} \right)^{\frac{1}{p}}$$

$$\leq \left(\int_{S^{s-1}} \int_0^R \left(\int_0^y \left| \frac{(y-t)^{r-1}}{(r-1)!} \partial^\alpha f(t\underline{u}) \right| dt \right)^p dy\, d\underline{u} \right)^{\frac{1}{p}}.$$

We now apply the Hölder inequality and get

$$\int_0^y \left| \frac{(y-t)^{r-1}}{(r-1)!} \partial^\alpha f(t\underline{u}) \right| dt \leq \|\partial^\alpha f\|_{L^p([0,y]\underline{u})} \left\| \frac{(y-t)^{r-1}}{(r-1)!} \right\|_{L^q([0,y])}$$

with $1/p + 1/q = 1$. $\|\partial^\alpha f\|_{L^p([0,y]\underline{u})} \leq \|\partial^\alpha f\|_{L^p([0,R]\underline{u})}$. Also,

$$\left\| \frac{(y-t)^{r-1}}{(r-1)!} \right\|_{L^q([0,y])} = \left(\int_0^y \left| \frac{(y-t)^{r-1}}{(r-1)!} \right|^q dt \right)^{\frac{1}{q}} = K|y|^{r-1+1/q},$$

with K a constant independent of f. We deduce

$$\left(\int_{S^{s-1}} \int_0^R \left(\int_0^y \left| \frac{(y-t)^{r-1}}{(r-1)!} \partial^\alpha f(t\underline{u}) \right| dt \right)^p dy \, d\underline{u} \right)^{1/p}$$

$$\leq K \int_0^R |y|^{p(s-1)+p/q} dy \left(\int_{S^{(s-1)}} \|\partial^\alpha f\|_{L^p([0,R]\underline{u})}^p d\underline{u} \right)^{1/p} = K'\|\partial^\alpha f\|_{L^p(G)} R^s.$$

The case $p = \infty$ follows in a similar but simpler way

$$\left\| \int_0^y \frac{(y-t)^{r-1}}{(r-1)!} \partial^\alpha f(t\underline{u}) dt \right\|_{L^\infty(G)} \leq \int_0^R \frac{(y-t)^{r-1}}{(r-1)!} dt \|\partial^\alpha f(t\underline{u})\|_{L^\infty(G)}$$

$$= \frac{R^r}{r!} \|\partial^\alpha f(t\underline{u})\|_{L^\infty(G)}$$

$$\|f - q_{r-1}\|_{L^\infty(D(R))} \leq C_0 \frac{R^r}{r!} \sum_{|\alpha|=r} \|\partial^\alpha f(t\underline{u}) dt\|_{L^\infty(D(R))}$$

$$\leq \frac{(sR)^r}{r!} \sum_{|\alpha|=r} \|\partial^\alpha f(t\underline{u}) dt\|_{L^\infty(D(R))}. \tag{6.3}$$

6.1.2 A Global Approximation Scheme

In this second scheme, we fix a lattice $\Lambda \subset V$ and a *model function* $B(x)$ with support on a compact set A. On the function $B(x)$ we make the basic assumption that its translates under Λ form a partition of unity:

$$1 = \sum_{\lambda \in \Lambda} B(x - \lambda).$$

This remarkable property is satisfied by the box spline associated to a list of vectors $X \subset \Lambda$ spanning V.

Let us anticipate a fundamental example in one variable. The simplest kind of family of box splines is $b_m(x)$, which, in the terminology of Section 7.1.1, is associated to the list $1, \ldots, 1$ of length $m + 1$. The proof of the following statements follow from the theory to be developed in Chapter 7.

Example 6.2. The box spline $b_m(x)$ can be defined through an auxiliary function $t_m(x)$ (the multivariate spline $T_{1^{m+1}}(x)$ cf. 7.1.1), as follows:

$$t_m(x) := \begin{cases} 0 & \text{if } x < 0, \\ x^m/m! & \text{if } x \geq 0, \end{cases} \qquad b_m(x) := \sum_{i=0}^{m+1} (-1)^i \binom{m+1}{i} t_m(x-i).$$

$$(6.4)$$

It can be easily seen that $b_m(x)$ is of class C^{m-1}. It is supported in the interval $[0, m+1]$, and given by the formula

$$b_m(x) := \sum_{i=0}^{k} (-1)^i \binom{m+1}{i} \frac{(x-i)^m}{m!}, \quad \forall x \in [k, k+1].$$

The box spline $b_m(x)$ is known as the *cardinal B-spline of Schoenberg*; cf. [95].

Notice in particular that when $x > m+1$, the formula for $b_m(x)$ is

$$\frac{1}{m!} \sum_{i=0}^{m+1} (-1)^i \binom{m+1}{i} \tau_1^i x^m = \frac{\nabla_1^{m+1}}{m!} x^m = 0$$

showing that $b_m(x)$ vanishes outside the interval $[0, m+1]$.

The approximation method we want to discuss is the following. Fix a scale n^{-1}, $n \in \mathbb{N}$, and use the scaled function $B_n(x) := B(xn)$ and its translates $B(n(x + \lambda/n)) = B(nx + \lambda)$ by elements of Λ/n to approximate a given function f. Of course, if $B(x)$ has support in A, then $B(nx + \lambda)$ has support in the set $(A - \lambda)/n$.

Clearly, we still have

$$1 = \sum_{\lambda \in \Lambda} B(nx - \lambda) = \sum_{\mu \in \Lambda/n} B_n(x - \mu).$$

We fix some compact set K over which we request to have a good approximation and want to estimate how well we can approximate a given function f with a linear combination of the translates $B_n(x - \mu)$ as n tends to infinity.

We start from the identity $f(x) = \sum_{\mu \in \Lambda/n} f(x) B_n(x - \mu)$. Assume, for instance, that $0 \in A$ and notice that, as n grows $f(x)$ tends to be nearly constant on the small set $A/n - \mu$. Thus we have that each function $f(x) B_n(x - \mu), \mu \in \Lambda/n$ can be well approximated by $f(\mu) B_n(x - \mu)$ from which we obtain the approximation

$$f(x) \cong \sum_{\mu \in \Lambda/n} f(\mu) B_n(x - \mu). \tag{6.5}$$

Better approximations of the form $\sum_{\mu \in \Lambda/n} c_\mu B_n(x - \mu)$ can be found by modulating the coefficients c_μ (cf. Section 18.1.6).

6.1.3 The Strang–Fix Condition

One can try to combine both approximation schemes and ask how good the approximation given by (6.5) is as n tends to infinity. The way we intend to measure this is the following. Let B be a model function in $\mathbb{R}^s$ as before (with compact support), and Λ a lattice in $\mathbb{R}^s$ such that $\sum_{\lambda \in \Lambda} B(x - \lambda) = 1$.

Definition 6.3.   (i) A function on Λ is called a *mesh-function*.
(ii) Let $a : \Lambda \to \mathbb{C}$ be a mesh-function. We set $B * a$ to be the function

$$(B * a)(x) := \sum_{\lambda \in \Lambda} B(x - \lambda)a(\lambda).$$

This is well-defined, since $B(x)$ has compact support; it is called the *discrete convolution* of B with a.
(iii) The vector space $\mathcal{S}_B^{\Lambda}$ formed by all the convolutions $B * a$, where a is a mesh-function, is called the *cardinal space* associated to B and Λ.
(iv) A function $f(x)$ on $\mathbb{R}^s$ restricts to a mesh-function $f_{|\Lambda}$ on Λ. We set

$$B *' f := B * f_{|\Lambda} \tag{6.6}$$

and call it the *semidiscrete convolution* of B with f.

Observe that, if a is a mesh function with compact support, we have the identity

$$(B * a) *' f = B *' (f * a). \tag{6.7}$$

Example 6.4. Consider the function from Example 6.2

$$b_1(x) := \begin{cases} x & \text{if } 0 \leq x \leq 1 \\ 2 - x & \text{if } 1 \leq x \leq 2 \\ 0 & \text{otherwise.} \end{cases}$$

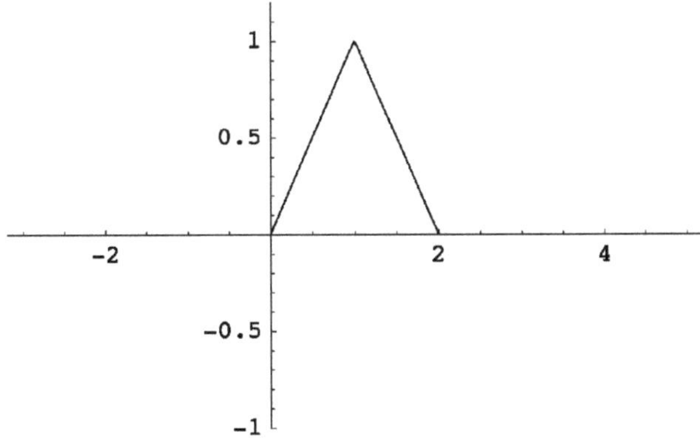

It is easily verified that its cardinal spline space is the space of all continuous functions on $\mathbb{R}$ that are linear on each interval $[n, n+1]$, $n \in \mathbb{Z}$.

For every positive real number h we define the linear *scale operator*

$$(\sigma_h f)(x) := f(x/h). \tag{6.8}$$

We apply it to functions in the cardinal space:

$$\sigma_h\left(\sum_{\lambda \in \Lambda} B(x - \lambda)a(\lambda)\right) = \sum_{\lambda \in \Lambda} B(x/h - \lambda)a(\lambda) = \sum_{\mu \in h\Lambda} (\sigma_h B)(x - \mu)a(\mu/h).$$

$$\tag{6.9}$$

In particular, we shall apply this when $h = n^{-1}$, $n \in \mathbb{N}$ so that $h\Lambda$ is a refinement of Λ.

We can consider $\sigma_h a$ as a function on $h\Lambda$. Formula (6.9) means that:

$$\sigma_h(B * a) = \sigma_h(B) * \sigma_h(a).$$

Therefore we have the following result:

Proposition 6.5. *The space $\mathcal{S}_h := \sigma_h(\mathcal{S}_B^\Lambda)$ equals the cardinal space for the scaled function $\sigma_h B$ and the lattice $h\Lambda$, i.e., $\mathcal{S}_h = \mathcal{S}_{\sigma_h(B)}^{h\Lambda}$.*

One of the ideas of the *finite element method* (see [105]) is to approximate functions by elements of $\mathcal{S}_{1/n}$ so that the approximation is more and more accurate as $n \in \mathbb{N}$ tends to ∞.

Definition 6.6. We say that B has approximation power $\geq r$ if for any bounded domain G and a smooth function f in a neighborhood of G, there exists a sequence $g_h \in \sigma_h(\mathcal{S}), h = 1/n, \ n \in \mathbb{N}$, such that for every $p \geq 1$ we have

$$\|g_h - f\|_{L^p(G)} = O(h^r).$$

That is, there exist g_h and a constant K with $\|g_h - f\|_{L^p(G)} < Kh^r, \ \forall h$.

The Strang–Fix conditions amount to a general statement, which we shall discuss only in the case of box splines, giving a usable criterion to check the approximation power of a function B. Denote by H^p the Sobolev space (for some fixed domain) of functions that are L^2 together with their derivatives up to order p and by

$$\|u\|_{H^p}^2 = \sum_{|\alpha| \leq p} \|\partial^\alpha u\|_{L^2}^2$$

the corresponding Hilbert norm. This is convenient due to Plancherel's theorem, which implies that $\|\partial^\alpha u\|_{L^2}^2 = \|x^\alpha \hat{u}\|_{L^2}^2$.

The Strang –Fix conditions are (see [106]):

Theorem 6.7. *Let B be a compactly supported function of class C^p and let $r \leq p + 1$. The following conditions are equivalent:*

(1) $\hat{B}(0) \neq 0$, but $\hat{B}$ has zeros of order at least r at all other points of $2\pi\mathbb{Z}^s$.
(2) For every polynomial q of degree $< r$ the semidiscrete convolution $B *' q$ is also a polynomial of the same degree.

(3) For each function $u \in H^r$ there exists an element $g_h \in \mathcal{S}_h$, associated to a sequence w_i^h with the property that,

$$\|u - g_h\|_{H^s} \leq c_s h^{r-s} \|u\|_{H^r}, \quad 0 \leq s \leq r - 1 \tag{6.10}$$

$$\sum_i |w_i^h|^2 \leq K \|u\|_{H^0}^2, \tag{6.11}$$

where the constants c_s, K are independent of u.

Condition (3) expresses the approximation power in terms of L^2 norms, but also L^∞ norms can be used.

As for condition (2), we need to study a stronger property.

Definition 6.8. We say that a function ψ with compact support is a *superfunction* of *strength* r for $\mathcal{S}$ if $\psi \in \mathcal{S}$ and $\psi *' f = f$ for all polynomials of degree $< r$.

In Section 18.1 we shall:

(i) See how to construct superfunctions associated to box splines B_X.
(ii) Prove that the approximation power of such a function $B = B_X$ is the maximum r such that the space of all polynomials of degree $< r$ is contained in the cardinal space $\mathcal{S}_{B_X}$.
(iii) Determine r and describe the space $D(X)$ of all polynomials contained in $\mathcal{S}_{B_X}$.
(iv) Finally, give an efficient and explicit approximation algorithm based on semidiscrete convolution with superfunctions.

Part II

The Differentiable Case

7

Splines

The theme of this chapter is a brief introduction to two classes of functions that have been used in numerical analysis: the box spline and the multivariate spline associated to a list of vectors X. They both compute the volume of a variable polytope.

The theory of splines plays an important role in numerical analysis, and it is outside the scope of this book to discuss it in any detail. For an introduction the reader can consult, for instance, [96].

7.1 Two Splines

7.1.1 The Box Spline

Take a finite list $X := (a_1, \ldots, a_m)$ of nonzero vectors $a_i \in \mathbb{R}^s$, thought of as the columns of an $s \times m$ matrix, still denoted by X. If X spans $\mathbb{R}^s$, one builds an important function for numerical analysis, the *box spline* $B_X(x)$, implicitly defined by the formula

$$\int_{\mathbb{R}^s} f(x) B_X(x) dx := \int_0^1 \cdots \int_0^1 f(\sum_{i=1}^m t_i a_i) dt_1 \cdots dt_m, \tag{7.1}$$

where $f(x)$ varies over a suitable set of test functions.

If 0 is not in the convex hull of the vectors a_i, then one has a simpler function $T_X(x)$, the *multivariate spline* (cf. [40]), characterized by the formula

$$\int_{\mathbb{R}^s} f(x) T_X(x) dx = \int_0^\infty \cdots \int_0^\infty f\left(\sum_{i=1}^m t_i a_i\right) dt_1 \cdots dt_m, \tag{7.2}$$

where $f(x)$ varies in a suitable set of test functions (usually continuous functions with compact support or more generally, exponentially decreasing on

C. De Concini and C. Procesi, *Topics in Hyperplane Arrangements, Polytopes and Box-Splines*, Universitext, DOI 10.1007/978-0-387-78963-7_7,
© Springer Science+Business Media, LLC 2010

the cone $C(X)$). From now on, in any statement regarding the multivariate spline $T_X(x)$ we shall tacitly assume that the convex hull of the vectors in X does not contain 0.

Remark 7.1. If X does not span V, we have to consider T_X and B_X as measures on the subspace spanned by X. In fact, it is best to consider them as tempered distributions, as one easily verifies (see also Proposition 7.17).

One then proves that T_X and B_X are indeed functions as soon as X spans V, i.e., when the support of the distribution has maximal dimension.

In the rest of this section we assume that X spans V.

Both B_X and T_X have a simple geometric interpretation as functions computing the volumes $V_X(x)$ and $V_X^1(x)$ of the variable polytopes $\Pi_X(x)$ and $\Pi_X^1(x)$ introduced in Section 1.3.

Let F be an L^1 function on $\mathbb{R}^m$. Given $v \in \mathbb{R}^s$, consider the variable subspace $V(v) := \{w \in \mathbb{R}^m \mid Xw = v\}$ and let $dw(v)$ be the Lebesgue measure in $V(v)$ induced by the Euclidean structure of $\mathbb{R}^m$.

Lemma 7.2. *For almost every $v \in \mathbb{R}^s$ we have that F restricted to the space $V(v)$ is L^1, and if $G(v) := \int_{V(v)} F(w)dw(v)$, we have that $G(v)$ is L^1 and*

$$\sqrt{\det(XX^t)} \int_{\mathbb{R}^m} F(t_1, \ldots, t_m)dt_1 \cdots dt_m = \int_{\mathbb{R}^s} G(x_1, \ldots, x_s)dx_1 \cdots dx_s. \qquad (7.3)$$

Proof. Denote by e_i the standard basis of $\mathbb{R}^m$. Choose an orthonormal basis $u_1, \ldots, u_m$ for $\mathbb{R}^m$ such that its last $m - s$ elements form a basis of the kernel of the matrix X. If Q denotes the orthogonal matrix with columns the u_i, we have $Qe_i = u_i$; hence XQ is a matrix of block form $(B, 0)$ with B an $s \times s$ invertible matrix. Take variables $z_1, \ldots, z_m$ with $\sum_i x_i e_i = \sum_j z_j u_j$. We have that for given $\underline{z} := \{z_1, \ldots, z_s\}$ (that we think of as a column vector), the integral $\int \cdots \int F(z_1, \ldots, z_m)dz_{s+1} \cdots dz_m$ consists in integrating the function F on the space $V(v)$ where $v = B\underline{z}$; thus it equals $G(B\underline{z})$. We change variables and apply Fubini's Theorem, and since Q is orthogonal, we have

$$\int_{\mathbb{R}^m} F(t_1, \ldots, t_m)dt_1 \cdots dt_m = \int_{\mathbb{R}^s} G(B\underline{z})dz_1 \cdots dz_s.$$

By a second change of variables,

$$\int_{\mathbb{R}^s} G(B\underline{z})dz_1 \cdots dz_s = |\det(B)|^{-1} \int_{\mathbb{R}^s} G(\underline{z})dz_1 \cdots dz_s.$$

From $XQ = (B, 0)$ we have $XX^t = BB^t$, and hence $|\det(B)| = \det(XX^t)^{\frac{1}{2}}$.

Theorem 7.3. $V_X(x) = \sqrt{\det(XX^t)}\, T_X(x)$, $V_X^1(x) = \sqrt{\det(XX^t)}\, B_X(x)$.

Proof. Let us give the proof in the case of T_X, the other case being similar. We want to apply Lemma 7.2 to the function $F = \chi_+ f(\sum_{i=1}^m t_i a_i)$ where χ_+ denotes the characteristic function of the quadrant $\mathbb{R}_+^m$ and f is any given test function.

The space that we denoted by $V(v)$ equals the subspace of t_i with $\sum_{i=1}^m t_i a_i = v$; on this space the function F equals $f(v)$ on the polytope $\Pi_X(v)$ and 0 elsewhere, and thus by definition, the associated function $G(v)$ equals $f(v)V_X(v)$, and the theorem follows from the formulas (7.3) and (7.2).

This proof defines the volume function only weakly, and thus the given equality holds a priori only almost everywhere. In order to characterize the volume, we still have to prove that V_X is continuous on $C(X)$. Furthermore, we shall find explicit formulas for T_X as a continuous function on $C(X)$. We shall do this in Proposition 7.19 by an explicit recursive formula, using the convexity of the cone.

Remark 7.4. Let M be an invertible $s \times s$ matrix. Then

$$\int_{\mathbb{R}^s} f(x)T_{MX}(x)dx = \int_{\mathbb{R}_+^m} f\Big(\sum_{i=1}^m t_i M a_i\Big)dt = \int_{\mathbb{R}^s} f(Mx)T_X(x)dx$$

$$= |\det(M)|^{-1} \int_{\mathbb{R}^s} f(x)T_X(M^{-1}x)dx,$$

so that

$$T_{MX}(x) = |\det(M)|^{-1}T_X(M^{-1}x). \tag{7.4}$$

7.1.2 *E*-splines

It is useful to generalize these notions, introducing a *parametric version* of the previously defined splines called *E*-splines, introduced by A. Ron in [92].

Fix parameters $\mu := (\mu_1, \ldots, \mu_m)$ and define the functions (or tempered distributions) $B_{X,\underline{\mu}}, T_{X,\underline{\mu}}$ on V by the implicit formulas

$$\int_V f(x)B_{X,\underline{\mu}}(x)dx := \int_0^1 \cdots \int_0^1 e^{-\sum_{i=1}^m t_i \mu_i} f\Big(\sum_{i=1}^m t_i a_i\Big)dt_1 \cdots dt_m \tag{7.5}$$

$$\int_V f(x)T_{X,\underline{\mu}}(x)dx := \int_0^\infty \cdots \int_0^\infty e^{-\sum_{i=1}^m t_i \mu_i} f\Big(\sum_{i=1}^m t_i a_i\Big)dt_1 \cdots dt_m. \tag{7.6}$$

The same proof as in Theorem 7.3 shows that these functions have a nice geometric interpretation as integrals on the polytopes $\Pi_X^1(x)$ and $\Pi_X(x)$:

$$B_{X,\underline{\mu}}(x) = \int_{\Pi_X^1(x)} e^{-\sum_{i=1}^m t_i \mu_i}dz; \quad T_{X,\underline{\mu}}(x) = \int_{\Pi_X(x)} e^{-\sum_{i=1}^m t_i \mu_i}dz, \tag{7.7}$$

where the measure dz is induced from the standard Lebesgue measure on $\mathbb{R}^m$ multiplied by the normalization constant $\sqrt{\det(XX^t)}^{-1}$. Of course for $\underline{\mu} = 0$, we recover the previous definitions.

Remark 7.5. Since the t_i restricted to the polytope $\Pi_X(x)$ span the linear functions, the function $T_{X,\mu}(x)$ may be interpreted as a generating series for integrals of polynomials on this polytope.

We come now to the main formula, which will allow us to change the computation of these functions into a problem of algebra. By applying formula (7.5) we obtain their Laplace transforms:

Theorem 7.6.

$$\int_V e^{-\langle x \mid y \rangle} B_{X,\underline{\mu}}(y)\,dy = \int_0^1 \cdots \int_0^1 e^{-\sum_{i=1}^m t_i(\langle x \mid a_i \rangle + \mu_i)}\,dt_1 \cdots dt_m \qquad (7.8)$$

$$= \prod_{a \in X} \frac{1 - e^{-a - \mu_a}}{a + \mu_a},$$

and

$$\int_V e^{-\langle x \mid y \rangle} T_{X,\mu}(y)\,dy = \int_0^\infty \cdots \int_0^\infty e^{-\sum_{i=1}^m t_i(\langle x \mid a_i \rangle + \mu_i)}\,dt_1 \cdots dt_m \qquad (7.9)$$

$$= \prod_{a \in X} \frac{1}{a + \mu_a}.$$

Proof. The result follows immediately from the elementary formulas

$$\int_0^1 e^{-tb}\,dt = \frac{1 - e^{-b}}{b} \qquad \text{and} \qquad \int_0^\infty e^{-tb}\,dt = \frac{1}{b}.$$

The second one is valid only when $b > 0$.

We have introduced the shorhand $a := \langle x \mid a \rangle$ for the linear function a on U.

In the course of this book we give an idea of the general algebraic calculus involving the spline functions that we have introduced. Under the Laplace transform one can reinterpret the calculus in terms of the structure of certain algebras of rational functions (or exponentials) as D-modules.

Remark 7.7. Due to its particular importance, we introduce the notation $d_{X,\mu} := \prod_{a \in X}(a + \mu_a)$. When $\underline{\mu} = 0$ we set $d_X := \prod_{a \in X} a$.

From the expressions for the Laplace transforms one gets that the box spline can be obtained from the multivariate spline by a simple combinatorial formula using the twisted difference operators ∇_a^μ (cf. formula (5.7)).

Proposition 7.8. *For every subset $S \subset X$, set $a_S := \sum_{a \in S} a$ and $\mu_S := \sum_{a \in S} \mu_a$. Then*

$$B_{X,\underline{\mu}}(x) = \prod_{a \in X} \nabla_a^{\mu_a} T_{X,\mu}(x) = \sum_{S \subset X} (-1)^{|S|} e^{-\mu_S} T_{X,\mu}(x - a_S). \qquad (7.10)$$

In particular, for $\underline{\mu} = 0$,

$$B_X(x) = \prod_{a \in X} \nabla_a T_X(x) = \sum_{S \subset X} (-1)^{|S|} T_X(x - a_S). \qquad (7.11)$$

Proof. The proposition follows from the two formulas of Theorem 7.6 by applying the rule (3.4) that gives the commutation relation between the Laplace transform and translations.

If we choose for T_X, B_X the functions that are continuous on $C(X)$ and $B(X)$ respectively, this identity is true pointwise only on strongly regular points (see definition 1.50). For instance, if $X = \{1\}$, then $C(X) = [0, \infty)$ and we can take T_X to be the characteristic function of $[0, \infty)$. Then $B_X(x) = T_X(x) - T_X(x - 1)$ is the characteristic function of $[0, 1)$ and not of $[0, 1]$.

But in case T_X is continuous on the entire space, formula 7.11 expresses B_X as a continuous function.

Example 7.9. Let us draw some of the functions of Example 6.2

$$t_m(x) := T_{1^{m+1}}(x), \quad b_m(x) := B_{1^{m+1}}(x).$$

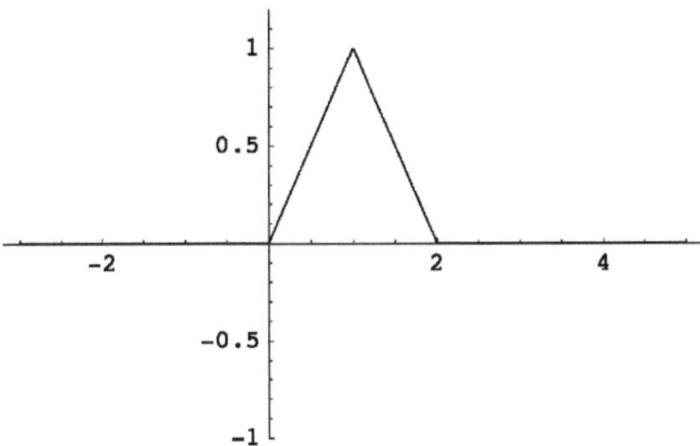

Fig. 7.1. The function $b_1(x)$

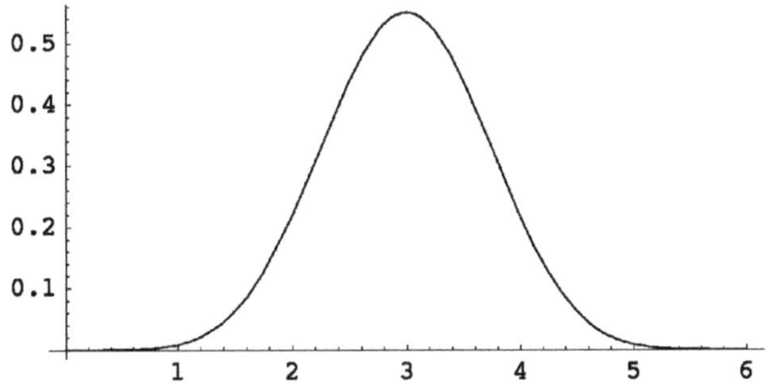

Fig. 7.2. The function $b_5(x)$

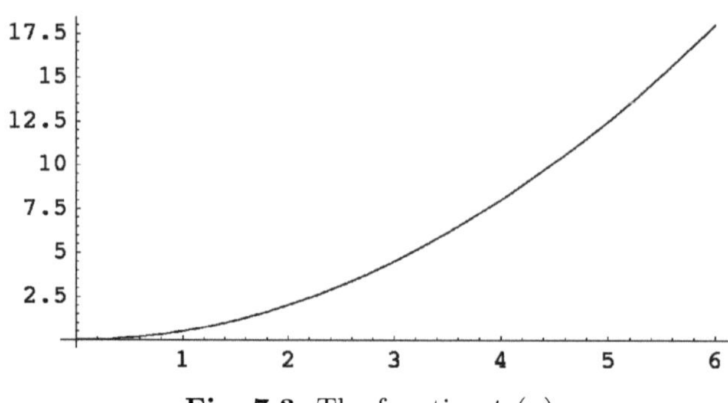

Fig. 7.3. The function $t_2(x)$

7.1.3 Shifted Box Spline

Let us make explicit some simple consequences in the case of the parameters $\mu = 0$. In this case, the formulas of the Laplace transform become:

$$LB_X = \prod_{a \in X} \frac{1 - e^{-a}}{a}, \quad LT_X = \prod_{a \in X} \frac{1}{a}. \tag{7.12}$$

It is useful to shift the box spline by $\rho_X := \frac{1}{2} \sum_{a \in X} a$ (as in Section 2.4.1) and consider $B_X(x + \rho_X)$ with support in the

$$\text{shifted box} \quad B(X) - \rho_X = \{\sum_{a \in X} t_a a \mid -\frac{1}{2} \leq t_a \leq \frac{1}{2}\}.$$

From formula (3.4) we have that the Laplace transform of $B_X(x + \rho_X)$ is

$$\prod_{a \in X} \frac{e^{a/2} - e^{-a/2}}{a}, \tag{7.13}$$

that is now seen to be symmetric with respect to the origin. In other words, $B_X(x + \rho_X) = B_X(-x + \rho_X)$.

This setting is particularly interesting when X is the set of positive roots of a root system (see Section 2.4).

Then formula (7.13), which expresses the Laplace transform of the shifted box spline, is invariant under the action of the Weyl group W.

Proposition 7.10. *The shifted box spline is symmetric under W.*

Proof. Let us recall the simple proof (cf. [89], [102]).

The Weyl group W is generated by the reflections s_i relative to the simple roots α_i. The reflection s_i maps α_i to $-\alpha_i$ and permutes the remaining positive roots. Thus

$$s_i\Big(\prod_{\alpha\in X}\frac{e^{\alpha/2}-e^{-\alpha/2}}{\alpha}\Big)=\prod_{\alpha\in X,\alpha\neq\alpha_i}\frac{e^{\alpha/2}-e^{-\alpha/2}}{a}\,\frac{e^{-\alpha_i/2}-e^{\alpha_i/2}}{-\alpha_i}$$

$$=\prod_{\alpha\in X}\frac{e^{\alpha/2}-e^{-\alpha/2}}{\alpha}.$$

Recall that we have already shown in Section 2.4 that in this case, the shifted box is the generalized permutohedron, the convex envelope of the W-orbit of ρ_X, and computed all its basic invariants.

As a consequence, we see that the *affine Weyl group* $\tilde{W}$ acts on the cardinal spline space translated by ρ_X. Where $\tilde{W}$ is the group of isometries generated by W and by the translations under the root lattice Λ, i.e., the lattice generated by the roots. The affine Weyl group is a semi-direct product $\tilde{W}:=\Lambda\rtimes W$.

Remark 7.11. If we take in the shifted cardinal space the finite combinations of translates of $B_X(x-\rho)$, we see that, as representation of $\tilde{W}$ it equals $\mathrm{Ind}_W^{\tilde{W}}(1)$. This is the representation induced by the trivial representation, from W to $\tilde{W}$.

7.1.4 Decompositions

We want to make a simple remark on a basic reduction computation for the functions T_X. This computation is based on the notion of *decomposition*, which will be analyzed more systematically in Section 20.1.

Assume that the list X is, up to reordering, the concatenation of lists $X_1,\ldots,X_k$ with $V=\langle X\rangle=\oplus_{j=1}^k\langle X_j\rangle$. In this case set $V_j=\langle X_j\rangle$ for each $j=1,\ldots,k$. The space V_j has a canonical Euclidean structure and thus a Lebesgue measure induced from that of V. As usual, write an element $x=(x_1,\ldots,x_k)\in V$ with the $x_j\in V_j$.

Proposition 7.12.

$$T_X(x)=c\prod_{j=1}^k T_{X_j}(x_j),\qquad B_X(x)=c\prod_{j=1}^k B_{X_j}(x_j),\qquad(7.14)$$

where the constant c is the ratio between the product of Lebesgue measures on the V_j's and the Lebesgue measure on V.

The constant c is the absolute value of the determinant of a basis of V, constructed by choosing for each i an orthonormal basis of V_i.

Proof. By formula (7.10), it is enough to prove the statement for T_X. We are going to apply the definition given by Formula (7.2) using as test function $f(x)$ a product $f(x)=\prod_{j=1}^k f_j(x_j)$. We get

$$\int_{\mathbb{R}^s} f(x)T_X(x)dx = \int_0^\infty \cdots \int_0^\infty f(\sum_{i=1}^m t_i a_i)dt_1 \cdots dt_m$$

$$= \prod_{j=1}^k \left[\int_0^\infty \cdots \int_0^\infty f_j\left(\sum_{a \in X_j} t_a a \right) \prod_{a \in X_j} dt_a \right],$$

$$= \prod_{j=1}^k \int_{V_j} f_j(x_j)T_{X_j}(x_j)dx_j.$$

Thus c is computed by $c\,dx = \prod_{j=1}^k dx_j$. This computation can be reduced easily to a computation of a determinant, by taking a basis $u_1, \ldots, u_s$ of V the union of orthonormal bases for all the subspaces V_j. Then $c = |\det(u_1, \ldots, u_s)|$.

Remark 7.13. For the support of the two functions we have

$$C(X) = \prod_{i=1}^k C(X_i), \quad B(X) = \prod_{i=1}^k B(X_i). \tag{7.15}$$

7.1.5 Recursive Expressions

The defining formula (7.2) implies immediately that, if $X = (Y, z)$

Proposition 7.14. *In the sense of distributions we have:*
1. *$D_z(T_X) = T_Y$.*
2. *$D_z(B_X) = \nabla_z B_Y$.*

Proof. 1. Given a function f in the Schwartz space, we have, by definition of distributional derivative,

$$\langle D_z(T_X) \mid f \rangle = \langle T_X \mid - D_z(f) \rangle$$

$$= -\int_0^\infty \cdots \int_0^\infty \left(\int_0^\infty D_z(f)\left(\sum_{i=1}^{m-1} t_i a_i + tz \right)dt \right)dt_1 \cdots dt_{m-1}$$

$$= \int_0^\infty \cdots \int_0^\infty f\left(\sum_{i=1}^{m-1} t_i a_i \right)dt_1 \cdots dt_{m-1} = \langle T_Y \mid f \rangle.$$

2. The proof is similar (use formula (5.5)),

$$\langle D_z(B_X) \mid f \rangle = \int_0^1 \cdots \int_0^1 \left[f\left(\sum_{i=1}^{m-1} t_i a_i \right) - f\left(\sum_{i=1}^{m-1} t_i a_i + z \right) \right]dt_1 \cdots dt_{m-1}$$

$$= \int_0^1 \cdots \int_0^1 (\nabla_{-z} f)\left(\sum_{i=1}^{m-1} t_i a_i \right)dt_1 \cdots dt_{m-1} = \langle \nabla_z B_Y \mid f \rangle.$$

It is customary to compute both T_X and B_X recursively as follows.

The initial computation is the following.

Proposition 7.15. *Let* $X = (a_1, \ldots, a_s)$ *be a basis of* V, $d := |\det(a_1, \ldots, a_s)|$. *Then* $B(X)$ *is the parallelepiped with edges the* a_i, $C(X)$ *is the positive quadrant generated by* X. *Furthermore,*

$$B_X = d^{-1}\chi_{B(X)}, \qquad T_X = d^{-1}\chi_{C(X)}, \tag{7.16}$$

where for any given set A, *we denote by* χ_A *its characteristic function.*

Proof. This follows immediately from the definitions (cf. also (7.4)).

It is convenient to think of B_X, in case $X = (a_1, \ldots, a_s)$ is a basis, not as the characteristic function of the compact parallelepiped (multiplied by d^{-1}) but rather that of the half-open parallelepiped given by $\{\sum_{i=1}^s t_i a_i, \ 0 \leq t_i < 1\}$.

There are two advantages in this definition. The first is that the identity (7.10) holds unconditionally, and the second appears when the vectors X are in $\mathbb{Z}^s$, since then the translates of B_X under this lattice give a partition of unity.

Let us apply formula (7.14) in a more systematic way. This tells us that we should concentrate our computations to the case in which X is *indecomposable*, that is, it cannot be written in the form $X = X_1, X_2$ with $\langle X \rangle = \langle X_1 \rangle \oplus \langle X_2 \rangle$. Under this assumption unless X is a single vector in a one dimensional space, X is automatically *nondegenerate*, (cf. Definition 2.23), i.e., the minimal length $m(X)$ of a cocircuit is ≥ 2.

Example 7.16. The simplest nondegenerate example is

$$X = (e_1, \ldots, e_s, \sum_{i=1}^s e_i).$$

Since we have taken as definition of T_X, B_X a weak one, we should think of these functions as being defined up to sets of measure 0.

It will be proved (cf. Theorem 7.24) that for nondegenerate X, both functions can be taken to be continuous on the whole space. We start by proving that they can be taken continuous on their support.

From (7.15) we are reduced to the nondegenerate case. Since the one-dimensional case is trivial we may assume $m > 0$, consider $Y := (a_1, \ldots, a_{m-1})$ and set $z := a_m$, so that $X = (Y, z)$. Since X is nondegenerate, Y still spans $V = \mathbb{R}^s$.

Proposition 7.17. *1.* B_X *is supported in the* box $B(X)$. *Similarly,* T_X *is supported in* $C(X)$. *They can both be represented by continuous functions on* $B(X)$ *and* $C(X)$ *respectively.*

2. We can choose T_X *and* B_X *to be given by*

$$T_X(x) = \int_0^\infty T_Y(x - tz)dt, \quad B_X(x) = \int_0^1 B_Y(x - tz)dt. \tag{7.17}$$

3. *The cone $C(X)$ can be decomposed into the union of finitely many polyhedral cones C_i so that T_X restricted to each C_i is a homogeneous polynomial of degree $m - s$.*

Proof. Working by induction on the dimension s of V and on the length m of X, we may assume that T_Y is continuous on its support $C(Y)$. For any Schwartz function f

$$\int_{\mathbb{R}_+^m} f\Big(\sum_{i=1}^m t_i a_i\Big) dt_1 \cdots dt_m = \int_0^\infty \Big(\int_V T_Y(x) f(x + tz) dx\Big) dt$$

$$= \int_0^\infty \Big(\int_V T_Y(x - tz) f(x) dx\Big) dt = \int_V \Big(\int_0^\infty T_Y(x - tz) dt\Big) f(x) dx.$$

We claim that $\int_0^\infty T_Y(x - tz) dt$ is well defined and continuous everywhere. Furthermore, it vanishes outside $C(X)$ and it is strictly positive in the interior of $C(X)$. This will allow us to identify it with T_X.

By definition, $C(X) = C(Y) + \mathbb{R}^+ z$, so for each point $x \in V$, the set $I_x := \{t \in \mathbb{R}^+ \,|\, x - tz \in C(Y)\}$ is empty unless $x \in C(X)$.

We claim that if $x \in C(X)$, then $I_x = [a(x), b(x)]$ is an interval and both $a(x), b(x)$ are continuous functions on $C(X)$.

Let $C(Y) := \{v \,|\, \langle \phi_i \,|\, v \rangle \geq 0\}$ for some elements $\phi_1, \ldots, \phi_M \in V^*$. We have that $I_x = \{t \geq 0 \,|\, \langle \phi_i \,|\, x \rangle \geq t \langle \phi_i \,|\, z \rangle, \forall i\}$.

These inequalities can be written as

$$t \geq \langle \phi_i \,|\, x \rangle \langle \phi_i \,|\, z \rangle^{-1}, \quad \forall i \,|\, \langle \phi_i \,|\, z \rangle < 0,$$

$$t \leq \langle \phi_i \,|\, x \rangle \langle \phi_i \,|\, z \rangle^{-1}, \quad \forall i \,|\, \langle \phi_i \,|\, z \rangle > 0.$$

When $\langle \phi_i \,|\, z \rangle = 0$ we have either that every t or no t satisfies the inequality. Since $C(X)$ is a pointed cone, there is at least one i such that $\langle \phi_i \,|\, z \rangle > 0$; otherwise, $-z \in C(Y) \subset C(X)$.

This proves that I_x is a bounded closed interval.

Set

$$a(x) := \max\{0, \langle \phi_i \,|\, x \rangle \langle \phi_i \,|\, z \rangle^{-1}\}, \qquad \forall i \,|\, \langle \phi_i \,|\, z \rangle < 0,$$
$$b(x) := \min\{\langle \phi_i \,|\, x \rangle \langle \phi_i \,|\, z \rangle^{-1}\}, \qquad \forall i \,|\, \langle \phi_i \,|\, z \rangle > 0.$$

The set I_x is empty as soon as $a(x) > b(x)$, and as we have seen, this happens exactly outside the cone $C(X)$.

Clearly, the two functions $a(x), b(x)$ are continuous and piecewise linear with respect to a decomposition into cones of the space.

The continuity of T_X on $C(X)$ now follows from the continuity of T_Y on $C(Y)$ and that of $a(x)$ and $b(x)$. This proves 1 and 2.

We claim now that the function T_X is piecewise polynomial of homogeneous degree $m - s$ with $s = \dim\langle X \rangle$.

Again proceed by induction: T_Y is continuous on $C(Y)$, which is decomposed into polyhedral cones $C_i(Y)$, so that T_Y is a polynomial $p_{i,Y}(x)$ homogeneous of degree $m-1-s$ on each $C_i(Y)$. Then as before, we have piecewise linear functions $a_i(x), b_i(x)$, so that when $a_i(x) \le b_i(x)$ the intersection $I_x^i := \{t \in \mathbb{R}^+ \,|\, x - tz \in C_i(Y)\}$ is the segment $[a_i(x), b_i(x)]$; otherwise it is empty.

Thus we can decompose the cone $C(X)$ into polyhedral cones on each of which all the functions a_i, b_i are linear.

When x varies over one of these cones C, the interval I_x is decomposed into intervals $[u_j(x), u_{j+1}(x)]$, $j = 1, \ldots, k$, so that all the $u_j(x)$ are linear, and when $t \in [u_j(x), u_{j+1}(x)]$ the point $x - tz$ lies in some given cone $C_j(Y)$. Therefore, on C the function T_X is a sum:

$$T_X(x) = \sum_{j=1}^{k-1} \int_{u_j(x)}^{u_{j+1}(x)} T_Y(x - tz)\,dt.$$

We know that on $C_j(Y)$ the function $T_Y(x) = p_{j,Y}(x)$ is a homogeneous polynomial of degree $m-1-s$. So $p_{j,Y}(x - tz) = \sum_{k=0}^{m-1-s} t^k p_{j,Y}^k(x)$ with $p_{j,Y}^k(x)$ homogeneous of degree $m-1-s-k$ and

$$\int_{u_j(x)}^{u_{j+1}(x)} t^k p_{j,Y}^k(x)\,dt = (k+1)^{-1} p_{j,Y}^k(x)[u_{j+1}(x)^{k+1} - u_j(x)^{k+1}].$$

This is a homogeneous polynomial of degree $m-1-s+1 = m-s$.

We leave to the reader to interpret the previous recursion as a statement on distributions. With $X = \{a_1, \ldots, a_m\}$.

Theorem 7.18. *The distribution T_X is the convolution*

$$T_X = T_{a_1} * \cdots * T_{a_m}.$$

A similar statement holds for $B_X = B_{a_1} * \cdots * B_{a_m}$.

We now relate T_X with the volume of $\Pi_X(x)$.

Proposition 7.19. *1. For any $x \in C(X)$, $T_X(x)$ equals the volume of the polytope $\Pi_X(x)$, while $B_X(x)$ equals the volume of the polytope $\Pi_X^1(x)$, both divided by the normalization factor $\sqrt{\det(XX^t)}$.*
2. T_X (resp. B_X) is strictly positive in the interior of $C(X)$ (resp. $B(X)$).

Proof. 1 Let us prove this for T_X. In the degenerate case, we can assume that z does not lie in the subspace $W = \langle Y \rangle$. We have a unique decomposition $x = y + tz, y \in W, t \in \mathbb{R}$. Since $x \in C(X)$, we must have $t \ge 0$ and $y \in C(Y)$. In this case, $\Pi_X(x) = \Pi_Y(y) \times \{tz\}$. The volume $\mathrm{Vol}(\Pi_X(x))$ of $\Pi_X(x)$ equals that of $\Pi_Y(y)$.

Let us now assume that X is nondegenerate. Consider on the polytope $\Pi_X(x)$ the function t_m. This ranges over the set of positive t such that $x - ta_m \in C(Y)$. If t_m is constant on the polytope $\Pi_X(x)$, this lies in a hyperplane and its volume is 0. By formula (7.17), it also follows that $T_X(x) = 0$. Otherwise, by Fubini's theorem $\mathrm{Vol}(\Pi_X(x)) = c_m \int_0^\infty \mathrm{Vol}(\Pi_Y(x - ta_m))dt$, for some positive constant c_m. By induction, since the recursive formulas are the same up to constants, we then see that $\mathrm{Vol}(\Pi_X(x))$ is continuous on $C(X)$, and the first claim follows from the weak equality proved in Theorem 7.3.

2 Follows from part 1 and Proposition 1.47.

Remark 7.20. In the parametric case we have the analogous recursive formulas

$$T_{X,\underline{\mu}}(x) = \int_0^\infty e^{-\mu_z t} T_{Y,\underline{\mu}}(x - tz)dt, \quad B_{X,\underline{\mu}}(x) = \int_0^1 e^{-\mu_z t} B_{Y,\underline{\mu}}(x - tz)dt.$$

$$(7.18)$$

We finish this section by introducing a fundamental space of polynomials $\tilde{D}(X)$.

Definition 7.21. We set $\tilde{D}(X)$ to be the space spanned by all the homogeneous polynomials that coincide on some open set with the multivariate spline T_X (cf. Proposition 7.17 3.) and all of the derivatives of any order of these polynomials.

Notice that these polynomials also describe B_X locally by formula (7.11).

The study of $\tilde{D}(X)$ will be one of our central themes (see Chapter 11). We shall describe it in various other ways, in particular as the space $D(X)$ of solutions of a remarkable set of differential equations (Definition 11.2). Only in Theorem 11.37 will we see that $\tilde{D}(X) = D(X)$.

7.1.6 Smoothness

Splines are used to interpolate and approximate functions. For this purpose it is important to understand the class of smoothness of a spline. We are going to show that $m(X)$ (the minimum number of elements in a cocircuit in X cf. Section 2.2.1), determines the class of differentiability of T_X.

Assume that X spans the ambient space V and $m(X) = 1$. Thus there is $a \in X$ such that $X \setminus \{a\}$ does not span. In this case we may apply formulas (7.14) and (7.16) that show that T_X is discontinuous.

Definition 7.22. Given a nonzero vector a we say that a function f on $\mathbb{R}^s$, is *continuous in the direction of a* if, for every $x \in \mathbb{R}^s$ the function $f(x - ta)$ is continuous in the variable t.

Lemma 7.23. *Let $X = \{Y, a\}$ with Y spanning $\mathbb{R}^s$, then T_X is continuous in the direction of a.*

Proof. We have that T_Y is given by a piecewise polynomial function and T_X is obtained from the recursive expression from which the claim follows.

Assume now $m(X) \geq 2$.

Theorem 7.24. *B_X and T_X are of class $C^{m(X)-2}$.*

Proof. Let $m(X) \geq 2$. Let us first prove that T_X is everywhere continuous. Since T_X is continuous on $C(X)$, and 0 outside $C(X)$ it is enough to prove that $T_X = 0$ on the boundary of T_X. The hypothesis $m(X) \geq 2$ implies that T_X is continuous in the direction of a for each $a \in X$ and these elements span $\mathbb{R}^s$. We claim that given a point x in the boundary of $C(X)$, we must have that for some $a \in X$, the points $x - ta$, $t > 0$, do not lie in $C(X)$. Otherwise, by convexity, all the points $x - ta$, $t \leq \epsilon$, would lie in $C(X)$ for $\epsilon > 0$ sufficiently small. But these points generate a convex set in $C(X)$ whose interior contains x, contradicting the hypothesis. For such an a, $T_X(x - ta)$ is continuous and is equal to 0 for $t > 0$. It follows that $T_X(x) = 0$.

Assume now that $m(X) \geq 3$. By Proposition 7.14, for each $a \in X$ we have that T_X has a derivative in the direction of a that is continuous and in fact of class at least $C^{m(X)-3}$ by induction. Since the elements of X span $\mathbb{R}^s$, the claim follows.

As for B_X, it is enough to apply formula (7.11) to T_X.

7.1.7 A Second Recursion

Let us discuss another recursive formula due to Micchelli for T_X.

From the previous reduction theorem we may assume that $m(X) \geq 2$, so that both T_X and $T_{X \setminus a_i}$ for all $a_i \in X$ are functions that are continuous on their support.

Theorem 7.25. *Let $b = \sum_{i=1}^{m} t_i a_i$. We then have that if either $m(X) \geq 3$ or b is in the open set where T_X coincides with some polynomial, then*

$$T_X(b) = \frac{1}{m-s} \sum_{i=1}^{m} t_i T_{X \setminus a_i}(b). \tag{7.19}$$

Proof. Let us fix coordinates. We first assume that $b = (b_1, \ldots, b_s)$ lies in an open set where T_X is a homogeneous polynomial of degree $m - s$. Thus by Euler's identity, in this open set,

$$T_X(x) = \frac{1}{m-s} \sum_{i=1}^{s} x_i \frac{\partial}{\partial x_i} T_X(x).$$

We have

$$\sum_{i=1}^{s} b_i \frac{\partial}{\partial x_i} = D_b = \sum_{i=1}^{m} t_i D_{a_i}.$$

Thus

$$T_X(b) = \frac{1}{m-s} \sum_{j=1}^{m} t_j D_{a_j} T_X(b) = \sum_{j=1}^{m} t_j T_{X \setminus a_j}(b).$$

If we make the further assumption that $m(X) \geq 3$, then all the functions involved in formula (7.19) are continuous, and thus the identity holds unconditionally.

When $m(X) = 2$ the above formula fails for special values of b. For instance, in the simple example of the three vectors $e_1, e_2, e_1 + e_2$ and the point $b := e_{\frac{1}{2}} + e_2/2$, we have $T_X(b) = \frac{1}{2}$, while for the right-hand side we get 1.

One can use this theorem together with the reduction to the case $m(X) \geq 2$ to effectively compute the function T_X. One of the main difficulties is that when we arrive at $m(X) = 1$, we have to do some complicated combinatorics.

8

R_X as a D-Module

In this chapter, the word D-module is used to denote a module over one of the two Weyl algebras $W(V), W(U)$ of differential operators with polynomial coefficients on V, U respectively. The purpose of this chapter is to determine an expansion in partial fractions of the regular functions on the complement of a hyperplane arrangement. This is essentially the theory of Brion–Vergne (cf. [28], [7]). We do it using the D-module structure of the algebra of regular functions. Finally, by inverse Laplace transform all this is interpreted as a calculus on the corresponding distributions.

8.1 The Algebra R_X

8.1.1 The Complement of Hyperplanes as Affine Variety

Let us return to the hyperplane arrangement $\mathcal{H}_{X,\underline{\mu}}$ in $U = V^*$ associated to a list of vectors X in V and parameters $\underline{\mu}$. Consider the open set $\mathcal{A}_{X,\underline{\mu}}$, the complement of the given hyperplanes. We usually start with a real vector space V but it is useful to define $\mathcal{A}_{X,\underline{\mu}}$ as the complement of the given hyperplanes in the complexified dual $U_{\mathbb{C}} = \hom(V, \mathbb{C})$.

The set $\mathcal{A}_{X,\underline{\mu}}$ is an affine algebraic variety with coordinate ring the algebra

$$R_{X,\underline{\mu}} = S[V_{\mathbb{C}}][\prod_{a \in X} (a + \mu_a)^{-1}], \tag{8.1}$$

obtained from the algebra $S[V_{\mathbb{C}}]$ of complex-valued polynomial functions by inverting the polynomial $\prod_{a \in X}(a + \mu_a)$.

The study of this algebra will be the main theme of this chapter. We shall do this by describing the algebra $R_{X,\underline{\mu}}$ as a module over the Weyl algebra.

In order not to have a very complicated notation, unless there is a risk of confusion we shall simply write $R = R_{X,\underline{\mu}}$.

C. De Concini and C. Procesi, *Topics in Hyperplane Arrangements, Polytopes and Box-Splines*, Universitext, DOI 10.1007/978-0-387-78963-7_8,
© Springer Science+Business Media, LLC 2010

As a preliminary step, take a subset $Y \subset X$ and set $\underline{\mu}(Y)$ to be the family $\underline{\mu}$ restricted to Y. We have thus the inclusion of algebras $R_{Y,\underline{\mu}(Y)} \subset R_{X,\underline{\mu}}$, and clearly this is an inclusion of $W(U)$ modules.

Let us now introduce a filtration in $R = R_{X,\underline{\mu}}$ by D-submodules that we will call the *filtration by polar order.*

This is defined algebraically as follows.

Definition 8.1 (filtration by polar order).

$$R_k = \sum_{Y \subset X, \; \dim\langle Y \rangle \leq k} R_{Y,\underline{\mu}(Y)}. \tag{8.2}$$

In other words, the subspace R_k of degree $\leq k$ is spanned by all the fractions $f \prod_{a \in X}(a + \mu_a)^{-h_a}$, $h_a \geq 0$, for which the vectors a with $h_a > 0$ span a space of dimension $\leq k$. Notice that $R_s = R$.

In the next subsection we shall start with a special case in order to introduce the building blocks of the theory. From now on, we assume that the spaces have been complexified and write V, U instead of $V_{\mathbb{C}}$, $U_{\mathbb{C}}$.

8.1.2 A Prototype D-module

We start by considering, for some $t \leq s$, the algebra of partial Laurent polynomials $L := \mathbb{C}[x_1^{\pm 1}, \ldots, x_t^{\pm 1}, x_{t+1}, \ldots, x_s]$ as a module over the Weyl algebra $W(s)$.

A linear basis of the space L is given by the monomials $M := x_1^{h_1} \ldots x_s^{h_s}$, $h_i \in \mathbb{Z}, \forall i \leq t, h_i \in \mathbb{N}, \forall i > t$. For such a monomial M set $p(M) := \{i \,|\, h_i < 0\}$, the *polar set* of M, and for any subset $A \subset \{1, \ldots, t\}$ let L_A be the subspace with basis the monomials M with $p(M) \subset A$.

Define furthermore $L_k := \sum_{|A| \leq k} L_A$. We see immediately that the following is true

Proposition 8.2. *The subspace L_A is a $W(s)$ submodule for each A.*
L_k/L_{k-1} has as basis the classes of the monomials M with $|p(M)| = k$.

For each A let W_A be the space defined by the linear equations $x_i = 0$, $\forall i \in A$, we shall denote by N_A the module N_{W_A} (defined in Section 4.1.2).

If $|A| = k$, define M_A as the subspace of L_k/L_{k-1} with basis the classes of the monomials M with $p(M) = A$.

Theorem 8.3. $L_k/L_{k-1} = \oplus_{|A|=k} M_A.$
The submodule M_A is canonically isomorphic to N_A by an isomorphism mapping the generator $\delta_A := \delta_{W_A}$ to the class of $1/\prod_{i \in A} x_i$.

Proof. All the statements are essentially obvious. We have already remarked that L_k/L_{k-1} has as basis the classes of the monomials M with $|p(M)| = k$. Thus $L_k/L_{k-1} = \oplus_{|A|=k} M_A.$

The class $[\frac{1}{\prod_{i \in A} x_i}] \in L_k/L_{k-1}$ clearly satisfies the equations

$$x_i \left[\frac{1}{\prod_{i \in A} x_i} \right] = \frac{\partial}{\partial x_j} \left[\frac{1}{\prod_{i \in A} x_i} \right] = 0, \quad \forall i \in A, \ j \notin A,$$

so by formula (4.1), it generates a submodule isomorphic to N_A.

Given $\prod_{j \notin A} x_j^{h_j} \prod_{i \in A} \frac{\partial^{k_i}}{\partial x_i}$, $h_i, k_i \geq 0$ we have explicitly

$$\prod_{j \notin A} x_j^{h_j} \prod_{i \in A} \frac{\partial^{k_i}}{\partial x_i} \frac{1}{\prod_{i \in A} x_i} = \prod_{j \notin A} x_j^{h_j} \prod_{i \in A} (-1)^{h_i} h_i! x_i^{-1-h_i}.$$

Th classes of these elements form a basis of M_A.

Corollary 8.4. *Let $\{a_1, \ldots, a_t\}$ be a set of linearly independent elements of V and take numbers $\mu_1, \ldots, \mu_t$. The successive quotients of the filtration by polar order, of the ring $S[V][\prod_{i=1}^t (a_i + \mu_i)^{-1}]$, are direct sums of the modules N_{W_A} associated to the affine linear subspaces W_A of equations $a_i + \mu_i = 0$, $i \in A$ for all subsets $A \subset \{1, \ldots, t\}$.*

Proof. This is an immediate consequence of Theorem 8.3 on applying an affine change of coordinates.

8.1.3 Partial Fractions

In this section we are going to deduce some identities in the algebra $R_{X,\mu}$. In the linear case, where all $\mu_a = 0$, we drop the $\underline{\mu}$ and write simply R_X. Related approaches can be found in [113] and [63]

Let us first develop a simple identity. Take vectors b_i, $i = 0, \ldots, k$. Assume that $b_0 = \sum_{i=1}^k \alpha_i b_i$. Choose numbers ν_i, $i = 0, \ldots, k$, and set

$$\nu := \nu_0 - \sum_{i=1}^k \alpha_i \nu_i. \tag{8.3}$$

If $\nu \neq 0$, we write

$$\frac{1}{\prod_{i=0}^k (b_i + \nu_i)} = \nu^{-1} \frac{b_0 + \nu_0 - \sum_{i=1}^k \alpha_i (b_i + \nu_i)}{\prod_{i=0}^k (b_i + \nu_i)}. \tag{8.4}$$

When we develop the right-hand side, we obtain a sum of $k+1$ terms in each of which one of the elements $b_i + \nu_i$ has disappeared.

Let us draw a first consequence on separation of variables. We use the notation $P(X, \underline{\mu})$ of Section 2.1.1 for the points of the arrangement, and given $p \in P(X, \underline{\mu})$, X_p for the set of $a \in X$ such that $a + \mu_a$ vanishes at p.

Proposition 8.5. *Assume that X spans V. Then:*

$$\prod_{a \in X} \frac{1}{a + \mu_a} = \sum_{p \in P(X,\underline{\mu})} c_p \prod_{a \in X_p} \frac{1}{a + \mu_a} = \sum_{p \in P(X,\underline{\mu})} c_p \prod_{a \in X_p} \frac{1}{a - \langle a \mid p \rangle} \tag{8.5}$$

with c_p given by the formula:

$$c_p = \prod_{a \in X \setminus X_p} \frac{1}{\langle a \mid p \rangle + \mu_a}.$$

Proof. This follows by induction applying the previous algorithm of separation of denominators.

Precisely, if X is a basis, there is a unique point of the arrangement and there is nothing to prove. Otherwise, we can write $X = (Y, z)$ where Y still spans V. By induction

$$\prod_{a \in X} \frac{1}{a + \mu_a} = \frac{1}{z + \mu_z} \prod_{a \in Y} \frac{1}{a + \mu_a} = \sum_{p \in P(Y, \underline{\mu})} c'_p \frac{1}{z + \mu_z} \prod_{a \in Y_p} \frac{1}{a + \mu_a}.$$

We need to analyze each product

$$\frac{1}{z + \mu_z} \prod_{a \in Y_p} \frac{1}{a + \mu_a}. \tag{8.6}$$

If $\langle z \mid p \rangle + \mu_z = 0$, then $p \in P(X, \underline{\mu})$, $X_p = \{Y_p, z\}$, and we are done. Otherwise, choose a basis $y_1, \ldots, y_s$ from Y_p and apply the previous proposition to the list $z, y_1, \ldots, y_s$ and the corresponding numbers μ_z, μ_{y_i}. The product (8.6) develops as a linear combination of products involving one fewer factor, and we can proceed by induction.

It remains to compute c_p. For a given $p \in P(X, \underline{\mu})$,

$$\prod_{a \in X \setminus X_p} \frac{1}{a + \mu_a} = c_p + \sum_{q \in P(X, \underline{\mu}),\ q \neq p} c_q \frac{\prod_{a \in X_p}(a + \mu_a)}{\prod_{a \in X_q}(a + \mu_a)}.$$

Hence, evaluating both sides at p yields

$$c_p = \prod_{a \in X \setminus X_p} \frac{1}{\langle a \mid p \rangle + \mu_a}.$$

Remark 8.6. In case X does not span the ambient space, we have that the minimal subspaces of the corresponding arrangement are all parallel to the subspace W defined by the equations $a = 0$, $a \in X$. One then has the same type of formula on replacing the points of the arrangement with these minimal subspaces.

We use this remark in order to start a development in partial fractions:

Proposition 8.7. *We can write a fraction*

$$\prod_{a \in X} (a + \mu_a)^{-h_a}, \quad h_a \geq 0,$$

as a linear combination of fractions in which the elements $a_i + \mu_i$ appearing in the denominator vanish on a subspace of the arrangement.

At this point we are still far from having a normal form for the partial fraction expansion. In fact, a dependency relation $\sum_{i=0}^{k} \alpha_i(b_i + \nu_i) = 0$ produces a relation

$$\sum_{j=0}^{k} \frac{\alpha_j}{\prod_{i=0,\ i\neq j}^{k}(b_i + \nu_i)} = \frac{\sum_{i=0}^{k} \alpha_i(b_i + \nu_i)}{\prod_{i=0}^{k}(b_i + \nu_i)} = 0. \tag{8.7}$$

So let us return to (8.3). If $\nu = 0$, we can write

$$\frac{1}{\prod_{i=0}^{k}(b_i + \nu_i)} = \frac{\sum_{j=1}^{k} \alpha_j(b_j + \nu_j)}{(b_0 + \nu_0)\prod_{i=0}^{k}(b_i + \nu_i)} = \sum_{j=1}^{k} \frac{\alpha_j}{(b_0 + \nu_0)^2 \prod_{i=1,\ i\neq j}^{k}(b_i + \nu_i)}.$$

We shall use Proposition 8.7, these relations, and the idea of unbroken bases, in order to normalize the fractions.

Theorem 8.8. *Consider a fraction $F := \prod_{i=1}^{k}(b_i + \nu_i)^{-h_i}$, $h_i > 0$, where the elements b_i form an ordered sublist extracted from the list $X = (a_1, \ldots, a_m)$, and ν_i denotes μ_{b_i}. Assume that the equations $b_i + \nu_i$ define a subspace W of the arrangement of codimension d. Then F is a linear combination of fractions $\prod_{r=1}^{d}(a_{i_r} + \mu_{i_r})^{-k_r}$, where:*

(a) The elements $a_{i_r} + \mu_{i_r}$, $r = 1, \ldots, d$ appearing in the denominator form an unbroken basis on W (see Definition 2.29).

(b) $k_r > 0$ for all r.

(c) $\sum_{r=1}^{d} k_r = \sum_{i=1}^{k} h_i$.

Proof. Assume that there is an element $a_i \in X_W$ breaking a sublist of $(b_1, \ldots, b_k)$ consisting of elements a_{i_j}, i.e., $a_i + \mu_i = \sum_j \alpha_j(a_{i_j} + \mu_{i_j})$ and $i < i_j$, $\forall j$. We multiply the denominator and the numerator by $a_i + \mu_i$ and then apply the previous identity (8.7). Thus the fraction $\prod_{j=1}^{k}(b_j + \nu_j)^{-1}$ equals a sum of fractions in which one of the terms $a_{i_j} + \mu_{i_j}$ has been replaced by $a_i + \mu_i$ and $i < i_j$. In an obvious sense, the fraction has been replaced by a sum of fractions that are *lexicographically strictly inferior*, but the elements appearing in the denominator still define W and span the same subspace spanned by the elements $b_1, \ldots, b_k$. The sum of the exponents also remains constant. The algorithm stops when each fraction appearing in the sum is of the form required in (a). The rest of the Theorem is clear.

In particular, R_X is spanned by the functions of the form $f/(\prod_i b_i^{k_i+1})$, with f a polynomial, $B = \{b_1, \ldots, b_t\} \subset X$ a linearly independent subset, and $k_i \geq 0$ for each $i = 1, \ldots, t$.

8.1.4 The Generic Case

Let us (by abuse of notation) denote by $\mathcal{B}_X$ the family of all subsets $\sigma := \{i_1, \ldots, i_s\} \subset \{1, \ldots, m\}$ with the property that the set $\underline{b}_\sigma := (a_{i_1}, \ldots, a_{i_s})$ is a basis extracted from the list X. Consider finally the corresponding point $p_\sigma : a_{i_k} + \mu_{i_k} = 0$, $k = 1, \ldots, s$.

Definition 8.9. We say that *the parameters are generic* if these points are all distinct.

In particular, we have that $\langle a_i \,|\, p_\sigma \rangle + \mu_i = 0, \iff i \in \sigma$. This condition holds on a nonempty open set of the space of parameters, in fact, on the complement of the following hyperplane arrangement:

For each sublist $a_{i_1}, \ldots, a_{i_{s+1}}$ of the list X consisting of $s + 1$ vectors spanning V, form the $(s + 1) \times (s + 1)$ matrix

$$
\begin{vmatrix}
a_{i_1} & \cdots & a_{i_{s+1}} \\
\mu_{i_1} & \cdots & \mu_{i_{s+1}}
\end{vmatrix}. \tag{8.8}
$$

Its determinant is a linear equation in the μ_i, and the hyperplanes defined by these linear equations give the nongeneric parameters.

If $\underline{b}_\sigma := (a_{i_1}, \ldots, a_{i_s})$ is a basis and v_σ is the determinant of the matrix with columns this basis, one easily sees that the determinant of the matrix (8.8) equals $v_\sigma(\langle a_{i_{s+1}} \,|\, p_\sigma \rangle + \mu_{i_{s+1}})$.

Set $d_\sigma := \prod_{k \in \sigma}(a_k + \mu_k)$. By Proposition 8.5 we get

$$
\prod_{i=1}^{m} \frac{1}{(a_i + \mu_i)} = \sum_\sigma \frac{c_\sigma}{d_\sigma}, \quad c_\sigma = \prod_{i=1,\ i \notin \sigma}^{m} (\langle a_i \,|\, p_\sigma \rangle + \mu_i)^{-1}. \tag{8.9}
$$

This formula has a simple geometric meaning, in the sense that the function $\prod_{i=1}^{m} 1/(a_i + \mu_i)$ has simple poles at the points p_σ and the residues have a fairly simple formula. When the parameters μ_i specialize or even become all 0, the poles coalesce and become higher-order poles. We will return on this point in Part IV.

8.1.5 The Filtration by Polar Order

We are now ready to analyze the filtration of the algebra R by polar order.

Theorem 8.10. R_k/R_{k-1} *is a direct sum of copies of the modules N_W as W runs over the subspaces of the arrangement of codimension k.*

For given W, the isotypic component of M_W is canonically isomorphic to the module $N_W \otimes \Theta_W$, where Θ_W is the vector space spanned by the elements $d_{\underline{c}}^{-1} := \prod_{a \in \underline{c}}(a + \mu_a)^{-1}$ as $\underline{c}$ runs over all the bases of X_W.

The space Θ_W has as basis the elements $d_{\underline{c}}^{-1}$ as $\underline{c}$ runs over all the unbroken bases in X_W.

Proof. We prove the statement by induction on k. For $k = 0$ we have as the only space of codimension 0 the entire space V. Then $R_0 = S[V] = N_V$ as desired.

Next, using the expansion (8.2) and Theorem 8.8, it follows that:

$$
R_k = \sum_{W, \underline{c}} R_{\underline{c}, \mu(\underline{c})}
$$

as W runs over the subspaces of the arrangement of codimension $\leq k$ and $\underline{c}$ over the unbroken bases in the subset X_W.

Consider, for each $W, \underline{c}$ with codimension $W = k$, the composed map $R_{\underline{c}, \mu(\underline{c})} \subset R_k \to R_k/R_{k-1}$.

Clearly, this factors to a map

$$i_{\underline{c}} : (R_{\underline{c}, \mu(\underline{c})})_k/(R_{\underline{c}, \mu(\underline{c})})_{k-1} \to R_k/R_{k-1}.$$

By the discussion in Section 8.1.2 we know that $(R_{\underline{c}, \mu(\underline{c})})_k/(R_{\underline{c}, \mu(\underline{c})})_{k-1}$ is identified with the irreducible module N_W. Thus either the map $i_{\underline{c}}$ is an injection or it is 0. In this last case, the module N_W appears in a composition series of the module R_{k-1}. By induction, R_{k-1} has a composition series for which each irreducible factor has as characteristic variety the conormal space to some subspace of codimension $\leq k-1$. If $i_{\underline{c}} = 0$, then N_W is isomorphic to one of these factors. This is a contradiction, since the characteristic variety of N_W is the conormal space to W, a space of codimension k. It follows that $i_{\underline{c}}$ is an injection. Let $\mathrm{Im}(i_{\underline{c}})$ denote its image.

As a consequence, we can deduce at least that

$$R_k/R_{k-1} = \sum_{W, \underline{c}} \mathrm{Im}(i_{\underline{c}})$$

as W runs over the subspaces of the arrangement of codimension k and $\underline{c}$ over the unbroken bases in the subset X_W.

Since for two different W_1, W_2 the two modules N_{W_1}, N_{W_2} are not isomorphic, it follows that for given W, the sum $\sum_{\underline{c}} \mathrm{Im}(i_{\underline{c}})$ as $\underline{c}$ runs over the unbroken bases in the subset X_W gives the isotypic component of type N_W of the module R_k/R_{k-1}, and this module is the direct sum of these isotypic components. Thus the first part of the theorem is proved, and in order to complete the proof, it is enough to verify that Θ_W is spanned by the elements $d_{\underline{c}}^{-1}$ and that the sum $\sum_{\underline{c}} \mathrm{Im}(i_{\underline{c}})$ is direct as $\underline{c}$ runs over the unbroken bases in the subset X_W.

The fact that the given elements $d_{\underline{c}}^{-1}$ span Θ_W follows by Theorem 8.8. In order to show that the sum $\sum_{\underline{c}} \mathrm{Im}(i_{\underline{c}})$ is direct, it is enough to verify, using the third part of Corollary 4.2 (which as noted extends to N_W), that the classes in R_k/R_{k-1} of the elements $\prod_{a \in \underline{c}}(a + \mu_a)^{-1}$ are linearly independent as $\underline{c}$ runs over the unbroken bases of X_W. This last point is nontrivial and requires a special argument, that we prove separately in the next proposition.

Proposition 8.11. *For given W of codimension k, the classes of the elements $\prod_{a \in \underline{c}}(a + \mu_a)^{-1}$ as $\underline{c} \subset X_W$ runs over the unbroken bases are linearly independent modulo R_{k-1}*

Proof. We first reduce to the case $k = s$ and $\mu = 0$.

Consider the algebra $R' := S[V][\prod_{a \in X_W}(a + \mu_a)^{-1}]$. Suppose we have already proved that the previous classes are linearly independent modulo R'_{k-1}.

We have certainly that $R'_{k-1} \subset R_{k-1}$, but in fact, $R'/R'_{k-1} \subset R/R_{k-1}$. Indeed, by what we have seen before, R'/R'_{k-1} is a direct sum of modules isomorphic to N_W and so with characteristic variety the conormal space of W. We have seen that N_W cannot appear in a composition series of R_{k-1}, so the mapping $R'/R'_{k-1} \to R/R_{k-1}$ must be injective.

In order to prove the proposition for R' we can further apply, to the algebra R', a translation automorphism (that preserves the filtration) and assume that $0 \in W$ or $\mu_a = 0$, $\forall a \in X_W$.

Finally, we can choose coordinates $x_1, \ldots, x_s$ such that all the equations $a \in X_W$ depend on the first k coordinates, and hence

$$R' = \mathbb{C}[x_1, \ldots, x_k][\prod_{a \in X_W} a^{-1}] \otimes \mathbb{C}[x_{k+1}, \ldots, x_s],$$

$$R'_{k-1} = \mathbb{C}[x_1, \ldots, x_k][\prod_{a \in X_W} a^{-1}]_{k-1} \otimes \mathbb{C}[x_{k+1}, \ldots, x_s],$$

$$R'/R'_{k-1}$$

$$= \mathbb{C}[x_1, \ldots, x_k]\Big[\prod_{a \in X_W} a^{-1}\Big]/\mathbb{C}[x_1, \ldots, x_k]\Big[\prod_{a \in X_W} a^{-1}\Big]_{k-1} \otimes \mathbb{C}[x_{k+1}, \ldots, x_s],$$

and we are reduced to the case $s = k$ and $\underline{\mu} = 0$, as desired. Let us thus assume we are in this case.

We are going to apply a method that actually comes from cohomology, so the reader may understand it better after reading the following chapters.

We need to work with top differential forms, rather than with functions. We need two facts:

(1) The forms $R_{s-1} dx_1 \wedge dx_2 \wedge \cdots \wedge dx_s$ are exact.
In fact, a form in $R_{s-1} dx_1 \wedge dx_2 \wedge \cdots \wedge dx_s$ is a linear combination of forms, each one of which, in suitable coordinates, is of type $\prod_{i=1}^{s} x_i^{h_i} dx_1 \wedge dx_2 \wedge \cdots \wedge dx_s$ where $h_i \in \mathbb{Z}$ and at least one of the h_i, say for instance h_s, is in $\mathbb{N}$. Then

$$\prod_{i=1}^{s} x_i^{h_i} dx_1 \wedge \cdots \wedge dx_s = (-1)^{s-1} d\Big((h_s + 1)^{-1} \prod_{i=1}^{s-1} x_i^{h_i} x_s^{h_s+1} dx_1 \wedge \cdots \wedge dx_{s-1}\Big).$$

(2) A residue computation.
Let $\underline{b} = (b_1, \ldots, b_s)$ be a basis extracted from X. Thus R_X is a subring of the ring of rational functions in the variables $b_1, \ldots, b_s$. To $\underline{b}$ we associate an injection

$$j_{\underline{b}} : R_X \to \mathbb{C}[[u_1, \ldots, u_s]][(u_1 \cdots u_s)^{-1}].$$

Here $\mathbb{C}[[u_1, \ldots, u_s]][(u_1 \cdots u_s)^{-1}]$ is the ring of formal Laurent series in the variables $u_1, \ldots, u_s$, defined by

$$j_{\underline{b}}(f(b_1, \ldots, b_s)) = f(u_1, u_1 u_2, \ldots, u_1 u_2 \cdots u_s).$$

Let us show that this is well-defined. Given $a \in X$, let $k = \gamma_{\underline{b}}(a)$ be the maximum index such that $a \in \langle b_k, \ldots, b_s \rangle$. We have

$$a = \sum_{j=k}^{s} \alpha_j b_j = u_1 \cdots u_k \left(\alpha_k + \sum_{j=k+1}^{s} \alpha_j \prod_{i=k+1}^{j} u_i \right) \tag{8.10}$$

with $\alpha_k \neq 0$, so that a^{-1} can be expanded as a Laurent series. Clearly, this map extends at the level of differential forms.

For a differential form ψ in $\mathbb{C}[[u_1, \ldots, u_s]][(u_1 \cdots u_s)^{-1}]du_1 \wedge \cdots \wedge du_s$ we define its residue $\mathrm{res}(\psi)$ to be the coefficient of $(u_1 \cdots u_s)^{-1}du_1 \wedge \cdots \wedge du_s$.

Lemma 8.12. *A form ψ with coefficients in $\mathbb{C}[[u_1, \ldots, u_s]][(u_1 \cdots u_s)^{-1}]$ is exact if and only if $\mathrm{res}(\psi) = 0$.*

Proof. It is clear that if we take a Laurent series and apply the operator $\frac{\partial}{\partial u_i}$ we get a Laurent series in which u_i never appears with the exponent -1. Conversely, such a Laurent series f is of the form $\frac{\partial g}{\partial u_i}$ for some easily computed g. The lemma follows by a simple induction.

As a consequence we have that the top forms with coefficients in R_{s-1} map to forms with zero residue.

Definition 8.13. Given a top differential form $\omega \in \Omega_X^s$ and an unbroken basis $\underline{b}$, we define $\mathrm{res}_{\underline{b}}(\omega)$ as the residue of $j_{\underline{b}}(\omega)$.

One can immediately extend this definition to the parametric case X, μ, a point $p \in P(X, \mu)$, and an unbroken basis $\underline{b}$ in X_p. The corresponding residue will be denoted by $\mathrm{res}_{\underline{b},p}(\omega)$.

Given a basis $\underline{b} := (b_1, \ldots, b_s)$, let us define

$$\omega_{\underline{b}} := d \log b_1 \wedge \cdots \wedge d \log b_k. \tag{8.11}$$

The main technical result from which all the next results follow is given by the following result of Szenes [108]:

Lemma 8.14. *Let $\underline{b}$ and $\underline{c}$ be two unbroken bases extracted from X. Then*

$$\mathrm{res}_{\underline{b}}(\omega_{\underline{c}}) = \delta_{\underline{b},\underline{c}}.$$

Proof. First notice that $j_{\underline{b}}(\omega_{\underline{b}}) = d \log u_1 \wedge \cdots \wedge d \log u_s$ so that $\mathrm{res}_{\underline{b}}(\omega_{\underline{b}}) = 1$.

With the notations of formula 8.10, notice that if $\underline{c} \neq \underline{b}$, there are two distinct elements $c, c' \in \underline{c}$ such that $\gamma_{\underline{b}}(c) = \gamma_{\underline{b}}(c')$. Indeed, since in an unbroken basis, the first element is always also the first element in the list X, $\gamma_{\underline{b}}(c_1) = 1$. If $\langle c_2, \ldots, c_s \rangle \neq \langle b_2, \ldots, b_s \rangle$, then there exists an index $i > 1$ with $\gamma_{\underline{b}}(c_i) = 1$. If $\langle c_2, \ldots, c_s \rangle = \langle b_2, \ldots, b_s \rangle$ then both $\underline{c} \setminus \{c_1\}$ and $\underline{b} \setminus \{b_1\}$ are unbroken bases of $\langle c_2, \ldots, c_s \rangle$ extracted from $X \cap \langle c_2, \ldots, c_s \rangle$ and everything follows by induction.

Given a c in $\underline{c}$ with $\gamma_{\underline{b}}(c) = k$, we write c in the form $(\prod_{i=1}^{k} u_i)f$ with $f(0) \neq 0$, so that $d\log c = d\log(\prod_{i=1}^{k} u_i) + d\log f$. Expand $j_{\underline{b}}(\omega_c)$ using the previous expressions. We get a linear combination of forms. All terms containing only factors of type $d\log(\prod_{i=1}^{k} u_i)$ vanish, since two elements are repeated. The others are a product of a closed form by a form $d\log(f)$ with $f(0) \neq 0$, which is exact. The product $a \wedge db$ of a closed form by an exact form equals $\pm d(a \wedge b)$, an exact form. So the residue is 0.

This clearly finishes the proof of Proposition 8.11, since the matrix $(\mathrm{res}_{\underline{b}}(\omega_{\underline{c}}))$ for $\underline{b}$ and $\underline{c}$ unbroken is the identity matrix.

Remark 8.15. In the case of generic parameters μ it is easy to see that for each subspace W of the arrangement, the module $\overline{N}_W$ appears in a composition series of R with multiplicity 1.

We now interpret the previous theorem as an expansion into partial fractions. Take a subspace W given in suitable coordinates $x_1, \ldots, x_s$ by the equations

$$W = \{x_1 - \gamma_1 = x_2 - \gamma_2 = \cdots = x_k - \gamma_k = 0\}$$

for suitable constants $\gamma_1, \gamma_2, \ldots, \gamma_k$. Generalizing Proposition 4.4, N_W is generated by an element δ_W satisfying

$$x_i \delta_W = \gamma_i \delta_W, \ i \leq k, \qquad \frac{\partial}{\partial x_i}\delta_W = 0, \ i > k.$$

Furthermore, as a module over $\mathbb{C}[\frac{\partial}{\partial x_1}, \ldots, \frac{\partial}{\partial x_k}, x_{k+1}, x_{k+2}, \ldots, x_s]$, N_W is free of rank 1 generated by δ_W.

The subring $A_W := \mathbb{C}[\frac{\partial}{\partial x_1}, \ldots, \frac{\partial}{\partial x_k}, x_{k+1}, x_{k+2}, \ldots, x_s] \subset W(U)$ does not depend only on W, but also on the choice of coordinates. Nevertheless, we may make such a choice for each W in the arrangement and obtain an explicit family of subrings indexed by the subspaces of the arrangement. Apply Theorem 8.10 and the previous remarks and notice that as δ_W we can take the class of the element $\prod_{a \in \underline{c}}(a + \mu_a)^{-1}$, we obtain the following result

Theorem 8.16. *We have a direct sum decomposition*

$$R_{X,\underline{\mu}} = \oplus_{W,\underline{c}} A_W \prod_{a \in \underline{c}}(a + \mu_a)^{-1}. \tag{8.12}$$

The explicit monomial basis for A_W produces, for each of the summands $A_W \prod_{a \in \underline{c}}(a + \mu_a)^{-1}$, a basis of explicit rational functions with denominators of type $\prod_{a \in \underline{c}}(a + \mu_a)^{h_a}, \ h_a < 0$.

This expansion is what we consider as an *expansion into partial fractions of the rational functions in* $R_{X,\underline{\mu}}$.

The filtration by polar order has the following interpretation.

Remark 8.17. $R_{X,k}$ is the maximal $W(U)$ submodule of R_X whose composition series is formed by irreducibles whose characteristic variety is the conormal bundle of subspaces of dimension $\geq s - k$.

The following proposition follows immediately:

Proposition 8.18. *If $Z \subset X$ is a sublist, then $R_{Z,k} = R_Z \cap R_{X,k}$, $\forall k$.*

8.1.6 The Polar Part

It is important to understand better the special case of $k = s$ and the module R/R_{s-1}, to which we sometimes refer as the *module of polar parts*. In this case, the irreducible modules appearing are associated to the subspaces of the arrangement of codimension s, that is, the points of the arrangement $P(X, \underline{\mu})$.

For $p \in P(X, \underline{\mu})$ and an unbroken basis $\underline{b} \subset X_p$, let us denote by $u_{\underline{b}}$ the class of the element $d_{\underline{b}}^{-1} = \prod_{a \in \underline{b}} (a + \mu_a)^{-1}$ in the quotient R/R_{s-1}.

We have that
$$f u_{\underline{b}} = f(p) u_{\underline{b}}, \ \forall f \in S[V],$$
and $u_{\underline{b}}$ generates a submodule isomorphic to N_p.

Lemma 8.19. *Given $v \in V$, the operator $v - \langle v \,|\, p \rangle$ is locally nilpotent on N_p.*

Proof. We can first translate p to 0 and reduce to prove that v is locally nilpotent on N_0. This is the Fourier transform of the algebra of polynomials, so the statement is equivalent to proving that a derivative is locally nilpotent on the space of polynomials, and this is obvious.

Recall that given a vector space V, a linear operator t, a number α, and a nonzero vector v, one says that v is a generalized eigenvector, of eigenvalue α, for t, if there is a positive integer k with $(t - \alpha)^k v = 0$. In a finite-dimensional vector space V, a set T of commuting operators has a canonical Fitting decomposition $V = \oplus_\alpha V_\alpha$, where the α are distinct functions on T and on V_α each $t \in T$ has only $\alpha(t)$ as an eigenvalue. The same statement is clearly true even if V is infinite-dimensional but the union of T stable finite-dimensional subspaces. Summarizing we obtain the results of Brion–Vergne [28]:

Theorem 8.20. *(1) The isotypic component of type N_p in R/R_{s-1} is the Fitting subspace for the commuting operators induced from V, where each vector $v \in V$ has generalized eigenvalue $\langle v \,|\, p \rangle$.*
(2) Any $S[V]$ submodule in R/R_{s-1} decomposes canonically into the direct sum of its intersections with the various isotypic components.
(3) Each isotypic component N_p is a free $S[U]$-module with basis the elements $u_{\underline{b}}$, as $\underline{b}$ runs over the unbroken bases of X_p.

Corollary 8.21. *Let Ψ_X be the linear subspace of R spanned by all the fractions $F := \prod_{i=1}^{k}(b_i + \mu_{b_i})^{-h_i}$, where the elements b_i form an ordered sublist extracted from the list $X = (a_1, \ldots, a_m)$, the elements b_i span V, and all $h_i > 0$.*

Then Ψ_X is a free $S[U]$-module with basis the element $d_{\underline{b}}^{-1}$ as $\underline{b}$ runs over the unbroken bases relative to some point of the arrangement.

As $S[U]$-module, $R = \Psi_X \oplus R_{s-1}$.

Proof. From Theorem 8.8, the space Ψ_X is also spanned by all the fractions $F := \prod_{i=1}^{k}(b_i + \mu_{b_i})^{-h_i}$ with all $h_i > 0$ and where the elements b_i form an unbroken basis. By the previous theorem, the classes of these elements give a basis of R/R_{s-1} therefore these elements must give a basis of Ψ_X and Ψ_X maps isomorphically to R/R_{s-1}.

8.1.7 Two Modules in Correspondence

The theory of the Laplace transform tells us how to transform some basic manipulations on distributions as an algebraic calculus. In our setting, this is best seen by introducing the following two D-modules in Fourier duality:

The first is the D-module $\mathcal{D}_{X,\underline{\mu}} := W(V)T_{X,\underline{\mu}}$ generated, in the space of tempered distributions on V, by $T_{X,\underline{\mu}}$ under the action of the algebra $W(V)$ (of differential operators on V with polynomial coefficients).

The second D-module is the algebra $R_{X,\underline{\mu}} := S[V][\prod_{a \in X}(a + \mu_a)^{-1}]$, of rational functions on U, obtained from the polynomials on U by inverting the element $d_{X,\underline{\mu}} := \prod_{a \in X}(a + \mu_a)$. As we have seen, this is a module under $W(U)$, and it is the coordinate ring of the open set $\mathcal{A}_{X,\underline{\mu}}$, complement of the union of the affine hyperplanes of U of equations $a = -\mu_a$, $a \in X$.

Theorem 8.22. *Under the Laplace transform, $\mathcal{D}_{X,\underline{\mu}}$ is mapped isomorphically onto $R_{X,\underline{\mu}}$. In other words, we get a canonical isomorphism of $\hat{\mathcal{D}}_{X,\underline{\mu}}$ with the algebra $R_{X,\underline{\mu}}$ as $W(U)$-modules.*

Proof. The injectivity of the Laplace transform on $\mathcal{D}_{X,\underline{\mu}}$ follows from the isomorphism of the Fourier transform on the Schwartz space and Proposition 3.5.

To see the surjectivity, notice, by definition and formula (7.9), that the image of $\mathcal{D}_{X,\underline{\mu}}$ under the Laplace transform is the smallest D-module in the field of rational functions on U containing $d_{X,\underline{\mu}}^{-1}$. Since $R_{X,\underline{\mu}}$ contains $d_{X,\underline{\mu}}^{-1}$, it suffices to see that $d_{X,\underline{\mu}}^{-1}$ generates $R_{X,\underline{\mu}}$ as a D-module.

From formula (8.12), since all the A_W are contained in $W(V)$, it suffices to see that the elements $\prod_{a \in \underline{c}}(a + \mu_a)^{-1}$ are in the $W(V)$-module generated by $d_{X,\underline{\mu}}^{-1}$. This is trivial, since $\prod_{a \in \underline{c}}(a + \mu_a)^{-1} = \prod_{a \notin \underline{c}}(a + \mu_a)d_{X,\underline{\mu}}^{-1}$.

From the basic properties of the Laplace transform and formula (8.12) it is easy to understand the nature of the tempered distributions in the space $\mathcal{D}_{X,\underline{\mu}} = \oplus_{W,\underline{c}} L^{-1}(A_W \prod_{a \in \underline{c}}(a + \mu_a)^{-1})$.

In fact, take a linearly independent (or unbroken) set $\underline{c} \subset X$. We know by formula (3.4) that $\prod_{a \in \underline{c}}(a + \mu_a)^{-1}$ is the Laplace transform of the distribution supported in the cone generated by the elements of $\underline{c}$ and that on this cone is the function e^ℓ, where ℓ is the unique linear function on the span of $\underline{c}$ such that $\ell(a) = -\mu_a$, $\forall a \in \underline{c}$.

In particular, when $\underline{c}$ is a basis, the equations $a + \mu_a = 0$, $a \in \underline{c}$, define a point $p \in P(X, \underline{\mu})$ of the arrangement, and $\ell(v) = \langle v | p \rangle$, $\forall v \in V$. So we think of p as a function on V and write equivalently e^p.

When we apply the operators of A_W, we either multiply by polynomials or take suitable derivatives.

In particular, for $\underline{\mu} = 0$ we get tempered distributions that are linear combinations of polynomial functions on the cones $C(A)$, $A \subset X$ a linearly independent subset and their distributional derivatives.

From the proof of Theorem 8.22 one deduces that the corresponding (under the inverse Laplace transform) filtration on $\mathcal{D}_X$ can be described geometrically as follows.

We cover $C(X)$ by the positive cones $C(A)$ spanned by linearly independent subsets of $A \subset X$. We define $C(X)_k$ to be the k-dimensional skeleton of the induced stratification, a union of k-dimensional cones. We then get the following

Proposition 8.23. $\mathcal{D}_{X,k}$ *consists of the tempered distributions in* $\mathcal{D}_X$ *whose support is contained in* $C(X)_k$.

9

The Function T_X

As the title says, this chapter is devoted to the study of the functions T_X, for which we propose formulas and algorithms that compute its values. The material is mostly a reinterpretation of results by Brion–Vergne (see [28]).

9.1 The Case of Numbers

9.1.1 Volume

Before we plunge into the combinatorial difficulties of the higher dimensional theory let us discuss the simple case in which $s = 1$. In other words, X is a row vector $\underline{h} := (h_1, \dots, h_m)$ of positive numbers. We want to compute $T_{\underline{h}}(x)$ or the volume function whose Laplace transform is $|\underline{h}| \prod h_i^{-1} y^{-m}$ with $|\underline{h}| = \sqrt{\sum_i^m h_i^2}$. The computation of the function whose Laplace transform is y^{-m} is rather easy. If χ denotes the characteristic function of the half-line $x \geq 0$, we have

$$L(\chi) = y^{-1}.$$

So we deduce

$$L((-x)^k \chi) = \frac{d^k y^{-1}}{dy^k} = (-1)^k k! y^{-k-1}.$$

Thus

$$T_{\underline{h}}(x) = \begin{cases} 0 & \text{if} \quad x < 0 \\ \frac{x^{m-1}}{(m-1)! \prod h_i} & \text{if} \quad x \geq 0. \end{cases} \tag{9.1}$$

- The polyhedron $\Pi_{\underline{h}}(x)$ is a *simplex*, given by:

$$\left\{ (t_1, \dots, t_m) \,\middle|\, t_i \geq 0, \ \sum_i t_i h_i = x \right\}.$$

- $\Pi_{\underline{h}}(x)$ is the convex envelope of the vertices $P_i = (0, 0, \dots, 0, x/h_i, 0, \dots, 0)$.

C. De Concini and C. Procesi, *Topics in Hyperplane Arrangements, Polytopes and Box-Splines*, Universitext, DOI 10.1007/978-0-387-78963-7_9,
© Springer Science+Business Media, LLC 2010

We could compute the volume of $\Pi_{\underline{h}}(x)$ directly but we can use the previous formula getting

Theorem 9.1. *The volume of $\Pi_{\underline{h}}(x)$ is given by*

$$\frac{\sqrt{\sum_i^m h_i^2}\, x^{m-1}}{(m-1)!\, \prod h_i}.$$

For the box spline one uses formula (7.10).

The special case $\underline{h} = \{1, \ldots, 1\}$, where 1 appears $m+1$-times has been discussed in Example 6.2, where we denoted by $t_m(x), b_m(x)$ the corresponding splines.

9.2 An Expansion

9.2.1 Local Expansion

With the notation of the previous chapters we have the following proposition, which allows us to reduce the computation of $T_{X,\underline{\mu}}$ to that of the various T_{X_p}, for $p \in P(X, \underline{\mu})$.

Proposition 9.2.

$$T_{X,\underline{\mu}}(x) = \sum_{p \in P(X,\underline{\mu})} c_p e^p T_{X_p}(x) \tag{9.2}$$

with c_p given by the formula

$$c_p = \prod_{a \in X \setminus X_p} \frac{1}{\langle a \mid p \rangle + \mu_a}.$$

Proof. We apply formula (8.5):

$$\prod_{a \in X} \frac{1}{a + \mu_a} = \sum_{p \in P(X,\underline{\mu})} c_p \prod_{a \in X_p} \frac{1}{a + \mu_a} = \sum_{p \in P(X,\underline{\mu})} c_p \prod_{a \in X_p} \frac{1}{a - \langle a \mid p \rangle}.$$

Using (7.9), we then see that $\prod_{a \in X_p} (a - \langle a \mid p \rangle)^{-1}$ is the Laplace transform of $e^p T_{X_p}$. From this the claim follows.

This together with formula (7.10) gives for the box spline

$$B_{X,\underline{\mu}}(x) = \sum_{p \in P(X,\underline{\mu})} c_p \sum_{S \subset X \setminus X_p} (-1)^{|S|} e^{-\mu_S + p} B_{X_p}(x - a_S).$$

Of course, this is also a reformulation of the identity of the Laplace transforms:

$$\prod_{a \in X} \frac{1 - e^{-a - \mu_a}}{a + \mu_a} = \sum_{p \in P(X,\underline{\mu})} c_p \prod_{a \in X \setminus X_p} (1 - e^{-a - \mu_a}) \prod_{a \in X_p} \frac{1 - e^{-a - \mu_a}}{a + \mu_a}.$$

As a consequence, of these formulas the essential problem is the determination of T_X in the nonparametric case.

9.2.2 The Generic Case

Before we start the general discussion, let us make explicit in the generic case the formulas that we have introduced in Section 8.1.4. Let us recall the formula (8.9), that

$$\prod_{i=1}^{m} \frac{1}{(a_i + \mu_i)} = \sum_{\sigma} \frac{1}{\prod_{i=1,\ i\notin\sigma}^{m}(\langle a_i \,|\, p_\sigma\rangle + \mu_i)} \prod_{k\in\sigma} \frac{1}{a_k + \mu_k}.$$

Since we are in the generic case, the set $P(X,\mu)$ coincides with the set of points p_σ as $\sigma \in \mathcal{B}_X$ runs over the subsets indexing bases. For such a p_σ we have that X_{p_σ} coincides with the basis a_i, $i \in \sigma$, and $T_{X_{p_\sigma}} = L^{-1}(\prod_{k\in\sigma} a_k^{-1})$ is equal to $v_\sigma^{-1}\chi_\sigma$, where χ_σ is the characteristic function of the positive cone C_σ generated by the vectors a_i, $i \in \sigma$ and v_σ the volume of the parallelepiped they generate. Finally, p_σ is given by the coordinates $a_i = -\mu_i$, $i \in \sigma$. We deduce the following explicit formula.

Proposition 9.3. *On a point x the function $T_{X,\mu}(x)$ equals the sum*

$$\sum_{\sigma\,|\ x\in C_\sigma} b_\sigma e^{\langle x\,|\,p_\sigma\rangle}, \quad b_\sigma := \frac{1}{v_\sigma \prod_{i=1,\ i\notin\sigma}^{m}(\langle a_i \,|\, p_\sigma\rangle + \mu_i)}.$$

We have thus the list of functions $b_\sigma e^{\langle x\,|\,p_\sigma\rangle}$, the set of σ such that $x \in C_\sigma$ is constant on each big cell. So on each big cell, $T_{X,\mu}(x)$ is a sum of a sublist of the functions $b_\sigma e^{\langle x\,|\,p_\sigma\rangle}$.

Observe that the points p_σ depend on the parameters μ_i and can be explicited by Cramer's rule. Formula (8.8) gives a determinantal expression for the terms $\langle a_i \,|\, p_\sigma\rangle + \mu_i$.

Now let us vary the generic parameters as $t\mu_i$ with t a parameter. We then have that the corresponding points p_σ vary as tp_σ.

The factors $v_\sigma \prod_{i=1,\ i\notin\sigma}^{m}(\langle a_i \,|\, p_\sigma\rangle + \mu_i)$ vary as

$$v_\sigma \prod_{i=1,\ i\notin\sigma}^{m} (\langle a_i \,|\, tp_\sigma\rangle + t\mu_i) = t^{m-s} v_\sigma \prod_{i=1,\ i\notin\sigma}^{m} (\langle a_i \,|\, p_\sigma\rangle + \mu_i)$$

while the numerator is given by

$$e^{\langle x\,|\,tp_\sigma\rangle} = \sum_{k=0}^{\infty} t^k \langle x \,|\, p_\sigma\rangle^k / k!.$$

From formula (7.7) it follows that $T_{X,t\mu}(x)$ is a holomorphic function in t around 0 and that its value for $t = 0$ is the (normalized) volume $T_X(x)$ of the polytope $\Pi_X(x)$. We deduce that

$$\sum_{\sigma,\ x\in C_\sigma} \frac{\langle x \,|\, p_\sigma\rangle^k}{v_\sigma \prod_{i=1,\ i\notin\sigma}^{m}(\langle a_i \,|\, p_\sigma\rangle + \mu_i)} = 0, \quad \forall k < m - s$$

and a *formula for the volumes,*

$$T_X(x) = \sum_{\sigma,\ x \in C_\sigma} \frac{\langle x \mid p_\sigma \rangle^{m-s}}{(m-s)! v_\sigma \prod_{i=1,\ i \notin \sigma}^{m} (\langle a_i \mid p_\sigma \rangle + \mu_i)}. \tag{9.3}$$

This formula (from [27]) contains parameters μ_i and the corresponding p_σ (determined by a generic choice of the μ_i), although its value $T_X(x)$ is independent of these parameters.

This formula, although explicit, is probably not the best, since it contains many terms (indexed by all bases σ, with $x \in C_\sigma$) and the not so easy to compute auxiliary p_σ. We shall discuss in the next section a direct approach to the volume.

Example 9.4. Take $X = \{1, \ldots, 1\}$, m times. Then the formula gives, for $x \geq 0$,

$$\sum_{j=1}^{m} \frac{(-x\mu_j)^{m-1}}{(m-1)! \prod_{i \neq j, i=1}^{m} (-\mu_j + \mu_i)} = \frac{x^{m-1}}{(m-1)!}$$

and the identity:

$$\sum_{j=1}^{m} \frac{\mu_j^{m-1}}{\prod_{i \neq j, i=1}^{m} (-\mu_j + \mu_i)} = (-1)^{m-1},$$

which one can prove directly by developing the determinant of the Vandermonde matrix along the first row $(\mu_1^{m-1}, \ldots, \mu_m^{m-1})$.

The other identities

$$\sum_{j=1}^{m} \frac{\mu_j^{k}}{\prod_{i \neq j, i=1}^{m} (-\mu_j + \mu_i)} = 0, \qquad \forall k < m - 1,$$

come by developing the determinant of the Vandermonde matrix once we substitute the first row with one of the following rows.

9.2.3 The General Case

We can now pass to the general case. By Proposition 9.2 our essential problem is the determination of T_X in the nonparametric case. In view of this, we assume that we indeed are in this case, and we are going to state and prove the main formula that one can effectively use for computing the function T_X. To be concrete, we may assume that we have chosen coordinates $x_1, \ldots, x_s$ for U.

Let us denote by $\mathcal{NB}$, or if needed $\mathcal{NB}_X$, the set of all unbroken bases extracted from X.

Theorem 9.5. *There exist uniquely defined polynomials $p_{\underline{b},X}(x)$, homogeneous of degree $|X| - s$ and indexed by the unbroken bases in X, with*

$$\frac{1}{d_X} = \sum_{\underline{b} \in \mathcal{NB}} p_{\underline{b},X}(\partial_x) \frac{1}{d_{\underline{b}}}, \qquad d_{\underline{b}} := \prod_{a \in \underline{b}} a. \tag{9.4}$$

Proof. The element d_X^{-1} lies in Ψ_X and by Corollary 8.21, the elements $\frac{1}{d_{\underline{b}}}$, with $\underline{b}$ unbroken bases, form a basis of Ψ_X as an $S[U]$-module. The claim follows.

Example 9.6. We write the elements of X as linear functions.
Take first

$$X = [x + y, x, y] = [x, y] \begin{vmatrix} 1 & 1 & 0 \\ 1 & 0 & 1 \end{vmatrix}$$

$$\frac{1}{(x + y)\, x\, y} = \frac{1}{y\, (x + y)^2} + \frac{1}{x\, (x + y)^2}$$

$$= -\frac{\partial}{\partial x} \frac{1}{y\, (x + y)} - \frac{\partial}{\partial y} \frac{1}{x\, (x + y)}.$$

Next take the more complicated list:

$$X = [x + y, x, -x + y, y] = [x, y] \begin{vmatrix} 1 & 1 & -1 & 0 \\ 1 & 0 & 1 & 1 \end{vmatrix},$$

$$\frac{1}{(x + y)\, x\, y\, (-x + y)} = -\frac{1}{y\, (x + y)^3} + \frac{4}{(x + y)^3(-x + y)} + \frac{1}{x\, (x + y)^3}$$

$$= \frac{1}{2}\left[-\frac{\partial^2}{\partial^2 x} \frac{1}{y\, (x + y)} + \left(\frac{\partial}{\partial x} + \frac{\partial}{\partial y}\right)^2 \left(\frac{1}{(x + y)(-x + y)}\right) + \frac{\partial^2}{\partial^2 y} \frac{1}{x\, (x + y)}\right].$$

9.3 A Formula for T_X

9.3.1 Jeffrey–Kirwan Residue Formula

We can now interpret Theorem 9.5 as follows:

Theorem 9.7. *Given a point x in the closure of a big cell Ω we have*

$$T_X(x) = \sum_{\underline{b}\,|\,\Omega \subset C(\underline{b})} |\det(\underline{b})|^{-1} p_{\underline{b},X}(-x), \tag{9.5}$$

where for each unbroken basis $\underline{b}$, $p_{\underline{b},X}(x)$ is the homogeneous polynomial of degree $|X| - s$ uniquely defined in Theorem 9.5.

Proof. We apply the inversion of the Laplace transform (cf. (7.16), (7.9)) to formula (9.4), obtaining

$$L^{-1}(d_A^{-1}) = \sum_{\underline{b} \in \mathcal{NB}} p_{\underline{b},A}(-x_1, \ldots, -x_n)| \det(\underline{b})|^{-1} \chi_{C(\underline{b})}.$$

This function is clearly continuous only on the set of regular points and a priori coincides with T_X only outside a subset of measure zero. By the continuity of T_X in $C(X)$ (part 3 of Proposition 7.17), it is sufficient to prove our claim in the interior of each big cell. This last fact is clear.

Example 9.8. In example 9.6 we have thus that for the list $X = [x+y, x, y]$,

$$\frac{1}{(x+y)\,x\,y} = \frac{1}{y\,(x+y)^2} + \frac{1}{x\,(x+y)^2}$$

$$= -\frac{\partial}{\partial x}\frac{1}{y\,(x+y)} - \frac{\partial}{\partial y}\frac{1}{x\,(x+y)}.$$

$$L^{-1}(d_X^{-1}) = x\chi_{C((0,1),(1,1))} + y\chi_{C((1,0),(1,1))}.$$

For the more complicated list $X = [x+y, x, -x+y, y]$, we have

$$L^{-1}(d_X^{-1}) = \frac{1}{2}[-x^2\chi_{C((0,1),(1,1))} + \frac{(x+y)^2}{2}\chi_{C((1,1),(-1,1))} + y^2\chi_{C((1,0),(1,1))}].$$

The corresponding geometric description is as follows

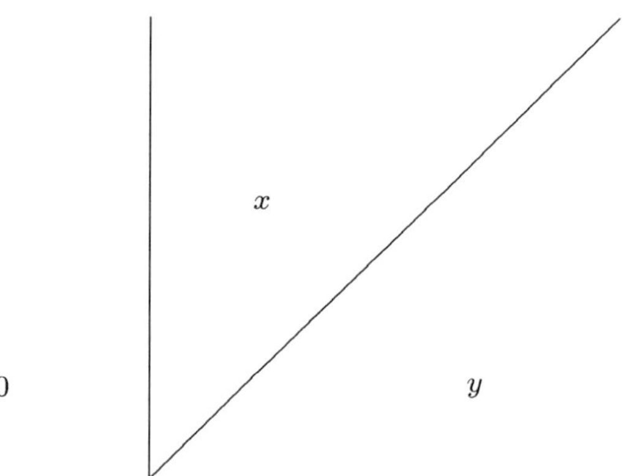

Fig. 9.1. The multivariate spline, Courant element

For the first list and for the second we have for $2T_X$ the picture (9.2)

The picture for the box splines is more complex. We can obtain it from T_X using formula (7.11). In the literature the box spline corresponding to the first list is called the *Courant element* or *hat function*, while the second is referred to as the *Zwart–Powell* or *ZP element*. Observe that the first list

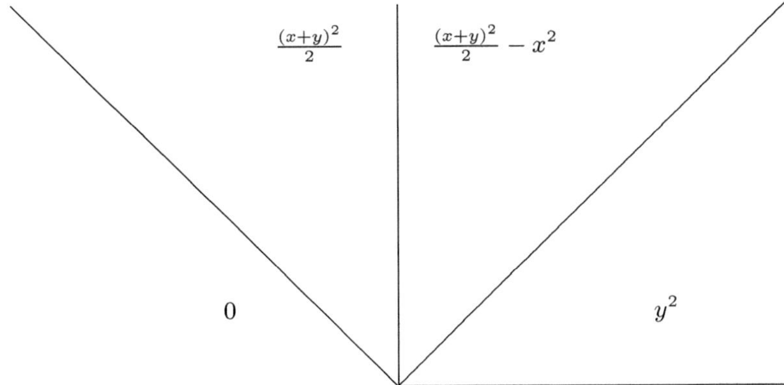

Fig. 9.2. The multivariate spline $2T_X$, ZP element

is (after a linear change of coordinates) the set of positive roots for type A_2, i.e., the vectors

$$(1,0), \ \left(\frac{1}{2}, \frac{\sqrt{3}}{2}\right), \ \left(-\frac{1}{2}, \frac{\sqrt{3}}{2}\right),$$

while the second is the set of positive roots for type B_2, and so we can apply the symmetry remarks of Section 2.4.1.

For A_2 the box becomes the regular hexagon, and after a shift that centers it at zero, the box spline becomes invariant under the full group S_3 of symmetries of the hexagon.

In fact, one could also consider the case of G_2, that has a symmetry group with 12 elements, and positive roots

$$(1,0), \ \left(\frac{1}{2}, \frac{\sqrt{3}}{2}\right), \ \left(-\frac{1}{2}, \frac{\sqrt{3}}{2}\right), \ \left(\frac{3}{2}, \sqrt{3}\right), \ \left(0, \sqrt{3}\right), \ \left(-\frac{3}{2}, \sqrt{3}\right).$$

In the first case we have the formula:

$$B_X(p) = T_X(p) - T_X(p - (1,0)) - T_X(p - (0,1)) - T_X(p - (1,1))$$

$$+ T_X(p - (1,1)) + T_X(p - (2,1)) + T_X(p - (1,2)) - T_X(p - (2,2))$$

$$= T_X(p) - T_X(p - (1,0)) - T_X(p - (0,1))$$

$$+ T_X(p - (2,1)) + T_X(p - (1,2)) - T_X(p - (2,2)).$$

Thus B_X is represented by Figure 9.3, (the regions are the regions of polynomiality on the box and inside each of them we write the corresponding polynomial).

In the second case we have the formula (after simplifications):

$$B_X(p) = T_X(p) - T_X(p - (1,0)) - T_X(p - (-1,1)) + T_X(p - (-1,2))$$

$$+ T_X(p - (2,1)) - T_X(p - (2,2)) - T_X(p - (0,3)) + T_X(p - (1,3)).$$

$$X = \begin{vmatrix} 1 & 0 & 1 & -1 \\ 0 & 1 & 1 & 1 \end{vmatrix}$$

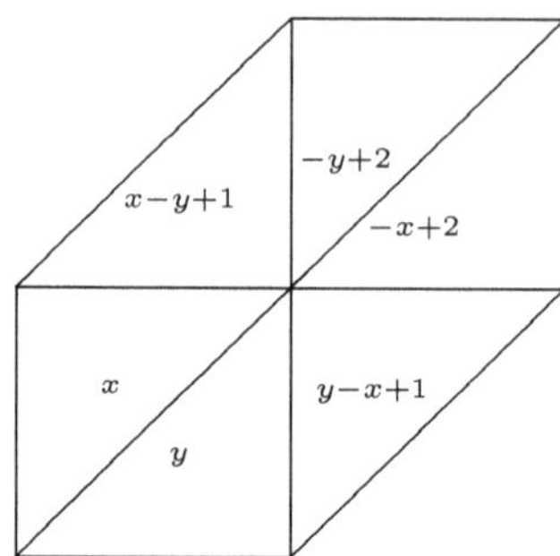

Fig. 9.3. The box spline $2B_X$, Courant element

Thus $2B_X$ is represented in figure 9.4 (the regions are the regions of polynomiality on the box and inside each of them we write the corresponding polynomial).

Remark 9.9. The symmetry in the picture (9.4) can best be understood by recalling that according to Section 2.4.1, the shifted box spline is invariant under the reflection group associated to the vectors in X that are the positive roots of B_2. This is the dihedral group with eight elements. The picture shows clearly the content of Theorem 2.56, that is, the extremal points of the shifted box are the (free) orbit under the dihedral group of the vertex ρ_X.

For G2 the computations have been done by Marco Caminati and are available on request. We get in Figure 9.5 a dodecahedron divided into 115 regions of polynomiality. From now on, unless there is a possibility of confusion, we shall write $p_{\underline{b}}$ instead of $p_{\underline{b},X}$.

Remark 9.10. (1) A residue formula for the polynomial $p_{\underline{b},X}$ will be given in formula (10.10).

(2) The polynomials $p_{\underline{b}}(x)$, with $\underline{b}$ an unbroken basis, will be characterized by differential equations in Section 11.10.

(3) We will also show (Corollary 17.7) that the polynomials $p_{\underline{b}}(x)$ are a basis of the space spanned by the polynomials coinciding with T_X on the big cells.

(4) It is easily seen that the number of big cells is usually much larger than the dimension of this last space. This reflects into the fact that the polynomials coinciding with T_X on the big cells satisfy complicated linear dependency relations that can be determined by the incidence matrix between big cells and cones generated by unbroken bases.

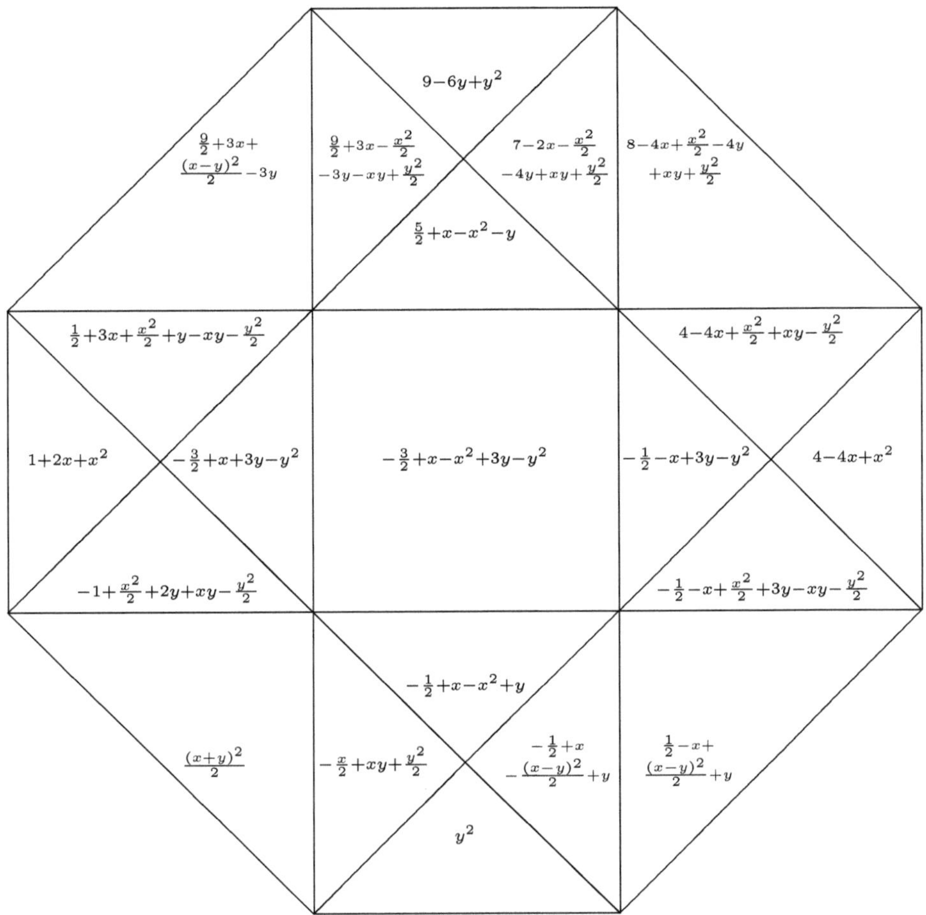

Fig. 9.4. The box spline $2B_X$, ZP element

Remark 9.11. The decomposition into big cells, which is essentially a geometric and combinatorial picture, corresponds to a decomposition of the cone $C(X)$ into regions in which the multivariate spline T_X has a polynomial nature (see 11.35). Moreover, the contributions to the polynomials on a given cell Ω come only from the unbroken bases $\underline{b}$ such that $\Omega \subset C(\underline{b})$.

This is in accordance with Theorem 9.16, that we will prove presently. This theorem states that one can use only the quadrants associated to unbroken bases in order to decompose the cone into big cells.

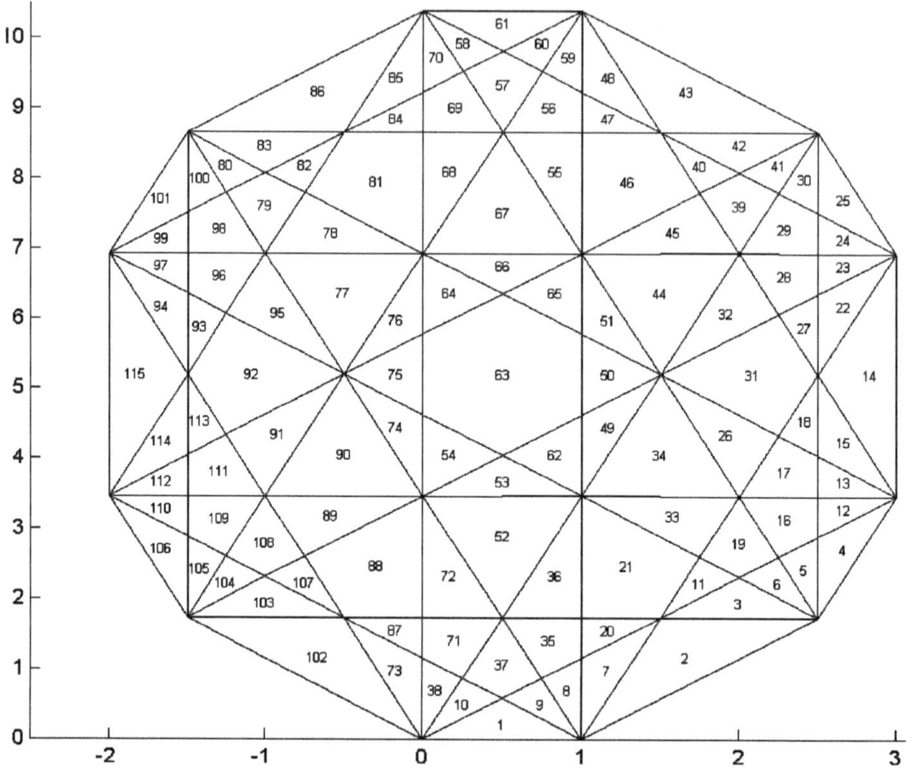

Fig. 9.5. The box spline for G_2

9.4 Geometry of the Cone

9.4.1 Big Cells

Formula (9.5) depends on the choice of an ordering of X and so on the corresponding unbroken bases, while T_X is clearly intrinsic. This suggests the possibility that the stratification of $C(X)$ induced by the big cells and their faces may be constructed by using only the stratifications associated to the cones $C(\underline{b})$ when $\underline{b}$ is unbroken. In order to prove this we need a few simple combinatorial lemmas on polytopes whose proof we recall for completeness.

First let us observe that the cone, the stratification, and the notion of unbroken do not change if we rescale the vectors in X by positive numbers. We may thus rescale, using our assumptions on X, each a_i to a vector v_i, such that all these vectors lie in an affine hyperplane Π of equation $\langle \phi, x \rangle = 1$ for some linear form ϕ.

The intersection of the cone $C(X)$ with Π is the convex polytope Σ, the envelope of the vectors v_i. Each cone, generated by $k+1$ independent vectors in X, intersects Π in a k-dimensional simplex. In particular, the strongly regular points of $C(X)$ are obtained by projecting the points in Σ that are not contained in any $(s-2)$-dimensional simplex generated by vectors in X.

Consider the following stratifications of Σ. Let us choose a family $\mathcal{I}$ of bases extracted from X. For each such basis $\underline{b}$ consider the family $\mathcal{F}_{\underline{b}}$ of relative interiors of faces of the simplex $\sigma_{\underline{b}}$ generated by $\underline{b}$. In other words we consider the relatively open simplices $\overset{\circ}{\sigma}_{\underline{c}}, \underline{c} \subset \underline{b}$, and set $\mathcal{F}_{\mathcal{I}} = \cup_{\underline{b} \in \mathcal{I}} \mathcal{F}_{\underline{b}}$. Given a point $v \in \Sigma$ we set $Z(v) = \{ f \in \mathcal{F}_{\mathcal{I}} | v \in f \}$ and define an equivalence relation $R_{\mathcal{I}}$ on Σ by setting v and w as equivalent (or belonging to the same stratum) if $Z(v) = Z(w)$.

We want to compare the equivalence relations $R_{\mathcal{I}}$ for various choices of $\mathcal{I}$. We start with a special case, assuming that the set X consists of $s+1$ vectors $v_0, \ldots, v_s$. We set $\mathcal{I}$ equal to the family of all bases formed by elements in X. Having chosen $0 \le j \le s$, we set $\mathcal{I}_j$ to be equal to the family of all bases formed by elements in X and that contain the vector v_j. By Theorem 1.49 $\Sigma = \cup_{\underline{b} \in \mathcal{I}} \sigma_{\underline{b}}$. This is refined as

Lemma 9.12. $\Sigma = \cup_{\underline{b} \in \mathcal{I}_j} \sigma_{\underline{b}}.$

Proof. By suitably reordering, we can assume that $j = s$. If $v_0, \ldots v_{s-1}$ are not linearly independent, the claim follows from Lemma 1.49.

Let us now suppose that $\underline{b} = \{ v_0, \ldots v_{s-1} \}$ is a basis of V. Take any $v \in \sigma_{\underline{b}} \setminus \{ v_s \}$. Consider the line ℓ joining v with v_s. This line intersects $\sigma_{\underline{b}}$ in a segment $[a, b]$, and (by Lemma 1.3) it intersects Σ in the convex envelope of the three points a, b, v_s. Then v is either in the segment $[a, v_s]$ or in $[b, v_s]$. Since a, b are in $(s-2)$-dimensional faces of $\sigma_{\underline{b}}$, we deduce that v lies in the convex envelope of an $(s-2)$-dimensional face τ of $\sigma_{\underline{b}}$ and v_s. If v_s is independent of τ, we have thus an $(s-1)$-dimensional simplex having v_s as a vertex in which v lies. Otherwise, by induction, v lies in any case in some simplex having v_s as a vertex, which is then contained in a larger $(s-1)$-dimensional simplex.

Lemma 9.13. *Let σ be a simplex, q a point in the interior of a face τ of σ, and p a point different from q. Assume that the segment $[p, q]$ intersects σ only in q. Then the convex hull of τ and p is a simplex having τ as a codimension-one face and meeting σ in τ.*

Proof. If p does not lie in the affine space spanned by σ, the statement is obvious.

Otherwise, we can assume that σ is the convex hull of the basis vectors $e_1, \ldots, e_m$, τ is the face of vectors with nonzero (positive) coordinates for $i \le k$, and p lies in the affine space generated by $e_1, \ldots, e_m$.

The condition that $\{ tq + (1-t)p \mid 0 \le t \le 1 \} \cap \sigma = \{ q \}$ is equivalent to the fact that there is an i larger than k such that the i-th coordinate of p is negative. This condition does not depend on the point $q \in \tau$ and shows that p is affinely independent of τ. Since when $0 \le t < 1$ this coordinate remains negative, our claim follows.

Lemma 9.14. *With the same notation as above, the equivalence relations $R_\mathcal{I}$ and $R_{\mathcal{I}_j}$ coincide.*

Proof. As before, assume that $j = s$. Observe that a simplex $\mathring{\sigma}_{\underline{c}} \in \mathcal{F}_\mathcal{I} \setminus \mathcal{F}_{\mathcal{I}_s}$ if and only if $\underline{c} \subset \underline{b}$ and v_s is in the affine envelope of $\underline{c}$. Thus, if $\underline{b} = (v_0, \ldots, v_{s-1})$ is not a basis, then $\mathcal{I} = \mathcal{I}_s$, and there is nothing to prove.

Otherwise, assume that $\underline{b} = (v_0, \ldots, v_{s-1})$ is a basis and take a such a simplex $\mathring{\sigma}_{\underline{c}} \in \mathcal{F}_\mathcal{I} \setminus \mathcal{F}_{\mathcal{I}_s}$; thus $\underline{c} \subset \underline{b}$ and v_s is in the affine hull of $\underline{c}$.

If two vectors $v, w \in \Sigma$ are congruent under $R_{\mathcal{I}_s}$ and $v \in \mathring{\sigma}_{\underline{c}}$, we need to show that also $w \in \mathring{\sigma}_{\underline{c}}$. Notice that by the equivalence $R_{\mathcal{I}_j}$, w is in the affine envelope of $\sigma_{\underline{c}}$, so if $\underline{c} \subsetneq \underline{b}$, we can work by induction on the dimension.

We may thus assume that $\underline{c} = \underline{b}$.

If v_s lies in the simplex $\sigma_{\underline{b}}$, the set Σ coincides with $\sigma_{\underline{b}}$. Then one easily sees that each face of $\sigma_{\underline{b}}$ is stratified by elements in $\mathcal{F}_{\mathcal{I}_s}$ and there is nothing to prove.

Assume now $v_s \notin \sigma_{\underline{b}}$. If $v_s = w$, the fact that $v \cong v_s$ for the equivalence $R_{\mathcal{I}_s}$ means that $v = v_s = w$ and we are done. So now assume $v_s \neq w$ and by contradiction $w \notin \mathring{\sigma}_{\underline{b}}$. The half-line starting from v_s and passing through w meets for the first time the simplex $\sigma_{\underline{b}}$ in a point u that is in the interior of some proper face τ of $\sigma_{\underline{b}}$ and w is in the half-open segment $(v_s, u]$. By the previous lemma, the convex hull of v_s and τ is a simplex ρ and meets $\sigma_{\underline{b}}$ exactly in τ. Both $\mathring{\rho}$ and $\mathring{\tau}$ belong to $\mathcal{F}_{\mathcal{I}_s}$ by construction, and $w \in \mathring{\rho} \cup \mathring{\tau}$. So the same is true for v and $v \notin \mathring{\sigma}_{\underline{b}}$, a contradiction.

Let us now return to our set of vectors X with a fixed ordering. Set $\mathcal{I}$ equal to the family of all bases that can be extracted from X, and as before, $\mathcal{N}B$ equal to the family of unbroken bases with respect to the chosen ordering.

Choose an element $a \in X$ and assume that the element b is the successor of a in our ordering. Define a new ordering by exchanging a and b. The following lemma tells us how the set $\mathcal{N}B$ changes.

Lemma 9.15. *An unbroken basis $\sigma := (a_1, \ldots, a_s)$ for the first order remains unbroken for the second unless all of the following conditions are satisfied:*

(i) $a = a_i$ appears in σ.

(ii) b does not appear in σ.

(iii) b is dependent on the elements $a_j, j \geq i$ in σ following a.

If all these conditions hold, then $\sigma' := (a_1, \ldots, a_{i-1}, b, a_{i+1}, \ldots, a_s)$ is an unbroken basis for the second order. All unbroken bases for the second order that are not unbroken for the first order are obtained in this way.

Proof. The proof is immediate and left to the reader.

Theorem 9.16. *The equivalence relations $R_{\mathcal{I}}$ and $R_{\mathcal{N}B}$ coincide.*

Proof. As before, choose an element $a \in X$ with a successor b in our ordering. Define a new ordering by exchanging a and b and denote by $\mathcal{N}B'$ the family of unbroken bases with respect to the new ordering. We claim that the equivalence relations $R_{\mathcal{N}B}$ and $R_{\mathcal{N}B'}$ coincide.

Since every basis extracted from X is an unbroken basis for a suitable ordering and we can pass from one ordering to another by a sequence of elementary moves consisting of exchanging an element with its successor, this will prove our theorem.

Set $\overline{\mathcal{N}B} = \mathcal{N}B \cup \mathcal{N}B'$. Take a basis $\underline{b} \in \overline{\mathcal{N}B} - \mathcal{N}B$. By Lemma 9.15 we have that $\underline{b} = \{c_1, \ldots, c_{k-1}, b, c_{k+1}, \ldots, c_r\}$ with $a, b, c_{k+1}, \ldots, c_r$ linearly dependent. Consider the set of vectors $\underline{b} \cup \{a\}$. To this set we can apply Lemma 9.14 and deduce that the equivalence relation induced by the family of all bases extracted from $\underline{b} \cup \{a\}$ coincides with the equivalence relation induced by the subfamily of all bases containing a. These are all easily seen to lie in $\mathcal{N}B$. We deduce that $R_{\mathcal{N}B}$ and $R_{\overline{\mathcal{N}B}}$ coincide. By symmetry, $R_{\mathcal{N}B'}$ and $R_{\overline{\mathcal{N}B}}$ coincide too, whence our claim.

Remark 9.17. We have thus proved that two strongly regular points x, y belong to the same big cell if and only if the set of unbroken bases $\underline{b}$ for which $x \in C(\underline{b})^0$ coincides with that for which $y \in C(\underline{b})^0$.

Example 9.18. A_3 ordered as $\alpha_1, \alpha_2, \alpha_3, \alpha_1 + \alpha_2, \alpha_2 + \alpha_3, \alpha_1 + \alpha_2 + \alpha_3$.

We have six unbroken bases, all of which necessarily contain α_1:

$$\{\alpha_1, \ \alpha_2, \ \alpha_3\}; \{\alpha_1, \ \alpha_2, \ \alpha_2 + \alpha_3\}; \{\alpha_1, \ \alpha_2, \ \alpha_1 + \alpha_2 + \alpha_3\};$$

$$\{\alpha_1, \ \alpha_3, \ \alpha_1 + \alpha_2\}; \{\alpha_1, \ \alpha_3, \ \alpha_1 + \alpha_2 + \alpha_3\}; \{\alpha_1, \ \alpha_1 + \alpha_2, \ \alpha_2 + \alpha_3\}.$$

The decomposition into big cells

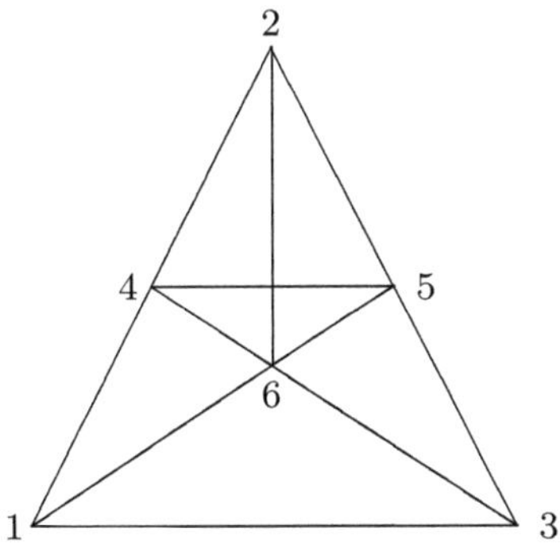

is obtained superposing the cones associated to unbroken bases:

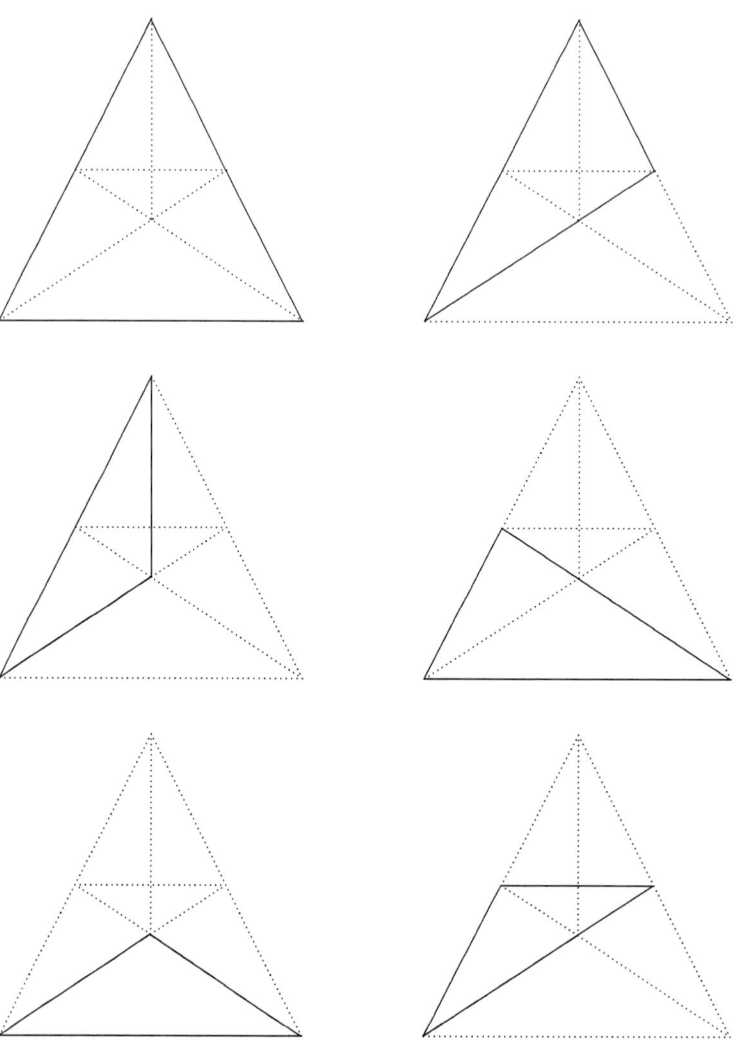

10

Cohomology

This chapter is independent of the remaining treatment and can be skipped without compromising the understanding of what follows.

10.1 De Rham Complex

10.1.1 Cohomology

The cohomology of the complement of a hyperplane arrangement has been computed by Orlik–Solomon [83] in general and over $\mathbb{Z}$. Here we want only to discuss how one computes it over $\mathbb{C}$ by using the algebraic de Rham complex. This is due to the fact that by a theorem of Grothendieck [57], we can compute the complex de Rham cohomology of a smooth affine variety by computing the cohomology of its algebraic de Rham complex.

So let us analyze the algebraic de Rham complex of the complement $\mathcal{A}_{X,\mu}$ of the hyperplane arrangement associated to a datum X, μ.

This complex is described by the graded algebra of algebraic differential forms $R_{X,\mu} \otimes \bigwedge dV$, where dV is the space of differentials of linear functions. In coordinates x_i we have that $\bigwedge dV$ is the exterior algebra generated by the elements dx_i.

We have already seen the (noncanonical) decomposition

$$R_{X,\mu} = \bigoplus_{W,\underline{c}} A_W \prod_{a \in \underline{c}} (a + \mu_a)^{-1}, \tag{10.1}$$

where W runs among the subspaces of the arrangement determined by the datum X, μ and $\underline{c}$ runs through the bases unbroken on W (see Definition 2.29).

The main point is the following:

Lemma 10.1. $A_W \prod_{a \in \underline{c}} (a + \mu_a)^{-1} \otimes \bigwedge dV$ *is a subcomplex.*

C. De Concini and C. Procesi, *Topics in Hyperplane Arrangements, Polytopes and Box-Splines*, Universitext, DOI 10.1007/978-0-387-78963-7_10,
© Springer Science+Business Media, LLC 2010

Proof. Choose coordinates such that $\mathbb{C}[\frac{\partial}{\partial x_1},\ldots,\frac{\partial}{\partial x_k},x_{k+1},\ldots,x_s] := A_W$, and $\underline{c} = \{x_1,\ldots,x_k\}$. In degree 0 we have clearly that

$$d\left(\frac{\partial^{h_1}}{\partial x_1}\cdots\frac{\partial^{h_k}}{\partial x_k}x_{k+1}^{h_{k+1}}x_{k+2}^{h_{k+2}}\cdots x_s^{h_s}\prod_{i=1}^{k}(x_i+\mu_i)^{-1}\right)$$

$$= \sum_{i=1}^{k}\frac{\partial^{h_1}}{\partial x_1}\cdots\frac{\partial^{h_i+1}}{\partial x_i}\cdots\frac{\partial^{h_k}}{\partial x_k}x_{k+1}^{h_{k+1}}x_{k+2}^{h_{k+2}}\cdots x_s^{h_s}\prod_{i=1}^{k}(x_i+\mu_i)^{-1}dx_i$$

$$+ \sum_{i=k+1}^{s}h_i\frac{\partial^{h_1}}{\partial x_1}\cdots\frac{\partial^{h_k}}{\partial x_k}x_{k+1}^{h_{k+1}}x_{k+2}^{h_{k+2}}\cdots x_i^{h_i-1}\cdots x_s^{h_s}\prod_{i=1}^{k}(x_i+\mu_i)^{-1}dx_i.$$

This formula extends to all degrees.

The previous decomposition of the de Rham complex into complexes thus gives a decomposition of the cohomology. We now have to understand the cohomology of a single summand $A_W\prod_{a\in\underline{c}}(a+\mu_a)^{-1}\otimes\bigwedge dV$. This is easily seen to be the tensor product of s basic complexes of the following two types:

1. The de Rham complex in one variable

$$\cdots 0 \cdots \xrightarrow{d} 0 \to \mathbb{C}[x] \xrightarrow{d} \mathbb{C}[x]dx \to 0 \xrightarrow{d} \cdots 0 \cdots .$$

Here $df = f'dx$ (where f' denotes the derivative with respect to x). Thus it is clear that the cohomology of this complex is 0 in all degrees except degree 0, where it is 1-dimensional, generated by the class of the cocycle 1.

2. The complex in one variable

$$\cdots 0 \to \mathbb{C}[\frac{d}{dx}]x^{-1} \xrightarrow{d} C[\frac{d}{dx}]x^{-1}dx \to 0 \cdots 0 \cdots .$$

Here $d(f(\frac{d}{dx})x^{-1}) = \frac{d}{dx}f(\frac{d}{dx})x^{-1}dx$. Thus it is clear that the cohomology of this complex is 0 in all degrees except degree 1, where it is 1-dimensional, generated by the class of the cocycle $d\log(x) = x^{-1}dx$.

In the first case we apply it to $x = x_i$ and in the second to $x = x_i + \mu_i$. The following is immediate

Theorem 10.2. *Let* $\underline{c} = \{c_1,\ldots,c_k\}$ *and set* $\mu_i := \mu_{c_i}$.

The cohomology of the complex $A_W\prod_{a\in\underline{c}}(a+\mu_a)^{-1}\otimes\bigwedge dV$ *is 0 except in dimension k where it is 1-dimensional generated by the class of the cocycle* $d\log(c_1+\mu_1)\wedge d\log(c_2+\mu_2)\wedge\cdots\wedge d\log(c_k+\mu_k)$.

Proof. This is immediate from the previous remarks and the Künneth formula.

For a subspace W of the arrangement and an unbroken basis $\underline{c} = \{c_1,\ldots,c_k\}$ on W, we set $\omega_{W,\underline{c}}$ equal to the cohomology class of the differential form $d\log(c_1+\mu_1)\wedge d\log(c_2+\mu_2)\wedge\ldots\wedge d\log[c_k+\mu_k)$. As a Corollary we get

Theorem 10.3. *The classes* $\omega_{W,\underline{c}}$ *form a basis of* $H^*(\mathcal{A}_{X,\underline{\mu}},\mathbb{C})$.

For any point $p \in P(X, \underline{\mu})$ we can consider the arrangement, centered at p, consisting of the hyperplanes of equation $a + \mu_a = 0$, $a \in X_p$. Let $\underline{\mu}_p$ be the restriction of $\underline{\mu}$ to X_p. Clearly, $\mathcal{A}_{X,\underline{\mu}} \subset \mathcal{A}_{X_p,\underline{\mu}_p}$ and these inclusions induce a map

$$f : \oplus_{p \in P(X,\underline{\mu})} H^*(\mathcal{A}_{X_p,\underline{\mu}_p}, \mathbb{C}) \to H^*(\mathcal{A}_{X,\underline{\mu}}, \mathbb{C}).$$

Corollary 10.4. *The map f is an isomorphism in top cohomology.*

10.1.2 Poincaré and Characteristic Polynomial

We are now ready to prove the theorem of Orlik–Solomon, stated in Section 2.2.4. We now assume $\underline{\mu} = 0$. Let us consider the Poincaré polynomial

$$p_X(t) = \sum_i \dim H^i(\mathcal{A}_X, \mathbb{C}) t^i.$$

Let $\chi_X(q)$ be the characteristic polynomial defined in Section 2.2.4. We have

Theorem 10.5 (Orlik–Solomon). $p_X(t) = (-t)^s \chi_X(-t^{-1})$.

Proof. Notice that there is an obvious bijection between the set of subspaces of the arrangement $\mathcal{H}_X$ and the set of subspaces in V spanned by elements in X mapping a subspace W of $\mathcal{H}$ to its annihilator $W^\perp$. Using this and the definition (2.8) of $\chi_X(q)$, we can write

$$\chi_X(q) = \sum_{W \text{ subspace of } \mathcal{H}_X} \mu(\hat{0}, W^\perp) q^{\dim W}.$$

On the other hand, by Theorem 10.3, we have that

$$p_X(t) = \sum_{W \text{ subspace of } \mathcal{H}_X} \nu(W) t^{s - \dim W},$$

where $\nu(W)$ is the number of unbroken bases on W. Thus everything will follow once we show that for any subspace W of $\mathcal{H}_X$,

$$\nu(W) = (-1)^{s - \dim W} \mu(\hat{0}, W^\perp) = (-1)^{\dim W^\perp} \mu(\hat{0}, W^\perp).$$

If X is the empty list in a space of dimension zero, then $\nu(0) = \mu(\hat{0}, \hat{0}) = 1$. We can thus assume that X is nonempty and proceed by induction on the dimension of $W^\perp$. In particular, our result is true for any list $X_{W^\perp} = X \cap W^\perp$ with $W \neq \{0\}$. From this we immediately deduce that for each such W, $\nu(W) = (-1)^{s - \dim W} \mu(\hat{0}, W^\perp)$.

Now notice that the fact that X is nonempty clearly implies that we have a free action of $\mathbb{C}^*$ on $\mathcal{A}_X$ by homotheties that induces a homeomorphism of $\mathcal{A}_X$ with $\mathbb{C}^* \times \overline{\mathcal{A}}_X$, where $\overline{\mathcal{A}}_X$ is the quotient of $\mathcal{A}_X$ by $\mathbb{C}^*$. In particular, the Euler characteristic $p_X(-1)$ of $\mathcal{A}_X$ is equal to zero. We thus deduce that

$$\sum_{W \text{ subspace of } \mathcal{H}_X} \nu(W)(-1)^{s-\dim W} = 0.$$

Using the inductive hypothesis yields

$$(-1)^s \nu(\{0\}) = -\sum_W \nu(W)(-1)^{s-\dim W}$$

$$= -\sum_W \mu(\hat{0}, W) = (-1)^s \mu(\hat{0}, V),$$

where the sum is over all nonzero subspaces W of the arrangement $\mathcal{H}_X$.

We should add that in the case of root systems, the theory is much more precise, but it would take us too far afield to explain it. In this case, by a result of Brieskorn [21], one has the remarkable factorization of the Poincaré polynomial as $\prod_{i=1}^{s}(1 + b_i q)$, where the positive integers b_i are the exponents of the root system (see [20]). This has been extended by the work of Terao [111], [112] on free arrangements. Moreover, in this case the Weyl group acts on the cohomology, and its character has also been object of study; see, for instance, Lehrer and Solomon [73].

10.1.3 Formality

We want to go one step forward and understand a basic fact, called *formality*, and the structure of cohomology as an algebra.

The formality of $\mathcal{A}_{X,\mu}$ means that there is a subalgebra of cocycles of the differential graded algebra of differential forms isomorphic to the cohomology algebra $H^*(\mathcal{A}_{X,\mu}, \mathbb{C})$.

Let then $\mathcal{H}_X$ be the subalgebra of the de Rham complex generated by all the elements $d\log(a + \mu_a)$, $a \in X$. Clearly, $\mathcal{H}_X$ consists of cocycles such that we have a algebra homomorphism from $\mathcal{H}_X$ to the cohomology algebra $H^*(\mathcal{A}_{X,\mu}, \mathbb{C})$.

We will transform the equations (8.7) into algebraic relations on the forms $d\log(a + \mu_a)$. Let us define $\omega_a := d\log(a + \mu_a)$.

Take a dependency relation $\sum_{i=0}^{k} s_i b_i = 0$, with all the s_i's different from zero. In particular, $\sum_{i=0}^{k} s_i d(b_i) = 0$. Set $\omega_i := d\log(b_i + \mu_{b_i})$. We have then

$$\omega_0 \wedge \cdots \wedge \cdots \wedge \omega_k = 0. \tag{10.2}$$

If, moreover, $\sum_{i=0}^{k} s_i(b_i + \mu_{b_i}) = 0$, we have a stronger relation of lower degree. Take two indices $h, j \leq k$ and multiply the identity $\sum_{i=0}^{k} s_i d(b_i) = 0$ by $d(b_0) \wedge \cdots \wedge d(\check{b}_h) \wedge \cdots \wedge d(\check{b}_j) \wedge \cdots \wedge d(b_k)$ we get[1]

---

[1] By $\check{x}$ we mean that we *remove* x.

$$0 = s_h d(b_h) \wedge d(b_0) \wedge \cdots \wedge d(\breve{b}_h) \wedge \cdots \wedge d(\breve{b}_j) \wedge \cdots \wedge d(b_k)$$
$$+ s_j d(b_j) \wedge d(b_0) \wedge \cdots \wedge d(\breve{b}_h) \wedge \cdots \wedge d(\breve{b}_j) \wedge \cdots \wedge d(b_k).$$

Thus we have the unique form $\gamma = (-1)^k d(b_0) \wedge \cdots \cdots \wedge d(\breve{b}_k) \wedge \cdots \wedge d(b_s) s_k^{-1}$ that does not depend on k. From this and the relation (8.7), we have that whenever we have the relation $\sum_{i=0}^{k} s_i(b_i + \mu_i) = 0$ we deduce

$$\sum_{i=0}^{k} (-1)^i \omega_0 \wedge \cdots \wedge \breve{\omega}_i \wedge \cdots \wedge \omega_k = 0. \tag{10.3}$$

In fact

$$\sum_{i=0}^{k} (-1)^i \omega_0 \wedge \cdots \wedge \breve{\omega}_i \wedge \cdots \wedge \omega_k = \Big(\sum_i \frac{s_i}{\prod_{j \neq i}(b_j + \mu_j)} \Big) \gamma = 0.$$

Theorem 10.6. *(1) The homomorphism from $\mathcal{H}_X$ to the cohomology algebra $H^*(\mathcal{A}_{X,\underline{\mu}}, \mathbb{C})$ is an isomorphism.*

(2) The products $d\log(a_{i_1} + \mu_{i_1}) \wedge \cdots \wedge d\log(a_{i_k} + \mu_{i_k})$ where $a_{i_1}, \ldots, a_{i_k}$ run through the unbroken bases for the corresponding spaces of the arrangement, are a linear basis of $\mathcal{H}_X$.

(3) The algebra $\mathcal{H}_X$ is presented as the exterior algebra in the generators ω_a, $a \in X$, modulo the relations (10.2), (10.3) deduced by the dependency relations.

Proof. Let us define the algebra $\tilde{\mathcal{H}}_X$ presented by the exterior generators ω_i and the relations 10.2, 10.3. We have a sequence of homomorphisms $\tilde{\mathcal{H}}_X \to \mathcal{H}_X \to H^*(\mathcal{A}_{X,\underline{\mu}}, \mathbb{C})$. From Theorem 10.3 we deduce that the composed homomorphism is surjective. Therefore, we need only show that the products corresponding to unbroken sets span $\tilde{\mathcal{H}}_X$.

First observe that by the relations (10.2) the algebra $\tilde{\mathcal{H}}_X$ is spanned by products of the ω_a involving only linearly independent elements of X. So let us take k linearly independent elements $b_i \in X$. Let W be the space of the arrangement given by $b_i + \mu_{b_i} = 0$. Take an $a \in X$ such that $a + \mu_a$ vanishes on W and a breaks the list b_i.

Using the relations (10.3) it is then clear that we can replace the product of the elements ω_{b_i} with a linear combination of products in which one of the ω_{b_i} has been replaced by ω_a. These terms are lexicographically strictly lower products, and the claim follows by recursion.

10.2 Residues

Inspired by the usual method in one complex variable for computing definite integrals, we want to develop a residue method to compute the function $T_{X,\underline{\mu}}$.

A similar approach will be later developed also for partition functions.

The results of this section are due to Brion–Vergne [27], [28], [25].

10.2.1 Local Residue

Proposition 9.2 can be interpreted as expressing $T_{X,\mu}$ as a sum of residues at the various points of the arrangement and it allows us to restrict our analysis to the case $\mu = 0$.

The description of the cohomology in top degree can be reformulated as follows. In Theorem 8.10 we have introduced the space Θ_X spanned by the elements $d_{\underline{b}} := \prod_{a \in \underline{b}} a^{-1}$, as $\underline{b}$ varies among the bases extracted from the list X, and we have proved that it has as basis the elements $d_{\underline{b}}$, as $\underline{b}$ varies among the unbroken bases. By the description of the top de Rham cohomology, we have the following result:

Proposition 10.7. *We have the direct sum decomposition*

$$R_X = \Theta_X \oplus \partial(R_X),$$

where $\partial(R_X)$ is the span of the partial derivatives of elements in R_X.

For any top differential form ψ, denote by $[\psi]$ its cohomology class. Recall that in Definition 8.13 we have defined the residue $\mathrm{res}_{\underline{b}}(\psi)$ for any unbroken basis $\underline{b}$. From the relations proved in Lemma 8.14 we get (recalling formula (8.11)) the following

Proposition 10.8.

$$[\psi] = \sum_{\underline{b} \in \mathcal{NB}_X} \mathrm{res}_{\underline{b}}(\psi)[\omega_{\underline{b}}]. \tag{10.4}$$

Consider the algebra $\mathcal{R}_X := L[d_X^{-1}]$, where L is the algebra of germs of functions holomorphic around 0. The algebra $\mathcal{R}_X$ is also a module over the Weyl algebra $W(V)$ and has a filtration by polar order. We have, for every k, a mapping $(R_X)_k/(R_X)_{k-1} \to (\mathcal{R}_X)_k/(\mathcal{R}_X)_{k-1}$.

Theorem 10.9. *The map*

$$R_X/(R_X)_{s-1} \to \mathcal{R}_X/(\mathcal{R}_X)_{s-1}$$

is an isomorphism.

The cohomology of the de Rham complex associated to $\mathcal{R}_X$ equals the cohomology of the complex associated to R_X.

Proof. First let us prove that the map is surjective. Take an element of $\mathcal{R}_X$. It can be written as linear combination of elements $f b_1^{-h_1} \cdots b_s^{-h_s}$, where $f \in L$ and the b_i give a basis of coordinates and $h_i \geq 0$. If we write f as a power series in the variables b_i we see that all the products $b_1^{k_1} \cdots b_s^{k_s} b_1^{-h_1} \cdots b_s^{-h_s}$ for which at least one $k_j > h_j$ lie in $(\mathcal{R}_X)_{s-1}$ therefore modulo this space, $f b_1^{-h_1} \cdots b_s^{-h_s}$ is in the image of R_X.

In order to prove that the map is an isomorphism, we can use the fact that it is a morphism of $W(V)$ modules. Then using Corollary 4.2 it is enough to prove that the images of the elements $d_{\underline{b}}^{-1}$ are still linearly independent in $\mathcal{R}_X/(\mathcal{R}_X)_{s-1}$. For this one can follow the same reasoning as in Proposition 8.11 (the residue computation).

As for the cohomology, we leave it to the reader.

We are now going to define the *total residue* of a function $f \in \mathcal{R}_X$. Choose once and for all a basis $x_1, \ldots, x_s$ of V.

Definition 10.10. Given $f \in \mathcal{R}_X$, its total residue $\mathrm{Tres}(f)$ is the cohomology class of the form $f\,dx_1 \wedge \cdots \wedge dx_s$.

We can now reformulate formula (10.4) as:

$$\mathrm{Tres}(f) = \sum_{\underline{b} \in \mathcal{NB}(X)} \mathrm{res}_{\underline{b}}(f)[\omega_{\underline{b}}].$$

We can compute the polynomials $p_{\underline{b},X}$ that appear in the expression (9.5) of the multivariate spline T_X from the following general formula. Recall that (Theorem 8.20) the space of polar parts is a free $S[U]$-module with basis the classes of the functions $d_{\underline{b}}^{-1} = \prod_{a \in \underline{b}} a^{-1}$ as $\underline{b} \in \mathcal{NB}(X)$:

Theorem 10.11 (Brion–Vergne). *For every $h \in \mathcal{R}_X$, write the class of h, in the space of polar parts, as $\sum_{\underline{b} \in \mathcal{NB}(X)} q_{\underline{b},X}[d_{\underline{b}}^{-1}]$ with $q_{\underline{b},X} \in S[U]$. We have*

$$q_{\underline{b},X}(-y) = det(\underline{b})\mathrm{res}_{\underline{b}}(e^{\langle y|x \rangle} h(x)). \tag{10.5}$$

Proof. We begin by observing that formula (10.5) makes sense, since if we expand $e^{\langle y|x \rangle} h(x)$ with respect to the variables $y = \{y_1, \ldots, y_s\}$, we get a power series whose coefficients lie in $\mathcal{R}_X$.

In order to prove this theorem we need some properties of Tres. The first property of Tres, that follows from the definition, is that given a function f and an index $1 \le i \le s$, we have $\mathrm{Tres}(\frac{\partial f}{\partial x_i}) = 0$; hence for two functions f, g

$$\mathrm{Tres}(\frac{\partial f}{\partial x_i} g) = -\mathrm{Tres}(f \frac{\partial g}{\partial x_i}).$$

In other words, for a polynomial P,

$$\mathrm{Tres}(P(\partial_x)(f)g) = \mathrm{Tres}(fP(-\partial_x)(g)). \tag{10.6}$$

We shall use the relation (10.6) for the function $f = e^{\langle y|x \rangle}$ for which we have

$$P(\partial_x)e^{\langle y|x \rangle} = P(y)e^{\langle y|x \rangle}. \tag{10.7}$$

The second simple property is that given a basis $\underline{b}$ extracted from X, that we think of as a system of linear coordinates on U, and a function f regular at 0, we have (by expanding in Taylor series in the variables $a \in \underline{b}$)

$$\mathrm{Tres}\Big(\frac{f}{\prod_{a\in\underline{b}}a}\Big) = f(0)\,\mathrm{Tres}\Big(\frac{1}{\prod_{a\in\underline{b}}a}\Big). \qquad (10.8)$$

We get

$$\mathrm{Tres}(e^{\langle y|x\rangle}h(x)) = \mathrm{Tres}\Big(\sum_{\underline{b}\in\mathcal{NB}(X)} e^{\langle y|x\rangle}q_{\underline{b},X}(\partial_x)\frac{1}{\prod_{a\in\underline{b}}a}\Big)$$

$$= \sum_{\underline{b}\in\mathcal{NB}(X)}\mathrm{Tres}\Big(\frac{q_{\underline{b},X}(-\partial_x)(e^{\langle y|x\rangle})}{\prod_{a\in\underline{b}}a}\Big) = \sum_{\underline{b}\in\mathcal{NB}(X)}q_{\underline{b},X}(-y)\mathrm{Tres}\Big(\frac{1}{\prod_{a\in\underline{b}}a}\Big)$$

$$= \sum_{\underline{b}\in\mathcal{NB}(X)}\frac{1}{det(\underline{b})}q_{\underline{b},X}(-y)[\omega_{\underline{b}}]. \qquad (10.9)$$

From this the theorem follows.

Thus applying this to $h(x) = d_X^{-1}$, we have

$$p_{\underline{b},X}(-y) = det(\underline{b})\mathrm{res}_{\underline{b}}\Big(\frac{e^{\langle y|x\rangle}}{d_X}\Big). \qquad (10.10)$$

Differential Equations

Most of this chapter is devoted to the discussion of the theory of Dahmen and Micchelli characterizing the space $D(X)$ of polynomials that satisfy the same differential equations satisfied by the polynomials that describe the box spline locally.

11.1 The First Theorem

11.1.1 The Space $D(X)$.

We shall use all the notation of Section 8.1 and assume that all the numbers μ_a are equal to zero.

By Corollary 8.20, R/R_{s-1} is a free $S[U]$-module with basis the elements $u_{\underline{b}}$, classes in R/R_{s-1} of the products $\prod_{a\in\underline{b}} a^{-1}$, as $\underline{b}$ runs over the unbroken bases of X.

Recall that $d_X = \prod_{a\in X} a$. We shall denote by u_X the class of d_X^{-1} in the space of polar parts R/R_{s-1}.

Lemma 11.1. *If Y is a cocircuit, we have $\prod_{a\in Y} a\, u_X = 0$.*

Proof. If $Y \subset X$ is a cocircuit, by definition $X \setminus Y$ does not span V so that $(\prod_{a\in Y} a)d_X^{-1} = \prod_{a\in X\setminus Y} a^{-1} \in R_{s-1}$.

If $Y \subset X$, let us define

$$D_Y := \prod_{a\in Y} D_a.$$

This element lies in the algebra $A = S[V]$ of polynomial differential operators with constant coefficients on V. Inspired by the previous lemma, let us give the following definition, of central importance in the work of Dahmen and Micchelli and in this book. Let us denote, as in Section 2.2.1, the set of cocircuits by $\mathcal{E}(X)$.

C. De Concini and C. Procesi, *Topics in Hyperplane Arrangements, Polytopes and Box-Splines*, Universitext, DOI 10.1007/978-0-387-78963-7_11,
© Springer Science+Business Media, LLC 2010

Definition 11.2. The space $D(X)$ is given by

$$D(X) := \{f \in S[U] \mid D_Y f = 0, \ \forall\, Y \in \mathcal{E}(X)\}. \tag{11.1}$$

The space $D(X)$ plays a fundamental role in the theory of B_X. In fact, in the course of this chapter we shall show that this space coincides with the space $\tilde{D}(X)$ introduced in Definition 7.21.

We shall often use a reduction to the irreducible case by the following proposition

Proposition 11.3. *Assume that X decomposes as $X_1 \cup \cdots \cup X_k$, so that the space $\mathbb{R}^s$ decomposes as $\mathbb{R}^s = \langle X_1 \rangle \oplus \cdots \oplus \langle X_k \rangle$.*

We identify the polynomials on a direct sum as the tensor product of the polynomials on the summands, and we have

$$D(X) = D(X_1) \otimes D(X_2) \otimes \cdots \otimes D(X_k). \tag{11.2}$$

Proof. It is enough to observe that a subset $Y = \cup_{i=1}^{k} Y_i$, $Y_i \subset X_i$ is a cocircuit if and only if at least one of the Y_i is a cocircuit, so the differential equations can be applied separately to the variables of each space $\langle X_i \rangle$.

Remark 11.4. When the irreducible X_i is reduced to just one vector, in the corresponding factor $\mathbb{R}$ the polynomials $D(X_i)$ reduce to the constants.

Recall that, in Definition 2.23, we have denoted by $m(X)$ the minimum length of elements in $\mathcal{E}(X)$. We have the following

Proposition 11.5. *The number $m(X)$ is the maximum among the numbers r such that all polynomials of degree $< r$ lie in $D(X)$.*

Proof. Every polynomial f of degree $p < m(X)$ satisfies the differential equations $D_Y f = 0, \forall\, Y \in \mathcal{E}(X)$, since D_Y is a product of more than p derivatives.

Conversely, if $Z \in \mathcal{E}(X)$ is a cocircuit of length $m(X)$, we can find a homogeneous polynomial f of degree $m(X)$ with $D_Z f = 1$ and $f \notin D(X)$.

Let us return to the basic formula (9.4):

$$\frac{1}{d_X} = \sum_{\underline{b} \in \mathcal{NB}} p_{\underline{b},X}(\partial_x)\frac{1}{d_{\underline{b}}}, \qquad d_{\underline{b}} := \prod_{a \in \underline{b}} a.$$

We shall apply this formula as follows. We work in the module P_X of polar parts and with the classes $u_X, u_{\underline{b}}$ of $d_X^{-1}, d_{\underline{b}}^{-1}$. We think of these classes as elements of the Fourier dual $\hat{P}_X$ of P_X, and for an element $m \in P_X$ we denote by $\hat{m}$ the same element thought of as an element of $\hat{P}_X$. By Theorem 8.20, each $\hat{u}_{\underline{b}}$ generates a submodule isomorphic to the module of polynomial functions on U, and

$$\hat{u}_X = \sum_{\underline{b} \in \mathcal{NB}} p_{\underline{b},X}(-x)\hat{u}_{\underline{b}}. \tag{11.3}$$

We can state the first theorem on differential equations satisfied by T_X:

Theorem 11.6. *The polynomials $p_{\underline{b},X}(-x)$ and hence the polynomials*

$$p_\Omega(x) := \sum_{\underline{b} \mid \Omega \subset C(\underline{b})} |\det(\underline{b})|^{-1} p_{\underline{b},X}(-x)$$

in formula (9.5) lie in $D(X)$.

Proof. The result follows from Lemma 11.1 on applying the algebraic Fourier duality to the module of polar parts and formula (11.3).

Using Proposition 8.23, the fact that the polynomials p_Ω that coincide with T_X in each big cell Ω lie in $D(X)$ has the following meaning in terms of distributions:

Corollary 11.7. *When Y is a cocircuit, the distribution $D_Y T_X$ is supported in the singular set of $C(X)$.*

11.2 The Dimension of $D(X)$

Since the elements D_Y are homogeneous, the space $D(X)$ is also graded, and it is interesting for each $k \geq 0$ to compute the dimension of its degree-k component $D_k(X)$. As usual, one can arrange these dimensions in a generating function,

$$H_X(q) := \sum_k \dim(D_k(X))q^k.$$

We will show a theorem proved in [41]:

Theorem 11.8. *The space $D(X)$ is finite-dimensional, of dimension the number $d(X)$ of linear bases of V that one can extract from X:*

$$H_X(q) := \sum_{\underline{b} \in \mathcal{B}_X} q^{m-s-e(\underline{b})} = q^{m-s} T(X, 1, q^{-1}). \tag{11.4}$$

Here $e(\underline{b})$ is the external activity of a basis $\underline{b} \in \mathcal{B}_X$ introduced in Section 2.2.2 (deduced from the Tutte polynomial). In particular, the top degree polynomials in $D(X)$ are of degree $m - s$.

We shall prove this theorem by exhibiting an explicit basis.

Moreover, from general facts we shall see that $D(X)$ is generated, under taking derivatives, by the homogeneous elements of top degree.

For the proof of these results we shall follow a direct approach. In fact, one can view these theorems as special cases of a general theory, the theory of Stanley–Reisner or face algebras.

The interested reader should read our paper *Hyperplane arrangements and box splines [43]* with an appendix by Anders Björner.

11.2.1 A Remarkable Family

In order to obtain a proof of Theorem 11.8, we start with the study of a purely algebraic geometric object.

For notational simplicity let us define $A := S[V]$.

We want to describe the *scheme* defined by the ideal I_X of A generated by the elements $d_Y := \prod_{a \in Y} a$ as Y runs over all the cocircuits.

Thus we are interested in the algebra

$$A_X := A/I_X. \tag{11.5}$$

We shall soon see that A_X is finite-dimensional. Formally A_X is the *coordinate ring* of the corresponding scheme.

The use of the word *scheme* may seem a bit fancy. What we really want to stress by this word is that we have a finite dimensional algebra (quotient of polynomials) whose elements as functions vanish exactly on some point p, but at least infinitesimally they are *not constant*. This can be interpreted analytically in the language of distributions and appears clearly in the dual picture that produces solutions of differential equations.

To warm us up for the proof let us verify that this scheme is supported at 0 (that implies that A_X is finite-dimensional).

For this, observe that the variety of zeros of a set of equations each one of which is itself a product of equations is the union of the subvarieties defined by selecting an equation out of each product. Thus what we need is the following:

Lemma 11.9. *Take one element b_Y from each cocircuit Y. Then the resulting elements b_Y span V, and hence the equations $b_Y = 0$, $Y \in \mathcal{E}(X)$ define the subvariety consisting of the point 0.*

Proof. If, by contradiction, these elements do not span V, their complement is a cocircuit. Since we selected an element from each cocircuit, this is not possible.

Let us extend the definition of our notions as follows. If $X = (a_1, \ldots, a_m)$ and $\underline{\mu} := (\mu_1, \ldots, \mu_m)$ are parameters, we define $I_X(\underline{\mu})$ to be the ideal given by the equations $\prod_{a_j \in Y}(a_j + \mu_j)$, $Y \in \mathcal{E}(X)$. This can be viewed either as an ideal in $S[V]$ depending on the parameters $\underline{\mu}$ or as an ideal in the polynomial ring

$$A(\underline{\mu}) := S[V][\mu_1, \ldots, \mu_m].$$

In this second setting it is clearly a homogeneous ideal, and we can apply to it the results of Section 5.1.2. This we shall do in the next section.

11.2.2 The First Main Theorem

Let us define, for a sublist Y of X, $d_Y(\underline{\mu}) := \prod_{a \in Y}(a + \mu_a)$.

We shall prove a stronger version of Theorem 11.8. Set

$$A_X(\underline{\mu}) := A(\underline{\mu})/I_X(\underline{\mu}).$$

We need some lemmas.

Lemma 11.10. *For generic μ, the ideal $I_X(\underline{\mu})$ defines $d(X)$ distinct points.*

Proof. Set theoretically, the variety defined by the ideal $I_X(\mu)$ is the union of the varieties described by selecting, for every cocircuit $Y \in \bar{\mathcal{E}}(X)$, an element $b_Y \in Y$ and setting the equation $b_Y + \mu_{b_Y} = 0$.

By Lemma 11.9, the vectors b_Y span V. Thus the equations $b_Y + \mu_{b_Y} = 0$ are either incompatible or define a point of coordinates $-\mu_{i_j}$, $j = 1,\ldots,s$, in some basis a_{i_j} extracted from X.

Conversely, given such a basis, every cocircuit must contain at least one of the elements a_{i_j}; hence the point given by the equations $a_{i_j} + \mu_{i_j} = 0$ must lie in the variety defined by $I_X(\underline{\mu})$.

As we have seen in Section 8.1.4, if the μ_i are generic, these $d(X)$ points are all distinct.

This lemma implies that for generic $\underline{\mu}$, $\dim(A_X(\underline{\mu})) \geq d(X)$ and equality means that all the points are reduced, that is, $A_X(\underline{\mu})$ coincides with the space of functions on these points and has no nilpotent elements.

We will proceed by induction. Let $X = \{Z, y\}$ and $\underline{\mu} = (\underline{\nu}, \mu_y)$. Choose a complement W to $\mathbb{C}y$ in V and write each element $a \in V$ as

$$a = \bar{a} + \lambda_a y, \ \bar{a} \in W, \ \lambda_a \in \mathbb{C},$$

according to the decomposition.

Finally, define $\overline{A}(\mu) := A(\mu)/A(\mu)(y + \mu_y)$. We can clearly identify $\overline{A}(\mu)$ with $S[W][\underline{\mu}]$.

Denote by $\pi : A(\mu) \to \overline{A}(\mu) = S[W][\underline{\mu}]$ the quotient homomorphism. If $a \in V$, we have that

$$\pi(a) = \bar{a} - \lambda_a \mu_y.$$

For $a \in Z$ set $\bar{\nu}_a := \mu_a - \lambda_a \mu_y$. Clearly,

$$S[W][\underline{\mu}] = S[W][\overline{\nu}][\mu_y].$$

Also, for any sublist $B \subset X$, set $\overline{B}$ equal to the list of nonzero vectors in W that are components of vectors in B.

Our first claim is the following

Lemma 11.11. *The image of $I_X(\underline{\mu})$ in $\overline{A}(\underline{\mu})$ is the ideal $I_{\overline{Z}}(\overline{\nu})[\mu_y]$.*

Proof. Indeed, the elements $d_Y(\underline{\mu}) = \prod_{a \in Y}(a + \mu_a)$, where Y runs over the set of cocircuits, generate the ideal $\bar{I}_X(\underline{\mu})$. A cocircuit is of the form $X \setminus H$, where H is a hyperplane spanned by a subset of X. If $y \notin H$, we have that $y + \mu_y$ divides $d_Y(\underline{\mu})$, and so $\pi(d_Y(\underline{\mu})) = 0$. On the other hand, the hyperplanes of previous type containing y are in one-to-one correspondence, by the projection map $H \mapsto W$, with the hyperplanes $\overline{H}$ of W generated by elements of $\overline{X} = \overline{Z}$. For such a hyperplane H we have clearly $\pi(d_Y(\underline{\mu})) = d_{\overline{Y}}(\overline{\nu})$ where $\overline{Y} := \overline{Z} \setminus \overline{H}$. Hence $\pi(I_X) = I_{\overline{Z}}(\overline{\nu})[\mu_y]$, as required.

Denote now by p_X the projection map $p_X : A(\underline{\mu}) \to A_X(\underline{\mu})$.

Lemma 11.12. *There exists a surjective map*

$$j : A_Z(\underline{\nu})[\mu_y] \to (y + \mu_y)A_X(\underline{\mu})$$

of $A(\underline{\mu})$-modules that makes the following diagram commute:

$$
\begin{array}{ccc}
A(\underline{\mu}) & \xrightarrow{\;y+\mu_y\;} & (y + \mu_y)A(\underline{\mu}) \\
{\scriptstyle p_Z}\downarrow & & {\scriptstyle p_X}\downarrow \\
A_Z(\underline{\nu})[\mu_y] & \xrightarrow{\;\;j\;\;} & (y + \mu_y)A_X(\underline{\mu})
\end{array}
$$

Proof. To see this it suffices to show that if $Y \in \mathcal{E}(Z)$ then

$$(y + \mu_y) \cdot \prod_{a \in Y}(a + \mu_a) \in I_X(\underline{\mu}).$$

We have two cases. If $y \in \langle Z \setminus Y \rangle$, then $Y = X \setminus \langle Z \setminus Y \rangle \in \mathcal{E}(X)$ and thus $\prod_{a \in Y}(a + \mu_a) \in I_X$.

If $y \notin \langle Z \setminus Y \rangle$, then $X \cap \langle Z \setminus Y \rangle = Z \setminus Y$ and $X \setminus \langle Z \setminus Y \rangle = X \setminus (Z \setminus Y)$ equals $(Y, y) \in \mathcal{E}(X)$, so that $(y + \mu_y) \prod_{a \in Y}(a + \mu_a) = \prod_{a \in (Y, y)}(a + \mu_a) \in I_X(\underline{\mu})$.

Our strategy is to construct an explicit basis of $A_X(\underline{\mu})$. Recall the definition of externally active elements $E(\underline{b})$ associated to a basis $\underline{b}$ (Definition 2.2). Let us consider then the elements

$$u_{\underline{b}}^X(\underline{\mu}) := d_{X \setminus (E(\underline{b}) \cup \underline{b})}(\underline{\mu}).$$

Theorem 11.13.   *(i) The ring $A_X(\underline{\mu})$ is a free module over the polynomial algebra $\mathbb{C}[\underline{\mu}]$ with basis the classes, modulo $I_X(\underline{\mu})$, of the elements $u_{\underline{b}}^X(\underline{\mu})$, as $\underline{b}$ runs over the bases extracted from X.*
  (ii) For all μ the ring $A_X(\underline{\mu}) := A/I_X(\underline{\mu})$ has dimension equal to the number $d(X)$ of bases that can be extracted from X.

Proof. Clearly, (ii) is a consequence of (i), so we need only to prove (i). We proceed by induction on the cardinality of X and on $\dim(V)$.

First look at the classes, modulo $y + \mu_y$, of the elements $u_{\underline{b}}^X(\underline{\mu})$, as $\underline{b}$ runs over the bases extracted from X with $y \in \underline{b}$. Then $\underline{c} := \overline{\underline{b}}$ is a basis (for W) extracted from $\overline{Z}$. Observe that since y is the last element of X, we have

$$\overline{X \setminus (E(\underline{b}) \cup \underline{b})} = \overline{Z} \setminus (E(\underline{c}) \cup \underline{c}),$$

so that

$$\pi(u_{\underline{b}}^X(\underline{\mu})) = u_{\underline{c}}^{\overline{Z}}(\overline{\nu}).$$

By induction, applying Lemma 11.11, we get that the elements $u_{\underline{c}}^{\overline{Z}}(\overline{\nu})$, with $\underline{c} \in \mathcal{B}_{\overline{Z}}$, give a basis of $\overline{A}_{\overline{Z}}(\nu)[\mu_y]$ over the polynomial algebra $\mathbb{C}[\underline{\nu}][\mu_y] = \mathbb{C}[\underline{\mu}]$.

It follows that given $F \in A_X(\underline{\mu})$, we can find polynomials $\pi_{F,\underline{b}} \in \mathbb{C}[\underline{\mu}]$, $\underline{b} \in \mathcal{B}_X$, $y \in X$ with $F - \sum_{\underline{b}} \pi_{F,\underline{b}} u_{\underline{b}}(\underline{\mu})$ lying in the ideal $(y + \mu_y)A_X(\underline{\mu})$.

Next let $\underline{b} \subset Z$ be a basis. Then

$$u_{\underline{b}}^X(\underline{\mu}) = (y + \mu_y)u_{\underline{b}}^Z(\underline{\mu}).$$

Thus again by induction, and by Lemma 11.12, the elements $u_{\underline{b}}^X(\underline{\mu})$ as $\underline{b}$ varies among the bases in Z span $(y + \mu_y)A_X(\underline{\mu})$ as a $\mathbb{C}[\underline{\mu}]$-module, as desired.

It remains to show the linear independence of the $u_{\underline{b}}^X(\underline{\mu})$. Assume that there is a linear combination

$$\sum_{\underline{b} \in \mathcal{B}_X} \pi_{\underline{b}} u_{\underline{b}}(\underline{\mu}) = 0,$$

with $\pi_{\underline{b}} \in \mathbb{C}[\underline{\mu}]$ not all equal to zero. By Lemma 11.10, we can find a value $\underline{\mu}_0$ of our parameters $\underline{\mu}$ with the property that not all the polynomials $\pi_{\underline{b}}$ vanish at $\underline{\mu}_0$ and the ring $A_X(\underline{\mu}_0)$ has dimension at least $d(X)$. On the other hand, the images of $u_{\underline{b}}^X(\underline{\mu})$ in $A_X(\underline{\mu}_0)$ span $A_X(\underline{\mu}_0)$ and are linearly dependent. Thus $\dim(A_X(\underline{\mu}_0)) < d(X)$, a contradiction.

Observe that as a consequence of the proof, we also have that $j : A_Z \to yA_X$ is bijective. We are going to use this fact later.

11.2.3 A Polygraph

As announced, we want to discuss the meaning of this theorem for the variety V_X given by $I_X(\mu)$ thought of as an ideal in $S[V][\mu_1, \ldots, \mu_m]$.

This variety is easily seen to be what is called a *polygraph*. It lies in $U \times \mathbb{C}^m$ and can be described as follows. Given a basis $\underline{b} := (a_{i_1}, \ldots, a_{i_s})$ of V, extracted from X, let $(\underline{b}^1, \ldots, \underline{b}^s)$ be the associated dual basis in U. Define a linear map $i_{\underline{b}} : \mathbb{C}^m \to U$ by

$$i_{\underline{b}}(\mu_1, \ldots, \mu_m) := -\sum_{j=1}^s \mu_{i_j} \underline{b}^j.$$

Let $\Gamma_{\underline{b}}$ be its graph.

Theorem 11.14. $V_X = \cup_{\underline{b}} \Gamma_{\underline{b}}$.

Proof. An element $(p, \mu_1^0, \ldots, \mu_m^0) \in V_X$ if and only if p is a point of the scheme defined by the ideal $I_X(\underline{\mu}^0)$, $\underline{\mu}^0 = (\mu_1^0, \ldots, \mu_m^0)$. That is, p is defined by the equations $a_{i_j} + \mu_{i_j} = 0$ for some basis $\underline{b} := (a_{i_1}, \ldots, a_{i_s})$ of V. This is equivalent to saying that $p = -\sum_{j=1}^s \mu_{i_j} \underline{b}^j$. Since all these steps are equivalences, the claim is proved.

The variety V_X comes equipped with a projection map ρ to $\mathbb{C}^m$ whose fibers are the schemes defined by the ideals $I_{\overline{X}}(\underline{\mu})$.

Remark 11.15. Theorem 11.13 implies that ρ is *flat* of degree $d(X)$ and $I_X(\underline{\mu})$ is the full ideal of equations of V_X. Furthermore, V_X is Cohen Macaulay.

This fact is remarkable since it is very difficult for a polygraph to satisfy these conditions. When it does, this usually has deep combinatorial implications (see for instance [60]).

One should make some remarks about the algebras $A_X(\underline{\mu})$ in general.

First let us recall some elementary commutative algebra. Take an ideal I of the polynomial ring $\mathbb{C}[x_1, \ldots, x_m]$. Then $\mathbb{C}[x_1, \ldots, x_m]/I$ is finite-dimensional if and only if the variety of zeros of I is a finite set of points $p_1, \ldots, p_k$. In this case moreover we have a canonical decomposition

$$\mathbb{C}[x_1, \ldots, x_m]/I = \oplus_{i=1}^k \mathbb{C}[x_1, \ldots, x_m]/I_{p_i},$$

where for each $p \in \{p_1, \ldots, p_k\}$, the ring $\mathbb{C}[x_1, \ldots, x_m]/I_p$ is the local ring associated to the point p.

Let p have coordinates $x_i = \mu_i$. The local ring $\mathbb{C}[x_1, \ldots, x_m]/I_p$ is characterized, in terms of linear algebra, as the subspace of $\mathbb{C}[x_1, \ldots, x_m]/I$ where the elements x_i have generalized eigenvalue μ_i. Thus the previous decomposition is just the Fitting decomposition of $\mathbb{C}[x_1, \ldots, x_m]/I$ into generalized eigenspaces for the commuting operators x_i.

In the case of the algebra $A_X(\underline{\mu})$, the quotient of $S[V]$ by the ideal $I_X(\underline{\mu})$ generated by the elements $\prod_{y \in Y}(y + \mu_y), Y \in \mathcal{E}(X)$ we see that if for a point $p \in P(X, \underline{\mu})$ we have that $\langle y \mid p \rangle + \mu_y \neq 0$, in the local ring of p the element $y + \mu_y$ is invertible, and so it can be dropped from the equations. We easily deduce the following result

Proposition 11.16. *For every point $p \in P(X, \underline{\mu})$, we have that the local component $A_X(\underline{\mu})(p)$ is identified to the algebra $\bar{A}_{X_p}(\underline{\mu}_p)$ defined by the sublist $X_p := (x \in X \mid \langle x \mid p \rangle + \mu_x = 0)$.*

Furthermore, by a change of variables $I_{X_p}(\underline{\mu}_p) = \tau_p(I_{X_p})$ where τ_p is the automorphism of A sending x to $x - \langle x \mid p \rangle, \forall x \in V$. Thus, the automorphism τ_p, induces an isomorphism between A_{X_p} and $A_{X_p}(\underline{\mu}_p)$.

11.2.4 Theorem 11.8

By Theorem 11.13, the ring $A_X = A/I_X$ has dimension $d(X)$. On the other hand, the space $D(X)$ is by the definition of the pairing (5.1) between $A = S[V]$ and $S[U]$ the space orthogonal to the ideal I_X. We deduce that the dimension of $D(X)$ equals the codimension of I_X, that is, $d(X)$, proving Theorem 11.8.

Let us now make some further considerations in the parametric case. Let us observe that for generic μ, that is, in the case in which $P(X, \underline{\mu})$ consists of $d(X)$ distinct points, the space of solutions of the system of differential equations defined by $I_X(\underline{\mu})$ has as a basis the functions e^p, $p \in P(X, \underline{\mu})$.

In the general case, Proposition 11.16 immediately implies the following

Theorem 11.17. *For each $p \in P(X, \underline{\mu})$ consider the sublist*

$$X_p := (x \in X \mid \langle x \mid p \rangle + \mu_x = 0).$$

Then the space $D(X, \underline{\mu})$ of solutions of the system of differential equations defined by $I_X(\underline{\mu})$ has the direct sum decomposition

$$D(X, \underline{\mu}) = \bigoplus_{p \in P(X, \underline{\mu})} e^p D(X_p).$$

Remark 11.18. In Definition 11.2 of $D(X)$, we can start by assuming that f is a tempered distribution. In fact, due to the property that the ideal generated by the elements D_Y contains all large enough products of derivatives (see Lemma 11.9), and applying Lemma 5.7, we deduce that any solution f is necessarily a polynomial.

11.3 A Realization of A_X

11.3.1 Polar Representation

Recall that we have introduced the notation $d_Y := \prod_{a \in Y} a$ for any list Y of vectors, in particular for a sublist $Y \subset X$. We shall use the notation u_Y for the class of d_Y^{-1} in the module of polar parts $P_X = R_X/(R_X)_{s-1}$. Notice that for any sublist $Y \subset X$ we have

$$d_{X \setminus Y} u_X = u_Y.$$

Definition 11.19. We define Q_X to be the $S[V]$ submodule of the space of polar parts P_X generated by the class u_X.

The spaces R_X and P_X are naturally graded (as functions). It is convenient to shift this degree by m so that u_X has degree 0 and the generators $u_{\underline{b}}$ have degree $m - s$. If $\underline{b}$ is an unbroken basis, these will be just the elements $u_{\underline{b}}$ introduced in Section 8.1.6 considered here when $\mu = 0$.

With these gradations the natural map $\pi : A \to Q_X$ defined by the formula $\pi(f) := f u_X$ preserves degrees.

We shall use the notation $E(\underline{b})$ of Definition 2.26 to denote the set of elements of X externally active with respect to a basis $\underline{b}$ extracted from X.

Theorem 11.20. *(1) The annihilator of u_X is the ideal I_X generated by the elements $d_Y = \prod_{a \in Y} a$, as Y runs over the cocircuits. Thus $Q_X \simeq A_X$ as graded A-modules.*

(2) The elements $u_{E(\underline{b}) \cup \underline{b}}$, as $\underline{b}$ runs over the bases extracted from X, are a basis of Q_X.

Proof. Since the classes, modulo I_X, of the elements $d_{X \setminus (E(\underline{b}) \cup \underline{b})}$, as $\underline{b}$ runs over the bases extracted from X, are a basis of A_X, we have that (1) and (2) are equivalent.

By Lemma 11.1, the annihilator of u_X contains the ideal I_X.

Thus, from Theorem 11.13, it is enough to see that $\dim Q_X \geq d(X)$.

We want to proceed by induction on s and on the cardinality of X.

If X consists of a basis of V, clearly both A_X and Q_X are 1-dimensional and the claim is clear. Assume that $X := (Z, y)$ and let us first discuss the special case in which Z spans a subspace V' of codimension-one in V, so y is a vector outside this subspace and $d(X)$ equals the number $d(Z)$ of bases of $V' = \langle Z \rangle$ extracted from Z.

We clearly have the inclusions

$$R_Z \subset R_X, \quad y^{-1} R_{Z,k-1} \subset R_{X,k} \quad \forall k.$$

Furthermore, the element $u_X \in P_X$ is killed by y. By induction on s, the theorem follows in this special case from the following lemma

Lemma 11.21. *(1) Multiplication by y^{-1} induces an isomorphism between P_Z and the kernel of multiplication by y in P_X.*

(2) Multiplication by y^{-1} induces an isomorphism between Q_Z and Q_X.

Proof. (1)   Since $y^{-1} R_{Z,k-1} \subset R_{X,k}$, it is clear that multiplication by y^{-1} induces a map from $P_Z = R_{Z,s-1}/R_{Z,s-2}$ to $P_X = R_{X,s}/R_{X,s-1}$. It is also clear that the image of this map lies in the kernel of the multiplication by y.

To see that this mapping gives an isomorphism to this kernel, order the elements of X so that y is the first element. An unbroken basis for X is of the form $(y, \underline{c})$ where $(\underline{c})$ is an unbroken basis for Z.

Now fix a set of coordinates $x_1, \ldots, x_s$ such that $x_1 = y$ and $x_2, \ldots, x_s$ is a basis of the span of Z. Denote by ∂_i the corresponding partial derivatives:

$$P_X = \oplus_{\underline{c}} \mathbb{C}[\partial_1, \ldots, \partial_s] u_{y,\underline{c}}, \quad P_Z = \oplus_{\underline{c}} \mathbb{C}[\partial_2, \ldots, \partial_s] u_{\underline{c}}.$$

We have that in each summand $\mathbb{C}[\partial_1, \ldots, \partial_s]u_{y,\underline{c}}$ the kernel of multiplication by x_1 coincides with $\mathbb{C}[\partial_2, \ldots, \partial_s]u_{(y,\underline{c})}$.

The claim then follows easily, since $\mathbb{C}[\partial_2, \ldots, \partial_s]u_{(y,\underline{c})}$ is the image, under multiplication by y^{-1} of $\mathbb{C}[\partial_2, \ldots, \partial_s]u_{\underline{c}}$.

(2) follows from part (1) and the formula $y^{-1}u_Z = u_X$.

Let us now assume that Z still spans V. We need thus to compare several of the objects under analysis in the case in which we pass from Z to X.

Let us consider the ring A/Ay, polynomial functions on the subspace of U where y vanishes, and denote by $\overline{Z}$ the set of nonzero vectors in the image of Z (or X) in A/Ay.

As in Theorem 11.13, the set $\mathcal{B}_X$ of bases extracted from X can be decomposed into two disjoint sets, $\mathcal{B}_Z$ and the bases containing y. This second set is in one-to-one correspondence with the bases of $V/\mathbb{C}y$ contained in $\overline{Z}$.

By Proposition 8.18, we have that under the inclusion $R_Z \subset R_X$ we obtain $R_{Z,k} = R_Z \cap R_{X,k}$, $\forall k$. Hence we get an inclusion of P_Z into P_X.

Let us consider in P_X the map of multiplication by y. We have clearly $yu_X = u_Z$, thus we obtain an exact sequence of A-modules:

$$0 \to K \to Q_X \xrightarrow{y} Q_Z \to 0,$$

where $K = Q_X \cap \ker(y)$.

We need to analyze K and prove that $\dim(K)$ is greater than or equal to the number $d_y(X)$ of bases extracted from X and containing y. Since $d(X) = d(Z) + d_y(X)$, this will prove the claim.

In order to achieve the inequality $\dim(K) \geq d_y(X)$, we will find inside K a direct sum of subspaces whose dimensions add up to $d_y(X)$.

Consider the set $\mathcal{S}_y(X)$ of all sublists Y of X that span a hyperplane $\langle Y \rangle$ not containing y and with $Y = X \cap \langle Y \rangle$ (we shall use the word *complete* for this last property).

For each $Y \in \mathcal{S}_y(X)$ we have, by Lemma 11.21 and Proposition 8.18, that multiplication by y^{-1} induces an inclusion $i_Y : P_Y \to P_{(y,Y)} \to P_X$ with image in the kernel $\ker(y) \subset P_X$. Thus we get a map

$$g := \bigoplus_{Y \in \mathcal{S}_y(X)} i_Y : \bigoplus_{Y \in \mathcal{S}_y(X)} P_Y \to \ker(y).$$

Lemma 11.22. *g is an isomorphism of $\oplus_{Y \in \mathcal{S}_y(X)} P_Y$ onto $\ker(y)$.*

Proof. As before, order the elements of X so that y is the first element. An unbroken basis for X is of the form $(y, \underline{c})$, where $(\underline{c})$ is an unbroken basis for $Y := X \cap \langle \underline{c} \rangle$.

By construction, $Y \in \mathcal{S}_y(X)$; and by Theorem 8.20, P_X is a free $S[U]$-module with basis the elements $u_{(y,\underline{c})}$ as $(y, \underline{c})$ run over the unbroken bases.

Similarly, $P_{(y,Y)}$ is a free $S[U]$-module over the elements $u_{(y,\underline{c})}$, such that $Y = \langle \underline{c} \rangle \cap X$. Thus we obtain the direct sum decomposition

$$P_X = \bigoplus_{Y \in \mathcal{S}_y(X)} P_{(y,Y)}.$$

Therefore

$$\ker(y) = \bigoplus_{Y \in \mathcal{S}(X)} P_{(y,Y)} \cap \ker(y).$$

By Lemma 11.21, for each $Y \in \mathcal{S}_y(X)$, the map i_Y gives an isomorphism of P_Y with $P_{(y,Y)} \cap \ker(y)$. This proves the lemma.

We can now finish the proof of Theorem 11.20.

(1) Notice that by part (2) of Lemma 11.21 and Lemma 11.22, we have an inclusion of $\oplus_{Y \in \mathcal{S}_y(X)} Q_Y$ into the space $K = Q_X \cap \ker(y)$. So we deduce that $\dim K \geq \sum_{Y \in \mathcal{S}_y(X)} \dim Q_Y = d_y(X)$. This gives the required inequality and implies also that we have a canonical exact sequence of modules:

$$0 \to \bigoplus_{Y \in \mathcal{S}_y(X)} Q_Y \to Q_X \xrightarrow{y} Q_Z \to 0. \tag{11.6}$$

We can easily prove as corollary a theorem by several authors, see [2], [51], [69].

Corollary 11.23. *Consider in $S[V]$ the subspace $\mathcal{P}(X)$ spanned by all the products $d_Y := \prod_{x \in Y} x$, $Y \subset X$, such that $X \setminus Y$ spans V.*

Then $S[V] = I_X \oplus \mathcal{P}(X)$, so that $\mathcal{P}(X)$ is in duality with $D(X)$. The elements $d_{X \setminus E(\underline{b}) \cup \underline{b}}$ give a basis of $\mathcal{P}(X)$.

Proof. Multiplying by d_X^{-1}, we see that $\mathcal{P}(X)d_X^{-1}$ is spanned by the polar parts $\prod_{x \in Z} x^{-1}$, $Z \subset X$, such that Z spans V. We have seen in Corollary 8.21 that this space of polar parts maps isomorphically to its image into P_X, and its image is clearly Q_X. This proves our claims.

Remark 11.24. The last theorem we have proved is equivalent to saying that in the algebra R_X the intersection $d_X^{-1}A \cap R_{X,s-1}$ is equal to $d_X^{-1}I_X$.

11.3.2 A Dual Approach

In this section we follow closely the paper [41] and characterize the space $\mathcal{P}(X)$, defined in the previous section, by differential equations. Notice that although an explicit basis for $\mathcal{P}(X)$ has been given using an ordering of X, the space $\mathcal{P}(X)$ is defined independently of this ordering. Moreover, if $Y \subset X$ is such that $X \setminus Y$ spans V, the same is true for any sublist of Y. It follows that by applying any directional derivative to d_Y, we still have a linear combination of monomials of the same type, hence an element in $\mathcal{P}(X)$. The fact that $\mathcal{P}(X)$

is stable under derivatives tells us, by Proposition 5.5, that it is defined by an ideal of differential equations with constant coefficients of codimension $d(X)$. Of course, now the elements of $\mathcal{P}(X)$ are thought of as polynomials on U, so that the differential operators are in $S[U]$. We want to determine this ideal. For this, given any hyperplane H spanned by a subset of X, let us choose a linear equation $\phi_H \in U$ for H and let us denote by $m(H, X) := |X \setminus H|$ the number of elements in X and not in H.

Theorem 11.25. *The ideal of differential equations satisfied by $\mathcal{P}(X)$ is generated by the elements $\phi_H^{m(H,X)}$ as H runs over the hyperplanes generated by subsets of X.*

Proof. Let K_X be the ideal generated by the elements $\phi_H^{m(H,X)}$ as H runs over the hyperplanes generated by subsets of X. Let $G(X)$ be the subspace of $S[V]$ of elements satisfying the differential equations in K_X. Let us begin by showing that the elements $\phi_H^{m(H,X)}$ vanish on $\mathcal{P}(X)$, so that $\mathcal{P}(X) \subset G(X)$ and K_X is contained in the ideal of differential equations satisfied by $\mathcal{P}(X)$.

To see this, let us compute a differential operator $\phi_H^{m(H,X)}$ on a monomial d_Y, where $X \setminus Y$ spans V. We split the elements of Y into two parts $Y_1 := Y \cap H$ and $Y_2 := Y \setminus Y_1$ and thus write $d_Y = d_{Y_1} d_{Y_2}$. By definition, the derivative associated to ϕ_H vanishes on the elements of Y_1. Therefore we compute $\phi_H^{m(H,X)}(d_Y) = d_{Y_1} \phi_H^{m(H,X)}(d_{Y_2})$. We claim that $\phi_H^{m(H,X)}(d_{Y_2}) = 0$. For this, it is enough to see that the cardinality of Y_2 is strictly less than $m(H, X)$. Since $Y_2 \subset X \setminus H$ and the complement of Y spans V, Y_2 is not equal to $X \setminus H$, and the claim follows.

At this point, since $\dim(\mathcal{P}(X)) = d(X)$, we are going to finish the proof by showing that $\dim(G(X)) \leq d(X)$.

We work by induction on the cardinality of X. If X is a basis, we have that K_X is the ideal generated by the basis dual to X and $G(X)$ is the 1-dimensional space of constant polynomials. Otherwise, we may assume that $X = (Z, y)$ and Z still spans V. Given a hyperplane H spanned by a subset of Z, we have that $m(H, X) = m(H, Z)$ if $y \in H$ and $m(H, X) = m(H, Z) + 1$ if $y \notin H$.

Notice that by the previous analysis and induction, we may assume that $\mathcal{P}(Z)$ is the subspace of $G(X)$ where all the operators $\phi_H^{m(H,X)-1}$ vanish, as H varies in the set $\mathcal{H}_y$ of all the hyperplanes spanned by subsets of Z that do not contain y. Each operator $\phi_H^{m(H,X)-1}$ applied to $G(X)$ has image in the space $G(X) \cap \ker(\phi_H)$. We thus have an exact sequence (that will turn out to be exact also on the right)

$$0 \to \mathcal{P}(Z) \to G(X) \to \oplus_{H \in \mathcal{H}_y} \phi_H^{m(H,X)-1} G(X).$$

For $H \in \mathcal{H}_y$ we want to estimate the dimension of $\phi_H^{m(H,X)-1} G(X)$. As usual, $\ker(\phi_H) = S[H]$.

As before, let us write $\mathcal{B}_X$ as the disjoint union of $\mathcal{B}_Z$ and the set $\mathcal{B}_X^{(y)}$. By definition, if $\underline{b} \in \mathcal{B}_X^{(y)}$, we have that $\underline{b} \setminus \{y\}$ is a basis of a hyperplane $H \in \mathcal{H}_y$ extracted from $Z \cap H$.

By induction, $d(H \cap Z) = \dim(G(H \cap Z))$. In order to complete our claim, since $d(X) = d(Z) + \sum_{H \in \mathcal{H}_y} d(H \cap Z))$, it is enough to see that $\phi_H^{m(H,X)-1} G(X) \subset G(H \cap Z))$.

Let us consider a hyperplane K of H generated by elements $Y \subset H \cap Z$. Let $\tilde{K}$ be the hyperplane in V spanned by Y and y. An equation $\phi_{\tilde{K}}$ for $\tilde{K}$ restricts on H to an equation for K. Thus our claim follows if we show that the elements of $\phi_H^{m(H,X)-1} G(X)$ lie in the kernel of the operator $\phi_{\tilde{K}}^{m(K,H \cap Z)}$. In other words, we need to show that for every such K, we have $\phi_H^{m(H,X)-1} \phi_{\tilde{K}}^{m(K,H \cap Z)} \in K_X$.

In order to see this, consider all the hyperplanes L_i generated by subsets of X and containing K. The corresponding forms ϕ_{L_i} lie in the subalgebra of polynomials constant on K, which is an algebra of polynomials in two variables, so we may think of each ϕ_{L_i} as a linear form $\phi_i := a_i x + b_i y$. Let us define $h_i := |(L_i \cap X) \setminus K|$. Note that each element $a \in X \setminus K$ lies in a unique L_i (equal to $\langle K, a \rangle$), so that $|X \setminus K| = \sum_i h_i$. Setting $h := \sum_i h_i$, we have that $m(L_i, X) = h - h_i$.

We may assume that $H = L_1$ and $\tilde{K} = L_2$. Also, since by construction clearly $(H \cap Z) \setminus K = (H \cap X) \setminus K$, we have $m(K, H \cap Z) = h_1$. Thus $\phi_H^{m(H,X)-1} \phi_{\tilde{K}}^{m(K,H \cap Z)} = \phi_1^{h-h_1-1} \phi_2^{h_1}$, and it is enough to show that this form is in the ideal generated by the forms $\phi_i^{h-h_i} \in K_X$.

Our claim now follows from the following analogue of Hermite's interpolation:

Lemma 11.26. *Let $\{\phi_i(x,y) = a_i x + b_i y\}$ be r linear forms, pairwise non-proportional, and let h_i, $i = 1, \ldots, r$, be positive integers. Set $h := \sum_{i=1}^{r} h_i$.*

Then all forms of degree $h - 1$ are contained in the ideal generated by the elements $\phi_i^{h-h_i}$.

Proof. Consider the space T of r-tuples of forms $f_1, \ldots, f_r$ with f_i of degree $h_i - 1$. Then T is an h-dimensional space and we have a linear map $(f_1, \ldots, f_r) \mapsto \sum_{i=1}^{r} f_i \phi_i^{h-h_i}$ from T to the h-dimensional space of forms of degree $h - 1$. Thus it is enough to show that this map is injective.

Take a relation $\sum_i f_i \phi_i^{h-h_i} = 0$. We want to show that all $f_i = 0$. By a change of variables we may assume that $\phi_1 = x$. When we apply the derivative $\frac{\partial}{\partial y}$ to this identity, we get

$$0 = \frac{\partial f_1}{\partial y} x^{(h-1)-(h_1-1)} + \sum_{i=2}^{r} [\frac{\partial f_i}{\partial y} \phi_i + f_i c_i] \phi_i^{(h-1)-h_i}, \quad c_i = (h - h_i) b_i.$$

Now we can apply induction (we have replaced h_1 with $h_1 - 1$) and deduce that

$$\frac{\partial f_1}{\partial y} = 0, \quad \frac{\partial f_i}{\partial y} \phi_i + f_i c_i = 0, \quad i = 2, \ldots, m.$$

In particular, $f_1 = cx^{h_1-1}$. The same argument applied to all the ϕ_i shows that $f_i\phi_i^{h-h_i} = c_i\phi_i^{h-1}$, $c_i \in \mathbb{C}$, and thus we have the relation $\sum_{i=1}^r c_i\phi_i^{h-1} = 0$ and its derivative $\sum_{i=2}^r c_ib_i(h-1)\phi_i^{h-2} = 0$.

We can finish the proof observing that, when $r \leq h$ and the ϕ_i are pairwise nonproportional the elements ϕ_i^{h-1} are linearly independent. This can be proved directly using the Vandermonde matrix (see page 187) but it also follows by induction using the previous argument.

Remark 11.27. In a recent paper [62], Holtz and Ron study a variation of this construction by also analyzing, in the case of X in a lattice, the quotient algebras modulo the ideal generated by $\phi_H^{m(H,X)+\epsilon}$ when $\epsilon = 1$ or -1. It turns out that these algebras, called *zonotopal* have combinatorial properties related to zonotopes. For $\epsilon = -1$ the dimension equals the number of integral interior points in the zonotope, while for $\epsilon = 1$, it is the total number of integral points in the zonotope.

11.3.3 Parametric Case

In the parametric case, we have the decomposition $d_X^{-1} = \sum_p c_p d_{X_p}^{-1}$ (cf. (8.5)). We set $Q_X(\underline{\mu})$ equal to the $S[V]$-module generated by u_X, and for each point $p \in P(X, \underline{\mu})$, we set $Q_X(p)$ equal to the $S[V]$-module generated by u_{X_p}.

Proposition 11.28. *1.*

$$Q_X(\underline{\mu}) = \oplus_{p \in P(X_p, \underline{\mu}_p)} Q_{X_p}(\underline{\mu}_p).$$

2. $Q_X(\underline{\mu})$ is isomorphic to $A_X(\underline{\mu})$, and the above decomposition coincides with the decomposition of $A_X(\underline{\mu})$ into its local components.

Proof. The first part follows immediately from Theorem 8.20.

As for the second, since the annihilator of u_X contains $I_{X,\underline{\mu}_p}$, we have a map $A_X(\underline{\mu}) \to Q_X(\underline{\mu})$, which by Fitting decomposition induces a map of the local summands. On each such summand this map is an isomorphism by the previous theorem, since we can translate p to 0. From this everything follows.

11.3.4 A Filtration

Set $L := d_X^{-1}A = d_X^{-1}S[V]$, and $L_k := L \cap R_{X,k}$. It is of some interest to analyze these deeper intersections.

If X does not necessarily span V, we define I_X as the ideal of $A = S[V]$ generated by the products $\prod_{x \in Y} x$, where $Y \subset X$ is any sublist such that the span of $X \setminus Y$ is strictly contained in the span of X. We set $A_X(V) = A/I_X$.

Of course, if we fix a decomposition $V := \langle X \rangle \oplus T$ we have an identification $A_X(V) = A_X \otimes S[T]$.

In particular, we can apply the previous definition to any subspace W of the arrangement and $X_W := \{a \in X \,|\, a = 0, \text{ on } W\}$.

Theorem 11.29. *For each k we have that L_k/L_{k-1} is isomorphic to the direct sum $\oplus A_{X_W}(V)$ as W varies over all the subspaces of codimension k of the arrangement.*

Proof. Consider the map $A \to L$ given by $f \to d_X^{-1} f$. By its very definition, and Theorem 11.20, the ideal I_X maps isomorphically onto L_{s-1}. So

$$L_{s-1} = \sum_H d_{X \cap H}^{-1} A$$

as H runs over the hyperplanes spanned by subsets of X. Set $L_H = d_{X \cap H}^{-1} A$. By Proposition 8.18, the filtration by polar order in R_X induces the same filtration in $R_{X \cap H}$ for each such hyperplane H. On the other hand, by definition, $R_{X,s-1} = \sum_H R_{X \cap H}$. We deduce that for each $k \le s - 1$,

$$L_k/L_{k-1} = \sum L_{H,k}/L_{H,k}.$$

We have $X \cap H = X_{H^\perp}$, and by induction we have that $L_{H,k}/L_{H,k-1}$ is the direct sum of all the pieces A_{X_W} as W varies over the spaces of codimension k in V of the arrangement defined by equations in $X \cap H$, in other words containing the line $H^\perp$. It follows that for each $k \le s$,

$$L_k/L_{k-1} = \sum_{W \text{ cod } W=k} A_{X_W}.$$

The fact that this sum is direct then follows from Theorem 8.10.

Remark 11.30. Using the Laplace transform, the previous discussion can be translated into an analysis of the distributional derivatives of the multivariate spline, i.e., into an analysis of the various discontinuities achieved by the successive derivatives on all the strata of the singular set.

11.3.5 Hilbert Series

When we compute the Hilbert series of $d_X^{-1} S[V]$ directly or as the sum of the contributions given by the previous filtration we get the identity:

$$\frac{q^{-|X|}}{(1-q)^s} = \sum_{k=0}^{s} \sum_{W \in \mathcal{W}_k} \frac{q^{-|X_W|}}{(1-q)^{s-k}} H_{X_W}(q), \qquad (11.7)$$

or

$$1 = \sum_{k=0}^{s} \sum_{W \in \mathcal{W}_k} q^{|X|-|X_W|}(1-q)^k H_{X_W}(q). \qquad (11.8)$$

Here $\mathcal{W}_k$ denotes the collection of subspaces of codimension k of the arrangement, and if $W \in \mathcal{W}_k$, we denote by X_W the set of elements in X vanishing on W.

In terms of external activity this formula is

$$q^{|X|} = \sum_{k=0}^{s}(q-1)^k \sum_{W \in \mathcal{W}_k} E_{X_W}(q). \tag{11.9}$$

Notice that this formula coincides with formula (2.10), which we proved combinatorially.

In the applications to the box spline it is interesting, given a set X of vectors that we list is some way, to consider for each $k \geq 0$ the list X^k in which every element $a \in X$ is repeated k times. Let us make explicit the relationship between $H_X(q)$ and $H_{X^k}(q)$.

First, the number of bases in the two cases is clearly related by the formula $d(X^k) = d(X)k^s$, since to each basis $\underline{b} := (b_1, \dots, b_s)$ extracted from X we associate k^s bases $\underline{b}(h_1, \dots, h_s)$, indexed by s numbers $1 \leq h_i \leq k$ each expressing the position of the corresponding b_i in the list of the k repeated terms in which it appears.

Now it is easy to see that

$$e(\underline{b}(h_1, \dots, h_s)) = ke(\underline{b}) + (h_1 - 1) + \cdots + (h_s - 1).$$

Thus we deduce the explicit formula

$$H_{X^k}(q) = \sum_{\underline{b} \in \mathcal{B}_X} \sum_{h_1, \dots, h_s = 1}^{k} q^{km - ke(\underline{b}) - (h_1 - 1) - \cdots - (h_s - 1)} = H_X(q^k)\left(\frac{q^k - 1}{q - 1}\right)^s.$$

11.4 More Differential Equations

11.4.1 A Characterization of the Polynomials $p_{\underline{b}}(-x)$

Our next task is to fully characterize the polynomials $p_{\underline{b}}(-x)$ appearing in formula (9.5) by differential equations. In Theorem 11.6, we have seen that these polynomials lie in $D(X)$.

For a given unbroken basis $\underline{b}$, consider the element $D_{X \setminus \underline{b}} := \prod_{a \in X,\ a \notin \underline{b}} D_a$.

Proposition 11.31. *1. The polynomials $p_{\underline{b}}(-x)$ satisfy the system*

$$D_{X \setminus \underline{b}} p_{\underline{c}}(-x_1, \dots, -x_s) = \begin{cases} 1 & \text{if } \underline{b} = \underline{c}, \\ 0 & \text{if } \underline{b} \neq \underline{c}. \end{cases} \tag{11.10}$$

2. The polynomials $p_{\underline{b}}(-x)$ for $\underline{b} \in \mathcal{NB}$ form a basis of $D(X)_{m-s}$. They are characterized, inside $D(X)$, by the equations (11.10).

Proof.   1 This follows from the identity $d_{X\setminus\underline{b}}u_X = u_{\underline{b}}$ and Formula (11.3). In fact, this last formula is based on the fact that the Fourier dual of the module of polar parts is a direct sum $\hat{P}_X = \oplus_{\underline{b}\in\mathcal{NB}}\mathbb{C}[x_1,\ldots,x_s]\hat{u}_{\underline{b}}$ of copies of the polynomial ring in which $\hat{u}_X = \sum_{\underline{b}\in\mathcal{NB}} p_{\underline{b}}(-x_1,\ldots,-x_s)\hat{u}_{\underline{b}}$. In the Fourier dual the identity $d_{X\setminus\underline{b}}u_X = u_{\underline{b}}$ becomes $D_{X\setminus\underline{b}}\hat{u}_X = \hat{u}_{\underline{b}}$.

2 The previous equations imply the linear independence. On the other hand, by Theorem 11.13, the dimension of $D(X)_{m-s}$ equals the number of unbroken bases, and the claim follows.

As a consequence, using formula (9.5), we can characterize by differential equations the multivariate spline $T_X(x)$ on each big cell Ω as the function T in $D(X)_{m-s}$ satisfying the equations

$$D_{X\setminus\underline{b}}T = \begin{cases} |\det(\underline{b})|^{-1} & \text{if}\quad \Omega \subset C(\underline{b}), \\ 0 & \text{otherwise.} \end{cases} \tag{11.11}$$

We have identified $D(X)$ with the dual of A_X and hence of Q_X. The basis $u_{\underline{b}}$, that we found in Q_X, thus defines a dual basis $u^{\underline{b}}$ in $D(X)$. In particular, for top degree we have the following:

Corollary 11.32. *When $\underline{b} \in \mathcal{NB}$, we have $u^{\underline{b}} = p_{\underline{b}}(-x)$.*

Proof. In order to avoid confusion, taking a basis x_i for V we write a typical vector of V as $\sum_i y_i x_i$ and consider the polynomial functions on V as polynomials in the variables y_i. The polynomial associated to $u^{\underline{b}}$ is by definition

$$\left\langle u^{\underline{b}} \,\middle|\, e^{\sum_i y_i x_i} u_X \right\rangle = \left\langle u^{\underline{b}} \,\middle|\, e^{\sum_i y_i x_i} \sum_{\underline{c}\in\mathcal{NB}} p_{\underline{c}}\!\left(\frac{\partial}{\partial x}\right) u_{\underline{c}} \right\rangle.$$

Let us compute $e^{\sum_i y_i x_i} \sum_{\underline{c}\in\mathcal{NB}} p_{\underline{c}}(\frac{\partial}{\partial x})u_{\underline{c}}$.

The commutation relation $[\frac{\partial}{\partial x_j}, e^{\sum_i x_i y_i}] = y_j e^{\sum_i x_i y_i}$ implies that for every polynomial f, we have $e^{\sum_i y_i x_i} f(\frac{\partial}{\partial x_i}) = f(\frac{\partial}{\partial x_i} - y_i)e^{\sum_i y_i x_i}$. Further, we have the fact that $e^{\sum_i y_i x_i}u_{\underline{c}} = u_{\underline{c}}$. So we get

$$e^{\sum_i y_i x_i} p_{\underline{c}}\!\left(\frac{\partial}{\partial x}\right) u_{\underline{c}} = p_{\underline{c}}\!\left(\frac{\partial}{\partial x_1} - y_1, \ldots, \frac{\partial}{\partial x_s} - y_s\right) u_{\underline{c}}.$$

Now, by degree considerations we have

$$\left\langle u^{\underline{b}} \,\middle|\, p_{\underline{c}}\!\left(\frac{\partial}{\partial x} - y\right) u_{\underline{c}} \right\rangle = \langle u^{\underline{b}} \,|\, p_{\underline{c}}(-y)u_{\underline{c}}\rangle = p_{\underline{c}}(-y)\delta^{\underline{b}}_{\underline{c}},$$

whence

$$\langle u^{\underline{b}} \,|\, e^{\sum_i y_i x_i}u_X\rangle = p_{\underline{b}}(-y_1,\ldots,-y_s).$$

We shall use a rather general notion of the theory of modules. Recall that the *socle* $s(M)$ of a module M is the sum of all its irreducible submodules. Clearly if $N \subset M$ is a submodule, $s(N) \subset s(M)$ while for a direct sum we have $s(M_1 \oplus M_2) = s(M_1) \oplus s(M_2)$. If M is an $S[V]$ module that is finite-dimensional as a vector space, then $s(M) \neq 0$ so that for a nonzero submodule N we must have $N \cap s(M) = s(N) \neq 0$.

Proposition 11.33. *The socle of the $S[V]$-module Q_X coincides with its top-degree part, with basis the elements $u_{\underline{b}}$.*

Proof. The socle of the algebra of polynomials, thought of as a module over the ring of constant coefficient differential operators, reduces clearly to the constants. By the Fourier transform, the socle of the algebra of constant coefficients differential operators thought of as a module over the polynomial ring, is also $\mathbb{C}$. It follows that the socle of P_X (as an $S[V]$-module), is the vector space with basis the elements $u_{\underline{b}}$. Since $u_{\underline{b}} \in Q_X$, we have $s(Q_X) = s(P_X)$, and the claim follows.

Theorem 11.34. *The space $D(X)$ is spanned by the polynomials $p_{\underline{b}}$ as $\underline{b}$ runs over the set of unbroken bases, and all of their derivatives.*

Proof. Consider in $D(X)$ the submodule N generated, under the action of $S[V]$, by the polynomials $p_{\underline{b}}$, as $\underline{b}$ runs over the set of unbroken bases.

Recall that $D(X)$ is in duality with A_X and so with Q_X. Consider the submodule $N^\perp$ of Q_X orthogonal to N. Since the polynomials $p_{\underline{b}}$, $\underline{b} \in \mathcal{NB}$, are in duality with the elements $u_{\underline{b}} \in Q_X$ and $s(Q_X)$ is generated by the $u_{\underline{b}}$, we have $N^\perp \cap s(Q_X) = \{0\}$. This implies that $N^\perp = \{0\}$, so that $D(X) = N$, as desired.

We finish this section observing that when we dualize the sequence (11.6), we get the exact sequence

$$0 \to D(Z) \longrightarrow D(X) \longrightarrow \bigoplus_{Y \in \mathcal{S}_y(X)} D(Y) \to 0.$$

By definition, an element ϕ of Q_X^* is the polynomial that, takes the value $\langle \phi \mid e^x u_X \rangle$ on an element $x \in V$. Thus, since the surjective map $Q_X \to Q_Z$ maps the generator u_X to the generator u_Z, the dual map corresponds to the inclusion $D(Z) \subset D(X)$. As for the other inclusion maps $Q_Y \to Q_X$, we have that the generator u_Y maps to $u_{y,Y} = \prod_{a \in X \setminus (y,Y)} a u_X$. Therefore, in the dual map, a polynomial $p \in D(X)$ maps, for each summand associated to Y, to the polynomial $D_{X \setminus (y,Y)} p$.

Observe that $D(Z)$ is also the image of $D(X)$ under the operator D_y.

11.4.2 Regions of Polynomiality

We already know, by Theorem 9.7 that the function T_X is a polynomial when restricted to any big cell. Conversely, we have the following result:

Theorem 11.35. *On two different big cells the function T_X coincides with two different polynomials.*

Proof. By Theorem 9.16 each big cell Ω inside $C(X)$, coincides with the intersection of the interior of the cones $C(\underline{b})$, as $\underline{b}$ varies among the unbroken bases such that $\Omega \subset C(\underline{b})$. Therefore, two different big cells correspond to two different subsets of $\mathcal{NB}$. On the other hand, by formula (9.5)

$$T_X(x) = \sum_{\underline{b} \mid \Omega \subset C(\underline{b})} |\det(\underline{b})|^{-1} p_{\underline{b}}(-x),$$

so our claim follows from the linear independence of the elements $p_{\underline{b}}$, proven in Proposition 11.10.

11.4.3 A Functional Interpretation

Call T_X^r the restriction of T_X to the open set $C^{\mathrm{reg}}(X)$ of strongly regular points. We note that T_X^r is a C^∞ function on $C^{\mathrm{reg}}(X)$. A differential operator P applied to T_X as a distribution, gives a distribution supported in the singular points if and only if we have $PT_X^r = 0$, on $C^{\mathrm{reg}}(X)$. If $P \in S[V]$ is polynomial with constant coefficients, this means that $Pd_X^{-1} \in R_{s-1}$ or that P is in the annihilator of u_X. Applying Theorem 11.20, we have the following:

Proposition 11.36. *The module Q_X can be identified with the space of functions on $C^{\mathrm{reg}}(X)$ obtained from T_X^r by applying derivatives.*

We can now complete our discussion on the relationship between $D(X)$ and the polynomials p_Ω coinciding with T_X in the interior of the big cells Ω.

Theorem 11.37. *The polynomials p_Ω coinciding with T_X in the interior of the big cells Ω span the top degree part of $D(X)$.*

Proof. We already know that these polynomials are linear combinations of the elements $p_{\underline{b}}$ that form a basis of the top degree part of $D(X)$. Thus, by duality, it is enough to show that every differential operator with constant coefficients that annihilates the elements p_Ω also annihilates the elements $p_{\underline{b}}$, in other words, that it is in I_X.

Now we use the previous proposition.

We have identified Q_X with a space of functions on $C^{\mathrm{reg}}(X)$. It is graded by homogeneity. In particular, in degree zero we have the locally constant functions.

Remark 11.38. It follows from Proposition 11.33 that the locally constant functions in Q_X have as basis the restriction to $C^{\mathrm{reg}}(X)$ of the characteristic functions of the cones $C(\underline{b})$, as $\underline{b}$ runs over the unbroken bases.

This theorem has an intriguing combinatorial corollary, whose proof we leave to the reader. We consider the incidence matrix between big cells Ω and unbroken base $\underline{b}$, where we put 1 or 0 in the $(\Omega, \underline{b})$ position according to wether $\Omega \subset C(\underline{b})$ or not. Then this matrix has maximum rank equal to the number of unbroken bases.

11.5 General Vectors

11.5.1 Polynomials

Let us first recall an elementary fact about polynomials. The space of homogeneous polynomials of degree k in s variables has as basis the set of monomials $x_1^{h_1} \cdots x_s^{h_s}$ indexed by s-tuples $h_1, \ldots, h_s$, $h_i \in \mathbb{N}$, $\sum_i h_i = k$. Let us establish a one-to-one correspondence between these sets of indices and subsets of $\{1, \ldots, k + s - 1\}$ consisting of $s - 1$ elements. We think of such a subset A as an ordered sequence $1 \le i_1 < i_2 < \cdots < i_s \le k + s - 1$. The correspondence associates to such an A the sequence $h_1, \ldots, h_s$ with the property that $h_1 + h_2 + \cdots + h_r$ counts the number of positive integers that do not belong to A and are less than the r-th element i_r of A (where $r \in \{1, \ldots, s - 1\}$).

In particular, the dimension of this space is $\binom{k+s-1}{s-1}$. The space of all polynomials in s variables of degree $\le k$ is isomorphic to the space of homogeneous polynomials of degree k in $s+1$ variables, and hence it has dimension $\binom{k+s}{s}$.

11.5.2 Expansion

The set of m-tuples of vectors $(a_1, \ldots, a_m)$ with $a_i \in \mathbb{R}^s$ clearly form a vector space of dimension sm. The condition that a subset A of $(a_1, \ldots, a_m)$ with s elements form a basis is expressed by the nonvanishing of the corresponding determinant. Thus the set of m-tuples of vectors with the property that every subset of s of them is a basis form a dense open set. A sequence with this property will be referred to as a *generic sequence*. For a generic sequence X many of our combinatorial theorems take a very special form.

First let us understand the unbroken bases.

Proposition 11.39. *A subsequence of s elements of a generic sequence X is an unbroken basis if and only if it contains a_1.*

Proof. Any unbroken basis contains a_1, so the condition is necessary. Conversely, a basis containing a_1 is unbroken, since every k elements extracted from X are linearly independent, provided that $k \le s$.

We thus have $\binom{m}{s}$ bases extracted from X and $\binom{m-1}{s-1}$ unbroken bases.

This implies readily the following theorem

Theorem 11.40. *If X is generic, the space $D(X)$ coincides with the space of all polynomials of degree $\le m - s$.*

Proof. By Theorem 11.13 we have $\dim(D(X)) = d(X)$, and if X is generic this equals $\binom{m}{s}$. Moreover, $D(X)$ is contained in the space of all polynomials of degree $\le m - s$ (in s variables), that has dimension $\binom{m}{s}$. The claim follows.

Corollary 11.41. *Consider the $\binom{m}{s-1}$ subsets of X formed by $m - s + 1$ elements. For each such subset A we have $d_A := \prod_{a \in A} a \in S^{m-s+1}(V)$. The elements d_A are linearly independent, and hence form a basis of $S^{m-s+1}(V)$ if and only if X is generic.*

Proof. Assume first X generic. By Theorem 11.40, the ideal I_X that is the orthogonal of $D(X)$ under the duality pairing, is just the ideal generated by $S^{m-s+1}(V)$. On the other hand the subsets of X of cardinality $m - s + 1$ are exactly the minimal cocircuits in X. It follows that the elements d_A as A runs over such subsets generate I_X. In particular, since they are homogeneous of degree $m - s + 1$, they span $S^{m-s+1}(V)$. Now we have $\dim S^{m-s+1}(V) = \binom{m}{s-1}$ thus these elements are necessarily linearly independent.

If on the other hand X is not generic there exist s linearly dependent elements Y in X, then the elements $d_{(a, X \setminus Y)}$, $a \in Y$ are linearly dependent.

Until the end of this chapter let us assume that X is generic. Let us now consider the expression of $1 / \prod_{a \in X} a = d_X^{-1}$.

Theorem 11.42. *There exist computable constants c_X^A with*

$$d_X^{-1} = \sum_{\substack{A \subset \{a_2, \ldots, a_m\}, \\ |A| = s-1}} \frac{c_X^A}{a_1^{m-s+1} \prod_{a \in A} a}. \tag{11.12}$$

Proof. By induction, on $m \geq s$. The case $m = s$ is trivial, so assume $m > s$ and set $Y := \{a_1, \ldots, a_{m-1}\}$. We have the following

$$d_X^{-1} = d_Y^{-1} a_m^{-1} = \sum_{\substack{A \subset \{a_2, \ldots, a_{m-1}\}, \\ |A| = s-1}} \frac{c_Y^A}{a_1^{m-s} \prod_{a \in A} a a_m}. \tag{11.13}$$

Fix a subset $A \subset \{a_2, \ldots, a_{m-1}\}$ with $s - 1$ elements. By assumption, the set $\overline{A} := A \cup \{a_m\}$ is a basis of V and we can write $a_1 = \sum_{b \in \overline{A}} d_A^b b$. If we now multiply numerator and denominator by a_1 and use this expression, we get

$$\frac{c_Y^A}{a_1^{m-s} \prod_{a \in A} a a_m} = \frac{a_1 c_Y^A}{a_1^{m-s+1} \prod_{a \in \overline{A}} a} = \sum_{b \in \overline{A}} d_A^b \frac{c_Y^A}{a_1^{m-s+1} \prod_{a \in \overline{A} \setminus \{b\}} a}.$$

Substituting into (11.13) one gets (11.12) and also a recursive expression for the constants c_X^A.

In order to compute the c_X^A we can alternatively use Corollary 11.41 in the following way. For each $A \subset \{a_2, \ldots, a_m\}$, $|A| = s - 1$, construct the element $q_A := \prod_{a \in \{a_2, \ldots, a_m\} \setminus A} a$. If we multiply 11.12 by $a_1^{m-s} d_X$, we get the expression

$$a_1^{m-s} = \sum_{\substack{A \subset \{a_2, \ldots, a_m\}, \\ |A| = s-1}} c_X^A q_A.$$

On the other hand, Corollary 11.41 applied to $\{a_2, \ldots, a_m\}$ tells us that the elements q_A form a basis of $S^{m-s}(V)$, so the constants c_X^A are the coefficients of the expansion of a_1^{m-s} in this basis.

11.5.3 An Identity

We want to deduce a simple identity from the previous discussion. Let us consider the vectors $a_i = \sum_{j=1}^{s} x_{i,j} e_j$ with variable coordinates, $i \in \{1, \ldots, m\}$. For a subset $A \subset \{1, \ldots, m\}$ with $s-1$ elements, set as before

$$q_A = \prod_{a \notin A} a = \sum_{h_1+h_2+\cdots+h_s=m-s+1} p_A^{h_1,\ldots,h_s}(x_{i,j}) \binom{m-s+1}{h_1 \, h_2 \cdots h_s} e_1^{h_1} \cdots e_s^{h_s}.$$

The polynomials $p_A^{h_1,\ldots,h_s}(x_{i,j})$ can be considered as entries of an $\binom{m}{s-1} \times \binom{m}{s-1}$ matrix Δ whose determinant is a homogeneous polynomial in the variables $x_{i,j}$ of degree $(m-s+1)\binom{m}{s-1}$, which we denote by $\Delta(x_{i,j})$. By Corollary 11.41, this polynomial is nonzero at $(a_1, \ldots, a_m)$ if, and only if, these vectors are generic. In other words, $\Delta(x_{i,j})$ vanishes exactly on the variety where one of the $\binom{m}{s}$ determinants $\det(B)$ of the matrix B, consisting of s columns extracted from X, vanish. These polynomials are all irreducible and each of degree s. Thus $\Delta(x_{i,j})$ is divisible by their product $\prod_B \det(B)$ that in turn is a polynomial of degree $s\binom{m}{s}$. Since $s\binom{m}{s} = (m-s+1)\binom{m}{s-1}$ we deduce that $\Delta(x_{i,j}) = c\prod_B \det(B)$ for some nonzero constant c. In particular, $\Delta(x_{i,j}) = 0$ defines the variety of $(a_1, \ldots, a_m)$ that are not generic.

11.5.4 The Splines

Formula (11.12) gives the explicit computation for the multivariate spline T_X. Given $A \subset \{a_2, \ldots, a_{m-1}\}$, with $|A| = s-1$, let b_A denote the element, dual to a_1, in the basis dual to the basis $A \cup \{a_1\}$, i.e.

$$\langle b_A \,|\, a_1 \rangle = 1, \quad \langle b_A \,|\, a \rangle = 0, \quad \forall a \in A.$$

Here b_A is a linear equation for the hyperplane generated by A (normalized using a_1). Thinking of b_A as a derivation, denote it by ∂_A. We have thus

$$d_X^{-1} = \sum_{\substack{A \subset \{a_2,\ldots,a_m\}, \\ |A|=s-1}} (-1)^{m-s} \partial_A^{m-s} \frac{c_X^A}{(m-s)! a_1 \prod_{a \in A} a}. \tag{11.14}$$

We deduce, from Proposition 11.31 the formula for the polynomials $p_{\underline{b}}$, where $\underline{b} = \{a_1, A\}$:

$$p_{\{a_1,A\}} = b_A^{m-s} \frac{c_X^A}{(m-s)!}.$$

As a corollary we get a variant of Corollary 11.41

Corollary 11.43. *Consider the $\binom{m}{s-1}$ subsets of X formed by $s-1$ elements. The elements b_A^{m-s+1} form a basis of $S^{m-s+1}(V)$.*

Proof. Apply Lemma 11.10.

Denote by χ_A the characteristic function of the cone generated by the basis (a_1, A) divided by the absolute value of the determinant δ_A of (a_1, A). Applying the inverse Laplace transform, we obtain

$$T_X = \sum_{\substack{A \subset \{a_2,\ldots,a_m\}, \\ |A| = s-1}} b_A^{m-s} \frac{c_X^A}{(m-s)!} \chi_A. \tag{11.15}$$

Given a big cell Ω, let us define

$$\mathcal{S}_\Omega := \{A \subset \{a_2,\ldots,a_m\}, \ |A| = s - 1, \ \Omega \subset C(a_1, A)\}.$$

We have thus by Theorem 9.7, when $x \in \overline{\Omega}$,

$$T_X(x) = \sum_{A \in \mathcal{S}_\Omega} b_A(x)^{m-s} \frac{c_X^A}{(m-s)!} \delta_A^{-1}. \tag{11.16}$$

We now want to take advantage of the fact that T_X is of class C^{m-s+1}, that is the maximal possible for a multivariate spline constructed with m vectors in $\mathbb{R}^s$. This implies that crossing a wall of the decomposition into big cells of $C(X)$, the two polynomials describing T_X in two consecutive cells must agree with all their derivatives of order $\leq m - s + 1$. Thus the difference of these two polynomials is a polynomial of degree $m - s$ vanishing with all these derivatives on the hyperplane generated by the wall.

By definition, each codimension-one face of a big cell Ω is part of an $(s-1)$-dimensional cone $C(A)$ generated by a subset A with $s - 1$ elements.

Assume now that $a_1 \notin A$. We have two possibilities: either a_1 lies in the same half-space as Ω or it lies in the opposite space. Using the expression (11.16) and the fact that the elements b_A^{m-s} are linearly independent, we see that in the first case the expression of T_X changes by dropping the term in b_A^{m-s}; in the second case, by adding it.

Of course, a_1 is auxiliary and one can make a different choice obtaining a different basis and rules for crossing walls. This observation appears to generate a number of interesting relations among the linear equations of these hyperplanes.

11.5.5 A Hyper-Vandermonde Identity

Consider again the vectors $a_i = \sum_{j=1}^s x_{i,j} e_j$ with variable coordinates, where $i \in \{1,\ldots,m\}$. For a subset $A = \{a_1,\ldots,a_{s-1}\} \subset \{1,\ldots,m\}$ with $s - 1$ elements, consider

$$f_A = a_1 \wedge \cdots \wedge a_{s-1} = \sum_{j=1}^s \ell_A^j(x_{i,j}) e_1 \wedge \cdots \wedge \check{e}_j \wedge \cdots \wedge e_s.$$

Denote by $u_j \in V^*$ the basis dual to e_i. We can identify, up to some determinant, $\bigwedge^{s-1} V = V^*$, and since $f_A \wedge a_i = 0$, $\forall i \in \{1,\ldots,s_1\}$, we think of

$f_A = \sum_{j=1}^{s}(-1)^{s-j}\ell_A^j(x_{i,j})u_j$ as a multiple of the linear form b_A. We now pass to the elements

$$f_A^{m-s+1} = \sum_{\sum_{i=1}^{s} h_i = m-s+1} s_A^{h_1,\ldots,h_s}(x_{i,j})u_1^{h_1}\cdots u_s^{h_s} \in S^{m-s+1}\left(\bigwedge^{s-1} V\right).$$

The polynomials $s_A^{h_1,\ldots,h_s}(x_{i,j})$ can be considered again as entries of a square $\binom{m}{s-1} \times \binom{m}{s-1}$ matrix Γ. We denote its determinant by $\Gamma(x_{i,j})$, this is a polynomial of degree $(m-s+1)\binom{m}{s-1}$. We apply the results of the previous paragraph (adding an auxiliary vector a_0 to the list that makes it of $m+1$ elements).

Lemma 11.44. *The polynomial $\Gamma(x_{i,j})$ is nonzero if and only if when we evaluate the $x_{i,j}$, the resulting vectors are generic.*

Proof. In one direction this is a consequence of corollary 11.43. On the other hand, if s elements in X are dependent, the s subsets of this sublist consisting of $s-1$ elements give rise either always to 0 or to proportional linear functions. In any case, the resulting elements b_A^{m-s+1} are no longer independent and thus cannot be a basis.

We have thus proved that the determinant $\Gamma(x_{i,j})$ vanishes exactly on the variety where one of the $\binom{m}{s}$ determinants $\det(B)$ of the matrix B, consisting of s columns extracted from X, vanish. These polynomials are all irreducible and each of degree s thus $\Gamma(x_{i,j})$ is divisible by their product $\prod_B \det(B)$, which in turn is a polynomial of degree $s\binom{m}{s}$. Since $s\binom{m}{s} = (m-s+1)\binom{m}{s-1}$, we deduce that $\Gamma(x_{i,j}) = c\prod_B \det(B)$ for some nonzero constant c. In particular, we have the following result:

Theorem 11.45. *For some nonzero constant c' we have $\Gamma(x_{i,j}) = c'\Delta(x_{i,j})$.*

Let us discuss in detail the case $m = 2$.

It is convenient to dehomogenize the coefficients and set $f_i = a_i x + y$, so that

$$f_i \wedge f_j = a_i - a_j \quad \text{and} \quad f_i^{m-1} = \sum_{j=0}^{m-1} a_i^{m-1-j}\binom{m-1}{j}x^{m-1-j}y^j.$$

The matrix Γ (in the basis $\binom{m-1}{j}x^{m-1-j}y^j$) is the Vandermonde matrix, and the resulting identity is the usual computation of its determinant.

For the matrix Δ we compute $\prod_{j\neq i}(a_j x + y) = \sum_{i=0}^{m-1} e_j^i x^j y^{m-1-j}$.

Thus the entry e_j^i of Δ is the j-th elementary symmetric function in the elements $a_1,\ldots,\breve{a}_i,\ldots,a_m$. By comparing leading terms, one finds that the constant c' in the previous identity equals 1.

We make explicit the discussion of the multivariate spline in dimension 2.

Let $X := \{a_1, \ldots, a_{m+1}\}$ be $m + 1$ pairwise nonproportional vectors, ordered in a clockwise form, in coordinates $a_i = (x_i, y_i)$. The cone generated by them is the quadrant $C(a_1, a_{m+1})$, and it is thus divided in m cells, the cones $C_i := C(a_i, a_{i+1})$, $i = 1, \ldots, m$. The unbroken bases for this ordering are the bases a_1, a_i, $i = 2, \ldots, m+1$ with determinant $d_i = x_1 y_i - y_1 x_i$. Each generates a cone $U_i := C(a_1, a_i)$.

A point in the closure of C_i lies exactly in the cells $U_j, j \geq i + 1$.

In order to apply formula (11.16) we first observe that the linear function b_i equals $d_i^{-1}(y_i x - x_i y)$, and then we only need to compute the coefficients of the expansion $a_1^{m-1} = \sum_{i=2}^{m+1} c_i \prod_{j \neq i, 1} a_i$. We finally get, for $x \in C_i$

$$T_X(x) = \sum_{j=i+1}^{m+1} b_i^{m-1} \frac{c_i}{(m-1)!|d_i|}.$$

Notice that when $i = 1$, the sum $\sum_{j=2}^{m+1} b_i^{m-1} \frac{c_i}{(m-1)!|d_i|}$ is also a multiple of the power a_1^{m-1}.

Problem: Using the duality pairing

$$S^{m-s+1}(\bigwedge^{s-1} V) \times S^{m-s+1}(V) \to (\bigwedge^{s} V)^{m+s-1},$$

we should compute the interesting matrix of pairings $\langle f_A^{m-s+1} \mid q_B \rangle$ indexed by pairs of subsets of X with $s - 1$ elements.

Part III

The Discrete Case

Integral Points in Polytopes

In this chapter we begin to study the problem of counting the number of integer points in a convex polytope, or the equivalent problem of computing a partition function. We start with the simplest case of numbers. We continue with the theorems of Brion and Ehrhart and leave the general discussion to the next chapters.

12.1 Decomposition of an Integer

12.1.1 Euler Recursion

Given positive integers $\underline{h} := (h_1, \ldots, h_m)$, the problem of counting the number of ways in which a positive integer n can be written as a linear combination $\sum_{i=1}^{n} k_i h_i$, with the k_i again positive integers, is a basic question in arithmetic. Its answer is quite complex. For instance, in the simple example in which we want to write

$$b = 2x + 3y, \quad x, y \in \mathbb{N},$$

we see directly that the answer depends on the class of b modulo 6:

Residue modulo 6	Number of solutions
0	$b/6 + 1$
1	$(b - 1)/6$
2	$(b + 4)/6$
3	$(b + 3)/6$
4	$(b + 2)/6$
5	$(b + 1)/6$

The following approach goes back at least to Euler, who showed that this function, that we denote by $\mathcal{T}_{\underline{h}}(n)$ or simply u_n, satisfies a simple recursive relation that at least in principle allows us to compute it. The relation is classically expressed as follows. First notice that $u_0 = 1$. Consider the polynomial $\prod_i (1 - x^{h_i})$, expand it, and then for any $n > 0$ substitute formally

C. De Concini and C. Procesi, *Topics in Hyperplane Arrangements, Polytopes and Box-Splines*, Universitext, DOI 10.1007/978-0-387-78963-7_12,
© Springer Science+Business Media, LLC 2010

for each x^r the expression u_{n-r} (setting $u_h = 0$ for $h < 0$). If we equate the resulting expression to 0, we get the required recursion (cf. formula 12.1).

We shall presently explain the meaning of these types of recursions in the general multidimensional case.

Let us start by remarking:

Lemma 12.1. *The number $\mathcal{T}_{\underline{h}}(n)$ is the coefficient of x^n in the power series expansion of*

$$S_{\underline{h}}(x) := \prod_{i=1}^{m} \frac{1}{1 - x^{h_i}} = 1 + \sum_{n=1}^{\infty} \mathcal{T}_{\underline{h}}(n) x^n.$$

Proof.

$$\prod_{i=1}^{m} \frac{1}{1 - x^{h_i}} = \prod_{i=1}^{m} \sum_{k=0}^{\infty} x^{k h_i} = \sum_{k_1=0,\ldots,k_m=0}^{\infty} x^{\sum_{i=1}^{m} k_i h_i} = 1 + \sum_{n=1}^{\infty} \mathcal{T}_{\underline{h}}(n) x^n.$$

If $\prod_{i=1}^{m}(1 - x^{h_i}) = 1 + \sum_{j=1}^{N} w_j x^j$, we have

$$(1 + \sum_{n=1}^{\infty} \mathcal{T}_{\underline{h}}(n) x^n)(1 + \sum_{j=1}^{N} w_j x^j) = 1,$$

that is, the recursion

$$\mathcal{T}_{\underline{h}}(n) + \sum_{j=1}^{n} w_j \mathcal{T}_{\underline{h}}(n - j) = 0, \ \forall n > 0, \quad \mathcal{T}_{\underline{h}}(0) = 1. \tag{12.1}$$

In general, given a sequence $\underline{v} := \{v_n\}$, $n \in \mathbb{Z}$, we shall consider its *generating function* $G_{\underline{v}}(x) := \sum_{n=-\infty}^{\infty} v_n x^n$.

Such expressions are not to be understood as functions or convergent series, but only symbolic expressions. They form a vector space but cannot in general be multiplied. Nevertheless, they form a module over the algebra of Laurent polynomials.

Let us now introduce the *shift operator* on sequences $\underline{v} := \{v_n\}$ defined by $(\tau\underline{v})_n := v_{n-1}$. We clearly have

$$G_{\tau\underline{v}}(x) = x G_{\underline{v}}(x).$$

Given a polynomial $p(x)$ and a constant c (that we think of as a sequence which is zero for $n \neq 0$), we say that $\underline{v}$ satisfies the recursion $(p(\tau), c)$ if $p(\tau)(\underline{v}) = c$.

In terms of generating functions this translates into $p(x)G_{\underline{v}}(x) = c$.

In our case, setting $p_{\underline{h}}(x) = \prod_{i=1}^{m}(1 - x^{h_i})$, the function $S_{\underline{h}}(x)$ obviously satisfies the recursion equation

$$p_{\underline{h}}(x)S_{\underline{h}}(x) = 1, \tag{12.2}$$

whence the recursion $(p_{\underline{h}}(\tau), 1)$. Notice that the series $S_{\underline{h}}(x)$ is clearly the unique solution to (12.2) that is a power series.

Remark 12.2. Observe (using (5.4)) that

$$p_{\underline{h}}(\tau) = \prod_{i=1}^{m} \nabla_{h_i}, \quad p_{\underline{h}}(\tau)f(x) = \sum_{S \subset \{1,\ldots,m\}} (-1)^{|S|} f\Big(x - \sum_{i \in S} h_i\Big).$$

Thus formula (12.2) gives the following recursive formula for $f(x) = S_{\underline{h}}(x)$:

$$f(x) = \sum_{S \subset \{1,\ldots,m\}, \, S \neq \emptyset} (-1)^{|S|+1} f\Big(x - \sum_{i \in S} h_i\Big), \tag{12.3}$$

subject to the initial conditions $f(0) = 1, f(j) = 0, \; \forall j \mid \; -\sum_{i=1}^{m} h_i < j < 0$.

We shall see a far-reaching generalization of these results in Section 13.2.2 and especially Theorem 13.52.

The number of ways in which n can be written as combination of numbers h_i was called by Sylvester a *denumerant,* and the first results on denumerants are due to Cayley and Sylvester, who proved that such a denumerant is a polynomial in n of degree $m - 1$ plus a periodic polynomial of lower degree called an *undulant*; see Chapter 3 of the book of Dickson [50] or the original papers of Cayley Sylvester, or in Kap. 3 of Bachmann [5]; and we present a variant of this in the next two sections. A different approach is also developed by Bell in 1943 [15]. A more precise description of the leading polynomial part and of the periodic corrections is presented in Section 16.3. Finally, a computational algorithm is presented in Section 16.4.

12.1.2 Two Strategies

Fix positive integers $\underline{h} := (h_1, \ldots, h_m)$. In order to compute for a given $n \geq 0$ the coefficient $T_{\underline{h}}(n)$ of x^n in the series expansion of the function

$$S_{\underline{h}}(x) = \prod_i \frac{1}{1 - x^{h_i}} = \sum_{n=0}^{\infty} T_{\underline{h}}(n)x^n,$$

or equivalently, of x^{-1} in the expansion of $x^{-n-1}S_{\underline{h}}(x)$, we can use two essentially equivalent strategies, that we are going to analyze separately from the algorithmic point of view.

1. Develop $S_{\underline{h}}(x)$ in partial fractions.

2. Compute the residue

$$\frac{1}{2\pi i} \oint \frac{x^{-n-1}}{\prod_i 1 - x^{h_i}} dx$$

around 0.

In both cases first we must expand in a suitable way the function $S_{\underline{h}}(x)$. Given k, let us denote by $\zeta_k := e^{\frac{2\pi i}{k}}$ a primitive k-th root of 1; we have then the identity

$$1 - x^k = \prod_{i=0}^{k-1}(1 - \zeta_k^i x) = (-1)^{k-1}\prod_{i=0}^{k-1}(\zeta_k^i - x).$$

Using this, we can write

$$S_{\underline{h}}(x) := (-1)^{\sum_i h_i - m}\prod_{i=1}^{m}\prod_{j=0}^{h_i-1}\frac{1}{\zeta_{h_i}^j - x} = \prod_{i=1}^{m}\prod_{j=0}^{h_i-1}\frac{1}{1 - \zeta_{h_i}^j x}. \tag{12.4}$$

Let μ be the least common multiple of the numbers h_i, and write $\mu = h_i k_i$. If $\zeta = e^{2\pi i/\mu}$, we have $\zeta_{h_i} = \zeta^{k_i}$; therefore we have the following:

Lemma 12.3.

$$S_{\underline{h}}(x) = \prod_i \frac{1}{1 - x^{h_i}} = \prod_{i=1}^{m}\prod_{j=0}^{h_i-1}\frac{1}{1 - \zeta^{k_i j} x} = \prod_{\ell=0}^{\mu-1}\frac{1}{(1 - \zeta^\ell x)^{b_\ell}},$$

where the integer b_ℓ is the number of k_i that are divisors of ℓ.

In particular, the function $x^{-n-1}S_{\underline{h}}(x)$, $n \geq 0$, has poles at 0 and at the μ-th roots of 1 (but not at ∞).

12.1.3 First Method: Development in Partial Fractions.

The classical method starts from the fact that there exist numbers c_i for which

$$\prod_{i=0}^{\mu-1}\frac{1}{(1 - \zeta^i x)^{b_i}} = \sum_{i=0}^{\mu-1}\sum_{k=1}^{b_i}\frac{c_{i,k}}{(1 - \zeta^i x)^k}. \tag{12.5}$$

In order to compute them, we can use recursively the simple identity, valid if $a \neq b$

$$\frac{1}{(1 - ax)(1 - bx)} = \frac{1}{a - b}\left[\frac{a}{(1 - ax)} - \frac{b}{(1 - bx)}\right]$$

and then (cf. Section 11.5.1)

$$\frac{1}{(1 - t)^k} = \sum_{h=0}^{\infty}\binom{k-1+h}{h}t^h \tag{12.6}$$

to get

$$\prod_{i=0}^{\mu-1}\frac{1}{(1 - \zeta^i x)^{b_i}} = \sum_{i=0}^{\mu-1}\sum_{k=1}^{b_i}c_{i,k}\left[\sum_{h=0}^{\infty}\binom{k-1+h}{h}(\zeta^i x)^h\right]. \tag{12.7}$$

We have thus obtained a formula for the coefficient

$$T_{\underline{h}}(n) = \sum_{i=0}^{\mu-1} \zeta^{in} \left[\sum_{k=1}^{b_i} c_{i,k} \binom{k-1+n}{n} \right].$$

Let us observe now that

$$\binom{k-1+n}{n} = \frac{(n+1)(n+2)\cdots(n+k-1)}{(k-1)!}$$

is a polynomial of degree $k-1$ in n, while the numbers ζ^{in} depend on the coset of n modulo μ. Given an $0 \le a < \mu$ and restricting us to the numbers $n = \mu k + a$, we have the following:

Theorem 12.4. *The function $T_{\underline{h}}(mk + a)$ is a computable polynomial in the variable k of degree $\le \max(b_i)$.*

The fact that $T_{\underline{h}}(n)$ is a polynomial on every coset means that it is a *quasipolynomial*, or *periodic polynomial*.

12.1.4 Second Method: Computation of Residues

Here the strategy is the following: shift the computation of the residue to the remaining poles, taking advantage of the fact that the sum of residues at all the poles of a rational function is 0.

From the theory of residues we have

$$\frac{1}{2\pi i} \oint \prod_i \frac{x^{-n-1}}{1 - x^{h_i}} dx = -\sum_{j=1}^{\mu-1} \frac{1}{2\pi i} \oint_{C_j} \prod_{t=1}^{\mu} \frac{x^{-n-1}}{(1 - \zeta^t x)^{b_t}} dx,$$

where C_j is a small circle around ζ^{-j}. In order to compute the term

$$\frac{1}{2\pi i} \oint_{C_j} \prod_{t=1}^{\mu} \frac{x^{-n-1}}{(1 - \zeta^t x)^{b_t}} dx, \tag{12.8}$$

we perform the change of coordinates $x = w + \zeta^{-j}$, obtaining

$$\frac{1}{2\pi i} \oint_{C_j - \zeta^{-j}} \prod_{t=1}^{\mu} \frac{(w + \zeta^{-j})^{-n-1}}{(1 - \zeta^{t-j} - \zeta^t w)^{b_t}} dw.$$

Now $\prod_{t=1,\ t\ne j}^{m} 1/(1 - \zeta^{t-j} - \zeta^t w)^{b_i}$ is holomorphic around 0, and we can explicitly expand it in a power series $\sum_{h=0}^{\infty} a_{j,h} w^h$, while

$$(w + \zeta^{-j})^{-n-1} = \zeta^{j(n+1)} \sum_{k=0}^{\infty} (-1)^k \binom{n+k}{k} (\zeta^j w)^k.$$

Finally, we have that (12.8) equals

$$\frac{(-1)^{b_j}}{2\pi i} \oint_{C_j - \zeta^{-j}} \zeta^{j(n+1-b_j)} \left(\sum_{k=0}^{\infty} (-1)^k \binom{n+k}{k} \zeta^{jk} w^k \right) \left(\sum_{h=0}^{\infty} a_{j,h} w^h \right) w^{-b_j} dw$$

$$= (-1)^{b_j} \zeta^{j(n+1-b_j)} \sum_{k+h=b_j-1} (-1)^k \zeta^{jk} \binom{n+k}{k} a_{j,h}.$$

Summing over j, we obtain an explicit formula for $\mathcal{T}_{\underline{h}}(n)$, again as a *quasipolynomial*.

Observe that in order to develop these formulas it suffices to compute a finite number of coefficients $a_{j,h}$.

In these formulas roots of unity appear, while the final partition functions are clearly integer-valued. An algorithmic problem remains. When we write out our expressions, we get, for each coset, a polynomial that takes integer values but that a priori is expressed with coefficients that are expressions in the root ζ. We need to know how to manipulate such an expression. This is an elementary problem, but it has a certain level of complexity and requires some manipulations on cyclotomic polynomials. We shall explain in Section 16.3 how to sum over roots of unity by computing so called Dedekind sums.

12.2 The General Discrete Case

12.2.1 Pick's Theorem and the Ehrhart Polynomial

As we shall see presently, the problem of counting integer points in a polytope is computationally difficult. One main point is how we describe the polytopes P and how to study this number as a function of the given data describing P. We have seen that there are basically two dual approaches to the description of convex polytopes. One is as the convex envelope of finitely many points, the other as the set of points satisfying finitely many linear inequalities. Most of our work is concerned with the second point of view, but we want first to recall some classical results on the first approach. We refer to [14] for details and several instructive examples.

There is a beautiful simple formula known as *Pick's theorem* in the 2-dimensional case. If P is a convex polygon whose vertices have integral coordinates, we call it an *integral polygon*. Let A be its area, and let I, B be the numbers of integer points in P and in its boundary, respectively. We have then

$$\text{Pick's theorem:} \qquad A = I + \frac{1}{2}B - 1.$$

The proof is by a simple subdivision argument, and we refer to [14] for details.

An attempt to generalize this result was made in several papers of Ehrhart [53], [54]. Starting from an integral polytope P he studied the number $p(n)$

of points of the homothetic polytopes nP with $n \in \mathbb{N}$ and showed that $p(n)$ is in fact a polynomial in n, the *Ehrhart polynomial*. For a rational polytope one instead obtains a quasipolynomial. Its leading term is clearly asymptotic to the volume of nP, and it is an interesting question to describe its various coefficients more geometrically. Not much is known about this, but we shall point out some properties.

12.2.2 The Space $\mathcal{C}[\Lambda]$ of Bi-infinite Series

In passing to the general case, we need to develop a small amount of general formalism that will help us in the proof of Brion's theorem.

Let Γ be a lattice of rank m, that by choosing a basis we can identify with $\mathbb{Z}^m$. As in Section 5.3.1 we denote by $\mathcal{C}[\Gamma]$ the space of complex-valued functions on Γ. This is the algebraic dual to the group algebra $\mathbb{C}[\Gamma]$. Indeed, given $f \in \mathcal{C}[\Gamma]$ and $c = \sum_{\gamma \in \Gamma} c_\gamma e^\gamma$, the pairing $\langle f \,|\, c \rangle$ is given by

$$\langle f \,|\, c \rangle = \sum_{\gamma \in \Gamma} f(\gamma) c_\gamma$$

This makes sense, since only finitely many coefficients c_γ are different from zero.

If $\gamma \in \Gamma$, we denote by δ_γ the delta function on Γ identically equal to 0 on Γ, except for $\delta_\gamma(\gamma) = 1$. Thus an element $f \in \mathcal{C}[\Gamma]$ can be expanded as

$$f = \sum_{\gamma \in \Gamma} f(\gamma) \delta_\gamma.$$

According to definition 5.20 this is also thought of as a distribution supported in Γ.

The *support* of a function $f \in \mathcal{C}[\Gamma]$ is the set of elements $\gamma \in \Gamma$ with $f(\gamma) \neq 0$.

Alternatively, if we consider the compact torus $T = \hom(\Gamma, S^1)$ dual to Γ (resp. the algebraic torus $T_{\mathbb{C}} = \hom(\Gamma, \mathbb{C}^*)$), the ring $\mathbb{C}[\Gamma]$ may be considered as a ring of C^∞ functions on T (resp. as the ring of regular functions on $T_{\mathbb{C}}$).

With this in mind we may also consider the space $\mathbb{C}[[\Gamma]]$ of formal unbounded power series

$$\sum_{\gamma \in \Gamma} g(\gamma) e^\gamma$$

as a space of formal generalized functions on T and think of the values $g(\gamma)$ as Fourier coefficients.

We have an obvious isomorphism of vector spaces

$$L : \mathcal{C}[\Gamma] \to \mathbb{C}[[\Gamma]]$$

defined by

$$Lf = \sum_{v \in \Gamma} f(\gamma)e^{-\gamma},$$

that the reader should think of as a formal Laplace transform.

Remark 12.5.   (i) Usually the functions f we are going to consider satisfy suitable growth conditions that ensure that $f(\gamma)$ can be considered as the Fourier coefficients of a function or sometimes a generalized function on T.

(ii) In general, $\mathcal{C}[\Gamma]$ is not an algebra. However, given $f_1, f_2 \in \mathcal{C}[\Gamma]$ with respective support S_1, S_2, if we assume that for any element $\gamma \in \Gamma$ we have only finitely many pairs $(\gamma_1, \gamma_2) \in S_1 \times S_2$ with $\gamma = \gamma_1 + \gamma_2$, then the convolution $f_1 * f_2$ is defined by

$$(f_1 * f_2)(\gamma) = \sum_{\gamma_1 + \gamma_2 = \gamma} f_1(\gamma_1)f_2(\gamma_2).$$

A related point of view is the following. Let $W := \Gamma \otimes \mathbb{R}$. If the function $f(v)$ is supported in a pointed cone $C \subset W$ and has a suitable polynomial growth, then $D := \sum_{v \in \Gamma} f(v)\delta_v$ is a tempered distribution and we have a discussion similar to that of Section 3.1.5. The Laplace transform, defined by extending formula (3.2) to distributions, is

$$Lf(y) = \sum_{v \in \Gamma} f(v)e^{-\langle v \mid y \rangle} = \langle D \mid e^{-\langle v \mid y \rangle} \rangle.$$

It converges on regions of the dual cone of C, i.e., where $\langle v \mid y \rangle > 0, \forall v \in C$. In fact, we have that as in Section 3.1.5, the Fourier transform of D is a boundary value of a function in a variable $z \in \mathrm{hom}(\Gamma, \mathbb{C})$ holomorphic in the region where the imaginary part of $z = x + iy$ lies in $\hat{C}$ and that gives the Laplace transform on the set $i\hat{C}$. The given formula shows that in fact, such a holomorphic function is periodic with respect to $\Gamma^* = \mathrm{hom}(\Gamma, 2\pi i\mathbb{Z})$ and thus it really defines a holomorphic function on some open set of the algebraic torus $T_{\mathbb{C}} = \mathrm{hom}(\Gamma, \mathbb{C})/\Gamma^*$ (see Proposition 5.13). In some of the main examples this holomorphic function coincides with a rational function. In that case, we shall identify the Laplace transform with this rational function.

The product of two elements Lf_1 and Lf_2 in $\mathbb{C}[[\Gamma]]$ is defined if and only if $f_1 * f_2$ is defined and by definition

$$L(f_1 * f_2) = Lf_1 Lf_2.$$

Consider $\mathbb{C}[\Gamma]$ as a subspace of $\mathbb{C}[[\Gamma]]$ and notice that the product of an element in $\mathbb{C}[\Gamma]$ and an arbitrary element is always well-defined, so that $\mathbb{C}[[\Gamma]]$ is a $\mathbb{C}[\Gamma]$ module.

As a preliminary to the analytic discussion of the Laplace transform of partition functions we may develop an algebraic formalism to which one can give an analytic content, as we shall see presently.

Definition 12.6. We say that an element $a \in \mathbb{C}[[\Gamma]]$ is *rational* if there is a nonzero $p \in \mathbb{C}[\Gamma]$ with $pa \in \mathbb{C}[\Gamma]$.

An element $a \in \mathbb{C}[[\Gamma]]$ is a *torsion element* if there is a nonzero $p \in \mathbb{C}[\Gamma]$ with $pa = 0$.

Let us denote by $\mathcal{R}$ the set of rational elements in $\mathbb{C}[[\Gamma]]$ and by $\mathcal{T}$ the set of *torsion* elements in $\mathbb{C}[[\Gamma]]$.

Proposition 12.7. *1. $\mathcal{R}$ and $\mathcal{T}$ are $\mathbb{C}[\Gamma]$ modules.*
2. $\mathcal{R}/\mathcal{T}$ is isomorphic, as a $\mathbb{C}[\Gamma]$ module, to the quotient field $\mathbb{C}(\Gamma)$ of $\mathbb{C}[\Gamma]$.

Proof. 1 is easy. Let us prove 2. First of all observe that if $a \in \mathcal{R}$ and we have two nonzero elements $p, q \in \mathbb{C}[\Gamma]$ such that $pa = r, qa = v \in \mathbb{C}[\Gamma]$, we have $pv = qr$ so that the rational function $i(a) := r/p = v/q$ is well-defined. Clearly, the map i is a $\mathbb{C}[\Gamma]$ linear map of $\mathcal{R}$ into $\mathbb{C}(\Gamma)$ with kernel $\mathcal{T}$. In order to finish the proof it is enough to see that i is surjective.

Take an injective homomorphism $u \in \hom(\Gamma, \mathbb{R})$. We are going to define a right inverse $s_u : \mathbb{C}(\Gamma) \to \mathcal{R}$ to i. Given any nonzero $p \in \mathbb{C}[\Gamma]$, write $p = \sum_\gamma c_\gamma e^\gamma \in \mathbb{C}[\Lambda]$. By assumption u takes its minimum for only one γ with $c_\gamma \neq 0$, call it γ_0. Thus p can be written as

$$p = c_{\gamma_0} e^{\gamma_0} \left(1 - \sum_{\gamma \neq \gamma_0} \frac{-c_\gamma}{c_{\gamma_0}} e^{\gamma - \gamma_0} \right),$$

with $c_{\gamma_0} \neq 0$ and $\langle u \,|\, \gamma - \gamma_0 \rangle > 0$.

We can then set

$$s_u(p^{-1}) = c_{\gamma_0}^{-1} e^{-\gamma_0} \left(\sum_{k=0}^{\infty} \Big(\sum_{\gamma \neq \gamma_0} \frac{-c_\gamma}{c_{\gamma_0}} e^{\gamma - \gamma_0} \Big)^k \right),$$

and for a fraction qp^{-1}, $p, q \in \mathbb{C}[\Gamma]$, we set $s_u(qp^{-1}) = q s_u(p^{-1})$. It is now easily verified that s_u is well defined and it has the required properties.

We shall denote by L_r the map from $L^{-1}(\mathcal{R})$ to $\mathbb{C}(\Gamma)$ given by the composition $i \circ L$ (this notation L_r, a *rational Laplace transform* can be justified by interpreting this map in the language of Laplace transforms, as we shall do in Section 14.3.2).

Remark 12.8. The section s_u has the following property, which is immediate to check. Given $h, k \in \mathbb{C}(\Gamma)$, the product $s_u(h) s_u(k)$ is well-defined and equals $s_u(hk)$.

We can now give an analytic content to this construction by looking only at those bi-infinite series that converge on some region of space. If we have fixed such a region and restricted to such series, we then have that if such a series is rational in the algebraic sense, it clearly converges to the restriction of its associated rational function. Conversely, given a region A of space and considering a Laurent polynomial p with the property that $|p(z)| < 1$ when $z \in A$, we clearly have that the series $\sum_{k=0} p(z)^k$ converges to the rational function $1/(1 - p(z))$.

In this sense the formal Laplace transform and the analytic one coincide.

12.2.3 Euler Maclaurin Sums

In this section the lattice Γ will be the lattice $\mathbb{Z}^m \subset \mathbb{R}^m$, so that we identify $\mathbb{C}[\Gamma] = \mathbb{C}[x_1^{\pm 1}, \ldots, x_m^{\pm 1}]$, $\mathbb{C}[[\Gamma]] = \mathbb{C}[[x_1^{\pm 1}, \ldots, x_m^{\pm 1}]]$ and $\mathbb{C}(\Gamma) = \mathbb{C}(x_1, \ldots, x_m)$.

We want to understand partition functions via the study of an analytic function associated to any polytope or even any region of space. We start by considering, for any bounded region K of $\mathbb{R}^m$, the finite set $\mathbb{Z}^m \cap K$ and the function

$$E_K = E_K(x) := \sum_{\underline{n} \in \mathbb{Z}^m \cap K} x^{\underline{n}}. \tag{12.9}$$

This is a Laurent polynomial, also known as an *Euler–Maclaurin sum*. It can be interpreted as the Laplace transform of the distribution $\sum_{v \in \mathbb{Z}^m \cap K} \delta_{-v}$.

Knowledge of the function E_K implies knowledge of the number of integral points in K that equals $E_K(1)$. Although this is a tautological statement, we shall see that a remarkable theorem of Brion [22] implies the existence of *very compact* expressions for the function E_K.

The meaning of this is best explained by the easiest example. When $K = [0, n]$, $n \in \mathbb{N}$, we have $E_K = \sum_{i=0}^{n} x^i$, a sum of $n + 1$ terms. On the other hand

$$\sum_{i=0}^{n} x^i = \frac{1 - x^{n+1}}{1 - x},$$

an expression as a rational function in which both numerator and denominator have a very short expression. Starting from this type of formula and adding other geometrical considerations, Barvinok has developed effective computational approaches to counting integral points (see for instance [10]).

There are several approaches to Brion's theorem, and we shall follow closely [65]. The idea behind these formulas is the following: we first extend formally the definition of E_K to any region in space; in general we have an infinite sum $E_K = \sum_{\underline{n} \in \mathbb{Z}^m \cap K} x^{\underline{n}}$, which in general has no analytic meaning but lies in the space of *formal bi-infinite series* defined in the previous section.

We have the obvious Grassmann formula

$$E_{A \cup B} + E_{A \cap B} = E_A + E_B. \tag{12.10}$$

If $w \in \mathbb{Z}^m$ and $A \subset \mathbb{R}^m$, we also have

$$E_{w+A}(x) = x^w E_A(x). \tag{12.11}$$

We want to apply this construction to polyhedral cones. The crucial result is the following.

Let $C = C(a_1, \ldots, a_t)$ be a polyhedral cone with $a_1, \ldots, a_t \in \mathbb{Z}^m$.

Lemma 12.9. *For every vector w we have $E_{w+C}(x) \in \mathcal{R}$.*

Proof. If $\dim C = 0$, then $C = \{0\}$ and $E_{w+C} = x^w$ or 0, depending on whether $w \in \mathbb{Z}^m$. We can thus assume $\dim C > 0$. Lemma 1.40 ensures that a cone can be decomposed into simplicial cones. So by formula (12.10) we reduce to the case $C = C(a_1, \ldots, a_t)$ with $a_1, \ldots, a_t \in \mathbb{Z}^m$ linearly independent vectors.

Take $v \in C$. Then $v = \sum_{i=1}^t \lambda_i a_i$ with $0 \leq \lambda_i$. Write for each $i = 1, \ldots, t$, $\lambda_i = [\lambda_i] + \gamma_i$, with $[\lambda_i]$ the integer part of λ_i and $0 \leq \gamma_i < 1$. Then

$$v - \sum_{i=1}^t [\lambda_i] a_i = \sum_{i=1}^t \gamma_i a_i \in \mathbb{Z}^m. \tag{12.12}$$

Consider the box $B = B(a_1, \ldots, a_t) = \{\sum_i^t t_i a_i,\ 0 \leq t_i < 1\}$. Formula (12.12) implies that each element of $(w+C) \cap \mathbb{Z}^m$ can be uniquely written in the form $a + b$, where $a \in \{\sum_{i=1}^t m_i a_i,\ m_i \in \mathbb{N}\}$ and $b \in (w + B) \cap \mathbb{Z}^m$.

Then E_{w+B} is a Laurent polynomial and moreover, we have the factorization $E_{w+C} = E_{w+B} \prod_{i=1}^t (\sum_{h=0}^\infty x^{ha_i})$, or

$$\prod_{i=1}^t (1 - x^{a_i}) E_{w+C} = E_{w+B} \in \mathbb{C}[x_1^{\pm 1}, \ldots, x_m^{\pm 1}], \tag{12.13}$$

and our claim follows.

Let us then define by $\varPi$ the subspace of $\mathbb{C}[[x_1^{\pm 1}, \ldots, x_m^{\pm 1}]]$ spanned by the elements E_K as K varies over the cones $w + C(a_1, \ldots, a_t)$ with $a_1, \ldots, a_t \in \mathbb{Z}^m$. From the previous discussion it follows that $\varPi$ is a $\mathbb{C}[x_1^{\pm 1}, \ldots, x_m^{\pm 1}]$ submodule of $\mathcal{R}$. We can thus associate, using the map i, to each such cone C and vector w the rational function $\psi(w + C) := i(E_{w+C})$. We have the basic properties of ψ.

Theorem 12.10. *(1) If $\{a_1, \ldots, a_t\}$ is part of a rational basis of $\mathbb{Z}^m$, then*

$$\psi(w + C(a_1, \ldots, a_t)) = \frac{E_{w+B}}{\prod_{i=1}^t (1 - x^{a_i})}, \tag{12.14}$$

with B the box generated by the a_i's.
(2) If $\{a_1, \ldots, a_t\}$ is part of an integer basis of $\mathbb{Z}^m$ and $a \in \mathbb{Z}^m$, we have
$$\psi(a + C(a_1, \ldots, a_t)) = \frac{x^a}{\prod_{i=1}^t (1 - x^{a_i})}.$$

Proof. (1) follows immediately from the formula (12.13) proved in Lemma 12.9.

As for (2), if the a_i are part of an integral basis and $a \in \mathbb{Z}^m$, we have $a + B \cap \mathbb{Z}^m = \{a\}$, and the claim follows from part (1).

Remark 12.11. $i(\varPi)$ is contained in the space of rational functions with poles on the hypersurfaces $x^a - x^b = 0$ for $a, b \in \mathbb{N}^m$.

12.2.4 Brion's Theorem

We start with some convex geometry.

Definition 12.12.   (i) If P is a convex polyhedron and $a \notin P$, we say that
a *sees* a point p of P if $[a, p] \cap P = \{p\}$.
(ii) We say that a sees a face F of P if a sees each point $p \in F$.
(iii) For any $a \notin P$ we set Σ_a to be the set of all points in P seen by a.

Lemma 12.13. Σ_a *is a union of faces.*

Proof. Given $p \in \Sigma_a$ there is a unique face F of P such that $p \in \mathring{F}$. Thus it
suffices to see that a sees F. If $F = \{p\}$ there is nothing to prove. Otherwise,
take any other point $b \in F$. Then there is a segment $[b, c] \subset F$ with $p \in (b, c)$.
Consider the triangle T with vertices a, b, c. We have that $T \cap P$ is convex,
$[b, c] \subset T \cap P$ and $[a, p] \cap T \cap P = \{p\}$. It follows from elementary plane
geometry that $T \cap P = [b, c]$ and hence $[a, b] \cap P = \{b\}$, as desired.

Notice that Σ_a is a (not necessarily convex) polyhedron. If P is a polytope
(i.e., compact), so also is Σ_a.

Given a polytope Q, its Euler characteristic $\chi(Q)$ is given by

$$\chi(Q) = \sum_{F \text{ face of } Q} (-1)^{\dim F}.$$

The Euler characteristic is a basic topological invariant, and we shall use only
the fact that for a contractible space, the Euler characteristic equals 1. One
could give an elementary direct proof of this fact for a convex polytope.

Lemma 12.14. *Given a convex polytope P and a point $a \notin P$, Σ_a is homeo-
morphic to a convex polytope. Thus Σ_a is contractible and $\chi(\Sigma_a) = 1$.*

Proof. By Theorem 1.8 there is a unique point $b \in P$ of minimal distance from
a. Take the segment $[a, b]$ and consider the hyperplane H orthogonal to $[a, b]$
and passing through b. Again by Theorem 1.8, a and P lie on two different
sides of H. Let us now consider the convex cone Γ with vertex a consisting
of all half-lines originating from a and meeting P. If we now take any point
$c \in \Sigma_a$, the segment $[a, c]$ lies in Γ and has to meet H, and thus meets $\Gamma \cap H$
in a unique point $\phi(c)$. Conversely, any half-line in Γ meets H for the first
time in a point b, and by Lemma 12.13, P in a point $c \in \Sigma_a$ with $b = \phi(c)$.
This gives (*by projection*) a homeomorphism between Σ_a and $\Gamma \cap H$. The fact
that $\Gamma \cap H$ is a convex polytope follows from the proof. In fact, $\Gamma \cap H$ is the
convex envelope of the projections $\phi(v)$ of the vertices of P contained in Σ_a.

Let the convex polyhedron P be defined by $\{p \,|\, \langle \phi_i \,|\, p \rangle \leq a_i, \ 1 \leq i \leq n\}$ and
take a face F. By Lemma 1.24, the subset $S_F \subset [1, n]$ of the indices $1, 2, \ldots, n$
for which $\langle \phi_i \,|\, p \rangle = a_i, \forall p \in F$, defines $F := \{p \in P \,|\, \langle \phi_i \,|\, p \rangle = a_i, \forall i \in S_F\}$
and $\mathring{F} = \{p \in P \,|\, \langle \phi_i \,|\, p \rangle = a_i$ if and only if $i \in S_F\}$.

Lemma 12.15. *1. The complement A of the set of points that see the face F equals the set*

$$C_F := \{p \in V \mid \langle \phi_i \mid p \rangle \leq a_i, \forall i \in S_F\}.$$

2. The set C_F is a cone with vertices the affine span of F.

Proof. A point p is in A if either $p \in P \subset C_F$ or there are an element $b \in F$ and a t with $0 < t < 1$ such that $tp + (1 - t)b \in P$. If $i \in S_F$, we have then $\langle \phi_i \mid tp + (1 - t)b \rangle = t\langle \phi_i \mid p \rangle + (1 - t)a_i \leq a_i$; hence $\langle \phi_i \mid p \rangle \leq a_i$, so $p \in C_F$. Conversely, if $p \in C_F$, we have $\langle \phi_i \mid p \rangle \leq a_i$ for $i \in S_F$. Take $b \in F$. We have that $sp + (1 - s)b \in P$ if and only if $\langle \phi_j \mid sp + (1 - s)b \rangle \leq a_j$ for all j. This condition is satisfied by all $0 \leq s \leq 1$ if $j \in S_F$. If $j \notin S_F$ for any $b \in \mathring{F}$ (the relative interior), we have $\langle \phi_j \mid b \rangle < a_j$ hence there is s_j with $0 < s_j < 1$ with $\langle \phi_j \mid sp + (1 - s)b \rangle = s\langle \phi_j \mid p \rangle + (1 - s)\langle \phi_j \mid b \rangle \leq a_j$ for $0 \leq s \leq s_j$. Taking the minimum s among such s_j, we see that $tp + (1 - t)b \in P$ for $0 \leq t \leq s$ and p does not see F.

The second part follows easily from the first upon noticing that Lemma 1.24 implies also that the affine span of F is the linear space defined by the equations $\langle \phi_i \mid p \rangle = a_i, \forall i \in S_F$. $\qquad \square$

Definition 12.16. A convex polytope P given by inequalities

$$P := \{p \in V \mid \langle \phi_i \mid p \rangle \leq a_i\}$$

will be called *pseudorational* if all the ϕ_i can be taken with integral coefficients. It will be called *rational* if also all the a_i can be taken integral.

Lemma 12.17. *If P is pseudorational, the element E_{C_F} lies in $\mathcal{R}$ for every face F of P and it is in $\mathcal{T}$ if $\dim(F) > 0$.*

Proof. By definition, $C_F := \{p \in V \mid \langle \phi_i \mid p \rangle \leq a_i, \forall i \in S_F\}$. The space $U := \{a \mid \langle \phi_i \mid a \rangle = 0, \forall i \in S_F\}$ is a space parallel to the affine span of F. Thus we have that $\dim(U) = \dim(F)$. By our assumption, U is defined by a system of linear equations with integer coefficients. If $\dim(F) > 0$, it follows that there is a nonzero element $\underline{a} \in U \cap \mathbb{Z}^m$. Then $C_F = C_F + \underline{a}$ so that $E_{C_F} = x^{\underline{a}} E_{C_F}$ or $(1 - x^{\underline{a}})E_{C_F} = 0$ hence $E_{C_F} \in \mathcal{T}$.

When $F = \{v\}$ is a vertex, setting $C := \{p \in V \mid \langle \phi_i \mid p \rangle \leq 0, \forall i \in S_F\}$, we have $C_F = v + C$. By hypothesis C is a cone pointed at 0, and by Remark 1.30, we have that $C = C(u_1, \dots, u_t)$ for some integral vectors u_i spanning V. We can then apply Theorem 12.10. $\qquad \square$

Proposition 12.18. *The following identity holds:*

$$\sum_{F \text{ face of } P} (-1)^{\dim F} E_{C_F} = E_P. \tag{12.15}$$

Proof. Given $\underline{n} \in \mathbb{Z}^m$, we need to compute the coefficient of $x^{\underline{n}}$ in

$$\sum_{F \text{ face of } P} (-1)^{\dim F} E_{C_F}.$$

If $\underline{n} \in P$, then $\underline{n} \in C_F$ for every face F. Thus the coefficient of $x^{\underline{n}}$ equals

$$\sum_{F \text{ face of } P} (-1)^{\dim F} = \chi(P) = 1,$$

since P is contractible.

If $\underline{n} \notin P$, then by definition $\underline{n} \in C_F$ if and only if $\underline{n}$ does not see F. It follows that the coefficient of $x^{\underline{n}}$ equals

$$\sum_{F \text{ face of } P} (-1)^{\dim F} - \sum_{\substack{F \text{ face of } P, \\ \underline{n} \text{ sees } F}} (-1)^{\dim F} = \chi(P) - \chi(\Sigma_{\underline{n}}) = 1 - 1 = 0$$

by Proposition 12.18.

We can now state and prove Brion's theorem [22] and [23]. We use the notation of Theorem 12.10, $\psi(C_{v_j}) = i(E_{C_{v_j}})$.

Theorem 12.19 (Brion). *Let P be a pseudorational convex polytope, and let $v_1, \ldots, v_h$ be its vertices. The following identity of rational functions holds*

$$E_P = \sum_{j=1}^{h} \psi(C_{v_j}). \tag{12.16}$$

Proof. Notice that $E_P \in \mathbb{C}[x_1^{\pm 1}, \ldots, x_m^{\pm 1}]$; hence $E_P = i(E_P)$.

Now let us apply i to the identity (12.15) and let us observe that by Lemma 12.17, if $\dim F > 0$, then $\psi(C_F) = 0$. It follows that

$$E_P = \sum_{F \text{ face of } P} (-1)^{\dim F} \psi(C_F) = \sum_{j=1}^{h} \psi(C_{v_j}), \tag{12.17}$$

as desired.

Remark 12.20. A particularly simple form of the previous formula occurs when all the cones C_{v_j} are simplicial (cf. definition 1.39), since in this case one can further apply the explicit formula (12.14).

Recall that in Section 1.3.4 we have called a polytope for which all the cones C_{v_j} are simplicial a simplicial polytope and have seen how they naturally arise as the variable polytopes associated to regular points.

One can extend the previous ideas to not necessarily convex polytopes. Let us define a pseudorational polytope as a union of convex pseudorational polytopes. Since the intersection of two convex pseudorational polytopes is still a convex pseudorational polytope, one easily sees that $E_P \in \mathcal{R}$ for all pseudo-rational polytopes. One can thus associate a rational function $\psi(P) := i(E_P)$ to any pseudorational polytope, and this assignment satisfies the Grassmann rule $\psi(P_1 \cup P_2) + \psi(P_1 \cap P_2) = \psi(P_1) + \psi(P_2)$, and the translation invariance $\psi(\underline{a} + P) = x^{\underline{a}}\psi(P)$ whenever $\underline{a} \in \mathbb{Z}^m$. Such assignments to polytopes (and also to polyhedra) have been studied systematically by McMullen (cf. [81], [80]).

12.2.5 Ehrhart's Theorem

To finish, we are going to show how to deduce the classical result of Ehrhart stated in Section 12.2.1 from Brion's theorem. Let P be a polytope with integral vertices $v_1, \ldots, v_h \in \mathbb{Z}^m$. Consider for any positive integer t the polytope tP, whose vertices are clearly $tv_1, \ldots, tv_h$. Set $F_P(t) = E_{tP}(1)$ (the number of integral points in tP).

Theorem 12.21 (Ehrhart). $F_P(t)$ *is a polynomial in* t.

Proof. We need a preliminary fact.

Lemma 12.22. *Let* $Q_j(x_1, \ldots, x_m, t) \in \mathbb{C}[x_1^{\pm 1}, \ldots, x_m^{\pm 1}, t]$, $j = 1, \ldots, h$ *and* $0 \neq P(x_1, \ldots, x_m) \in \mathbb{C}[x_1^{\pm 1}, \ldots, x_m^{\pm 1}]$. *Take* $w_1, \ldots, w_h \in \mathbb{Z}^m$. *If*

$$L_t[x_1, \ldots, x_m] := \frac{\sum_{j=1}^{h} Q_j(x_1, \ldots, x_m, t) x^{tw_j}}{P(x_1, \ldots, x_m)} \in \mathbb{C}[x_1^{\pm 1}, \ldots, x_m^{\pm 1}], \ \forall t \in \mathbb{N},$$

$$(12.18)$$

then there is a polynomial $p(t) \in \mathbb{C}[t]$ *with*

$$p(t) = L_t[1, \ldots, 1], \ \forall t \in \mathbb{N}.$$

Proof. Take an auxiliary variable z and exponents $n_1, \ldots, n_m \in \mathbb{N}$ with $P(z^{n_1}, \ldots, z^{n_m}) \neq 0$. We can substitute in formula (12.18) for x_i the element z^{n_i} and reduce to the one-variable case. If $P(1) \neq 0$, we evaluate for $z = 1$ and $L_t(1) = P(1)^{-1} \sum_{j=1}^{h} Q_j(1, t) \in \mathbb{C}[t]$. Otherwise, write $P(z) = (1 - z)^m R(z)$ with $R(1) \neq 0$. We now proceed by induction on $m > 0$ by applying l'Hopital's rule:

$$L_t(1) = \lim_{z \to 1} \frac{\sum_{j=1}^{h} Q_j(z, t) z^{tw_j}}{P(z)} = \lim_{z \to 1} \frac{\sum_{j=1}^{h} \frac{d[Q_j(z,t) z^{tw_j}]}{dz}}{\frac{dP(z)}{dz}}.$$

We have that the fraction of derivatives satisfies the same hypothesis of the lemma, but now $dP(z)/dz$ has a zero of order $m - 1$ at 1, thus allowing us to proceed by induction.

We return to the proof of Ehrhart's Theorem. For each $j = 1, \ldots, h$, set $C_j = C_{v_j} - v_j = C_{tv_j} - tv_j$. Then C_j is a cone with vertex the origin and $\psi(C_{tv_j}) = x^{tv_j}\psi(C_j)$. It follows from (12.16) that

$$E_{tP} = \sum_{j=1}^{h} x^{tv_j} \psi(C_j) = \frac{\sum_{j=1}^{h} H_j(x_1, \ldots, x_m) x^{tv_j}}{K(x_1, \ldots, x_m)}, \qquad (12.19)$$

where $K(x_1, \ldots, x_m)$ and the $H_j(x_1, \ldots, x_m)$ are polynomials. Thus, since $E_{tP} \in \mathbb{C}[x_1^{\pm 1}, \ldots, x_m^{\pm 1}]$, $\forall t \in \mathbb{N}$, the theorem follows from Lemma 12.22.

The Partition Functions

The main purpose of this chapter is to discuss the theory of Dahmen–Micchelli describing the difference equations that are satisfied by the quasipolynomials that describe the partition function $\mathcal{T}_X$ on the big cells. These equations allow also us to develop possible recursive algorithms.

Most of this chapter follows very closely the paper we wrote with M. Vergne [44]; see also [45].

13.1 Combinatorial Theory

As usual, in this chapter Λ is a lattice in a real vector space V and X a list of vectors in Λ.

13.1.1 Cut-Locus and Chambers

In this section we assume that X generates a pointed cone $C(X)$ and recall some definitions from Section 1.3.3. The set of singular points $C^{\mathrm{sing}}(X)$ is defined as the union of all the cones $C(Y)$ for all the subsets of X that do not span the space V.

The set of strongly regular points $C^{\mathrm{reg}}(X)$ is the complement of $C^{\mathrm{sing}}(X)$. The big cells are the connected components of $C^{\mathrm{reg}}(X)$. The big cells are the natural open regions over which the multivariate spline coincides with a polynomial (Theorem 9.7).

Let us recall some properties of the cut locus (Definition 1.54). We have seen in Proposition 1.55 that the translates $C^{\mathrm{sing}}(X)+\Lambda$ give a periodic hyperplane arrangement. The union of the hyperplanes of this periodic arrangement is the cut locus, and each connected component of its complement is a chamber. Each chamber of this arrangement, by Theorem 2.7, is the interior of a bounded polytope. The function B_X is a polynomial on each chamber. The set of chambers is invariant under translation by elements of Λ. If X is a basis of the lattice Λ, we have that the chambers are open parallelepipeds.

C. De Concini and C. Procesi, *Topics in Hyperplane Arrangements, Polytopes and Box-Splines*, Universitext, DOI 10.1007/978-0-387-78963-7_13,
© Springer Science+Business Media, LLC 2010

We introduce a very convenient notation.

Definition 13.1. Given two sets A, B in V, we set

$$\delta(A \,|\, B) := \{\alpha \in \Lambda \,|\, (A - \alpha) \cap B \neq \emptyset\} = (A - B) \cap \Lambda. \qquad (13.1)$$

Notice that given $\beta, \gamma \in \Lambda$, we have that

$$\delta(A \,|\, B) + \beta - \gamma = \delta(A + \beta \,|\, B + \gamma).$$

We shall apply this definition mostly to zonotopes and so usually write $\delta(A \,|\, X)$ instead of $\delta(A \,|\, B(X))$. Also, if $A = \{p\}$, we write $\delta(p \,|\, B)$ instead of $\delta(\{p\} \,|\, B)$

Remark 13.2. If r is a regular vector (i.e., a vector that does not lie in the cut locus), the set $\delta(r \,|\, X)$ depends only on the chamber $\mathfrak{c}$ in which r lies and equals $\delta(\mathfrak{c} \,|\, X)$.
    That is, $\gamma \in \delta(\mathfrak{c} \,|\, X)$ if and only if $\mathfrak{c} \subset \gamma + B(X)$.

The following proposition is an immediate generalization of Proposition 2.50 (which depends on the decomposition of the zonotope into parallelepipeds).

Proposition 13.3. *If r is a regular vector, then $\delta(r \,|\, X)$ has cardinality $\delta(X)$. Furthermore, if we consider the sublattice Λ_X of Λ generated by X, every coset with respect to Λ meets $\delta(r \,|\, X)$.*

Is is often useful to perform a reduction to the nondegenerate case (see Definition 2.23). If X spans Λ and $X = X_1 \cup X_2$ is a decomposition, we have that $\Lambda = \Lambda_1 \oplus \Lambda_2$, where Λ_1, Λ_2 are the two lattices spanned by X_1, X_2. We have then $C(X) = C(X_1) \times C(X_2)$, $B(X) = B(X_1) \times B(X_2)$, and the big cells for X are products $\Omega_1 \times \Omega_2$ of big cells for the two subsets. The same is also true for the chambers. If we normalize the volumes so that $\Lambda, \Lambda_1, \Lambda_2$ all have covolume 1 we have

$$\delta(X) = \delta(X_1)\delta(X_2) = \mathrm{vol}(B(X_1))\mathrm{vol}(B(X_2)) = \mathrm{vol}(B(X)).$$

Finally, for a chamber $\mathfrak{c}_1 \times \mathfrak{c}_2$ we have

$$\delta(\mathfrak{c}_1 \times \mathfrak{c}_2 | X) = \delta(\mathfrak{c}_1 | X_1) \times \delta(\mathfrak{c}_2 | X_2).$$

    These formulas allow us to restrict most of our proofs to the case in which X is nondegenerate.

13.1.2 Combinatorial Wall Crossing

Here we are going to assume that X generates V and use the notation of Section 13.1.1. Let us recall a few facts from Section 2.1.2. Given a *chamber*, its closure is a compact convex polytope and we have a decomposition of the entire space into faces. We shall use fraktur letters as $\mathfrak{f}$ for these faces.

Let $\mathfrak{c}, \mathfrak{g}$ be two chambers whose closures intersect in an $(s-1)$-dimensional face $\mathfrak{f}$. The face $\mathfrak{f}$ generates a hyperplane K that is the translate of some hyperplane H generated by a subset of elements in X. If this is the case, we shall say that $\mathfrak{c}$ and $\mathfrak{g}$ are H-separated.

Choose an equation $u \in U$ for H. We can then decompose X into the disjoint union of three sets:

A where u takes positive values, B where u takes negative values, and C where it vanishes.

Recall then that $B(X)$ has two faces parallel to H, namely $a_A + B(C)$ and $a_B + B(C)$. For any $D \subset A \cup B$, D nonempty, set

$$a_{D,u} := \sum_{a \in D \cap A} a - \sum_{b \in D \cap B} b.$$

Theorem 13.4. *Let $\mathfrak{c}, \mathfrak{g}$ be two H separated chambers, $\alpha \in \delta(\mathfrak{g} \,|\, X) \setminus \delta(\mathfrak{c} \,|\, X)$. Then either for all $D \neq \emptyset$, and $D \subset A \cup B$ we have that $\alpha + a_{D,u} \in \delta(\mathfrak{c} \,|\, X)$ or for all $D \neq \emptyset, D \subset A \cup B$ we have $\alpha - a_{D,u} \in \delta(\mathfrak{c} \,|\, X)$.*

Proof. We can translate and assume that $\alpha = 0$. This implies that $\mathfrak{g}$ intersects $B(X)$, and hence it is contained in $B(X)$, while $\mathfrak{c}$ is disjoint from $B(X)$. It follows that either $\mathfrak{f}$ lies in $a_A + B(C)$ or $\mathfrak{f}$ lies in $a_B + B(C)$.

Assume that we are in the first case. Since $D \neq \emptyset$, u is strictly positive on $a_{D,u}$. On the other hand, $a_A + B(C)$ is the set of points in $B(X)$ where u attains its maximum. We deduce that for each $\epsilon > 0$, $x \in \mathfrak{f}$, $x + \epsilon a_{D,u}$ does not lie in $B(X)$. Thus if ϵ is small enough, $z := x + \epsilon a_{D,u}$ lies in $\mathfrak{c}$. Now

$$z - a_{D,u} = x - (1 - \epsilon)a_{D,u} \in B(C) + a_A - (1 - \epsilon)\left(\sum_{a \in D \cap A} a - \sum_{b \in D \cap B} b \right)$$

$$= B(C) + \sum_{a \in A \setminus D} a + \epsilon\left(\sum_{a \in D \cap A} a \right) + (1 - \epsilon)\left(\sum_{b \in D \cap B} b \right) \subset B(X).$$

It follows that $a_{D,u} \in \delta(\mathfrak{c} \,|\, X)$, as desired.

The second case follows in exactly the same way using $-u$ instead of u (notice that $a_{D,u} = -a_{D,-u}$).

13.1.3 Combinatorial Wall Crossing II

Since the cut locus is invariant under translation under elements of Λ, if $\mathfrak{c}$ is a chamber and $\lambda \in \Lambda$ we have that $\mathfrak{c} - \lambda$ is also a chamber.

If $X = \underline{b}$ is a basis, the chambers associated to $\underline{b}$ are the parallelepipeds obtained by translation from the interior of $B(\underline{b})$.

Lemma 13.5. *Given a basis $\underline{b}$ in X, each chamber $\mathfrak{c}$ can be translated into $B(\underline{b})$.*

We have seen that if r is regular, $\delta(r\,|\,X)$ is formed by $\delta(X)$ points and is independent of r as long as r varies in a chamber $\mathfrak{c}$. We denote it by $\delta(\mathfrak{c}\,|\,X)$. We observe that we have

$$\delta(\mathfrak{c} - \lambda\,|\,X) = \delta(\mathfrak{c}\,|\,X) - \lambda, \ \forall \lambda \in \Lambda. \tag{13.2}$$

We now want to understand how $\delta(r\,|\,X)$ changes when we pass from one chamber to another.

Let $\mathfrak{c}, \mathfrak{g}$ be two chambers whose closures intersect in an $(s-1)$-dimensional face $\mathfrak{f}$. The face $\mathfrak{f}$ generates a hyperplane K that is the translate of some hyperplane H generated by a subset $C := H \cap X$ of elements in X. With the previous notation, we take the two opposite faces $F_0 = a_A + B(C)$, and $F_1 = a_B + B(C)$ of $B(X)$. The zonotope $B(X)$ lies entirely in the region bounded by the hyperplanes $a_A + H, a_B + H$ and contains the prism, with basis F_0, F_1 (join of the two). We start with a simple geometric lemma whose proof is immediate.

Lemma 13.6. *We have one and only one of the following three possibilities:*

1. *$\mathfrak{f}$ is in the interior of $B(X)$. In this case $\mathfrak{c}, \mathfrak{g}$ are also in the interior of $B(X)$.*
2. *$\mathfrak{f}$ is in the exterior of $B(X)$. In this case $\mathfrak{c}, \mathfrak{g}$ are also in the exterior of $B(X)$.*
3. *$\mathfrak{f}$ is in one of the two faces F_0, F_1. In this case the one of $\mathfrak{c}, \mathfrak{g}$ that lies on the same side of K as $B(X)$ is in the interior of $B(X)$; the other is in the exterior.*

The set C also determines a cut locus in H (and hence in each of its translates) that is usually properly contained in the intersection of the cut locus for X with H. Thus any chamber $\mathfrak{f}$ of the cut locus for X contained in H is also contained in a unique chamber $\mathfrak{f}'$ for the cut locus of C.

We can thus first translate $\mathfrak{c}$ so that its face $\mathfrak{f}$ lies in the unique hyperplane K tangent to $B(X)$, parallel to H, and such that $\mathfrak{c}$ lies to the opposite side of $B(X)$ (using one of the two possible opposite faces of $B(X)$ parallel to $\mathfrak{f}$). Let $F_0 := B(X) \cap K$ and u_{F_0} the corresponding vector in U. The set X decomposes as $X = A \cup B \cup C$ according to the sign $+, -, 0$ taken by u_{F_0}. Thus $F_0 = \sum_{a \in A} a + B(C) = a_A + B(C)$.

By translating in K and using Lemma 13.5 we may assume that $\mathfrak{f}$ lies in F_0.

By our previous remarks, there is a unique chamber $\mathfrak{f}'$ for the arrangement generated by C in H such that $\mathfrak{f}$ is contained in $\mathfrak{f}' + a_A$. We then have that $\mathfrak{c}$ is disjoint from $B(X)$, while $\mathfrak{g}$ lies inside $B(X)$. Their common face $\mathfrak{f}$ lies in the interior of F_0. We then have, with all this notation, the following theorem

Theorem 13.7 (Combinatorial wall crossing formula). *Assume that $\mathfrak{g}$ lies in the same half-space cut by K as $B(X)$ then*

$$\delta(\mathfrak{g}\,|\,X) \setminus \delta(\mathfrak{c}\,|\,X) = \delta(\mathfrak{f}\,|\,F_0) = \delta(\mathfrak{f}'\,|\,C) - a_A,$$

$$\delta(\mathfrak{c}\,|\,X) \setminus \delta(\mathfrak{g}\,|\,X) = \delta(\mathfrak{f}\,|\,F_1) = \delta(\mathfrak{f}'\,|\,C) - a_B.$$

Proof. If $\alpha \in \delta(\mathfrak{g}\,|\,X) \cup \delta(\mathfrak{c}\,|\,X)$, we must have that $\mathfrak{f} - \alpha$ lies in $B(X)$. Thus by the previous lemma, either $\mathfrak{f} - \alpha$ lies in the interior of $B(X)$, and then $\alpha \in \delta(\mathfrak{g}\,|\,X) \cap \delta(\mathfrak{c}\,|\,X)$, or in one of two faces F_0, F_1. If $\mathfrak{f} - \alpha \subset F_0$, that is, if $\alpha \in \delta(\mathfrak{f}\,|\,F_0)$, we have performed a translation that preserves K. Thus $\mathfrak{g} - \alpha$ still lies in the same half-space cut by K as $B(X)$ and we have $\alpha \in \delta(\mathfrak{g}\,|\,X) \setminus \delta(\mathfrak{c}\,|\,X)$.

If, on the other hand, $\mathfrak{f} - \alpha \subset F_1$, then by the same reasoning, we deduce that $\alpha \in \delta(\mathfrak{c}\,|\,X) \setminus \delta(\mathfrak{g}\,|\,X)$. So we have seen that

$$\delta(\mathfrak{g}\,|\,X) \setminus \delta(\mathfrak{c}\,|\,X) = \delta(\mathfrak{f}\,|\,F_0), \quad \delta(\mathfrak{c}\,|\,X) \setminus \delta(\mathfrak{g}\,|\,X) = \delta(\mathfrak{f}\,|\,F_1).$$

Finally,

$$\delta(\mathfrak{f}\,|\,F_0) = \delta(\mathfrak{f}'\,|\,C) - a_A, \quad \delta(\mathfrak{f}\,|\,F_1) = \delta(\mathfrak{f}'\,|\,C) - a_B.$$

Remark 13.8. Notice that we have, in particular,

$$F_0 + a_B = F_1 + a_A, \quad \delta(\mathfrak{f}\,|\,F_1) + a_B = \delta(\mathfrak{f}\,|\,F_0) + a_A.$$

Notice that this theorem is clearly related to Theorem 13.4.

13.2 The Difference Theorem

The main purpose of this chapter is to prove that a partition function $\mathcal{T}_X$ is a quasipolynomial on the regions $\Omega - B(X)$, for each big cell Ω. The fact that $\mathcal{T}_X$ is a quasipolynomial not just on the big cells but in fact in the larger regions $\Omega - B(X)$ may be considered as a discrete analogue of the differentiability properties of the multivariate spline T_X.

In fact, we have a more precise statement that identifies the quasipolynomial f coinciding with $\mathcal{T}_X$ on $\Omega - B(X)$ with an element of a canonical space, that we denote by $DM(X)$, uniquely determined by its *initial values*, that is, the values it takes on $\delta(u\,|\,X)$, where $u \in \Omega$ is close to zero. In this case $\delta(u\,|\,X) \cap C(X) = \{0\}$, so that, in order to coincide with $\mathcal{T}_X$ on $\Omega - B(X)$ we must have that $f(0) = 1$ while $f(a) = 0$ for all nonzero elements of $\delta(u\,|\,X)$. These are the defining initial conditions.

The partition function is well-defined only when the vectors in X lie in a pointed cone. However, the reader can understand that many of the results in this section are independent of this assumption.

13.2.1 Topes and Big Cells

Definition 13.9. By the word *tope* we mean a connected component of the complement in V of the union of the hyperplanes generated by subsets of X.

Remark 13.10. Usually big cells are larger than topes. For example, let V be of dimension 3 with basis the vectors a, b, c, $X = (a, b, c, d)$ with $d = a + b + c$. The cone $C(X)$ is the cone generated by the three vectors a, b, c. It is the union of the (closure of) 3 big cells the cones $C(a, b, d)$, $C(a, c, d)$, $C(b, c, d)$, each of which is the union of two topes.

For example, $C(a, b, d) = C(a, a + b, d) \cup C(a + b, b, d)$.

Lemma 13.11. *Each tope τ contains a unique chamber $\mathfrak{c}$, with $0 \in \bar{\mathfrak{c}}$ and conversely, one such chamber is contained in a unique tope.*

Proof. The union of the chambers contained in τ is dense in τ, so there is at least one such chamber, but the walls of this chamber passing through 0 define the tope τ as the one containing $\mathfrak{c}$.

Definition 13.12. We call a chamber *initial* if $0 \in \bar{\mathfrak{c}}$.

Proposition 13.13. *Given any initial chamber $\mathfrak{c}$, we have that $\delta(\mathfrak{c} \mid X)$ contains the points of Λ in the interior of $-B(X)$.*

Proof. If $\gamma \in \overset{\circ}{B}(X)$ then $\gamma + \mathfrak{c} \subset B(X)$, that is, $\mathfrak{c} \subset -\gamma + B(X)$.

Next take a face F of $B(X)$. We want to see whether the points of Λ in the interior of F lie in $\delta(\mathfrak{c} \mid X)$. If $\gamma \in \overset{\circ}{F}$ then $\gamma + \mathfrak{c} \subset B(X)$, that is, $\mathfrak{c} \subset -\gamma + B(X)$ if and only if $\bar{\mathfrak{c}} \subset -\gamma + B(X)$.

13.2.2 A Special System

In order to understand and develop our ideas, let us start with a simple heuristic argument. If Y is a cocircuit, we have $\prod_{y \in Y}(1 - e^y)\mathcal{T}_X = \mathcal{T}_{X \setminus Y}$, and this last function, when expanded in a power series, is supported in the cone generated by $X \setminus Y$, which, by definition does not intersect any big-cell. Since at the level of coefficients, multiplication by $(1 - e^y)$ is equivalent to applying the difference operator ∇_y, we begin to suspect that a quasipolynomial function coinciding with $\mathcal{T}_X$ on $\Omega - B(X)$ if it exists must satisfy the difference equations $\nabla_Y f = 0$, where $\nabla_Y := \prod_{v \in Y} \nabla_v$, as $Y \in \mathcal{E}(X)$ runs over the cocircuits (for the definition see Section 2.2.1).

In this chapter we wish to work in an arithmetic way. Thus let us denote by $\mathcal{C}[\Lambda]$ the abelian group of $\mathbb{Z}$-valued functions on Λ. When we consider the space of $\mathbb{C}$-valued functions on Λ, we shall denote it by $\mathcal{C}_{\mathbb{C}}[\Lambda]$. In fact, the reader will easily understand that our results would be valid for functions with values in any commutative ring A.

Definition 13.14. Given X a list in Λ we define

$$DM(X) := \{f \in \mathcal{C}[\Lambda] \mid \nabla_Y f = 0\} \tag{13.3}$$

as Y runs over all cocircuits of X.

Observe that Λ acts under translation on $\mathcal{C}[\Lambda]$ and $DM(X)$ is a $\mathbb{Z}[\Lambda]$ submodule.

Example 13.15. (i)   When $s = 1$ the only cocircuit is X itself. In the special case in which $X = \{1, 1, \ldots, 1\}$, m times 1, the equation is $\nabla_1^m f = 0$. Its space of solutions has as basis the binomial coefficients $\binom{n}{i}$, $i = 0, \ldots, m-1$.

In this case $\mathcal{T}_X(n)$ coincides with $\binom{n+m-1}{m-1}$ on the integers $n > -m$.

(ii)   Assume now that $X = \{a_1, \ldots, a_s\}$ are linearly independent, so they form a basis of a sublattice Λ_X. Then each $\{a_i\}$ is a cocircuit and $DM(X)$ is the abelian group of functions on Λ that are constant on the cosets of Λ_X.

Notice that the action of ∇_Y on functions is given as follows. For every sublist $S \subseteq Y$ set $a_S := \sum_{a \in S} a$. Then

$$\nabla_Y f(b) = \sum_{S \subseteq Y} (-1)^{|S|} f(b - a_S). \tag{13.4}$$

The equation $\nabla_Y f(b) = 0$ can be also written as

$$f(b) = \sum_{S \subseteq Y,\ S \neq \emptyset} (-1)^{|S|+1} f(b - a_S). \tag{13.5}$$

Equation (13.5) shows that such a difference equation is in fact a recursion relation. The entire system associated to cocircuits will be the basic system of recursions.

Let us for the moment discuss the simple case of dimension-one, i.e., when X is a list of positive numbers $a_1, \ldots, a_m$. In this case, $B(X)$ is the interval $[0, \sum_{i=1}^{m} a_i]$, and we take for Ω the set of strictly positive numbers so that $\Omega - B(X)$ is the open half-line $(-\sum_{i=1}^{m} a_i, \infty)$. Clearly, $\mathcal{T}_X(n)$ equals 0 if $n < 0$ and $\mathcal{T}_X(0) = 1$. Here we have a unique difference equation, that is, $\prod_{i=1}^{m} \nabla_{a_i} F = 0$, which we can use in the form of formula (13.5). If $n > 0$, we have, for every nonempty S, that $n - a_S \in [n - \sum_{i=1}^{m} a_i, n) \subset (-\sum_{i=1}^{m} a_i, \infty)$, and that allows us to compute $\mathcal{T}_X(n)$ subject to the initial conditions

$$\mathcal{T}_X(0) = 1,\ \mathcal{T}_X(n) = 0,\ \forall\, n \in \mathbb{Z},\ \text{such that} \ -\sum_{i=1}^{m} a_i < n < 0.$$

Using the fact that this is a quasipolynomial, only finitely many computations are needed to develop a formula explicitly (cf. Section 16.4).

So we start by studying the set of difference equations given in (13.3). The theory is quite rich and can be approached from several directions, as we will see presently. Let $J_X \subset \mathbb{Z}[\Lambda]$ be the ideal generated by the elements $\prod_{a \in Y} (1 - e^a)$ as Y runs over the cocircuits. For $a \in \Lambda$ denote by $[e^a]$ the class of e^a modulo J_X. We have, as in Theorem 5.22, the following result

Proposition 13.16. *Under the pairing between the abelian groups $DM(X)$ and $\mathbb{Z}[\Lambda]/J_X$ defined by*

$$(f, [e^a]) := f(a)$$

for $f \in DM(X)$ and $a \in \Lambda$, we have that $DM(X)$ is identified with its dual (over $\mathbb{Z}$) of $\mathbb{Z}[\Lambda]/J_X$.

Our next task is thus to study the algebra $\mathbb{Z}[\Lambda]/J_X$.

13.2.3 On $DM(X)$

As in Section 11.2.1, we can generalize our setting as follows. Take variables $\underline{t} = (t_1, \ldots, t_m)$. Consider the algebraic torus $(\mathbb{C}^*)^m$ whose coordinate ring is the ring of Laurent polynomials $\mathbb{C}[t_1^{\pm 1}, \ldots, t_m^{\pm 1}]$ and the algebraic torus $G := U_{\mathbb{C}}/\hom(\Lambda, \mathbb{Z})$ whose character group is Λ and whose coordinate ring is $\mathbb{C}[\Lambda] = \mathbb{C}[x_1^{\pm 1}, \ldots, x_s^{\pm 1}]$. They both have integral forms with coordinate rings $R := \mathbb{Z}[t_1^{\pm 1}, \ldots, t_m^{\pm 1}]$ and $\mathbb{Z}[\Lambda] = \mathbb{Z}[x_1^{\pm 1}, \ldots, x_s^{\pm 1}]$.

In the ring $R[\Lambda]$ we take the ideal $J_X(\underline{t})$ generated by the functions $N_Y(\underline{t}) := \prod_{a_i \in Y}(1 - t_i e^{-a_i})$ as Y runs over the cocircuits. Sometimes given $a = a_i$ in X, we are going to write, by abuse of notation, t_a for t_i. In coordinates, if $a_i = (a_{i,1}, \ldots, a_{i,s})$ we have $1 - t_i e^{-a_i} = 1 - t_i \prod_{j=1}^{s} x_j^{-a_{i,j}}$.

Thus the ideal $J_X(\underline{t})$ defines a subscheme Z of the torus whose geometric points are $G \times (\mathbb{C}^*)^m$.

We set $\mathbb{A} = R[\Lambda]/J_X(\underline{t})$.

If we take the projection $\pi : Z \to (\mathbb{C}^*)^m$, then clearly the subscheme defined by J_X is the fiber of π over the identity. In general, if we take any list $\underline{\mu} = (\mu_1, \ldots, \mu_m)$, the fiber of π over the point of coordinates e^{μ_a} is the subscheme in G defined by the ideal $J_X(\underline{\mu})$ generated by the functions $N_Y(\underline{\mu}) := \prod_{a \in Y}(1 - e^{-a+\mu_a})$ as Y runs over the cocircuits.

Given a basis $\underline{b} = \{b_1, \ldots, b_s\}$ for V extracted from X, consider the lattice $\Lambda_{\underline{b}} \subset \Lambda$ that $\underline{b}$ generates in Λ.

The group $\Lambda/\Lambda_{\underline{b}}$ is finite of order $[\Lambda : \Lambda_{\underline{b}}] = |\det(\underline{b})|$, and its character group is the finite subgroup $T(\underline{b})$ of G that is the intersection of the kernels of the characters e^b for $b \in \underline{b}$, that is the kernel of the surjective homomorphism

$$\pi_{\underline{b}} : G \to (\mathbb{C}^*)^s, \quad \pi_{\underline{b}} : g \mapsto (e^{b_1}(g), \ldots, e^{b_s}(g)).$$

The set $T_{\underline{\mu}}(\underline{b}) := \{g \in G \mid \pi_{\underline{b}}(g) = (e^{\mu_1}, \ldots, e^{\mu_s})\}$ is a coset of $T(\underline{b})$.

Recall that by Proposition 2.15, we have the following formula for the volume $\delta(X)$ of the zonotope $B(X)$:

$$\delta(X) = \sum_{\underline{b} \in \mathcal{B}_X} |\det(\underline{b})|. \tag{13.6}$$

Define $P_{\underline{\mu}}(X) := \cup_{\underline{b} \in \mathcal{B}_X} T_{\underline{\mu}}(\underline{b})$. We call this set *the set of points of the arrangement associated to X and μ.*

For $p \in P_{\underline{\mu}}(X)$ set $X_p := \{a \in X \mid e^a(p) = e^{\mu_a}\}$. Furthermore, set $d(X_p)$ to be the number of bases $\underline{b} \in \mathcal{B}_X$ extracted from X_p.

Proposition 13.17.

$$\sum_{p \in P_{\underline{\mu}}(X)} d(X_p) = \delta(X). \tag{13.7}$$

Proof. Consider the set S of pairs $(p, \underline{b})$ such that $p \in T_{\underline{\mu}}(\underline{b})$, or equivalently $\underline{b} \subset X_p$. We count its cardinality in two ways. First we have seen that $T_{\underline{\mu}}(\underline{b})$ has $|\det(\underline{b})|$ elements, so $|S| = \sum_{\underline{b} \in \mathcal{B}_X} |\det(\underline{b})|$.

On the other hand, $|S| = \sum_{p \in P_{\underline{\mu}}(X)} d(X_p)$. By the paving of the zonotope we know that $\sum_{\underline{b} \in \mathcal{B}_X} |\det(\underline{b})| = \delta(X)$ is the volume of the zonotope, and the claim follows.

Proposition 13.18. *1. The variety defined by the ideal $J_X(\mu)$ is $P_{\underline{\mu}}(X)$.*

2. For generic values of the parameters $\nu_i = e^{\mu_i}$, the set $P_{\underline{\mu}}(X)$ consists of $\delta(X)$ distinct points.

Proof. The proof of 1 is similar to that of Lemma 11.9 and the first part of Theorem 11.13. The only difference is the fact that, when we extract a basis $\underline{b}$ from X, the equations $e^{-b + \mu_b} - 1 = 0$ for $b \in \underline{b}$ define $T_{\underline{\mu}}(\underline{b})$. By its definition (14.3), $P_{\underline{\mu}}(X) = \cup_{\underline{b}} T_{\underline{\mu}}(\underline{b})$.

Part 2 is trivial if X consists of a basis of V. Otherwise, write for a basis $\underline{b}$ in X an element a in $X \setminus \underline{b}$, $ka = \sum_{b \in B} n_b b$ with $k, n_b \in \mathbb{Z}$. Consider the subgroup in $(\mathbb{C}^*)^m$ that is kernel of the character $t_a^{-k} \prod_{b \in \underline{b}} t_b^{n_b}$. It is then immediate that if the point of coordinates $\nu_i = e^{\mu_i}$ lies outside the union of these subgroups as $\underline{b}$ and a vary, the set $P_{\underline{\mu}}(X)$ consists of $\delta(X)$ distinct points.

Choose any chamber $\mathfrak{c}$. The main theorem we plan to prove is the following

Theorem 13.19. *The classes modulo $J_X(\underline{t})$ of the elements e^a, $a \in \delta(\mathfrak{c}|X)$ give a basis over of $\mathbb{A} = R[\Lambda]/J_X(\underline{t})$ as an R-module.*

We start with a lemma

Lemma 13.20. *i) The classes modulo $J_X(\underline{t})$ of the elements e^a, $a \in \delta(\mathfrak{c}|X)$ span $\mathbb{A}$ as an R-module.*

Proof. Denote by B the span of the classes modulo $J_X(\underline{t})$ of the elements e^a, $a \in \delta(\mathfrak{c}|X)$. Observe that $\Lambda = \cup_{\mathfrak{g}} \delta(\mathfrak{g}|X)$ as $\mathfrak{g}$ varies over all the chambers. So it is enough to show that given a chamber $\mathfrak{g}$ and $a \in \delta(\mathfrak{g}|X)$, we have that e^a is in B.

Given such a chamber $\mathfrak{g}$, we define $d(\mathfrak{g})$ to be the minimum d such that there is a sequence $\mathfrak{c} = \mathfrak{g}_0, \ldots, \mathfrak{g}_d = \mathfrak{g}$ with $\overline{\mathfrak{g}_i}, \overline{\mathfrak{g}_{i+1}}$ intersecting in an $(s-1)$-dimensional face. Notice that $d(\mathfrak{g}) = 0$ if and only if $\mathfrak{g} = \mathfrak{c}$. So we can proceed by induction on $d(\mathfrak{g})$ and assume that the class of e^a is in B for each point $a \in \delta(\mathfrak{g}_{d-1}|X)$.

The closures of $\mathfrak{g}_d$ and $\mathfrak{g}_{d-1}$ intersect in an $(s-1)$-dimensional face $\mathfrak{f}$. Let H be the hyperplane through the origin parallel to $\mathfrak{f}$ and let $C = H \cap X$. If we choose an equation $u \in U$ for H we can divide $Y = X \setminus C$ into the disjoint union $A \cup B$, where A (resp. B) consists of the vectors on which u is positive (resp. negative).

Let $\alpha \in \delta(\mathfrak{g}|X)$. We claim that we can express the class of e^α modulo J_X as a linear combination with coefficients in R of the classes modulo J_X of the elements e^a, $a \in \delta(\mathfrak{g}_{d-1}|X)$. This will obviously imply our claim. If $\alpha \in \mathfrak{g}_{d-1}$, there is nothing to prove. So let us assume $\alpha \in \delta(\mathfrak{g}|X) \setminus \delta(\mathfrak{g}_{d-1}|X)$.

By Theorem 13.4 we have, up to changing the sign of u, that for all nonempty $D \subset Y$, $\alpha + a_{D,u} \in \mathfrak{g}_{d-1}$. On the other hand,

$$\prod_{a \in A}(1 - t_a^{-1}e^a) \prod_{b \in B}(1 - t_b e^{-b}) = (-1)^{|A|} \prod_{a \in A} t_a^{-1} e^{-a} \prod_{x \in Y}(1 - t_x e^{-x}) \in J_X(\underline{t}).$$

So $e^\alpha \prod_{a \in A}(1 - t_a^{-1}e^a) \prod_{b \in B}(1 - t_b e^{-b}) \in J_X(\underline{t})$. Developing we get

$$e^\alpha = \sum_{D \subset Y, D \neq \emptyset} t_D e^{\alpha + a_{D,u}}, \quad \text{modulo } J_X(\underline{t}),$$

with $t_D = (-1)^{|D|} \prod_{a \in A} t_a^{-1} \prod_{b \in B} t_b \in R$ which implies our claim.

We can now give a proof of Theorem 13.19.

Proof. By Lemma 13.20, it suffices to see that the classes modulo $J_X(\underline{t})$ of the elements e^a, $a \in \delta(\mathfrak{c}|X)$, are linearly independent over R.

By contradiction assume that there is a nonzero linear combination

$$\sum_{a \in \delta(\mathfrak{c}|X)} p_a e^a \in J_X(\underline{t})$$

with $p_a \in R$. Choose, using Proposition 13.18, a point of coordinates $\nu_i = e^{\mu_i}$ such that $P_{\underline{\mu}}(X)$ consists of $\delta(X)$ distinct points and such that at least one of the coefficients p_a is not zero at this point.

It follows that on the one hand, $\mathbb{C}[\Lambda]/J_X(\underline{\mu})$ has dimension at least $\delta(X)$, on the other it has dimension strictly smaller that $\delta(X)$ giving a contradiction.

Notice that the proof of the lemma furnishes an explicit algorithm of recursion to compute the value of a function in $DM(X)$ at any point once we know its values on a set $\delta(\mathfrak{c}|X)$.

We shall use mainly the following consequence of this theorem:

Theorem 13.21. *$DM(X)$ is a free abelian group of dimension $\delta(X)$ consisting of quasipolynomials.*

For any chamber $\mathfrak{c}$, evaluation of the functions in $DM(X)$ on the set $\delta(\mathfrak{c}\,|\,X)$ establishes a linear isomorphism of $DM(X)$ with the abelian group of all $\mathbb{Z}$-valued functions on $\delta(\mathfrak{c}\,|\,X)$.

Proof. Recall that the abelian group $DM(X)$ is the dual to the ring $\mathbb{Z}[\Lambda]/J_X$. Since the ideal defines the set $P(X)$ that consists of torsion points, we deduce from Theorem 5.32 that $DM(X)$ consists of quasipolynomials.

All the other statements follow from Theorem 13.19.

13.2.4 A Categorical Interpretation

It would be interesting to give a direct proof of Theorem 13.19 by showing that the recursive equations can be solved starting from the initial values. This approach requires us to prove that if we reach the same point by a different path of recursions, we indeed obtain the same value. Although in the end this is true we do not know a direct proof of this statement.

One may interpret the previous discussion as follows. We associate to X an infinite connected oriented graph whose vertices are the chambers; an oriented edge is associated to a common codimension-one face $\mathfrak{f}$ of two chambers $\mathfrak{c}, \mathfrak{g}$. We orient the edge from $\mathfrak{g}$ to $\mathfrak{c}$ if $B(X)$ and $\mathfrak{g}$ lie on the same side of the hyperplane spanned by $\mathfrak{f}$.

The lattice Λ acts on this graph with finitely many orbits on vertices and edges.

Given a chamber $\mathfrak{c}$, we associate to it the finite set $\delta(\mathfrak{c} \,|\, X)$ and denote by $D(\mathfrak{c})$ the abelian group of $\mathbb{Z}$-valued functions on $\delta(\mathfrak{c} \,|\, X)$.

An edge $\mathfrak{g} \xrightarrow{\mathfrak{f}} \mathfrak{c}$ identifies the rational hyperplane H, as well as the sets A, B, C and the bijective correspondence

$$\delta(\mathfrak{f}) : \delta(\mathfrak{c} \,|\, X) \to \delta(\mathfrak{g} \,|\, X)$$

given by

$$\delta(\mathfrak{f})(y) := \begin{cases} y & \text{if } y \in \delta(\mathfrak{c} \,|\, X) \cap \delta(\mathfrak{g} \,|\, X), \\ y - a_A + a_B & \text{if } y \in \delta(\mathfrak{c} \,|\, X) \setminus \delta(\mathfrak{g} \,|\, X). \end{cases}$$

Finally, to the same oriented edge we can also associate the linear map

$$D(\mathfrak{f}) : D(\mathfrak{c}) \to D(\mathfrak{g})$$

given by the matrix defined implicitly in Lemma 13.20 or its inverse.

This finite oriented graph incorporates all of the combinatorics of the abelian group $DM(X)$, that can be reconstructed from these data. A direct proof of the theorem should consist in verifying directly that when we compose the matrices on two given paths with the same endpoints we obtain the same result.

13.3 Special Functions

The purpose of this and the following section is to exhibit explicit elements in $DM(X)$. In order to achieve this, we develop first some formalism.

13.3.1 Convolutions and the Partition Function

We now return to the formalism of Section 12.2.2, where we introduced bi-infinite series $\mathbb{C}[[\Lambda]] := \{\sum_{v \in \Lambda} g(v) e^v\}$, that by abuse of notation we identify with $\mathcal{C}_{\mathbb{C}}[\Lambda]$ via the map $L : \mathcal{C}_{\mathbb{C}}[\Lambda] \to \mathbb{C}[[\Lambda]]$.

As in Section 12.2.2, we can then consider the subspace $\mathcal{R} \subset \mathbb{C}[[\Lambda]]$ of rational elements consisting of those elements $g \in \mathbb{C}[[\Lambda]]$ such that there exists $p \in \mathbb{C}[\Lambda], p \neq 0$ with $pg \in \mathbb{C}[\Lambda]$ together with its torsion subspace $\mathcal{T}$ of those $g \in \mathbb{C}[[\Lambda]]$ such that there exits $p \in \mathbb{C}[\Lambda], p \neq 0$ with $pg = 0$.

Notice that under the identification of the space of functions $\mathcal{C}_{\mathbb{C}}[\Lambda]$ with $\mathbb{C}[[\Lambda]]$, we have $DM(X) \subset \mathcal{T}$.

Recall that by Proposition 12.7, $\mathcal{R}/\mathcal{T}$ is isomorphic, under a map L_r, to the field of fractions $\mathbb{C}(\Lambda)$ of $\mathbb{C}[\Lambda]$.

If $\gamma \in \Lambda$ is nonzero, we set

$$\mathcal{T}_\gamma := \sum_{k=0}^{\infty} \delta_{k\gamma}, \quad L\mathcal{T}_\gamma = \sum_{k=0}^{\infty} e^{-k\gamma}.$$

The function $\mathcal{T}_\gamma$ is supported on the "half-line" $\mathbb{Z}_{\geq 0}\gamma$ and is the discrete analogue of the Heaviside function. In fact, $\nabla_\gamma \mathcal{T}_\gamma = \delta_0$ or $(1 - e^{-\gamma})L\mathcal{T}_\gamma = 1$.

Thus $L\mathcal{T}_\gamma \in \mathcal{R}$, or by abuse of notation $\mathcal{T}_\gamma \in \mathcal{R}$, and we simply write $L_r\mathcal{T}_\gamma := L_r L\mathcal{T}_\gamma$, getting

$$L_r\mathcal{T}_\gamma = \frac{1}{1 - e^{-\gamma}}.$$

Moreover, if $u \in \hom(\Lambda, \mathbb{R})$ is an injective homomorphism with $\langle u \,|\, a \rangle > 0$ and $s_u : \mathbb{C}(\Lambda) :\to \mathcal{R}$ the corresponding section, we have

$$L\mathcal{T}_\gamma = s_u\left(\frac{1}{1 - e^{-\gamma}}\right).$$

Notice that if instead $\langle u \,|\, a \rangle < 0$, then we have

$$s_u\left(\frac{1}{1 - e^{-\gamma}}\right) = -e^{\gamma} s_u\left(\frac{1}{1 - e^{\gamma}}\right) = -e^{\gamma} L\mathcal{T}_{-\gamma} \cong -\delta_{-\gamma} * \mathcal{T}_{-\gamma}. \tag{13.8}$$

Now let $X := (a_1, \ldots, a_m)$ be a finite list of vectors in Λ that spans a pointed cone in V. This implies the existence of a generic linear form u with $\langle u \,|\, a_i \rangle > 0$ for all i. Recall that in 1.45, we have defined the partition function on Λ by

$$\mathcal{T}_X(\lambda) := \#\{(n_1, \ldots, n_m) \,|\, \sum n_i a_i = \lambda, \ n_i \in \mathbb{N}\}.$$

Consider

$$C(X) := \{\sum_{i=1}^{n} m_i a_i, \ m_i \in \mathbb{Z}_{\geq 0}\} \subset \Lambda.$$

Clearly, the partition function is supported in $C(X)$.

Also the convolution $\mathcal{T}_{a_1} * \mathcal{T}_{a_2} * \cdots * \mathcal{T}_{a_m}$ is well-defined, in $i_u \mathcal{R}$ and we immediately verify the following

Proposition 13.22.

$$\mathcal{T}_X := \mathcal{T}_{a_1} * \mathcal{T}_{a_2} * \cdots * \mathcal{T}_{a_m}.$$

We shall thus think of $\mathcal{T}_X$ as a discrete analogue of the multivariate spline T_X (see Theorem 7.18). [1] We also have an analogue of B_X. Namely, setting

$$Q_X(v) := \#\Big\{(n_1,\ldots,n_m)\,\Big|\, \sum n_i a_i = v,\ n_i \in \{0,1\}\Big\},$$

where

$$Q_X := \sum_{v \in \Lambda} Q_X(v)\delta_v. \tag{13.9}$$

Applying our previous considerations we obtain the following

Lemma 13.23.   *(i) $\mathcal{T}_X \in \mathcal{R}$ and*

$$L_r \mathcal{T}_X = \prod_{a \in X} \frac{1}{1 - e^{-a}}. \tag{13.10}$$

(ii) $Q_X \in \mathcal{C}[\Lambda]$ and

$$L_r Q_X = \prod_{a \in X} (1 + e^{-a}) = \prod_{a \in X} \frac{1 - e^{-2a}}{1 - e^{-a}}. \tag{13.11}$$

If X is any list of nonzero vectors and u is a generic linear form, we can still construct $i_u(\prod_{a \in X} \frac{1}{1-e^{-a}}) = *_{a \in X} i_u(\frac{1}{1-e^{-a}})$. Now, though, we have to apply formula (13.8) and obtain that $i_u(\prod_{a \in X} \frac{1}{1-e^{-a}})$ is the convolution product of the functions $\mathcal{T}_{a_i}$ for the a_i with $\langle u \,|\, a_i \rangle > 0$ and of $-\delta_{-a_i} * \mathcal{T}_{-a_i}$ for the a_i with $\langle u \,|\, a_i \rangle < 0$.

The reader can verify that also in this setting, $i_u(\prod_{a \in X} \frac{1}{1-e^{-a}})$ can be interpreted as a suitable partition function that now depends on u.

13.3.2 Constructing Elements in $DM(X)$

Definition 13.24. A subspace of V spanned by a subset of X will be called a *rational subspace* (relative to X).

We shall denote by S_X the set of all rational subspaces and by $S_X^{(i)}$ the rational subspaces of dimension i.

Given a rational subspace $\underline{r}$, $X \setminus \underline{r}$ defines a hyperplane arrangement in the space $\underline{r}^{\perp} \subset U$ orthogonal to $\underline{r}$. Take an open face $F_{\underline{r}}$ in $\underline{r}^{\perp}$ with respect to this hyperplane arrangement.

We get a decomposition of $X \setminus \underline{r}$ into the two lists A, B of elements that are positive (resp. negative) on $F_{\underline{r}}$. We denote by $C(F_{\underline{r}}, X)$ the cone $C(A, -B)$ generated by the list $[A, -B]$. Then $C(A, -B)$ is a pointed cone.

---

[1] Some authors use the term *truncated multivariate spline.*

Notice that if $\underline{r} = \{0\}$, then $A \cup B = X$. In this we shall denote an open face $F_{\{0\}}$ simply by F and call it a regular face for X.

Notice that given $a \in \Lambda$, we have $\nabla_a = -\tau_a \nabla_{-a}$.

Lemma 13.25. *There exists a unique element $\mathcal{T}_X^{F_{\underline{r}}} \in \mathcal{C}[\Lambda]$ such that*

(i) $(\prod_{a \in X \setminus \underline{r}} \nabla_a) \mathcal{T}_X^{F_{\underline{r}}} = \delta_0$.

(ii) $\mathcal{T}_X^{F_{\underline{r}}}$ is supported in $-b_B + C(A, -B)$.

Proof. We define $\mathcal{T}_X^{F_{\underline{r}}}$ by the convolution product of partition functions:

$$\mathcal{T}_X^{F_{\underline{r}}} = (-1)^{|B|} \delta_{-b_B} * \mathcal{T}_A * \mathcal{T}_{-B}, \tag{13.12}$$

where $b_B = \sum_{b \in B} b$. Since

$$\prod_{a \in X \setminus \underline{r}} \nabla_a = (-1)^{|B|} \tau_{b_B} \Big(\prod_{a \in A} \nabla_a \Big) \Big(\prod_{b \in B} \nabla_{-b} \Big) \tag{13.13}$$

and $C(A, -B)$ is a pointed cone, $\mathcal{T}_X^{F_{\underline{r}}}$ is well-defined, satisfies the two properties, and is unique.

We can generalize this as follows. Choose an orientation of each $\underline{r} \in S_X$. Take a pair of rational subspaces $\underline{r} \in S_X^{(i)}$ and $\underline{t} \in S_X^{(i+1)}$ with $\underline{r} \subset \underline{t}$. We say that a vector v in $\underline{t} \setminus \underline{r}$ is positive, if the orientation on $\underline{t}$ induced by v and the orientation on $\underline{r}$, coincides with the chosen orientation of $\underline{t}$. Set

$$A = \{a \in X \cap \underline{t} \,|\, a \text{ is positive}\}, \quad B = \{b \in X \cap \underline{t} \,|\, b \text{ is negative}\}.$$

We define

$$\mathcal{T}_{\underline{r}}^{t+} := (-1)^{|B|} \delta_{-b_B} * \mathcal{T}_A * \mathcal{T}_{-B}, \tag{13.14}$$

$$\mathcal{T}_{\underline{r}}^{t-} := (-1)^{|A|} \delta_{-a_A} * \mathcal{T}_B * \mathcal{T}_{-A}. \tag{13.15}$$

The function $\mathcal{T}_{\underline{r}}^{t+}$ is supported on the cone $-\sum_{b \in B} b + C(A, -B)$ while $\mathcal{T}_{\underline{r}}^{t-}$ is supported on the cone $-\sum_{a \in A} a + C(-A, B)$.

Definition 13.26. Take a pair of oriented rational subspaces $\underline{r} \in S_X^{(i)}$ and $\underline{t} \in S_X^{(i+1)}$ with $\underline{r} \subset \underline{t}$. We define

$$\mathcal{Q}_{\underline{r}}^{t}(X) = \mathcal{T}_{\underline{r}}^{t+} - \mathcal{T}_{\underline{r}}^{t-}.$$

If X is fixed, we will write simply $\mathcal{Q}_{\underline{r}}^{t}$ instead of $\mathcal{Q}_{\underline{r}}^{t}(X)$.

We identify $\mathcal{C}[\Lambda \cap \underline{r}]$ to the subgroup in $\mathcal{C}[\Lambda]$ consisting of the elements supported in $\Lambda \cap \underline{r}$. If f is a function on $\Lambda \cap \underline{r}$, the hypotheses of Remark 12.5 are satisfied for the pairs $f, \mathcal{T}_{\underline{r}}^{t\pm}$, and we can perform the convolutions $\mathcal{T}_{\underline{r}}^{t\pm} * f$. So convolution by $\mathcal{Q}_{\underline{r}}^{t}$ induces a map

$$\Pi_{\underline{r}}^{t} : \mathcal{C}[\Lambda \cap \underline{r}] \to \mathcal{C}[\Lambda \cap \underline{t}], \quad f \mapsto \mathcal{Q}_{\underline{r}}^{t} * f.$$

Given a rational subspace $\underline{r}$, let us consider the abelian group $DM(X \cap \underline{r})$ inside $\mathcal{C}[\Lambda \cap \underline{r}] \subset \mathcal{C}[\Lambda]$.

Proposition 13.27. $\Pi_{\underline{r}}^{\underline{t}}$ *maps* $DM(X \cap \underline{r})$ *to* $DM(X \cap \underline{t})$.

Proof. Let $Y := X \cap \underline{r}$, $Y' := X \cap \underline{t}$. If $T \subset Y'$ is a cocircuit in Y', we need to see that $\nabla_T(\mathcal{Q}_{\underline{r}}^{\underline{t}} * f) = 0$ for each $f \in DM(X \cap \underline{r})$.

By definition,

$$\nabla_{A \cup B} \mathcal{T}_{\underline{r}}^{\underline{t}+} = \nabla_{A \cup B} \mathcal{T}_{\underline{r}}^{\underline{t}-} = \delta_0,$$

so that if $T = A \cup B = Y' \setminus Y$, then

$$\nabla_T(\mathcal{Q}_{\underline{r}}^{\underline{t}} * f) = (\nabla_T \mathcal{Q}_{\underline{r}}^{\underline{t}}) * f = 0.$$

Otherwise, the set $Y \cap T$ is a cocircuit in Y and then $\nabla_{Y \cap T} f = 0$. So

$$\nabla_T(\mathcal{Q}_{\underline{r}}^{\underline{t}} * f) = (\nabla_{T \setminus Y} \mathcal{Q}_{\underline{r}}^{\underline{t}}) * (\nabla_{Y \cap T} f) = 0.$$

This proposition gives a way to construct explicit elements of $DM(X)$. Take a flag ϕ of oriented rational subspaces $0 = \underline{r}_0 \subset \underline{r}_1 \subset \underline{r}_2 \subset \cdots \subset \underline{r}_s = V$ with $\dim(\underline{r}_i) = i$. Set $\mathcal{Q}_i := \mathcal{T}_{\underline{r}_{i-1}}^{\underline{r}_i+} - \mathcal{T}_{\underline{r}_{i-1}}^{\underline{r}_i-}$.

Proposition 13.28.

$$\mathcal{Q}_{\phi}^X := \mathcal{Q}_1 * \mathcal{Q}_2 * \cdots * \mathcal{Q}_s \tag{13.16}$$

lies in $DM(X)$.

13.4 A Remarkable Space

In this section we follow very closely a paper with M. Vergne [44].

13.4.1 A Basic Formula

Choose two rational spaces $\underline{r}, \underline{t}$ and a regular face $F_{\underline{r}}$ for $X \setminus \underline{r}$ in $\underline{r}^{\perp}$.

The image of $F_{\underline{r}}$ modulo $\underline{t}^{\perp}$ is contained in a unique regular face for the set $(X \cap \underline{t}) \setminus \underline{r}$. To simplify notation, we still denote this face by $F_{\underline{r}}$.

Proposition 13.29. *1.* $\nabla_{(X \setminus \underline{t}) \setminus \underline{r}} \mathcal{T}_{X \setminus \underline{r}}^{F_{\underline{r}}} = \mathcal{T}_{(X \setminus \underline{r}) \cap \underline{t}}^{F_{\underline{r}}}$.
2. For $g \in \mathcal{C}[\Lambda \cap \underline{r}]$,

$$\nabla_{X \setminus \underline{t}}(\mathcal{T}_{X \setminus \underline{r}}^{F_{\underline{r}}} * g) = \mathcal{T}_{(X \setminus \underline{r}) \cap \underline{t}}^{F_{\underline{r}}} * (\nabla_{(X \cap \underline{r}) \setminus \underline{t}} g). \tag{13.17}$$

Proof. 1 From equation (13.12), we see that

$$\nabla_{(X \setminus \underline{t}) \setminus \underline{r}} \mathcal{T}_X^{F_{\underline{r}}} = (-1)^{|B|} \delta_{-b_B} * \nabla_{(A \setminus \underline{t}) \setminus \underline{r}} (\mathcal{T}_A) * \nabla_{(B \setminus \underline{t}) \setminus \underline{r}} (\mathcal{T}_{-B}).$$

Using formula (13.13) we get

$$\nabla_{(X\setminus\underline{t})\setminus\underline{r}}\mathcal{T}_X^{F_{\underline{r}}} = (-1)^{|B\cap\underline{t}|}\delta_{-b_{B\cap\underline{t}}} * \mathcal{T}_{A\cap\underline{t}} * \mathcal{T}_{-B\cap\underline{t}} = \mathcal{T}_{(X\setminus\underline{r})\cap\underline{t}}^{F_{\underline{r}}},$$

which is our claim.

2 Let $g \in \mathcal{C}[\Lambda\cap\underline{r}]$. Take any rational subspace $\underline{t}$. We have the factorization $\nabla_{X\setminus\underline{t}} = \nabla_{(X\cap\underline{r})\setminus\underline{t}}\nabla_{(X\setminus\underline{t})\setminus\underline{r}}$; thus

$$\nabla_{X\setminus\underline{t}}(\mathcal{T}_{X\setminus\underline{r}}^{F_{\underline{r}}} * g) = (\nabla_{(X\setminus\underline{t})\setminus\underline{r}}\mathcal{T}_{X\setminus\underline{r}}^{F_{\underline{r}}}) * (\nabla_{(X\cap\underline{r})\setminus\underline{t}}g).$$

Since $\nabla_{(X\setminus\underline{t})\setminus\underline{r}}\mathcal{T}_{X\setminus\underline{r}}^{F_{\underline{r}}} = \mathcal{T}_{(X\setminus\underline{r})\cap\underline{t}}^{F_{\underline{r}}}$ from part 1, we obtain formula (13.17), that is the origin of all other formulas of this section.

13.4.2 The Abelian Group $\mathcal{F}(X)$.

The following abelian group will play a key role in what follows.

Definition 13.30.

$$\mathcal{F}(X) := \{f \in \mathcal{C}[\Lambda] \mid \nabla_{X\setminus\underline{r}}f \in \mathcal{C}[\Lambda\cap\underline{r}], \text{ for all } \underline{r} \in S_X\}. \qquad (13.18)$$

Notice that if $f \in \mathcal{F}(X)$, then f must in particular satisfy the relation corresponding to the space $\underline{r} = \{0\}$), that is,

$$\nabla_X f = c\delta_0,$$

or the equivalent relation

$$\prod_{a\in X}(1 - e^a)Lf = c.$$

with $c \in \mathbb{Z}$. In particular, $\mathcal{F}(X) \subset \mathcal{R}$ (the rational elements) and its image $L_r(\mathcal{F}(X))$ in the field $C(\Lambda)$ is the one dimensional abelian group spanned by

$$\prod_{a\in X}\frac{1}{1 - e^{-a}}.$$

Example 13.31. Let us give a simple example. Let $\Lambda = \mathbb{Z}$ and $X = [2, -1]$. Then it is easy to see that $\mathcal{F}(X)$ has as integral basis

$$\theta_1 = \sum_{n\in\mathbb{Z}}\delta_n, \qquad\qquad \theta_2 = \sum_{n\in\mathbb{Z}}n\delta_n,$$

$$\theta_3 = \sum_{n\in\mathbb{Z}}(\frac{n}{2} + \frac{1 - (-1)^n}{4})\delta_n, \qquad \theta_4 = \sum_{n\geq 0}(\frac{n}{2} + \frac{1 - (-1)^n}{4})\delta_n.$$

Here $\theta_1, \theta_2, \theta_3$ is an integral basis of $DM(X)$.

The first important fact on this abelian group is the following:

Lemma 13.32.   *(i) If F is a regular face for X, then $\mathcal{T}_X^F$ lies in $\mathcal{F}(X)$. (ii) The abelian group $DM(X)$ is contained in $\mathcal{F}(X)$.*

Proof. (i) Indeed, $\nabla_{X\setminus\underline{r}}\mathcal{T}_X^F = \mathcal{T}_{X\cap\underline{r}}^F \in \mathcal{C}[\Lambda \cap \underline{r}]$.
(ii) is clear from the definitions.

Each $\mathcal{T}_X^F$ is a partition function. In particular, if X generates a pointed cone, then the partition function $\mathcal{T}_X$ equals $\mathcal{T}_X^F$ for the face F that is positive on X.

In fact, there is a much more precise statement of which Lemma 13.32 is a very special case and that will be the object of Theorem 13.35.

13.4.3 Some Properties of $\mathcal{F}(X)$.

Let $\underline{r}$ be a rational subspace and $F_{\underline{r}}$ a regular face for $X \setminus \underline{r}$.

Proposition 13.33. *1. The map $g \mapsto \mathcal{T}_{X\setminus\underline{r}}^{F_{\underline{r}}} * g$ gives an injection from the group $\mathcal{F}(X \cap \underline{r})$ to $\mathcal{F}(X)$. Moreover,*

$$\nabla_{X\setminus\underline{r}}(\mathcal{T}_{X\setminus\underline{r}}^{F_{\underline{r}}} * g) = g, \ \forall g \in \mathcal{F}(X \cap \underline{r}). \tag{13.19}$$

2. $\nabla_{X\setminus\underline{r}}$ maps $\mathcal{F}(X)$ surjectively to $\mathcal{F}(X \cap \underline{r})$.

3. If $g \in DM(X \cap \underline{r})$, then $\nabla_{X\setminus\underline{t}}(\mathcal{T}_{X\setminus\underline{r}}^{F_{\underline{r}}} * g) = 0$ for any rational subspace $\underline{t}$ such that $\underline{t} \cap \underline{r} \neq \underline{r}$.

Proof. 1 If $g \in \mathcal{F}(X \cap \underline{r})$, then for every rational subspace $\underline{t}$ we have $\nabla_{(X\cap\underline{r})\setminus\underline{t}}g \in \mathcal{C}[\Lambda \cap \underline{t} \cap \underline{r}]$. By formula (13.17) we have that $\nabla_{X\setminus\underline{t}}(\mathcal{T}_{X\setminus\underline{r}}^{F_{\underline{r}}} * g)$ is the convolution of two functions supported in $\underline{t}$; hence it lies in $\mathcal{C}[\Lambda \cap \underline{t}]$, so that $\mathcal{T}_{X\setminus\underline{r}}^{F_{\underline{r}}} * g \in \mathcal{F}(X)$, as desired.

Formula (13.19) follows from the fact that $\nabla_{X\setminus\underline{r}}\mathcal{T}_{X\setminus\underline{r}}^{F_{\underline{r}}} = \delta_0$.

2 If $f \in \mathcal{F}(X)$, we have $\nabla_{X\setminus\underline{r}}f \in \mathcal{F}(X \cap \underline{r})$. In fact, taking a rational subspace $\underline{t}$ of $\underline{r}$, we have that $\nabla_{(X\cap\underline{r})\setminus\underline{t}}\nabla_{X\setminus\underline{r}}f = \nabla_{X\setminus\underline{t}}f \in \mathcal{C}[\Lambda \cap \underline{t}]$. The fact that $\nabla_{X\setminus\underline{r}}$ is surjective is a consequence of formula (13.19).

3 Follow from formula (13.17).

Proposition 13.33 allows us to associate to a rational space $\underline{r}$ and a regular face $F_{\underline{r}}$ for $X \setminus \underline{r}$ the operator

$$\Pi^{F_{\underline{r}}} : f \mapsto \mathcal{T}_{X\setminus\underline{r}}^{F_{\underline{r}}} * (\nabla_{X\setminus\underline{r}}f)$$

on $\mathcal{F}(X)$. From formula (13.19), it follows that the operator $\Pi^{F_{\underline{r}}}$ is a projector with image $\mathcal{T}_{X\setminus\underline{r}}^{F_{\underline{r}}} * \mathcal{F}(X \cap \underline{r})$.

13.4.4 The Main Theorem

Recall that $S_X^{(i)}$ is the set of rational subspaces of dimension i. Define the abelian groups

$$\mathcal{F}(X)_i := \cap_{\underline{t} \in S_X^{(i-1)} } \ker(\nabla_{X \setminus \underline{t}}) \cap \mathcal{F}(X).$$

Notice that by definition, $\mathcal{F}(X)_{\{0\}} = \mathcal{F}(X)$, that $\mathcal{F}(X)_{\dim V}$ is the abelian group $DM(X)$, and that $\mathcal{F}(X)_{i+1} \subseteq \mathcal{F}(X)_i$.

Choose, for every rational space $\underline{r}$, a regular face $F_{\underline{r}}$ for $X \setminus \underline{r}$.

Lemma 13.34. *Let $\underline{r} \in S_X^{(i)}$.*

(i) *The image of $\nabla_{X \setminus \underline{r}}$ restricted to $\mathcal{F}(X)_i$ is contained in the abelian group $DM(X \cap \underline{r})$.*

(ii) *If f is in $DM(X \cap \underline{r})$, then $T_{X \setminus \underline{r}}^{F_{\underline{r}}} * f \in \mathcal{F}(X)_i$.*

Proof. (i) First, we know by the definition of $\mathcal{F}(X)$, that $\nabla_{X \setminus \underline{r}} \mathcal{F}(X)_i$ is contained in the abelian group $\mathcal{C}[\Lambda \cap \underline{r}]$. Let $\underline{t}$ be a rational hyperplane of $\underline{r}$, so that $\underline{t}$ is of dimension $i - 1$. By construction, we have that for every $f \in \mathcal{F}(X)_i$,

$$0 = \nabla_{X \setminus \underline{t}} f = \nabla_{(X \cap \underline{r}) \setminus \underline{t}} \nabla_{X \setminus \underline{r}} f.$$

This means that $\nabla_{X \setminus \underline{r}} f$ satisfies the difference equations given by the cocircuits of $X \cap \underline{r}$, that is, it lies in $DM(X \cap \underline{r})$.

(ii) Follows from Proposition 13.33 3. $\qquad \square$

The following is the main theorem of this section.

Theorem 13.35. *With the previous choices, we have*

$$\mathcal{F}(X) = \oplus_{\underline{r} \in S_X} T_{X \setminus \underline{r}}^{F_{\underline{r}}} * DM(X \cap \underline{r}). \tag{13.20}$$

Proof. Consider the map $\mu_i : \mathcal{F}(X)_i \to \oplus_{\underline{r} \in S_X^{(i)}} DM(X \cap \underline{r})$ given by

$$\mu_i f := \oplus_{\underline{r} \in S_X^{(i)}} \nabla_{X \setminus \underline{r}} f$$

and the map $\mathbf{P}_i : \oplus_{\underline{r} \in S_X^{(i)}} DM(X \cap \underline{r}) \to \mathcal{F}(X)_i$ given by

$$\mathbf{P}_i(\oplus g_{\underline{r}}) := \sum T_{X \setminus \underline{r}}^{F_{\underline{r}}} * g_{\underline{r}}.$$

We claim that the sequence

$$0 \longrightarrow \mathcal{F}(X)_{i+1} \longrightarrow \mathcal{F}(X)_i \xrightarrow{\mu_i} \oplus_{\underline{r} \in S_X^{(i)}} DM(X \cap \underline{r}) \longrightarrow 0$$

is exact and the map $\mathbf{P}_i$ provides a splitting.

By definition, $\mathcal{F}(X)_{i+1}$ is the kernel of μ_i. Thus we only need to show that $\mu_i \mathbf{P}_i = \mathrm{Id}$. Given $\underline{r} \in S_X^{(i)}$ and $g \in DM(X \cap \underline{r})$, by formula (13.19) we have $\nabla_{X \backslash \underline{r}}(\mathcal{T}^{F_{\underline{r}}}_{X \backslash \underline{r}} * g) = g$. If instead we take $\underline{t} \neq \underline{r}$ another subspace of $S_X^{(i)}$, then $\underline{r} \cap \underline{t}$ is a proper subspace of $\underline{t}$. Proposition 13.33 *iii)* says that for $g \in DM(X \cap \underline{r})$, $\nabla_{X \backslash \underline{t}}(\mathcal{T}^{F_{\underline{r}}}_{X \backslash \underline{r}} * g) = 0$. Thus, given a family $g_{\underline{r}} \in DM(X \cap \underline{r})$, the function $f = \sum_{\underline{t} \in S_X^{(i)}} \mathcal{T}^{F_{\underline{t}}}_{X \backslash \underline{t}} * g_{\underline{t}}$ is such that $\nabla_{X \backslash \underline{r}} f = g_{\underline{r}}$ for all $\underline{r} \in S_X^{(i)}$. This proves our claim that $\mu_i \mathbf{P}_i = \mathrm{Id}$.

At this point, putting together these facts, Theorem 13.35 follows.

Definition 13.36. A collection $\mathbf{F} = \{F_{\underline{r}}\}$ of faces $F_{\underline{r}} \subset \underline{r}^{\perp}$ regular for $X \setminus \underline{r}$ indexed by the rational subspaces $\underline{r} \in S_X$ will be called an *X-regular collection.*

Given an X-regular collection $\mathbf{F}$, we can write, using Theorem 13.35, an element $f \in \mathcal{F}(X)$ as

$$f = \sum_{\underline{r} \in S_X} f_{\underline{r}} \quad \text{with} \quad f_{\underline{r}} \in \mathcal{T}^{F_{\underline{r}}}_{X \backslash \underline{r}} * DM(X \cap \underline{r}).$$

This expression for f will be called the $\mathbf{F}$ decomposition of f. In this decomposition, we always have $F_V = \{0\}$, $\mathcal{T}^{F_V}_{X \backslash V} = \delta_0$ and the component f_V is in $DM(X)$.

The abelian group $\mathcal{T}^{F_{\underline{r}}}_{X \backslash \underline{r}} * DM(X \cap \underline{r})$ will be referred to as *the $F_{\underline{r}}$-component* of $\mathcal{F}(X)$.

From Theorem 13.35, it follows that the operator $\mathrm{Id} - \mathbf{P}_i \mu_i$ projects $\mathcal{F}(X)_i$ to $\mathcal{F}(X)_{i+1}$ with kernel $\oplus_{\underline{r} \in S_X^{(i)}} \mathcal{T}^{F_{\underline{r}}}_{X \backslash \underline{r}} * DM(X \cap \underline{r})$ (this operator depends of $\mathbf{F}$). Thus the ordered product

$$\mathbf{\Pi}_i^{\mathbf{F}} := (\mathrm{Id} - \mathbf{P}_{i-1} \mu_{i-1})(\mathrm{Id} - \mathbf{P}_{i-2} \mu_{i-2}) \cdots (\mathrm{Id} - \mathbf{P}_0 \mu_0)$$

projects $\mathcal{F}(X)$ to $\mathcal{F}(X)_i$. Therefore, we have the following result

Proposition 13.37. *Let $\mathbf{F}$ be an X-regular collection and $\underline{r}$ a rational subspace of dimension i. The operator $P_{\underline{r}}^{\mathbf{F}} = \Pi^{F_{\underline{r}}} \mathbf{\Pi}_i^{\mathbf{F}}$ is the projector of $\mathcal{F}(X)$ to the $F_{\underline{r}}$-component $\mathcal{T}^{F_{\underline{r}}}_{X \backslash \underline{r}} * DM(X \cap \underline{r})$ of $\mathcal{F}(X)$.*

In particular, if $\dim(V) = s$, the operator

$$P_V := (\mathrm{Id} - \mathbf{P}_{s-1} \mu_{s-1})(\mathrm{Id} - \mathbf{P}_{s-2} \mu_{s-2}) \cdots (\mathrm{Id} - \mathbf{P}_0 \mu_0)$$

is the projector $\mathcal{F}(X) \to DM(X)$ associated to the direct sum decomposition

$$\mathcal{F}(X) = DM(X) \oplus \left(\oplus_{\underline{r} \in S_X \,|\, \underline{r} \neq V} \mathcal{T}^{F_{\underline{r}}}_{X \backslash \underline{r}} * DM(X \cap \underline{r}) \right).$$

Let $\mathbf{F} = \{F_{\underline{r}}\}$ be an X-regular collection. If $\underline{t}$ is a rational subspace and for each $\underline{r} \in S_{X \cap \underline{t}}$, we take the face containing the image of $F_{\underline{r}}$ modulo $\underline{t}^{\perp}$, we get an $X \cap \underline{t}$-regular collection. We denote this collection of faces for $X \cap \underline{t}$ by $\mathbf{F_{\underline{t}}}$ in the next proposition. The proof of this proposition is skipped, since it is very similar to preceding proofs.

Proposition 13.38. *Let $\underline{t}$ be a rational subspace. Take $f \in \mathcal{F}(X)$ and let $f = \sum_{\underline{r} \in S(X)} f_{\underline{r}}$ be the **F** decomposition of f and $\nabla_{X \backslash \underline{t}} f = \sum_{\underline{t} \in S_{X \cap \underline{t}}} g_{\underline{r}}$ be the $\mathbf{F}_{\underline{t}}$ decomposition of $\nabla_{X \backslash \underline{t}} f$. Then*

- $\nabla_{X \backslash \underline{s}} f_{\underline{r}} = 0$ *if $\underline{r} \notin S_{X \cap \underline{t}}$,*
- $\nabla_{X \backslash \underline{s}} f_{\underline{r}} = g_{\underline{r}}$ *if $\underline{r} \in S_{X \cap \underline{t}}$.*

13.4.5 Localization Theorem

In this section we are going to show that every element $f \in \mathcal{F}(X)$ coincides with a quasipolynomial on the sets $(\tau - B(X)) \cap \Lambda$ as τ varies over all topes (we simply say that f is a quasipolynomial on $\tau - B(X)$).

Definition 13.39. Let τ be a tope and $\underline{r}$ a proper rational subspace. We say that a regular face $F_{\underline{r}}$ for $X \backslash \underline{r}$ is nonpositive on τ if there exist $u_{\underline{r}} \in F_{\underline{r}}$ and $x_0 \in \tau$ such that $\langle u_{\underline{r}}, x_0 \rangle < 0$.

Given $x_0 \in \tau$, it is always possible to choose a regular face $F_{\underline{r}} \subset \underline{r}^{\perp}$ for $X \backslash \underline{r}$ such that x_0 is negative on some vector $u_{\underline{r}} \in F_{\underline{r}}$, since the projection of x_0 on $V/\underline{r}$ is not zero.

Definition 13.40. Let $\mathbf{F} = \{F_{\underline{r}}\}$ be an X-regular collection. We shall say that $\mathbf{F}$ is nonpositive on τ if each $F_{\underline{r}}$ is nonpositive on τ.

Let $f \in \mathcal{F}(X)$ and let $f = \sum f_{\underline{r}}$ be the **F** decomposition of f.

The content of the following result is quite similar to Paradan's localization theorem [87].

Theorem 13.41 (Localization theorem). *Let τ be a tope. Let $\mathbf{F} = \{F_{\underline{r}}\}$ be an X-regular collection nonpositive on τ.*

*The component f_V of the **F** decomposition $f = \sum_{\underline{r} \in S_X} f_{\underline{r}}$ is a quasipolynomial function in $DM(X)$ such that $f = f_V$ on $(\tau - B(X)) \cap \Lambda$.*

Proof. Let $\underline{r}$ be a proper rational subspace and write $f_{\underline{r}} = \mathcal{T}_{X \backslash \underline{r}}^{F_{\underline{r}}} * k_{\underline{r}}$, where $k_{\underline{r}} \in DM(X \cap \underline{r})$. In the notation of Lemma 13.25, the support of $f_{\underline{r}}$ is contained in the polyhedron $\underline{r} - \sum_{b \in B} b + C(F_{\underline{r}}, X \backslash \underline{r}) \subset \underline{r} + C(F_{\underline{r}}, X \backslash \underline{r})$.

This last polyhedron is convex, and by construction, it has a boundary limited by hyperplanes that are rational with respect to X. Thus, either we have $\tau \subset \underline{r} + C(F_{\underline{r}}, X \backslash \underline{r})$ or $\tau \cap (\underline{r} + C(F_{\underline{r}}, X \backslash \underline{r})) = \emptyset$.

Take $u_{\underline{r}} \in F_{\underline{r}}$ and $x_0 \in \tau$ such that $\langle u_{\underline{r}}, x_0 \rangle < 0$. Since $u_{\underline{r}} \geq 0$ on the set $\underline{r} + C(F_{\underline{r}}, X \backslash \underline{r})$ it follows that τ is not a subset of $\underline{r} + C(F_{\underline{r}}, X \backslash \underline{r})$, so that $\tau \cap (\underline{r} + C(F_{\underline{r}}, X \backslash \underline{r})) = \emptyset$.

In fact, we claim that the set $\tau - B(X)$ does not intersect the support $\underline{r} - \sum_{b \in B} b + C(F_{\underline{r}}, X \backslash \underline{r})$ of $f_{\underline{r}}$. Indeed, otherwise, we would have an equation $v - \sum_{x \in X} t_x x = s + \sum_{a \in A} k_a a + \sum_{b \in B} h_b(-b)$ in which $v \in \tau$, $0 \leq t_x \leq 1$, while $k_a \geq 0$, $h_b \geq 1$, $s \in \underline{r}$. This would imply that $v \in \underline{r} + C(F_{\underline{r}}, X \backslash \underline{r}))$, a contradiction. Thus f coincides with the quasipolynomial f_V on the set $(\tau - B(X)) \cap \Lambda$.

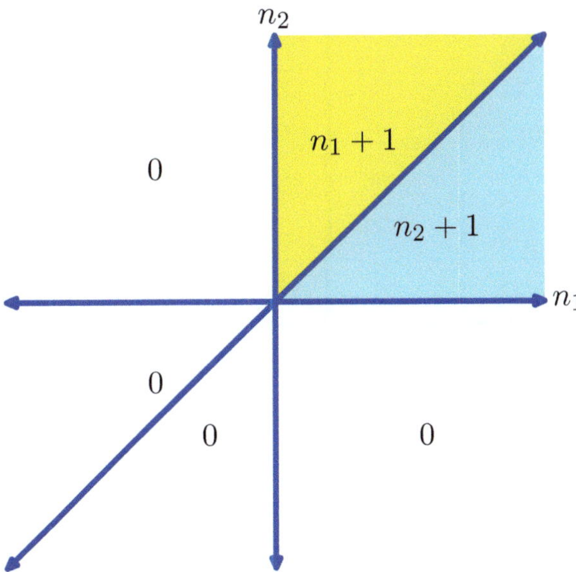

Fig. 13.1. The partition function of $X := (a, b, c)$.

Example 13.42. Figure 13.1 describes the partition function $\mathcal{T}_X$ for the list $X := (a, b, c)$ with $a := \omega_1, b := \omega_2, c := \omega_1 + \omega_2$ in the lattice $\Lambda := \mathbb{Z}\omega_1 \oplus \mathbb{Z}\omega_2$. Figure 13.2 describes the **F** decomposition relative to the tope containing x_0. Notice how the choice of **F** has the effect of pushing the supports of the elements $f_{\underline{r}}$ $(\underline{r} \neq V)$ away from τ.

A quasipolynomial is completely determined by the values that it takes on $(\tau - B(X)) \cap \Lambda$. Thus f_V is independent of the construction, so we may make the following definition

Definition 13.43. We shall denote by f^{τ} the quasipolynomial coinciding with f on $(\tau - B(X)) \cap \Lambda$.

The open subsets $\tau - B(X)$ cover V when τ runs over the topes of V (with possible overlapping). Thus the element $f \in \mathcal{F}(X)$ is entirely determined by the quasipolynomials f^{τ}.

Let us choose a scalar product on V, and identify V and U with respect to this scalar product. Given a point β in V and a rational subspace $\underline{r}$ in S_X, we write

$$\beta = p_{\underline{r}}\beta + p_{\underline{r}^{\perp}}\beta$$

with $p_{\underline{r}}\beta \in \underline{r}$ and $p_{\underline{r}^{\perp}}\beta$ in $\underline{r}^{\perp}$.

Definition 13.44. We say that $\beta \in V$ is generic with respect to $\underline{r}$ if $p_{\underline{r}}\beta$ is in a tope $\tau(p_{\underline{r}}\beta)$ for the sequence $X \cap \underline{r}$, and $p_{\underline{r}^{\perp}}\beta \in \underline{r}^{\perp}$ is regular for $X \setminus \underline{r}$.

We clearly have the following

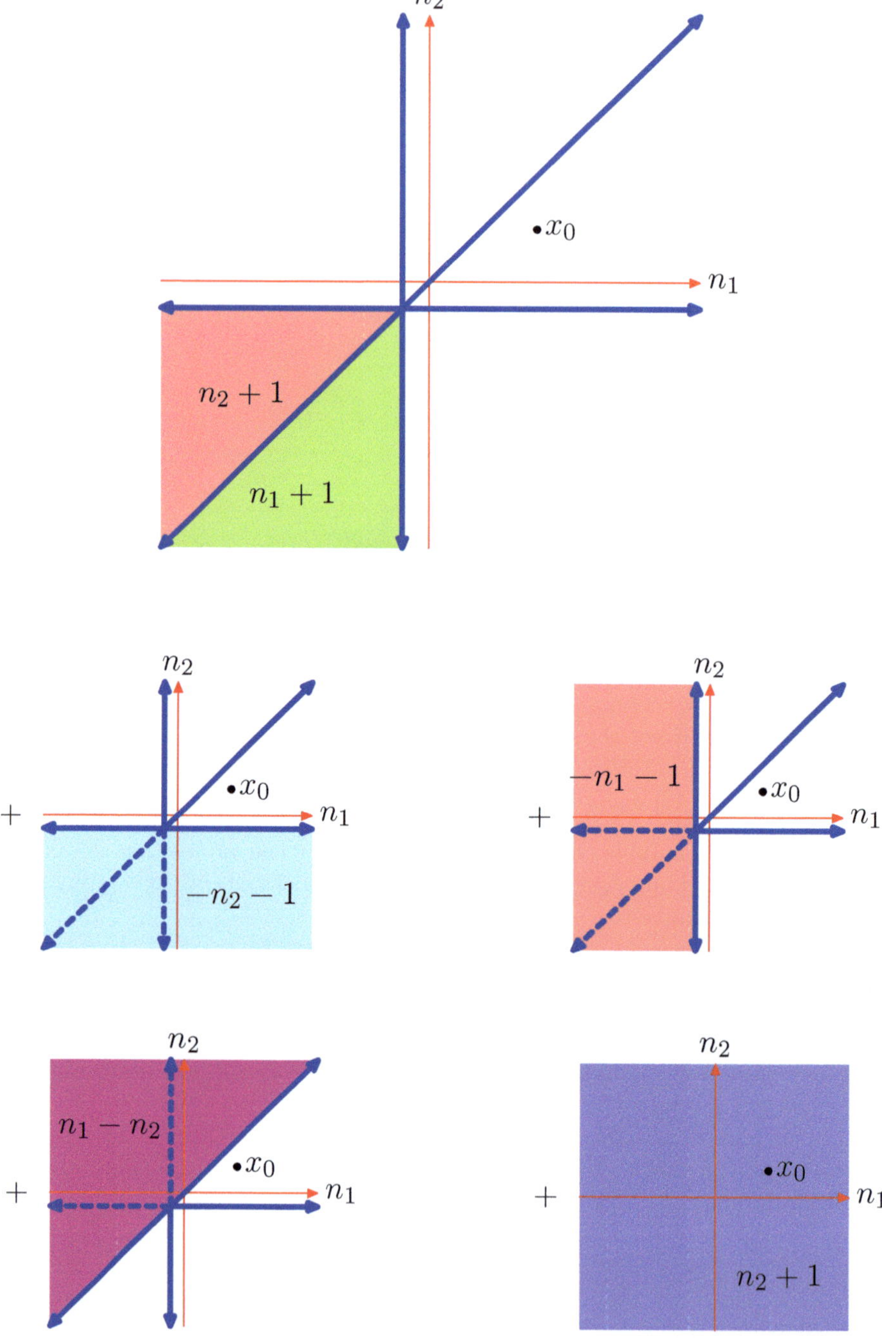

Fig. 13.2. F decomposition of the partition function of $X := (a, b, c)$ for **F** nonpositive on τ.

Proposition 13.45. *The set of β that are not generic with respect to $\underline{r}$ is a union of finitely many hyperplanes.*

By Theorem 13.41, if $f \in \mathcal{F}(X)$, then the element $\nabla_{X \setminus \underline{r}} f$ is in $\mathcal{F}(X \cap \underline{r})$ and coincides with a quasipolynomial $(\nabla_{X \setminus \underline{r}} f)^\tau \in DM(X \cap \underline{r})$ on each tope τ for the system $X \cap \underline{r}$.

Theorem 13.46. *Let $\beta \in V$ be generic with respect to all the rational subspaces $\underline{r}$. Let $F_{\underline{r}}^\beta$ be the unique regular face for $X \setminus \underline{r}$ containing $p_{\underline{r}^\perp} \beta$.*
Then

$$f = \sum_{\underline{r} \in S_X} \mathcal{T}_{X \setminus \underline{r}}^{-F_{\underline{r}}^\beta} * (\nabla_{X \setminus \underline{r}} f)^{\tau(p_{\underline{r}} \beta)}.$$

Proof. By the hypotheses made on β, the collection $\mathbf{F} = \{F_{\underline{r}}\}$ is an X-regular collection. Set $f = \sum_{\underline{r} \in S_X} \mathcal{T}_{X \setminus \underline{r}}^{F_{\underline{r}}} * q_{\underline{r}}$ with $q_{\underline{r}} \in DM(X \cap \underline{r})$. We apply Proposition 13.38. It follows that $q_{\underline{r}}$ is the component in $DM(X \cap \underline{r})$ in the decomposition of $\nabla_{X \setminus \underline{r}} f \in \mathcal{F}(X \cap \underline{r})$ with respect to the $X \cap \underline{r}$-regular collection induced by $\mathbf{F}$. Set $u_{\underline{t}} = -p_{\underline{t}^\perp} \beta$. Observe that for $\underline{t} \subset \underline{r}$, we have $\langle u_{\underline{t}}, p_{\underline{r}} \beta \rangle = -\|u_{\underline{t}}\|^2$, so that each $u_{\underline{t}}$ is negative on $p_{\underline{r}} \beta$. Thus the formula follows from Theorem 13.41.

13.4.6 Wall-Crossing Formula

We first develop a general formula describing how the functions f^τ change when crossing a wall. Then we apply this to the partition function $\mathcal{T}_X$ and deduce that it is a quasipolynomial on $\Omega - B(X)$, where Ω is a big cell.

Let H be a rational hyperplane and $u \in H^\perp$ a nonzero element. Then the two open faces in $H^\perp$ are half-lines $F_H = \mathbf{R}_{>0} u$ and $-F_H$. By Proposition 13.27, if $q \in DM(X \cap H)$, then $w := (\mathcal{T}_{X \setminus H}^{F_H} - \mathcal{T}_{X \setminus H}^{-F_H}) * q$ is an element of $DM(X)$.

Remark 13.47. In [28], a one-dimensional residue formula is given for w, allowing us to compute it.

Assume that τ_1, τ_2 are two adjacent topes, namely $\overline{\tau}_1 \cap \overline{\tau}_2$ spans a hyperplane H. The hyperplane H is a rational subspace. Let τ_{12} be the unique tope for $X \cap H$ such that $\overline{\tau}_1 \cap \overline{\tau}_2 \subset \overline{\tau_{12}}$.

Example 13.48. Let C be the cone generated by the vectors $a := \omega_3 + \omega_1$, $b := \omega_3 + \omega_2$, $c := \omega_3 - \omega_1$, $d := \omega_3 - \omega_2$ in a three-dimensional space $V := \mathbb{R}\omega_1 \oplus \mathbb{R}\omega_2 \oplus \mathbb{R}\omega_3$. Figure 13.3 represents the section of C cut by the affine hyperplane containing a, b, c, d. We consider $X := (a, b, c, d)$.

On the left of the picture we show the intersection of C with the two topes τ_1, τ_2 adjacent along the hyperplane H generated by b, d, and on the right, that with the tope τ_{12}. The list $X \cap H$ is $[b, d]$. The closure of the tope τ_{12} is *twice as big* as $\overline{\tau}_1 \cap \overline{\tau}_2$.

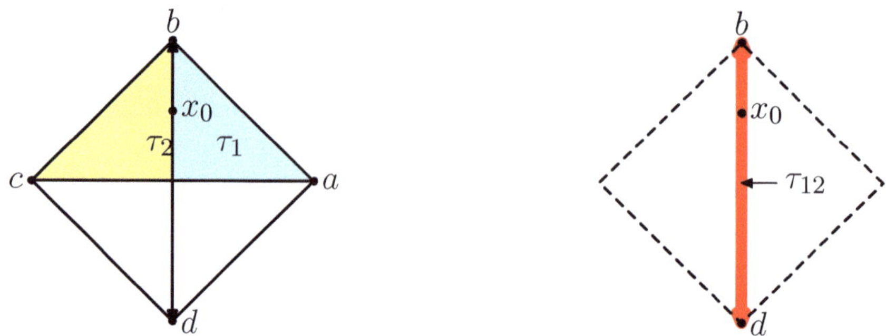

Fig. 13.3. Two adjacent topes of $X := (a, b, c, d)$.

Let $f \in \mathcal{F}(X)$. The function $\nabla_{X \setminus H} f$ is an element of $\mathcal{F}(H \cap X)$; thus by Theorem 13.41, there exists a quasipolynomial $(\nabla_{X \setminus H} f)^{\tau_{12}}$ on H such that $\nabla_{X \setminus H} f$ agrees with $(\nabla_{X \setminus H} f)^{\tau_{12}}$ on τ_{12}.

Theorem 13.49. *Let $\tau_1, \tau_2, H, \tau_{12}$ be as before and $f \in \mathcal{F}(X)$. Let F_H be the half-line in $H^\perp$ positive on τ_1. Then*

$$f^{\tau_1} - f^{\tau_2} = (\mathcal{T}_{X \setminus H}^{F_H} - \mathcal{T}_{X \setminus H}^{-F_H}) * (\nabla_{X \setminus H} f)^{\tau_{12}}. \tag{13.21}$$

Proof. Let x_0 be a point in the relative interior of $\overline{\tau}_1 \cap \overline{\tau}_2$ in H. Then x_0 does not belong to any X-rational hyperplane different from H (see Figure 13.3). Therefore, we can choose a regular vector $u_{\underline{r}}$ for $X \setminus \underline{r}$ for every rational subspace $\underline{r}$ different from H, V such that $u_{\underline{r}}$ is negative on x_0. By continuity, there are points $x_1 \in \tau_1$ and $x_2 \in \tau_2$ sufficiently close to x_0 and where these elements $u_{\underline{r}}$ are still negative. We choose $u \in H^\perp$ positive on τ_1. Consider the sequences $\mathbf{u}^1 = (u_{\underline{r}}^1)$, where $u_{\underline{r}}^1 = u_{\underline{r}}$ for $\underline{r} \neq H$ and $u_H^1 = -u$, and $\mathbf{u}^2 = (u_{\underline{r}}^2)$, where $u_{\underline{r}}^2 = u_{\underline{r}}$ for $\underline{r} \neq H$ and $u_H^2 = u$. Correspondingly, we get two X-regular collections $\mathbf{F}^1$ and $\mathbf{F}^2$.

For $i = 1, 2$ let $f = f_V^i + f_H^i + \sum_{\underline{r} \neq H, V} f_{\underline{r}}^i$ be the $\mathbf{F}^i$ decomposition of f.

We write $f_H^1 = \mathcal{T}_{X \setminus H}^{-F_H} * q^{(1)}$ with $q^{(1)} \in DM(X \cap H)$. Now the sequence $\mathbf{u}^1$ takes a negative value at the point x_1 of τ_1. Thus by Theorem 13.41, the component f_V^1 is equal to f^{τ_1}.

By Proposition 13.38,

$$\nabla_{X \setminus H} f = q^{(1)} + \sum_{\underline{r} \subset H, \underline{r} \neq H} \nabla_{X \setminus H} f_{\underline{r}}^1$$

is the $\mathbf{F}^1$ decomposition of $\nabla_{X \setminus H} f$, so that again by Theorem 13.41, we have $q^{(1)} = (\nabla_{X \setminus H} f)^{\tau_{12}}$ so that $f_H^1 = \mathcal{T}_{X \setminus H}^{-F_H} * (\nabla_{X \setminus H} f)^{\tau_{12}}$.

Similarly, $\mathbf{u}^2$ takes a negative value at the point x_2 of τ_2, so $f_V^2 = f^{\tau_2}$ and $f_H^2 = \mathcal{T}_{X\backslash H}^{F_H} * (\nabla_{X\backslash H} f)^{\tau_{12}}$.

Now from Proposition 13.37, when $\dim(\underline{r}) = i$,

$$f_{\underline{r}}^1 = \Pi^{F_{\underline{r}}^1} \mathbf{\Pi}_i^{\mathbf{F}^1} f, \qquad f_{\underline{r}}^2 = \Pi^{F_{\underline{r}}^2} \mathbf{\Pi}_i^{\mathbf{F}^2} f,$$

and for any $i < \dim V$, the operators $\Pi^{F_{\underline{r}}^1}\mathbf{\Pi}_i^{\mathbf{F}^1}$ and $\Pi^{F_{\underline{r}}^2}\mathbf{\Pi}_i^{\mathbf{F}^2}$ are equal. Thus $f_{\underline{r}}^1 = f_{\underline{r}}^2$ for $\underline{r} \neq V, H$. So we obtain $f_V^1 + f_H^1 = f_V^2 + f_H^2$, and our formula.

Consider now the case that X spans a pointed cone. Let us interpret formula (13.21) in the case in which $f = \mathcal{T}_X$. By Theorem 13.41, for a given tope τ, $\mathcal{T}_X$ agrees with a quasipolynomial $\mathcal{T}_X^\tau$ on $\tau - B(X)$. Recall that $\nabla_{X\backslash H}(\mathcal{T}_X) = \mathcal{T}_{X\cap H}$, as we have seen in Lemma 13.32. It follows that given two adjacent topes τ_1, τ_2 as above, $(\nabla_{X\backslash H} f)^{\tau_{12}}$ equals $(\mathcal{T}_{X\cap H})^{\tau_{12}}$ (extended by zero outside H). So we deduce the identity

$$\mathcal{T}_X^{\tau_1} - \mathcal{T}_X^{\tau_2} = (\mathcal{T}_{X\backslash H}^{F_H} - \mathcal{T}_{X\backslash H}^{-F_H}) * \mathcal{T}_{X\cap H}^{\tau_{12}}. \tag{13.22}$$

This is Paradan's formula ([86], Theorem 5.2).

Example 13.50. Assume $X = (a, b, c)$ as in Remark 13.42. We write $v \in V$ as $v = v_1\omega_1 + v_2\omega_2$. Let $\tau_1 = \{v_1 > v_2 > 0\}$, $\tau_2 = \{0 < v_1 < v_2\}$. Then one easily sees (see Figure 13.1) that

$$\mathcal{T}_X^{\tau_1} = (n_2 + 1), \qquad \mathcal{T}_X^{\tau_2} = (n_1 + 1), \qquad \mathcal{T}_{X\cap H}^{\tau_{12}} = 1.$$

Equality (13.22) is equivalent to the following identity of series, which is easily checked:

$$\sum_{n_1,n_2} (n_2 - n_1)x_1^{n_1} x_2^{n_2} = \left(- \sum_{n_1 \geq 0, n_2 < 0} x_1^{n_1} x_2^{n_2} + \sum_{n_1 < 0, n_2 \geq 0} x_1^{n_1} x_1^{n_2} \right)\left(\sum_h x_1^h x_2^h \right).$$

13.4.7 The Partition Function

We assume that $C(X)$ is a pointed cone. Let us now consider a big cell Ω. We need a lemma

Lemma 13.51. *Given a big cell Ω, let $\tau_1, \ldots, \tau_k$ be all the topes contained in Ω. Then*

$$\Omega - B(X) = \cup_{i=1}^k (\tau_i - B(X)).$$

Proof. Notice that $\cup_{i=1}^k \tau_i$ is dense in Ω. Given $v \in \Omega - B(X)$, $v + B(X)$ has nonempty interior, and thus its nonempty intersection with the open set Ω has non empty interior. It follows that $v + B(X)$ meets $\cup_{i=1}^k \tau_i$, proving our claim.

Now in order to prove the statement for big cells, we need to see what happens when we cross a wall between two adjacent topes.

Theorem 13.52. *On $(\Omega - B(X)) \cap \Lambda$, the partition function $\mathcal{T}_X$ agrees with a quasipolynomial $\mathcal{T}_X^{\Omega} \in DM(X)$.*

Proof. By Lemma 13.51, it suffices to show that given two adjacent topes τ_1, τ_2 in Ω, $\mathcal{T}_X^{\tau_1} = \mathcal{T}_X^{\tau_2}$.

But now notice that the positive cone spanned by $X \cap H$, the support of $\mathcal{T}_{X \cap H}$, is formed of singular vectors, and therefore it is disjoint from Ω by definition of big cells. Therefore, $\mathcal{T}_{X \cap H}$ vanishes on τ_{12}. Thus $\mathcal{T}_{X \cap H}^{\tau_{12}} = 0$, and our claim follows from formula (13.22).

This theorem was proven [37] by Dahmen–Micchelli for topes, and by Szenes–Vergne [110] for cells.

There is an important point of the theory of Dahmen–Micchelli that we should point out. Take a chamber $\mathfrak{c}$ inside $C(X)$ such that $0 \in \bar{\mathfrak{c}}$.

Lemma 13.53. $\delta(\mathfrak{c} \,|\, X) \cap C(X) = \{0\}$.

Proof. Assume that $a \in (\mathfrak{c} - B(X)) \cap \Lambda \cap C(X)$. We have $a = p - \sum_i t_i a_i$, $0 \le t_i \le 1$, $p \in \mathfrak{c}$, and thus $p \in a + C(X)$. The boundary of $a + C(X)$ is in the cut locus, and if $a \in C(X), a \ne 0$, we have that $0 \notin a + C(X)$. Therefore the chamber $\mathfrak{c}$ cannot be contained in $a + C(X)$ and hence it is disjoint from it, a contradiction.

From this we have, using Theorem 13.52 and Theorem 13.19, the following

Theorem 13.54. *Let Ω be a big cell contained in $C(X)$, $\mathfrak{c} \subset \Omega$ a chamber such that $0 \in \bar{\mathfrak{c}}$.*

Then $\mathcal{T}_X^{\Omega}$ is the unique element $f \in DM(X)$ such that $f(0) = 1$ and $f(a) = 0$, $\forall a \in \delta(\mathfrak{c} \,|\, X)$, $a \ne 0$.

Proof. By Theorem 13.52, $\mathcal{T}_X^{\Omega}$ coincides with $\mathcal{T}_X$ on $\delta(\mathfrak{c} \,|\, X) \subset (\Omega - B(X)) \cap \Lambda$. Now $\mathcal{T}_X$ is 0 outside $C(X)$, so its values on $\delta(\mathfrak{c} \,|\, X)$ are exactly the ones prescribed, and these determine $\mathcal{T}_X^{\Omega}$ uniquely by Theorem 13.19.

Remark 13.55. If Ω is a big cell contained in the cone $C(X)$, the open set $\Omega - B(X)$ contains $\overline{\Omega}$, so that the quasipolynomial $\mathcal{T}_X^{\Omega}$ coincides with $\mathcal{T}_X$ on $\overline{\Omega}$.

This is usually not so for $f \in \mathcal{F}(X)$ and a tope τ : the function f does not usually coincide with f^{τ} on $\overline{\tau}$.

We finally give a formula for $\mathcal{T}_X$ due to Paradan.

Let us choose a scalar product on V. We use the notation of Theorem 13.46. Since $\nabla_{X \backslash \underline{r}} \mathcal{T}_X = \mathcal{T}_{X \cap \underline{r}}$, we obtain as a corollary of Theorem 13.46 the following

Theorem 13.56. *(Paradan).* *Let $\beta \in V$ be generic with respect to all the rational subspaces $\underline{r}$. Then we have*

$$\mathcal{T}_X = \sum_{\underline{r} \in S_X} \mathcal{T}_{X \backslash \underline{r}}^{-F_{\underline{r}}^{\beta}} * (\mathcal{T}_{X \cap \underline{r}})^{\tau(p_{\underline{r}}\beta)}.$$

Remark 13.57. The set of $\beta \in V$ generic with respect to all the rational subspaces $\underline{r}$ decomposes into finitely many open polyhedral cones.

The decomposition depends only on the cone in which β lies.

In this decomposition, the component in $DM(X)$ is the quasipolynomial that coincides with $\mathcal{T}_X$ on the cell containing β.

Finally, if X spans a pointed cone and β has negative scalar product with X, the decomposition reduces to $\mathcal{T}_X = \mathcal{T}_X$, the component for $\underline{r} = \{0\}$.

13.4.8 The space $\tilde{\mathcal{F}}(X)$

When X does not span a pointed cone we can still construct several partition functions $\mathcal{P}_X^F := (-1)^{|B|}\delta_{-b_B} * H_A * H_{-B}$ for each decomposition $X = A \cup B$ into positive and negative vectors associated to every open face F (13.12).

It is thus interesting to develop a Theory which treats all these functions. This is best done introducing a new $\mathbb{Z}[\Lambda]$ module.

While $DM(X)$ is a $\mathbb{Z}[\Lambda]$ module, in general $\mathcal{F}(X)$ is not stable under $\mathbb{Z}[\Lambda]$, so we consider the $\mathbb{Z}[\Lambda]$ submodule $\tilde{\mathcal{F}}(X)$ in $\mathcal{C}[\Gamma]$ generated by $\mathcal{F}(X)$ and similarly for $\tilde{\mathcal{F}}_i(X)$.

Remark 13.58. It is easy to see that, if X' is deduced from X by changing sign to some of its vectors, then $\tilde{\mathcal{F}}(X) = \tilde{\mathcal{F}}(X')$.

Definition 13.59. Define $DM^{(G)}(X \cap \underline{r})$ to be the $\mathbb{Z}[\Lambda]$ submodule of $\mathcal{C}[\Lambda]$ generated by the image of $DM(X \cap \underline{r})$ inside $\mathcal{C}[\Lambda]$.

The group $\Lambda_{\underline{r}} = \Lambda \cap \underline{r}$ is a direct summand in Λ so $\mathbb{Z}[\Lambda]$ is isomorphic (in a non canonical way) to

$$\mathbb{Z}[\Lambda] = \mathbb{Z}[\Lambda_{\underline{r}}] \otimes \mathbb{Z}[\Lambda/\Lambda_{\underline{r}}].$$

It is then immediate to verify that

$$DM^{(G)}(X \cap \underline{r}) = \mathbb{Z}[\Lambda] \otimes_{\mathbb{Z}[\Lambda_{\underline{r}}]} DM(X \cap \underline{r}) = \mathbb{Z}[\Lambda/\Lambda_{\underline{r}}] \otimes DM(X \cap \underline{r}). \quad (13.23)$$

Since both the maps μ_i and the convolutions $\Theta_{X \setminus \underline{r}}^{F_{\underline{r}}}$ obviously extend to maps of $\mathbb{Z}[\Lambda]$ modules, we easily deduce

Corollary 13.60. *The sequence*

$$0 \to \tilde{\mathcal{F}}_{i+1}(X) \to \tilde{\mathcal{F}}_i(X) \overset{\mu_i}{\to} \bigoplus_{\underline{r} \in S_X^{(i)}} DM^{(G)}(X \cap \underline{r}) \to 0$$

is split exact with splitting:

$$\tilde{\mathcal{F}}_i(X) = \tilde{\mathcal{F}}_{i+1}(X) \oplus \bigoplus_{\underline{r} \in S_X^{(i)}} \mathcal{P}_{X \setminus \underline{r}}^{F_{\underline{r}}} * DM^{(G)}(X \cap \underline{r}).$$

From this, we obtain a decomposition dependent upon the choices of the faces $F_{\underline{r}}$

$$\tilde{\mathcal{F}}(X) = DM(X) \oplus \left(\oplus_{\underline{r} \in S_X \mid \underline{r} \neq V} \, \mathcal{P}^{F_{\underline{r}}}_{X \backslash \underline{r}} * DM^{(G)}(X \cap \underline{r}) \right). \qquad (13.24)$$

One easily verifies that $\tilde{\mathcal{F}}(X)$ can also be intrinsically defined as:

Definition 13.61. The space $\tilde{\mathcal{F}}(X)$ is the space of functions $f \in \mathcal{C}[\Lambda]$ such that $\nabla_{X \backslash \underline{r}} f$ is supported on a finite number of Λ translates of $\underline{r}$ for every proper rational subspace $\underline{r}$.

Remark 13.62. By definition, the filtration $\tilde{\mathcal{F}}_i(X)$ is defined by a condition on torsion or in the language of modules by support. One can verify that in particular $DM(X)$ is the part supported in dimension 0, that is of maximal possible torsion.

13.4.9 Generators of $\tilde{\mathcal{F}}(X)$

In this subsection, we assume that $\bar{a} \neq 0$ for any $a \in X$. Thus every open face F produces a decomposition $X = A \cup B$ into positive and negative vectors and can define as in (13.12):

$$\mathcal{P}^F_X := (-1)^{|B|} \delta_{-b_B} * H_A * H_{-B}.$$

Theorem 13.63. *The elements* $\mathcal{P}^F_X$, *as F runs on all open faces, generate* $\tilde{\mathcal{F}}(X)$ *as* $\mathbb{Z}[\Lambda]$ *module.*

Proof. Denote by M the $\mathbb{Z}[\Lambda]$ module generated by the elements $\mathcal{P}^F_X$, as F runs on all open faces. In general, from the description of $\tilde{\mathcal{F}}(X)$ given in Formula (13.24), it is enough to prove that elements of the type $\mathcal{P}^{F_{\underline{r}}}_X * g$ with $g \in DM(X \cap \underline{r})$ are in M. As $DM(X \cap \underline{r}) \subset \mathcal{F}(X \cap \underline{r})$, it is sufficient to prove by induction that each element $\mathcal{P}^{F_{\underline{r}}}_X * \mathcal{P}^K_{X \cap \underline{r}}$ is in M, where K is any open face for the system $X \cap \underline{r}$. We choose a linear function u_0 in the face $F_{\underline{r}}$. Thus u_0 vanishes on $\underline{r}$ and is non zero on every element $a \in X$ not in $\underline{r}$. We choose a linear function u_1 such that the restriction of u_1 to $\underline{r}$ lies in the face K. In particular, u_1 is non zero on every element $a \in X \cap \underline{r}$. We can choose ϵ sufficiently small such that $u_0 + \epsilon u_1$ is non zero on every element $a \in X$. Then $u_0 + \epsilon u_1$ defines an open face F in the arrangement $\mathcal{H}_X$. We see that $\mathcal{P}^{F_{\underline{r}}}_X * \mathcal{P}^K_{X \cap \underline{r}}$ is equal to $\mathcal{P}^F_X$.

13.4.10 Continuity

The space $D(X)$ being formed of polynomials can be viewed as restriction of polynomial functions on V to the lattice. Similarly for $DM(X)$ once we choose a representative for the exponential factor associated to each point of the arrangement.

There is a rather remarkable case in which the elements of the discrete space $\tilde{\mathcal{F}}(X)$ appear as restrictions of continuous functions to the lattice Λ, in fact piecewise polynomial or quasi–polynomial functions in the corresponding complex space $DM_V(X)$ of functions on V. We first need a simple

Lemma 13.64. *Let X be a list of m vectors in the lattice $\mathbb{Z}^n$. There exists an integer R such that, for every $f \in DM(X)$ and every $v \in \mathbb{Z}^n$, f is determined by the values that it takes in $C_{v,R} := v + [0, R]^n \cap \mathbb{Z}^n$.*

Proof. The elements of $DM(X)$ are quasi–polynomials, thus there exists some $k \in \mathbb{N}$ so that each f is a polynomial in the cosets of $k\mathbb{Z}^n$ in $\mathbb{Z}^n$ of some degree $\leq h$ for some h. Now a polynomial of degree h is completely determined if we know its value on a product of n sets $S_1 \times S_2 \ldots \times S_n$ with the property that each S_i has at least $h + 1$ elements. Finally if R is large enough each coset on $k\mathbb{Z}^n$ in $\mathbb{Z}^n$ intersected with $C_{v,R}$ contains such a product.

Theorem 13.65. *If X is a list of vectors in Λ, such that 0 is in the interior of the zonotope $B(X)$, then each function f in the space $\tilde{\mathcal{F}}(X)$ is the restriction to Λ of a continuous function $\underline{f}$ with the property that, on each affine tope, the function $\underline{f}$ lies in $DM_V(X)$.*

Proof. It is enough to do this for a function $f \in \mathcal{F}(X)$ since, after that, we can translate and take linear combinations which preserve the property.

If 0 is in the interior of the zonotope, for every tope τ we have $\bar{\tau} \subset \tau - B(X)$. Where $\bar{\tau}$ denotes the closure.

Now let f^τ be the quasi–polynomial agreeing by the *localization theorem* 13.41 with f on $\Lambda \cap (\tau - B(X))$.

We need to show that if τ_1, τ_2 are two topes the two functions f^{τ_1}, f^{τ_2} coincide on the intersection of the two closures.

This intersection is a rational polyhedral cone C spanning some linear subspace W defined over $\mathbb{Q}$, and the two functions induce two quasi–polynomials on this space.

The intersection $\Lambda \cap W$ is a lattice in W. We can now apply Lemma 13.64 since it is clear that the cone C has a non–empty interior and hence (from the cone property) it contains sets of type $C_{v,R}$ for any given R.

Next we know that the two quasi–polynomials f^{τ_1}, f^{τ_2} agree (with f) on $\Lambda \cap \bar{\tau}_1 \cap \bar{\tau}_2$ and (having fixed uniformly the periodic parts) it is clear that they agree on W.

13.5 Reciprocity

13.5.1 The Reciprocity Law

Let us discuss now a simplified version of the proof of Dahmen–Micchelli of the reciprocity law (generalizing the theorem for the Ehrhart polynomial).

Let us consider, for a given big cell Ω, the function $\mathcal{T}_X^{\Omega} \in DM(X)$, which coincides with the partition function on $\delta(\Omega \mid X) = (\Omega - B(X)) \cap \Lambda$.

Theorem 13.66.

$$\mathcal{T}_X^{\Omega}(\alpha) = (-1)^{|X|-\dim\langle X\rangle}\mathcal{T}_X^{\Omega}(-\alpha - 2\rho_X), \qquad (13.25)$$

where as usual, $\rho_X = \frac{1}{2}\sum_{x\in X} x$.

Proof. It is immediately seen that we can reduce to the case of X nondegenerate. Start with the simple case in which $s = 1$, $X = \{a\}$. Here $\Omega = \mathbb{R}^+$ and $\mathcal{T}_X^{\Omega}$ is the function on $\mathbb{Z}$ that is 1 on the multiples of a and zero otherwise. This function clearly satisfies (13.25).

Passing to the general case, we have that for any $y \in X$, $X \setminus \{y\}$ spans V. We proceed by induction on $|X|$.

Set $Q(\alpha) := \mathcal{T}_X^{\Omega}(\alpha) - (-1)^{|X|-\dim\langle X\rangle}\mathcal{T}_X^{\Omega}(-\alpha - 2\rho_X)$. We need to show that $Q(\alpha) = 0$.

We start by showing that $\nabla_y(Q) = 0$ for every $y \in X$.

Recall that $\nabla_y \mathcal{T}_X = \mathcal{T}_{X\setminus\{y\}}$. If Ω' is the unique big cell for $X \setminus \{y\}$ containing Ω it follows that $\nabla_y(\mathcal{T}_X^{\Omega}) = \mathcal{T}_{X\setminus\{y\}}^{\Omega'}$. Moreover,

$$\nabla_y(\mathcal{T}_X^{\Omega}(-\alpha - 2\rho_X)) = \mathcal{T}_X^{\Omega}(-\alpha - 2\rho_X) - \mathcal{T}_X^{\Omega}(-\alpha - 2\rho_X + y)$$
$$= -\nabla_y(\mathcal{T}_X^{\Omega})(-\alpha - 2\rho_X + y) = -\mathcal{T}_{X\setminus\{y\}}^{\Omega'}(-\alpha - 2\rho_{X\setminus\{y\}}).$$

If $y \in X$ is such that $\Omega \subset C(X \setminus \{y\})$, we have that

$$\nabla_y(Q)(\alpha) = \mathcal{T}_{X\setminus\{y\}}^{\Omega'}(\alpha) - (-1)^{|X|-\dim\langle X\rangle}(-\mathcal{T}_{X\setminus\{y\}}^{\Omega'}(-\alpha - 2\rho_{X\setminus\{y\}}))$$

$$= \mathcal{T}_{X\setminus\{y\}}^{\Omega'}(\alpha) - (-1)^{|X\setminus\{y\}|-\dim\langle X\setminus\{y\}\rangle}(\mathcal{T}_{X\setminus\{y\}}^{\Omega'}(-\alpha - 2\rho_{X\setminus\{y\}})) = 0$$

by the inductive hypothesis.

If $\Omega \cap C(X \setminus \{y\}) = \emptyset$, we have that $\nabla_y(\mathcal{T}_X^{\Omega})$ is identically zero. In particular, $\nabla_y(Q) = 0$. We deduce that $Q(\alpha)$ is constant on the lattice Λ^0 spanned by X. Since by definition $\mathcal{T}_X^{\Omega}$ and hence Q is 0 outside Λ^0, in order to prove our claim we have to show that there is an $\alpha \in \Lambda^0$ such that $Q(\alpha) = 0$.

We know by Theorem 13.54 that if we take a chamber $\mathfrak{c} \subset \Omega$ with $0 \in \mathfrak{c}$, then for every $\alpha \in \delta(\mathfrak{c} \mid X)$, $\alpha \neq 0$, we have $\mathcal{T}_X^{\Omega}(\alpha) = 0$.

Thus it suffices to find a nonzero element $\alpha \in \delta(\mathfrak{c} \mid X) \cap \Lambda^0$ with the property that also $-\alpha - 2\rho_X \in \delta(\mathfrak{c} \mid X) \setminus \{0\}$.

Notice now that $B(X) = 2\rho_X - B(X)$. It follows that for any vector u, $-\alpha - 2\rho_X \in u - B(X)$ if and only if $-\alpha \in u + 2\rho_X - B(X) = u + B(X)$, if and only if $\alpha \in -u - B(X)$.

Taking $u \in \mathfrak{c}$, we deduce that $-\alpha - 2\rho_X \in \delta(\mathfrak{c}\,|\,X)$ if and only if we have $\alpha \in \delta(-\mathfrak{c}\,|\,X)$.

In order to complete the proof, we have shown that we only need to find an element $\alpha \in \Lambda^0 \cap \delta(\mathfrak{c}\,|\,X) \cap \delta(-\mathfrak{c}\,|\,X)$ with $\alpha \neq 0$, $\alpha + 2\rho_X \neq 0$. Since both $\delta(\mathfrak{c}\,|\,X)$ and $\delta(-\mathfrak{c}\,|\,X)$ clearly contain the points in the lattice in the interior of $-B(X)$, everything will follow from the following lemma

Lemma 13.67. *(1) If for every $y \in X$, we have $\langle X \setminus \{y\}\rangle = V$, the interior $\overset{\circ}{B}(X)$ of $B(X)$ contains a point in Λ^0.*

(2) For any such point $\alpha \in \overset{\circ}{B}(X)$, $\alpha + 2\rho_X \neq 0$.

Proof. One can derive this lemma directly from the paving of the zonotope, but we can give a direct proof.

Part (2) is trivial, since if $\alpha \in B(X)$, we have always $\alpha + 2\rho_X \neq 0$.

Let us prove part (1). X contains a basis $e_1, \ldots, e_s$ of V, and by our assumption, for each $i = 1, \ldots, s$ there is an element $y_i \in X \setminus \{e_1, \ldots, e_s\}$ with the property that in the expansion $y_i = \sum_{j=1}^{s} a_{i,j} e_j$ we have $a_{i,i} \neq 0$. Thus there are distinct elements $x_1, \ldots, x_t \in X \setminus \{e_1, \ldots, e_s\}$ and numbers $(\lambda_1, \ldots, \lambda_t)$ in the cube $[0,1]^t$ such that the vector $v = \sum_{i=1}^{t} \lambda_i x_i = \sum_{j=1}^{s} b_j e_j$ has the property that all the coordinates b_j are nonzero for each $j = 1, \ldots, s$. By construction, v is in $B(X)$ as is every element $v + \sum_j t_j e_j$, $0 \leq t_j \leq 1$. Up to reordering we can assume that there is a $1 \leq k \leq s$ such that $b_j > 0$ exactly for $1 \leq j \leq k$ (notice that, since we have a linear function that is positive on X, $k \geq 1$).

Set $w = e_1 + \cdots + e_k$. We claim that w lies in the interior of $B(X)$. This is equivalent to the fact that 0 lies in the interior of $B(X) - w$. To see this, notice that if $1 \leq i \leq k$, then $-e_i = e_1 + \cdots + e_{i-1} + e_{i+1} + \cdots + e_k - w \in B(X) - w$. Also if $k < i \leq s$, $e_i = w + e_i - w \in B(X) - w$. Finally, $v = w + v - w \in B(X) - w$. It follows that $-e_1, \ldots, -e_k, e_{k+1}, \ldots, e_s, v$ are all in $B(X) - w$. Since $B(X) - w$ is convex, their convex hull Δ is also contained in $B(X) - w$. It is now clear that $0 = v - \sum_j b_j e_j = v + \sum_{j=1}^{k} b_j(-e_j) + \sum_{j=k+1}^{s}(-b_j)e_j$ lies in the interior of Δ so everything follows.

It is important also to understand the geometric meaning of the function $\mathcal{T}_X(b - 2\rho_X)$. In fact the set of solutions of $\sum_i n_i a_i = b - 2\rho_X$, $n_i \in \mathbb{N}$, is in one-to-one correspondence with the set of solutions of $\sum_i n_i a_i = b$, $n_i \in \mathbb{N}$, $n_i > 0$. In terms of polytopes, these last points are the integral points in the *interior* of the polytope $\Pi_X(b)$. Thus the reciprocity law can also be interpreted as follows

Theorem 13.68. *The number of integral points contained in the interior of a polytope $\Pi_X(b)$ for $b \in \Omega$ equals $(-1)^{|X| - \dim\langle X\rangle} \mathcal{T}_X^{\Omega}(-b)$.*

Proof. A point with coordinates n_a, $a \in \mathbb{N}$, and $\sum_{a \in X} n_a a = b$ is in the interior of $\Pi_X(b)$ if and only if $n_a > 0$, $\forall a$ (cf. Section 1.3.2). This means that the point with coordinates $n_a - 1$, $a \in \mathbb{N}$, such that $\sum_{a \in X}(n_a - 1)a = b - 2\rho_X$ lies in $\Pi_X(b - 2\rho_X)$. In other words, the number of integral points in the interior of $\Pi_X(b)$ equals the number of integral points in $\Pi_X(b - 2\rho_X)$ that equals $\mathcal{T}_X^{\Omega}(b - 2\rho_X)$ by Theorem 13.52 (since $2\rho_X \in B(X)$).

By reciprocity, this is $(-1)^{|X| - \dim\langle X \rangle} \mathcal{T}_X^{\Omega}(-b)$.

13.6 Appendix: a Complement

13.6.1 A Basis for $DM(X)$

This section is not needed elsewhere in this book and can be skipped.

We are going to explain how we can get an integral basis of $DM(X)$ from our previous construction. For this purpose, in order to perform a correct induction, we need to generalize it. We therefore consider, instead of a lattice Λ, a finitely generated abelian group Γ and a list of elements $X \subset \Gamma$. To these data we can associate again the space of integer-valued functions on Γ, that we denote by $\mathcal{C}[\Gamma]$ and its subspace $DM(X)$ defined by the equations $\nabla_Y f = 0$ as Y runs over the cocircuits. Here a cocircuit in X is a sublist such that $X \setminus Y$ generates a subgroup of infinite index in Γ, i.e., of smaller rank and that is minimal with this property.

Let Γ_t denote the torsion subgroup of Γ, $\Lambda := \Gamma/\Gamma_t$ the associated lattice and $V := \Gamma \otimes \mathbb{R} = \Lambda \otimes \mathbb{R}$. Let furthermore $\overline{X}$ denote the image of X in $\Lambda \subset V$.

We claim that $DM(X)$ is a free $\mathbb{Z}[\Gamma_t]$ module of rank $\delta(\overline{X})$ and we are going to construct an explicit basis.

We assume that X is totally ordered and spans a subgroup of finite index in Γ.

Take a basis $\underline{b}$ extracted from X. By this we mean that $\underline{b}$ is an integral basis for a subgroup of finite index in Γ, or equivalently that their images in V give a basis of V as a vector space.

Given $c \in X$, set $\underline{b}_{\geq c} = \{b \in \underline{b} \mid c \leq b\}$ (in the given ordering). Define $X_{\underline{b}}$ equal to the set of elements $c \in X$ such that their image in V is nonzero and linearly dependent on the images of $\underline{b}_{\geq c}$. These elements, according to Definition 2.26 consist of the elements in $\underline{b}$ plus all the elements in X, whose image in V is nonzero and externally active with respect to the image of $\underline{b}$.

We have that $X_{\underline{b}}$ is a sublist of X spanning also a subgroup of finite index in Γ; therefore we have that $DM(X_{\underline{b}})$ is a subgroup of $DM(X)$.

Now the reader should verify that the discussion leading to Proposition 13.28 holds in this more general case of the complete flag $\phi_{\underline{b}}$ of rational subspaces $\underline{r}_i$ generated by the images $\{b_1, \ldots, b_i\}$, $i = 1, \ldots, s$. We obtain the element $\mathcal{Q}_{\phi_{\underline{b}}}^{X_{\underline{b}}} \in DM(X_{\underline{b}}) \subset DM(X)$.

Choose a complete set $C_{\underline{b}} \subset \Lambda$ of coset representatives modulo the sublattice $\Lambda_{\underline{b}}$ spanned by $\underline{b}$. To every pair $(\underline{b}, \gamma)$ with $\underline{b} \in \mathcal{B}_X$ and $\gamma \in C_{\underline{b}}$, we associate the element

$$\mathcal{Q}^{X_{\underline{b}}}_{\phi_{\underline{b}}, \gamma} = \tau_\gamma \mathcal{Q}^{X_{\underline{b}}}_{\phi_{\underline{b}}}$$

in $DM(X)$. We can now state our next result.

Theorem 13.69. *The elements $\mathcal{Q}^{X_{\underline{b}}}_{\phi_{\underline{b}}, \gamma}$, as $\underline{b}$ varies in $\mathcal{B}_X$ and γ in $C_{\underline{b}}$, form a basis of $DM(X)$.*

Proof. We shall proceed by induction on the cardinality of X, the case in which X is empty being trivial.

First observe that if X' is obtained from X by removing all elements of finite order we have that $DM(X) = DM(X')$. The proposed basis does not involve the elements of finite order in X, therefore, we may assume that X does not contain any element of finite order. Let a in X be the least element. Write $X = [a, Z]$ and set $\tilde{Z}$ to be the image of Z in $\tilde{\Gamma} := \Gamma/\mathbb{Z}a$.

Observe first that the kernel of ∇_a is the set of functions that are constant on the cosets of $\mathbb{Z}a$ inside Γ, and therefore we may identify $\ker \nabla_a$ with $\mathcal{C}[\tilde{\Gamma}]$. Moreover, we claim that under this identification, that we denote by j_a, we have $\ker \nabla_a \cap DM(X) = j_a DM(\tilde{Z})$. This is clear, since a cocircuit in $\tilde{Z}$ is the image of a cocircuit in X not containing a.

It is also clear that ∇_a maps $DM(X)$ to $DM(Z)$. In fact, let $T \subset Z$ be a cocircuit in Z. Then $T \cup \{a\}$ is a cocircuit in X. Thus $\nabla_T(\nabla_a f) = 0$ if $f \in DM(X)$.

We now claim that the sequence

$$0 \to DM(\tilde{Z}) \xrightarrow{j_a} DM(X) \xrightarrow{\nabla_a} DM(Z) \to 0 \tag{13.26}$$

is exact.

By our previous consideration, in order to show the exactness of (13.26), it remains only to show that the map $\nabla_a : DM(X) \to DM(Z)$ is surjective. This we prove by induction.

If $\{a\}$ is a cocircuit, that is, if Z is degenerate, then $DM(Z) = 0$ and there is nothing to prove.

Otherwise, take a basis $\underline{b} \in \mathcal{B}_X$ such that a does not lie in $\underline{b}$. Then $\underline{b} \in \mathcal{B}_Z$. Observe that $Z_{\underline{b}} = X_{\underline{b}} \setminus \{a\}$ and that for every $\gamma \in C_{\underline{b}}$,

$$\mathcal{Q}^{Z_{\underline{b}}}_{\phi_{\underline{b}}, \gamma} = \nabla_a(\mathcal{Q}^{X_{\underline{b}}}_{\phi_{\underline{b}}, \gamma}).$$

Thus, by induction, the elements $\nabla_a(\mathcal{Q}^{X_{\underline{b}}}_{\phi_{\underline{b}}, \gamma})$, as $\underline{b}$ varies in $\mathcal{B}_Z$ and γ in $C_{\underline{b}}$, form a basis of $DM(Z)$, and the surjectivity of ∇_a follows.

Let us now take a basis $\underline{b}$ containing a. Consider the corresponding basis $\tilde{\underline{b}}$ inside $\tilde{Z}$. Take the quotient map $p : \Gamma \to \tilde{\Gamma}$. It is clear that p maps $C_{\underline{b}}$ onto a set $C_{\tilde{\underline{b}}}$ of representatives of the cosets of $\tilde{\Gamma}/\tilde{\Gamma}_{\tilde{\underline{b}}}$ in $\tilde{\Gamma}$. Define an equivalence relation on $C_{\underline{b}}$ setting $\gamma \simeq \gamma'$ if $p(\gamma - \gamma') \in \tilde{\Gamma}_{\underline{b}}$. If we choose an element in each equivalence class, we obtain a subset of $\tilde{\Gamma}$, which we can take as $C_{\tilde{b}}$.

By induction, the set of elements $\mathcal{Q}^{\tilde{Z}_{\underline{b}}}_{\phi_{\tilde{b}}, p(\gamma)}$, as $\underline{b}$ varies in $\mathcal{B}_{\tilde{Z}}$ and γ in $C_{\underline{b}}$, is a basis of $DM(\tilde{Z})$ as a $\mathbb{Z}[\Gamma_t]$ module. Now observe that

$$i_a\left(\mathcal{Q}^{\tilde{Z}_{\tilde{b}}}_{\phi_{\tilde{b}}, p(\gamma)}\right) = \mathcal{Q}^{X_b}_{\phi_{\underline{b}}, \gamma}. \tag{13.27}$$

In fact, whenever one multiplies a function $f = j_a(g) \in \ker \nabla_a$ by any element δ_c (or by a series formed by these elements), one sees that

$$\delta_c * j_a(g) = j_a(\delta_{p(c)} * g), \quad \forall c \in \Gamma, \ \forall g \in \mathcal{C}[\tilde{\Gamma}]. \tag{13.28}$$

Furthermore, the elements $(X_{\underline{b}} \setminus \{a\}) \cap \underline{r}_i$, $i \geq 2$, map surjectively under p to the elements of $\tilde{Z}_{p(\underline{b})} \cap \underline{r}_i / \underline{r}_1$, and thus the formulas defining the two functions on both sides of equation (13.27) coincide, factor by factor, for all steps of the flag from formula (13.28).

At this point everything follows from the exact sequence (13.26).

Example 13.70. Let us analyze this basis for $s-1$, that is, $X = (a_1, \ldots, a_m)$ is a list of numbers. Let us assume that they are positive for simplicity $a_i \in \mathbb{N}$. The bases extracted from X are just the numbers a_i and then the list $X_{\{a_i\}}$ equals $(a_1, \ldots, a_i)$. The flags $\phi_{\underline{b}}$ are just the trivial flag, and the function $\mathcal{Q}^{X_{\{a_i\}}}_{\phi_{\underline{b}}}$ is nothing but the quasipolynomial coinciding with the partition function $\mathcal{T}_{X_{\{a_i\}}}$ on the positive integers. The sublattice spanned by a_i is $\mathbb{Z}a_i$ and as coset representatives we can take the numbers $0, 1, \ldots, a_i - 1$. So the basis is formed by the quasipolynomials expressing the partition functions $\mathcal{T}_{X_{\{a_i\}}}$ and some of their translates.

Toric Arrangements

This chapter is devoted to the study of the periodic analogue of a hyperplane arrangement, that we call a toric arrangement. Thus the treatment follows the same strategy as in Chapter 8, although with several technical complications. We shall then link this theory with that of the partition functions in a way similar to the treatment of T_X in Chapter 9.

14.1 Some Basic Formulas

14.1.1 Laplace Transform and Partition Functions

In order to motivate the treatment of this chapter, we start by discussing the analogue, for partition functions, of Theorem 7.6 for splines.

Let us return to the main data, a lattice $\Lambda \subset V$ of rank s and a finite list $X = (a_1, \ldots, a_m)$ of elements in Λ. We set $U = V^*$, the dual over $\mathbb{R}$, and get that $U_{\mathbb{C}} = \hom(\Lambda, \mathbb{C})$. We also fix a list $\mu = (\mu_1, \ldots, \mu_m) \in \mathbb{C}^m$ of complex numbers, that we shall think of as a map $\mu : X \to \mathbb{C}$, so that sometimes if $a_i = a$ we will write μ_a for μ_i (notice that if we take two distinct elements a, b in the list X, although we could have that as elements they are equal, still it can very well be that $\mu_a \neq \mu_b$). Given $\lambda \in \Lambda$, we set

$$P(\lambda) = \Pi_X(\lambda) \cap \mathbb{N}^m := \left\{ (n_1, \ldots, n_m) \mid \sum_{i=1}^{m} n_i a_i = \lambda, n_i \in \mathbb{N} \right\},$$

and define Euler–Maclaurin sums as in (12.9) (setting $\nu_i := e^{-\mu_i}$) (cf. [25], [24]) by

$$\mathcal{T}_{X,\mu}(\lambda) := \sum_{(n_i) \in P(\lambda)} e^{-\sum_{i=1}^{m} n_i \mu_i} = \sum_{(n_i) \in P(\lambda)} \prod_{i=1}^{m} \nu_i^{n_i}.$$

We then set $\mathcal{T}_{X,\mu} = \sum_{\lambda \in \Lambda} \mathcal{T}_{X,\mu}(\lambda) \delta_\lambda$. One can verify that $\mathcal{T}_{X,\mu}$ is a tempered distribution, supported at the points of the lattice Λ lying in $\overline{C}(X)$.

C. De Concini and C. Procesi, *Topics in Hyperplane Arrangements, Polytopes and Box-Splines*, Universitext, DOI 10.1007/978-0-387-78963-7_14,
© Springer Science+Business Media, LLC 2010

Reasoning as in Section 12.2.2, we see that the Laplace transform of $\mathcal{T}_{X,\underline{\mu}}$ extends to a rational function on the torus with coordinate ring $\mathbb{C}[\Lambda]$, which, by abuse of notation, we still indicate with $L\mathcal{T}_{X,\underline{\mu}}$

$$L\mathcal{T}_{X,\underline{\mu}} = \frac{1}{\prod_{a \in X}(1 - e^{-a-\mu_a})}, \tag{14.1}$$

with the notable special case

$$L\mathcal{T}_X = \frac{1}{\prod_{a \in X}(1 - e^{-a})}. \tag{14.2}$$

14.1.2 The Coordinate Algebra

Formula (14.2) suggests that we work with the algebra

$$S_X := \mathbb{C}[\Lambda]\Big[\prod_{a \in X}(1 - e^{-a})^{-1}\Big].$$

More generally, we shall work with the algebra

$$S_{X,\underline{\mu}} := \mathbb{C}[\Lambda]\Big[\prod_{a \in X}(1 - e^{-a-\mu_a})^{-1}\Big].$$

The algebra $S_{X,\underline{\mu}}$ has a precise geometric meaning. As we have seen in Section 5.2.1, $\mathbb{C}[\Lambda]$ is the coordinate ring of the s-dimensional torus $T_{\mathbb{C}} = U_{\mathbb{C}}/\Lambda^*$ with $\Lambda^* := \hom(\Lambda, 2\pi i\mathbb{Z})$. The equation $\prod_{a \in X}(1 - e^{-a-\mu_a}) = 0$ defines a hypersurface $\mathcal{D}$ in $T_{\mathbb{C}}$, the union of the subvarieties H_a defined by the equations $e^{-a-\mu_a} = 1$, $a \in X$. The algebra S_X is the coordinate ring of the complement of $\mathcal{D}$ in $T_{\mathbb{C}}$, itself an affine variety, that we denote by $\mathcal{A}_{X,\underline{\mu}}$ or simply $\mathcal{A}$.

Remark 14.1. It is often convenient to think of $\mathbb{C}[\Lambda]$ as equal to $\mathbb{C}[x_1^{\pm 1}, \ldots, x_s^{\pm 1}]$ and of e^a as a monomial in the x_i with integer exponents.

Definition 14.2. The finite set whose elements are all the connected components of any nonempty subvariety in $T_{\mathbb{C}}$ of the form $H_Y := \cap_{a \in Y} H_a$ for some $Y \subset X$ will be called the (complete) toric arrangement associated to $(X, \underline{\mu})$ and denoted by $\mathcal{H}_{X,\underline{\mu}}$.

Remark 14.3. 1. Notice that a toric arrangement can be partially ordered by reverse inclusion. With respect to this order the torus $T_{\mathbb{C}}$ considered as the empty intersection is a unique minimum element.

2. Any nonempty intersection $H_Y = \cap_{a \in Y} H_a$ is a coset for the closed subgroup $K_Y := \cap_{a \in Y} \ker e^a \subset T_{\mathbb{C}}$ of $T_{\mathbb{C}}$. The connected component of K_Y through 1 is a subtorus T_Y; if H_Y is nonempty, the connected components of H_Y are cosets of T_Y.

In what follows, a special role will be played by the so called *points of the arrangement*, i.e., the elements of the arrangement of dimension zero. We have already encountered these points in Section 13.2.3 and denoted this set by $P_{\underline{\mu}}(X)$. Recall that, if we denote by $\mathcal{B}_X$ the family of subsets $\sigma \subset \{1, \dots, m\}$ such that $\underline{b}_\sigma := \{a_i | i \in \sigma\}$ is a basis, then

$$P_{\underline{\mu}}(X) := \cup_{\sigma \in \mathcal{B}_X} T_{\underline{\mu}}(\underline{b}_\sigma), \tag{14.3}$$

where $T_{\underline{\mu}}(\underline{b}_\sigma) = \cap_{b \in \underline{b}_\sigma} H_b$.

The points of the arrangement form the zero-dimensional pieces (or maximal elements under reverse inclusion) of the entire *toric arrangement*.

For any point in $P_{\underline{\mu}}(X)$ we choose once and for all a representative $\phi \in U_{\mathbb{C}}$ so that the given point equals e^ϕ and e^ϕ can be thought of as an exponential function on V. We will denote by $\tilde{P}_{\underline{\mu}}(X)$ the corresponding set of representatives.

For toric arrangements we can generalize the notions introduced in Section 2.1.1 for hyperplane arrangements. Given a subspace $p\mathcal{T}$ of the arrangement associated to X (where $\mathcal{T}$ is one of the tori $T_Y, Y \subset X$), with $p = e^\phi$, we consider the sublist $X_{p\mathcal{T}}$ of characters in $a \in X$ that take the value $e^{a+\mu_a} = 1$ on $p\mathcal{T}$.

A *basis $\underline{b}$ relative to $p\mathcal{T}$* is a basis extracted from the sublist $X_{p\mathcal{T}}$ of the span of $X_{p\mathcal{T}}$. For such a basis we denote by $\langle \underline{b} \rangle$ the linear span and by $\Lambda_{\underline{b}}$ the lattice that they span. We introduce the set

$$R_{\underline{b}} = \left\{ \lambda \in \Lambda \mid \lambda = \sum_{b \in \underline{b}} p_b b, \text{ with } 0 \le p_b < 1 \right\}. \tag{14.4}$$

Notice that this is a set of coset representatives of $\Lambda \cap \langle \underline{b} \rangle / \Lambda_{\underline{b}}$.

The formulas we are aiming at are built starting from some basic functions. These are analogues in the discrete case of the Laplace transforms of the characteristic functions of the cones generated by bases extracted from X. With $\underline{b}, \phi, p = e^\phi, p\mathcal{T}$ as before, set

$$\xi_{\underline{b},\phi} = \xi_{\underline{b},p\mathcal{T}} := \sum_{v \in \Lambda \cap C(\underline{b})} e^{\langle \phi | v \rangle} \delta_v.$$

Then $\xi_{\underline{b},\phi}$ is a tempered distribution supported on the cone $C(\underline{b})$ and satisfying the hypotheses of Section 12.2.2, so that its Laplace transform has an analytic meaning.

Proposition 14.4. *The Laplace transform of $\xi_{\underline{b},\phi}$ equals*

$$|R_{\underline{b}}| \frac{e(\phi)}{\prod_{a \in \underline{b}} (1 - e^{-a-\mu_a})},$$

where

$$e(\phi) := \frac{\sum_{\lambda \in R_{\underline{b}}} e^{\langle \phi | \lambda \rangle} e^{-\lambda}}{|R_{\underline{b}}|}. \tag{14.5}$$

Proof. Notice that if $v = \sum_{b \in \underline{b}} n_b b$, then $e^{\sum_b \mu_b n_b} = e^{\langle \phi \mid v \rangle}$. Also by our choice of $R_{\underline{b}}$, see (14.4), every element in $\Lambda \cap C(\underline{b})$ can be uniquely written as $\lambda + v$ with $\lambda \in R_{\underline{b}}$ and $v \in \Lambda_{\underline{b}} \cap C(\underline{b})$. Expanding, we get

$$|R_{\underline{b}}| \frac{e(\phi)}{\prod_{a \in \underline{b}}(1 - e^{-a-\mu_a})} =$$
$$= \left(\sum_{\lambda \in R_{\underline{b}}} e^{\langle \phi \mid \lambda \rangle} e^{-\lambda} \right) \sum_{v \in \Lambda_{\underline{b}} \cap C(\underline{b})} e^{\langle \phi \mid v \rangle - v} = \sum_{v \in \Lambda \cap C(\underline{b})} e^{\langle \phi \mid v \rangle} e^{-v},$$

that is the Laplace transform of $\xi_{\underline{b}, \phi}$, as desired.

Remark 14.5. 1. The normalization that we have chosen for the element $e(\phi)$ will be justified in Section 14.2.3.

2. If we choose a basis for $\Lambda \cap \langle \underline{b} \rangle$ and write the elements b_i as integral vectors in this basis, we have that $|R_{\underline{b}}| = |\det(\underline{b})|$.

The heart of our work will consist in developing a suitable formalism that will allow us to express in a way, similar to that of Section 9.3, the Laplace transform of the partition function from these special elements applied to the points of the arrangement. The result of this analysis will be presented in formulas (14.28) and (16.4).

14.2 Basic Modules and Algebras of Operators

14.2.1 Two Algebras as Fourier Transforms

As we have already stated, our main goal in this chapter is the study of the algebra $\mathbb{C}[\Lambda][\prod_{a \in X}(1 - e^{-a})^{-1}]$ as a *periodic analogue* of the algebras R_X. In other words, we shall describe it as a module under the action of a suitable algebra of differential operators. This we shall do in Section 14.3. As a preliminary step we develop the algebraic tools that we shall need for this purpose.

We shall use two noncommutative algebras that we describe directly as algebras of operators. In particular, we want to use the translation operators τ_a, $a \in \Lambda$, acting on functions on Λ or on $\Lambda \otimes \mathbb{R}$.

The definitions that we shall give are a formalization of the basic formulas (3.3), (3.4) for the Laplace transform.

Under the Laplace transform a polynomial becomes a differential operator with constant coefficients, while the translation τ_a becomes multiplication by e^{-a}.

Consider first the group algebra $\mathbb{C}[\Lambda]$ with basis the elements e^a, $a \in \Lambda$, as an algebra of functions on $U_{\mathbb{C}}$ (periodic with respect to Λ^*).

Definition 14.6. We set $\tilde{W}(\Lambda)$ to be the algebra of differential operators on $U_{\mathbb{C}}$ with coefficients the exponential functions e^λ, $\lambda \in \Lambda$.

In order to understand the structure of $\tilde{W}(\Lambda)$, let us first make explicit some commutation relations among the generators:

$$D_u(e^a) = \langle u \,|\, a \rangle e^a \implies [D_u, e^a] = \langle u \,|\, a \rangle e^a \quad \text{for } u \in U_{\mathbb{C}}, \ a \in \Lambda. \quad (14.6)$$

The algebra $\tilde{W}(\Lambda)$ can be also thought of as algebra of differential operators on the corresponding torus $T_{\mathbb{C}}$. For this choose a basis $\theta_1, \ldots, \theta_s$ of Λ with dual basis $e_1, \ldots, e_s$ of $U_{\mathbb{C}}$ and set $x_i := e^{\theta_i}$.

The operators $D_{e_i} = \frac{\partial}{\partial \theta_i}$, on the functions on $T_{\mathbb{C}}$, coincide with $x_i \frac{\partial}{\partial x_i}$. Therefore, the algebra $\tilde{W}(\Lambda)$ also equals

$$\mathbb{C}\left[x_1^{\pm 1}, \ldots, x_s^{\pm 1}, \frac{\partial}{\partial x_1}, \ldots, \frac{\partial}{\partial x_s}\right].$$

In particular, $\tilde{W}(\Lambda)$ contains (in fact it is a localization) the standard Weyl algebra $W(s) = \mathbb{C}[x_1, \ldots, x_s, \frac{\partial}{\partial x_1}, \ldots, \frac{\partial}{\partial x_s}]$. Recall that $S[U_{\mathbb{C}}]$ is identified with the algebra of differential operators with constant coefficients.

Lemma 14.7. $\mathbb{C}[\Lambda] = \mathbb{C}[x_1^{\pm 1}, \ldots, x_s^{\pm 1}]$ *is an irreducible module over* $\tilde{W}(\Lambda)$.
Additively, we have

$$\tilde{W}(\Lambda) = S[U_{\mathbb{C}}] \otimes \mathbb{C}[\Lambda] = \mathbb{C}[\frac{\partial}{\partial x_1}, \ldots, \frac{\partial}{\partial x_s}] \otimes \mathbb{C}[x_1^{\pm 1}, \ldots, x_s^{\pm 1}].$$

Proof. Take a nonzero submodule $N \subset C[\Lambda] = \mathbb{C}[x_1^{\pm 1}, \ldots, x_s^{\pm 1}]$ under the action of $\tilde{W}(\Lambda)$ and a nonzero element $f \in N$. By multiplying f by a suitable monomial in the x_i's, we can cancel denominators and assume $f \in \mathbb{C}[x_1, \ldots, x_s]$. Since $\tilde{W}(\Lambda) \supset W(s)$ and $\mathbb{C}[x_1, \ldots, x_s]$ is an irreducible $W(s)$-module, we get that $N \supset \mathbb{C}[x_1, \ldots, x_s]$. But this immediately implies that $N = \mathbb{C}[\Lambda]$, proving our first claim.

The commutation relations imply that we have a surjective linear map $S[U_{\mathbb{C}}] \otimes \mathbb{C}[\Lambda] \to \tilde{W}(\Lambda)$. The proof that this is a linear isomorphism is similar to that of the structure of the usual Weyl algebra, and is left to the reader. $\qquad \square$

The previous lemma tells us in particular that $\tilde{W}(\Lambda)$ is the algebra generated by the two commutative algebras $S[U_{\mathbb{C}}]$, $\mathbb{C}[\Lambda]$ modulo the ideal generated by the commutation relations (14.6).

The second algebra we are going to consider is the algebra $W^{\#}(\Lambda)$ of difference operators with polynomial coefficients on Λ.

This algebra is generated by $S[U_{\mathbb{C}}]$, that is now thought of as complex-valued polynomial functions on V and by the translation operators τ_a, $a \in \Lambda$. Again we have the commutation relations

$$[u, \tau_a] = \langle u \,|\, a \rangle \tau_a \quad \text{for } u \in U_{\mathbb{C}}, \ a \in \Lambda. \quad (14.7)$$

In a way similar to the previous case we have additively

$$W^{\#}(\Lambda) = S[U_{\mathbb{C}}] \otimes \mathbb{C}[\Lambda].$$

The fact that these algebras are determined by the relations (14.6) and (14.7) tell us also that $W^{\#}(\Lambda)$ is isomorphic to $\tilde{W}(\Lambda)$ under the isomorphism ϕ defined by

$$\phi(\tau_a) = e^{-a}, \quad \phi(u) = -D_u, \tag{14.8}$$

for $a \in \Lambda$, $u \in U$. Thus, given a module M over $W^{\#}(\Lambda)$, we shall denote by $\hat{M}$ the same space considered as a module over $\tilde{W}(\Lambda)$ and defined through the isomorphism ϕ^{-1} and think of it as a *formal Fourier transform* (similarly when we start from a module over $\tilde{W}(\Lambda)$).

The main example we have in mind is the following.

Define $S_X := \mathbb{C}[\Lambda][\prod_{a \in X}(1 - e^{-a})^{-1}]$ to be the localization of $\mathbb{C}[\Lambda]$, obtained by inverting $\delta_X := \prod_{a \in X}(1 - e^{-a})$, and consider S_X as a module over $\tilde{W}(\Lambda)$. From the previous remark we can and will thus consider S_X as the image under the Laplace transform of a $W^{\#}(\Lambda)$ module, that will turn out to be a module of distributions (cf. Section 14.3.2).

Remark 14.8. Using the coordinates $x_i = e^{\theta_i}$ we can write $e^{-a} = \prod_{i=1}^{s} x_i^{-m_i(a)}$, where $a = \sum_i m_i(a)\theta_i$. The algebra S_X is obtained from $\mathbb{C}[x_1, \ldots, x_s]$ by inverting the element $\prod_i x_i \prod_{a \in X}(1 - \prod_{i=1}^{s} x_i^{m_i(a)})$.

For the algebra S_X we want to describe its structure as a $\tilde{W}(\Lambda)$-module by mimicking the theory of R_X developed in Section 8.1.

This can be done, but there are some notable differences. The first is that a list $(a_1, \ldots, a_k)$ of linearly independent elements in Λ in general is not part of an integral basis for Λ. This is reflected in the fact that the subgroup that is the common kernel of the characters e^{a_i} is not necessarily connected. At the level of modules, this implies that the isotypic components are not indexed by the subgroups which are common kernels of lists of characters of X but rather by all their connected components. This creates some technical difficulties that make this theory more complex than the one developed for hyperplane arrangements.

14.2.2 Some Basic Modules

We keep the notation of the previous section,

$$\tilde{W}(\Lambda) = \mathbb{C}\left[x_1^{\pm 1}, \ldots, x_s^{\pm 1}, \frac{\partial}{\partial x_1}, \ldots, \frac{\partial}{\partial x_s}\right],$$

and extend the constructions of Section 8.1.2.

One can develop in exactly the same way as for $W(s)$ (cf. Section 4.1.3) a Bernstein filtration for the algebra $\tilde{W}(\Lambda)$ using the degree in the operators $\frac{\partial}{\partial x_i}$. The associated graded algebra $\mathrm{gr}(\tilde{W}(\Lambda))$ is then the commutative algebra $\mathbb{C}[x_1^{\pm 1}, \ldots, x_s^{\pm 1}, \xi_1, \ldots, \xi_s]$. This should be thought of as the coordinate ring of the cotangent bundle of the torus $(\mathbb{C}^*)^s$ (or more intrinsically of the torus $T_{\mathbb{C}}$).

For a finitely generated $\tilde{W}(\Lambda)$-module P we can then define a Bernstein filtration whose associated graded module is a module over the algebra $\mathrm{gr}(\tilde{W}(\Lambda))$ whose support is independent of the chosen filtration and is, by definition, the characteristic variety of P, a subvariety of $(\mathbb{C}^*)^s \times \mathbb{C}^s$ (or more intrinsically of the cotangent bundle of $T_{\mathbb{C}}$).

The algebra of Laurent polynomials is an irreducible module (by Lemma 14.7). Its characteristic variety is $(\mathbb{C}^*)^s \times 0$.

For some $k \le s$ fix constants $\underline{\mu} = (\mu_1, \ldots, \mu_k)$ and consider the localized algebra

$$S_{\underline{\mu}} = \mathbb{C}[\Lambda]\Big[\prod_{i=1}^k (1 - e^{-\theta_i - \mu_i})^{-1}\Big] = \mathbb{C}[x_1^{\pm 1}, \ldots, x_s^{\pm 1}]\Big[\prod_{i=1}^k (1 - \lambda_i x_i)^{-1}\Big],$$

$\lambda_i = e^{\mu_i}$ as a $\tilde{W}(\Lambda)$-module.

Consider first the case $k = s$ and in $S_{\underline{\mu}}$ the submodule $S_{\underline{\mu}, s-1}$ generated by all the elements that do not have all the factors $1 - \lambda_i x_i$ in the denominator. Let $p \in T_{\mathbb{C}}$ be the element with coordinates λ_i^{-1}.

Theorem 14.9. *The module $S_{\underline{\mu}}/S_{\underline{\mu}, s-1}$ is free of rank 1 over the algebra $\mathbb{C}[\frac{\partial}{\partial x_1}, \ldots, \frac{\partial}{\partial x_s}]$ generated by the class $\overline{u}$ (modulo $S_{\underline{\mu}, s-1}$) of the element $u := \prod_{i=1}^s (1 - \lambda_i x_i)^{-1}$.*

As a $\tilde{W}(\Lambda)$-module $S_{\underline{\mu}}/S_{\underline{\mu}, s-1}$ is irreducible. It is canonically isomorphic to the irreducible module N_p generated by the Dirac distribution at p via an isomorphism mapping $\overline{u}$ to δ_p.

Its characteristic variety is the cotangent space to $T_{\mathbb{C}}$ at p.

Proof. Since we have that $T_{\mathbb{C}}$ acts as a group of translations we can translate by p^{-1} or equivalently make a change of coordinates $y_i = \lambda_i x_i$ and assume that $p = 1, \lambda_i = 1$ for each $i = 1, \ldots s$. We are in fact in a special case of what was discussed in Section 8.1.3.

For each $i = 1, \ldots, s$, we have $(1 - x_i)u \in S_{\underline{\mu}, s-1}$, and hence $x_i \overline{u} = \overline{u}$. The fact that $\overline{u} \ne 0$ follows as in Proposition 8.11. Thus, by Proposition 4.4, $\overline{u}$ generates under the algebra $W(s) = \mathbb{C}[x_1, \ldots, x_s, \frac{\partial}{\partial x_1}, \ldots, \frac{\partial}{\partial x_s}] \subset \tilde{W}(\Lambda)$ an irreducible module N_1 isomorphic to the module generated by the Dirac distribution in 1. This proves the irreducibility of $S_{\underline{\mu}}/S_{\underline{\mu}, s-1}$ as a $\tilde{W}(\Lambda)$-module.

Furthermore, Proposition 4.4 also tells us that as a module over the algebra $\mathbb{C}[\frac{\partial}{\partial x_1}, \ldots, \frac{\partial}{\partial x_s}]$, $S_{\underline{\mu}}/S_{\underline{\mu}, s-1}$ is free with generator $\overline{u}$. The conclusion on the characteristic variety is then clear.

Remark 14.10. In the module $S_{\underline{\mu}}/S_{\underline{\mu}, s-1}$ the element $\overline{u}$ satisfies the equations

$$e^a \overline{u} = e^a(p)\overline{u}, \qquad \forall a \in \Lambda. \tag{14.9}$$

An element in $S_{\underline{\mu}}/S_{\underline{\mu}, s-1}$ satisfying the previous equations is a multiple of $\overline{u}$.

In any $\tilde{W}(\Lambda)$-module P, a nonzero element satisfying (14.9) generates a submodule isomorphic to N_p.

From now on, N_p will denote the irreducible $\tilde{W}(\Lambda)$ module generated by the Dirac distribution at p.

14.2.3 Induction

As we have already explained, the theory for lattices is more complex than that for hyperplane arrangements, essentially because a subgroup M of Λ does not have in general a complement M' with $\Lambda = M \oplus M'$. This creates some new phenomena in module theory, and we begin by analyzing the main technical point.

Let $M \subset \Lambda$ be a subgroup. We define the induction functor from $\tilde{W}(M)$-modules to $\tilde{W}(\Lambda)$-modules. Given a $\tilde{W}(M)$-module P, set

$$\mathrm{Ind}_M^\Lambda P := \mathbb{C}[\Lambda] \otimes_{\mathbb{C}[M]} P. \tag{14.10}$$

$\mathrm{Ind}_M^\Lambda P$ has a natural structure of a $\mathbb{C}[\Lambda]$ module.

Lemma 14.11. $\mathbb{C}[\Lambda] \otimes_{\mathbb{C}[M]} P$ *has a unique* $\tilde{W}(\Lambda)$*-module structure extending the natural* $\mathbb{C}[\Lambda]$ *module structure and such that:*

(1) On $1 \otimes P$ *the action of* $\tilde{W}(M)$ *coincides with the given one.*
(2) If ∂ *is a derivative that is zero on* $\mathbb{C}[M]$ *then it is also zero on* $1 \otimes P$*.*

Proof. The previous conditions define uniquely the action of $\mathbb{C}[\Lambda]$ and of the derivatives by $\partial(a \otimes p) = \partial a \otimes p + a \otimes \partial(p)$. It is easy to verify that these actions satisfy the commutation relations defining $\tilde{W}(\Lambda)$.

We have a general fact about these modules.

Proposition 14.12. *If* $N \subset M \subset \Lambda$ *is an inclusion of lattices and* P *is a* $\tilde{W}(N)$*-module, then we have*

$$\mathrm{Ind}_M^\Lambda P = \mathrm{Ind}_M^\Lambda \mathrm{Ind}_N^M P.$$

The functor $P \mapsto \mathrm{Ind}_M^\Lambda P$ *is exact.*[1]

Proof. The first part is clear from the definition; as for the second, it follows from the fact that $\mathbb{C}[\Lambda]$ is a free module over $\mathbb{C}[M]$. We may in fact choose as basis the elements e^λ, where λ runs over coset representatives of Λ/M.

---

[1] A functor is exact if it transforms exact sequences into exact sequences.

We want to apply this in particular to the $\tilde{W}(M)$-module N_p. We are going to split this discussion into two steps. Let us consider

$$\overline{M} := \{a \in \Lambda \,|\, \exists \ m \in \mathbb{N}^+,\ ma \in M\}.$$

From the theorem of elementary divisors (Section 5.2.2), it follows that we can split Λ as $\Lambda = \overline{M} \oplus M'$ and that M has finite index in $\overline{M}$.

The inclusions $M \subset \overline{M} \subset \Lambda$ induce surjective homomorphisms

$$T_{\mathbb{C}}^{\Lambda} \xrightarrow{f} T_{\mathbb{C}}^{\overline{M}} \xrightarrow{k} T_{\mathbb{C}}^{M}, \quad \pi = kf.$$

Let us set

$$K := \ker(\pi), \quad G := \ker(k), \quad \mathcal{T} := \ker(f).$$

the kernel $\mathcal{T} := \ker(f)$ is a torus with character group $\Lambda/\overline{M} = M'$. We can identify $T_{\mathbb{C}}^{\overline{M}}$ with a subtorus of $T_{\mathbb{C}}$ and write $T_{\mathbb{C}} = T_{\mathbb{C}}^{\overline{M}} \times \mathcal{T}$.

On the other hand, the kernel $G \subset T_{\mathbb{C}}^{\overline{M}}$ of k is a finite group (the dual of $\overline{M}/M$, Section 5.2.1). Finally $K = G \times \mathcal{T}$, and thus we have a bijection between the elements of G and the irreducible (or connected) components of $K = \ker \pi$. Given $g \in G$, the corresponding connected component is the coset $g\mathcal{T}$.

If $p \in T_{\mathbb{C}}$ and $q := \pi(p)$, we shall define

$$N_{pK} := \operatorname{Ind}_M^{\Lambda}(N_q). \tag{14.11}$$

By Proposition 14.12 we shall analyze separately the induction functors when M splits and when it is a subgroup of finite index. We start from the case in which M has finite index, so that $\overline{M} = \Lambda$ and $K = G$ is finite. In this case, we have the following theorem:

Theorem 14.13. $\operatorname{Ind}_M^{\Lambda}(N_q) \simeq \oplus_{p \in \pi^{-1}(q)} N_p$.

Proof. Let us take the two ideals $J \subset \mathbb{C}[M]$ and $I = J\mathbb{C}[\Lambda] \subset \mathbb{C}[\Lambda]$ generated by the same elements $e^b - e^b(q)$, $b \in M$. Then J is the ideal of the point q in the torus $T_{\mathbb{C}}^M$, while I is the ideal of definition of $\pi^{-1}(q)$ in $T_{\mathbb{C}}^{\Lambda}$. We have $\mathbb{C}[M]/J = \mathbb{C}$ and $I = \mathbb{C}[\Lambda] \otimes_{\mathbb{C}[M]} J$. It follows that a basis of $\mathbb{C}[\Lambda]$ over $\mathbb{C}[M]$ can be chosen by lifting a linear basis of $\mathbb{C}[\Lambda]/I$.

Since $\mathbb{C}[\Lambda]/I$ is the algebra of functions on $\pi^{-1}(q)$ we can decompose $\mathbb{C}[\Lambda]/I = \oplus_{p \in \pi^{-1}(q)} \mathbb{C}\chi_p$, where χ_p is the characteristic function of p. For each $p \in \pi^{-1}(q)$ let us choose a representative $c(p)$ in $\mathbb{C}[\Lambda]$ of χ_p in such a way that

$$1 = \sum_{p \in \pi^{-1}(q)} c(p).$$

By Proposition 14.12, we have $\operatorname{Ind}_M^{\Lambda}(P) = \oplus_{p \in \tilde{\pi}^{-1}(q)} c(p) \otimes P$, for any module P.

Let us apply this to N_q and its generating element u. Write

$$1 \otimes u = \sum_{p \in \pi^{-1}(q)} c(p) \otimes u.$$

We claim that if $a \in \Lambda$, then $e^a c(p) \otimes u = e^a(p) c(p) \otimes u$, so that by remark 14.10, $c(p) \otimes u$ generates a copy of N_p. To see our claim, notice that we have $e^a c(p) = e^a(p) c(p) + h$, $h \in J$, and since $hu = 0$, the claim follows.

Since the different N_p's are nonisomorphic, having distinct characteristic varieties, we deduce that $\mathrm{Ind}_M^\Lambda(N_q)$ contains a submodule isomorphic to $\oplus_{p \in \pi^{-1}(q)} N_p$. By construction this module contains the element $1 \otimes u$, which clearly generates $\mathrm{Ind}_M^\Lambda(N_q)$ as a $\tilde{W}(\Lambda)$-module, so everything follows.

It is convenient to make the elements $c(p)$ explicit as follows. Use exponential notation and assume that $p = e^\phi \in G$, so that $e^a(p) = e^{\langle \phi \mid a \rangle}$. By Proposition 5.16, $G = \ker(\pi)$ is in duality with Λ/M. The finite-dimensional group algebra $\mathbb{C}[\Lambda/M]$ is identified with the ring of functions on G.

We choose a set R of representatives in Λ for the cosets Λ/M. We now use the basic elements of formula (14.5) and set

$$c(p) = e(\phi) := |\Lambda/M|^{-1} \sum_{\lambda \in R} e^{\langle \phi \mid \lambda \rangle} e^{-\lambda}. \tag{14.12}$$

Character theory tells us that the element $e(\phi)$ has the property of taking the value 1 at the point e^ϕ and 0 on all the other points of G.

Moreover, it also tells us that $\sum_{e^\phi \in \pi^{-1}(q)} e^{\langle \phi \mid \lambda \rangle} = 0$, $\forall \lambda \neq 0$, in our set of representatives.

Therefore, the elements $e(\phi)$ satisfy

$$\sum_{e^\phi \in \pi^{-1}(q)} e(\phi) = 1. \tag{14.13}$$

Let us now consider the case in which M splits Λ, i.e., there exists a (noncanonical) M' with $\Lambda = M \oplus M'$. We have then a decomposition

$$\tilde{W}(\Lambda) \simeq \tilde{W}(M) \otimes \tilde{W}(M'), \quad \mathbb{C}[\Lambda] \simeq \mathbb{C}[M] \otimes \mathbb{C}[M'].$$

From this we deduce that if $q \in T_M$, we have

$$\mathrm{Ind}_M^\Lambda(N_q) \simeq N_q \otimes \mathbb{C}[M'] \tag{14.14}$$

as $\tilde{W}(M) \otimes \tilde{W}(M')$ modules. The preimage $\pi^{-1}(q)$ of q under the morphism $\pi : T \to T_M$ is a coset $p\mathcal{T}$, the torus $\mathcal{T}$ being the kernel of π, so that using the notation of formula (14.11), $\mathrm{Ind}_M^\Lambda(N_q) = N_{p\mathcal{T}}$.

Let us choose coordinates in such a way that $\mathbb{C}[M] = \mathbb{C}[x_1^{\pm 1}, \ldots, x_k^{\pm 1}]$ and $\mathbb{C}[M'] = \mathbb{C}[x_{k+1}^{\pm 1}, \ldots, x_s^{\pm 1}]$. In these coordinates for T, the variety $p\mathcal{T}$ is given by the equations $x_i = \alpha_i$, $i = 1, \ldots, k$, with the α_i some roots of 1.

Set $A := \mathbb{C}[\frac{\partial}{\partial x_1}, \ldots, \frac{\partial}{\partial x_k}, x_{k+1}^{\pm 1}, \ldots, x_s^{\pm 1}]$,

Proposition 14.14. *As a module over the algebra A, $N_{p\mathcal{T}}$ is free of rank 1, generated by an element (unique up to a constant factor) u satisfying the equations*

$$x_i u = \alpha_i u, \ \ i = 1, \ldots, k, \qquad \frac{\partial}{\partial x_j} u = 0, \ \ j = k+1, \ldots, s. \qquad (14.15)$$

Then module $N_{p\mathcal{T}}$ is irreducible with characteristic variety the conormal bundle to the coset $p\mathcal{T}$.

Proof. The first part follows immediately from formula (14.14) and Theorem 14.9. For the second part, arguing as in Lemma 14.7, one easily reduces to show the irreducibility of the $W(s)$-module generated by u, that was shown in Proposition 4.4.

The part on the characteristic variety follows from part 1.

Remark 14.15. We shall use the previous proposition as follows. Whenever we find in a module over $\tilde{W}(\Lambda)$ a nonzero element $\bar{u}$ satisfying the equations (14.15), we can be sure that the submodule generated by $\bar{u}$ is irreducible and isomorphic to $N_{p\mathcal{T}}$.

We pass now to the general case. Using the notation introduced above, we easily get the following result

Proposition 14.16. *Let $p \in T_{\mathbb{C}}$ and $q := \pi(p)$. Then*

$$N_{pK} := \mathrm{Ind}_M^\Lambda(N_q) = \mathrm{Ind}_{\overline{M}}^\Lambda \mathrm{Ind}_M^{\overline{M}}(N_q) \simeq \oplus_{g \in G} N_{pg\mathcal{T}}. \qquad (14.16)$$

Each $N_{pg\mathcal{T}}$ is an irreducible module with characteristic variety the conormal bundle to $pg\mathcal{T}$, and the summands are pairwise nonisomorphic.

14.2.4 A Realization

Take k linearly independent characters $\underline{b} := \{a_1, \ldots, a_k\}$ in Λ. Denote by M the sublattice of Λ that they generate, K the joint kernel of the characters e^{a_i}. In what follows we are going to use the notation of the previous section.

We want to analyze next the algebra $S_{\underline{\mu}} := \mathbb{C}[\Lambda][\prod_{i=1}^k (1 - e^{-a_i - \mu_i})^{-1}]$ as a $\tilde{W}(\Lambda)$-module and perform the usual filtration by submodules $S_{\underline{\mu},h}$. For each h, $S_{\underline{\mu},h}$ is the submodule of fractions whose denominator contains at most h among the factors $1 - e^{-a_i - \mu_i}$. Together with $S_{\underline{\mu}}$, let us also consider the algebra

$$R_{\underline{\mu}} := \mathbb{C}[M][\prod_{i=1}^k (1 - e^{-a_i - \mu_i})^{-1}]$$

as a $\tilde{W}(M)$ module.

Proposition 14.17. *For every h we have*

$$S_{\underline{\mu},h} = \mathrm{Ind}_M^\Lambda R_{\underline{\mu},h}.$$

Proof. We apply Theorem 14.9 and the exactness of the induction functor. It is clear that $S_{\underline{\mu},h} = \mathbb{C}[\Lambda] \otimes_{\mathbb{C}[M]} R_{\underline{\mu},h}$, at least as $\mathbb{C}[\Lambda]$-modules, but the uniqueness of the module structure ensured by Lemma 14.11 implies our claim.

We are now going to study $S_{\underline{\mu}}/S_{\underline{\mu},k-1}$. Set $q \in T_{\mathbb{C}}^M$ equal to the element such that $e^{-a_i}(q) = e^{\mu_i}$. Take $p \in \pi^{-1}(q)$. We deduce, using the notation of the previous section, the following corollary

Corollary 14.18. $\pi^{-1}(q) = pK = \cup_{g \in G}\, pg\mathcal{T}, \quad S_{\underline{\mu}}/S_{\underline{\mu},k-1} = \oplus_{g \in G} N_{pg\mathcal{T}}.$

We shall need to make more explicit certain canonical generators for the summands $N_{pg\mathcal{T}}$.

From the proof of Theorem 14.13 we have our next result

Proposition 14.19. *Let $p \in \pi^{-1}(q)$ and $\tilde{u} \in \mathbb{C}[\Lambda]$ be an element whose restriction to pK is the characteristic function of $p\mathcal{T}$.*

The class of $\tilde{u}/\prod_{i=1}^k(1 - e^{-a_i-\mu_i})$ in $S_{\underline{\mu}}/S_{\underline{\mu},k-1}$ does not depend on the choice of $\tilde{u}$ and is characterized, up to scalar multiples, as the unique generator of $N_{p\mathcal{T}}$ satisfying equations (14.15).

Given an element $\tilde{u}$ as in the previous proposition, we shall denote by $\omega_{\underline{b},p\mathcal{T}}$ the class of $\tilde{u}/\prod_{i=1}^k(1 - e^{-a_i-\mu_i})$ in $S_{\underline{\mu}}/S_{\underline{\mu},k-1}$.

If $p = e^\phi$, we can choose as $\tilde{u}$ the element $e(\phi)$ defined in formula (14.5), so that the elements $d_{\underline{b},\phi} \in S_{\underline{\mu}}$ defined as

$$d_{\underline{b},\phi} := \frac{e(\phi)}{\prod_{i=1}^k(1 - e^{-a_i-\mu_i})}. \tag{14.17}$$

give representatives of the classes $\omega_{\underline{b},p\mathcal{T}}$.

As a result, we are recovering essentially the elements introduced in Section 14.1.2 as special Laplace transforms.

14.3 The Toric Arrangement

14.3.1 The Coordinate Ring as a $\tilde{W}(\Lambda)$ Module

As in Section 8.1.7, for every $k \le s$ let us consider the $\tilde{W}(\Lambda)$ submodule $S_{X,\underline{\mu},k} \subset S_{X,\underline{\mu}}$ spanned by the elements

$$\frac{f}{\prod_i(1 - e^{-a_i-\mu_i})^{h_i}}, \quad f \in \mathbb{C}[\Lambda],$$

such that the vectors a_i that appear in the denominator with positive exponent span a subspace of dimension $\le k$.

We are going to describe $S_{X,\underline{\mu}}$ as a $\tilde{W}(\Lambda)$ module and deduce an expansion into partial fractions.

We start with a very elementary remark.

Lemma 14.20. *For each $n \geq 0$, the following identities hold:*

$$\frac{x^n}{1-x} = \frac{1}{1-x} - \sum_{i=0}^{n-1} x^i, \qquad \frac{x^{-n}}{1-x} = \frac{1}{1-x} + \sum_{i=1}^{n} x^{-i}.$$

Proof. We give the proof of the first identity, the proof of the second being completely analogous. For $n = 0$ there is nothing to prove. Proceeding by induction, we have

$$\frac{x^n}{1-x} = x\frac{x^{n-1}}{1-x} = x\left(\frac{1}{1-x} - \sum_{i=0}^{n-2} x^i\right) = \frac{x}{1-x} - \sum_{i=1}^{n-1} x^i = \frac{1}{1-x} - \sum_{i=0}^{n-1} x^i.$$

As a consequence, we get the following

Lemma 14.21. *Given linearly independent elements $a_1, \ldots, a_k \in \Lambda$, integers $h_1, \ldots, h_k$, and constants $\mu_1, \ldots, \mu_k$, we have that*

$$\frac{e^{\sum_i h_i a_i}}{\prod_{i=1}^{k}(1 - e^{-a_i-\mu_i})} = \frac{e^{-\sum_i h_i \mu_i}}{\prod_{i=1}^{k}(1 - e^{-a_i-\mu_i})} + K,$$

where K is a sum of fractions with numerator in $\mathbb{C}[\Lambda]$ and denominator a product of a proper subset of the set $\{1 - e^{-a_i-\mu_i}\}$, $i = 1, \ldots, k$.

Proof. In order to see this, we can first apply the previous lemma to each of the factors

$$\frac{e^{h_i a_i}}{1 - e^{-a_i-\mu_i}} = e^{-h_i\mu_i}\frac{e^{h_i(a_i+\mu_i)}}{1 - e^{-a_i-\mu_i}}$$

with $x = e^{-a_i-\mu_i}$ and then substitute.

Take a sequence $\underline{b} := (b_1, \ldots, b_k)$ of linearly independent elements in Λ and fix a constant μ_b for each $b \in \underline{b}$. Let $M \subset \Lambda$ be the lattice they generate and let $\overline{M}$ be defined as in Section 14.2.3. Choose as set of coset representatives in $\overline{M}$ of $\overline{M}/M$ the set $R_{\underline{b}}$ defined by (14.4).

Choose a point $p = e^\phi \in T_{\mathbb{C}}$ with the property that $e^{\langle\phi\,|\,b\rangle} = e^{-\mu_b}$ for each $b \in \underline{b}$ and define $e(\phi)$ as in (14.5).

Applying Lemma 14.21 to $d_{\underline{b},\phi} = e(\phi)/\prod_{b\in\underline{b}}(1 - e^{-b-\mu_b})$ (cf. (14.17)) we get the following result

Proposition 14.22. *For given $c \in \overline{M}$ we have*

$$e^c d_{\underline{b},\phi} = e^{\langle\phi\,|\,c\rangle} d_{\underline{b},\phi} + K,$$

where K is a sum of fractions with numerator in $\mathbb{C}[\Lambda]$ and denominator a product of a proper subset of the set $\{1 - e^{-b-\mu_b}\}$, $b \in \underline{b}$.

Proof. We have

$$e^c(d_{\underline{b},\phi}) = e^{\langle \phi \,|\, c\rangle}\left[\frac{\sum_{\lambda \in R_{\underline{b}}} e^{-\langle \phi \,|\, \lambda+c\rangle} e^{\lambda+c}}{|R_{\underline{b}}|\prod_{b\in\underline{b}}(1 - e^{-b-\mu_b})}\right].$$

Each $\lambda+c$ can be rewritten as $\lambda' + \sum_{b\in\underline{b}} h_b b$, $h_b \in \mathbb{Z}$ where λ' is the unique element in $R_{\underline{b}}$ congruent to $\lambda + c$ modulo $\overline{M}$. We have

$$e^{-\langle \phi \,|\, \lambda+c\rangle} = e^{-\langle \phi \,|\, \lambda'\rangle} e^{-\langle \phi \,|\, \sum_{b\in\underline{b}} h_b b\rangle} = e^{-\langle \phi \,|\, \lambda'\rangle} e^{\sum_{b\in\underline{b}} h_b \mu_b}.$$

By Lemma 14.21,

$$\frac{e^{\sum_{b\in\underline{b}} h_b b}}{\prod_{b\in\underline{b}}(1 - e^{-b-\mu_b})} = \frac{e^{-\sum_{b\in\underline{b}} h_b \mu_b}}{\prod_{b\in\underline{b}}(1 - e^{-b-\mu_b})} + K'.$$

Thus

$$\frac{e^{-\langle \phi \,|\, \lambda+c\rangle} e^{\lambda+c}}{|R_{\underline{b}}|\prod_{b\in\underline{b}}(1 - e^{-b-\mu_b})} = \frac{e^{-\langle \phi \,|\, \lambda'\rangle} e^{\sum_{b\in\underline{b}} h_b \mu_b} e^{\lambda'} e^{-\sum_{b\in\underline{b}} h_b \mu_b}}{|R_{\underline{b}}|\prod_{b\in\underline{b}}(1 - e^{-b-\mu_b})} + K''.$$

$$= \frac{e^{-\langle \phi \,|\, \lambda'\rangle} e^{\lambda'}}{|R_{\underline{b}}|\prod_{b\in\underline{b}}(1 - e^{-b-\mu_b})} + K''.$$

As λ runs over $R_{\underline{b}}$ also λ' runs over $R_{\underline{b}}$. We thus obtain

$$e^c(d_{\underline{b},\phi}) = e^{\langle \phi \,|\, c\rangle} d_{\underline{b},\phi} + K \tag{14.18}$$

where K' and hence K'', K, are sums of terms in which not every $b \in \underline{b}$ appears in the denominators, as desired.

Take $b_0 \in \overline{M}$, $\mu_0 \in \mathbb{C}$, and consider the element

$$\frac{e(\phi)}{(1 - e^{-b_0-\mu_0}) \prod_{b\in\underline{b}}(1 - e^{-b-\mu_b})} = \frac{d_{\underline{b},\phi}}{(1 - e^{-b_0-\mu_0})}.$$

Lemma 14.23. *Write formula (14.18) as*

$$(1 - e^{-b_0-\mu_0}) d_{\underline{b},\phi} = (1 - e^{-\langle \phi \,|\, b_0\rangle - \mu_0}) d_{\underline{b},\phi} - K. \tag{14.19}$$

(1) If $e^{\langle \phi \,|\, b_0\rangle + \mu_0} = 1$, we have

$$\frac{e(\phi)}{(1 - e^{-b_0-\mu_0}) \prod_{b\in\underline{b}}(1 - e^{-b-\mu_b})} = -\frac{K}{(1 - e^{-b_0-\mu_0})^2}. \tag{14.20}$$

(2) If $e^{\langle \phi \,|\, b_0\rangle + \mu_0} \neq 1$, we have

$$\frac{e(\phi)}{(1 - e^{-b_0-\mu_0}) \prod_{b\in\underline{b}}(1 - e^{-b-\mu_b})} \tag{14.21}$$

$$= (1 - e^{-\langle \phi \,|\, b_0\rangle - \mu_0})^{-1}\left[\frac{e(\phi)}{\prod_{b\in\underline{b}}(1 - e^{-b-\mu_b})} + \frac{K}{(1 - e^{-b_0-\mu_0})}\right].$$

We draw as an important conclusion a first expansion into partial fractions.

Proposition 14.24. *Let $(b_0, \ldots, b_r)$ be any list of elements in Λ, spanning a sublattice M and $(\mu_0, \ldots, \mu_r)$ a list of complex numbers. The fraction*

$$A = \prod_{i=0}^{r}(1 - e^{-b_i - \mu_i})^{-1}$$

can be written as a linear combination of fractions with numerators in $\mathbb{C}[M]$ and denominators $\prod_{i=0}^{r}(1 - e^{-b_i - \mu_i})^{h_i}$, $h_i \geq 0$, with the property that the elements b_i for which $h_i > 0$ are linearly independent.

Proof. When we encounter in the denominator of a fraction a product of type $\prod_{i=0}^{k}(1 - e^{-b_i - \mu_i})$ with the b_i linearly dependent, we have to make a replacement so as to arrive finally at linearly independent elements. This we do as follows. First we may assume that the elements $b_1, \ldots, b_k$ are linearly independent. Let M be the lattice that they span, so that $b_0 \in \overline{M}$, and $\underline{b} := (b_1, \ldots, b_k)$ is a basis for M. We may also assume that b_0 is not dependent on any proper subset of $b_1, \ldots, b_k$. Consider the unique point $q = e^\psi \in T_M$ with $e^{\langle \psi \mid b_0 \rangle + \mu_0} = 1$. For each $p = e^\phi \in \pi^{-1}(q) \in T_{\overline{M}}$ take the corresponding $e(\phi)$. We now apply the identity (14.13) so that

$$\frac{1}{\prod_{i=0}^{k}(1 - e^{-b_i - \mu_i})} = \sum_{e^\phi \in \pi^{-1}(q)} \frac{e(\phi)}{\prod_{i=0}^{k}(1 - e^{-b_i - \mu_i})}.$$

To each summand we can apply the previous identities (14.20), (14.21), and obtain a sum in which only linearly independent sets appear in the denominators. Iterating this algorithm we arrive at the desired expansion.

Corollary 14.25. $S_{X,\mu,k}$ *is the sum of the $S_{Y,\mu}$, where Y runs over all the linearly independent sublists of X consisting of k elements.*

Let $\underline{b} := (b_1, \ldots, b_k)$ be a linearly independent sublist of X generating some sublattice M and use the notation of Section 14.2.3 for $K, G, \mathcal{T}$. Let

$$S_{\underline{b},\mu} := \mathbb{C}[\Lambda]\left[\prod_{b \in \underline{b}}(1 - e^{-b - \mu_b})^{-1}\right].$$

According to Corollary 14.18 and Proposition 14.19, $S_{\underline{b},\mu}/S_{\underline{b},\mu,k-1}$ is a semisimple module, the direct sum of irreducibles $N_{p\mathcal{T}}$ indexed by the elements of $p = e^\phi \in T_{\overline{M}}$ such that $e^{-\langle \phi \mid b_i \rangle - \mu_i} = 1$. Each $N_{p\mathcal{T}}$ is generated by the class $\omega_{\underline{b},p\mathcal{T}}$ of the element $d_{\underline{b},\phi}$.

Proposition 14.26. *If $\underline{b} := (b_1, \ldots, b_k)$ is a linearly independent sublist of X then $S_{\underline{b},\mu}/S_{\underline{b},\mu,k-1}$ maps injectively into $S_{X,\mu,k}/S_{X,\mu,k-1}$.*

$S_{X,\mu,k}/S_{X,\mu,k-1}$ is the sum of all the images of the spaces $S_{\underline{b},\mu}/S_{\underline{b},\mu,k-1}$ as $\underline{b} := (b_1, \ldots, b_k)$ runs over the linearly independent sublists of X.

Proof. The second part follows from Corollary 14.25. As for the first, we know from Corollary 14.18 that $S_{\underline{b},\mu}/S_{\underline{b},\mu,k-1}$ is a direct sum of nonisomorphic irreducible modules of type $N_{p\mathcal{T}}$ with characteristic variety the conormal bundle to $p\mathcal{T}$, a subvariety of codimension k. Since $N_{p\mathcal{T}}$ is irreducible, it is enough to prove that the image of $N_{p\mathcal{T}}$ in $S_{X,\mu,k}/S_{X,\mu,k-1}$ is nonzero. The argument now is identical to that we carried out in Proposition 8.11. We have that by induction, a composition series of $S_{X,\mu,k-1}$ is formed by modules of type $N_{p'\mathcal{T}'}$, where $h\mathcal{T}'$ has codimension $\leq k-1$, hence with a characteristic variety different from that of $N_{p\mathcal{T}}$, and so $N_{p\mathcal{T}}$ cannot appear in a composition series of $S_{X,\mu,k-1}$.

Corollary 14.27. *If $Y \subset X$ is a sublist, we have, for all k,*

$$S_{Y,\mu} \cap S_{X,\mu,k} = S_{Y,\mu,k}.$$

We want to apply the previous proposition to start obtaining an expansion into partial fractions of the elements of $S_{X,\mu}$ and describe a composition series for $S_{X,\mu}$.

Definition 14.28. If $p\mathcal{T}$ is an element of the arrangement, we denote by $X_{p\mathcal{T}}$ the sublist of X consisting of those elements $a \in X$ such that $e^a = e^{-\mu_a}$ on $p\mathcal{T}$.

Assume that $p\mathcal{T}$ has codimension k. A maximal linearly independent list $\underline{b} := (b_1, \ldots, b_k)$ in $X_{p\mathcal{T}}$ will be called a *basis relative to $p\mathcal{T}$*.

If the element of the arrangement is a point $p = e^{\phi}$, $\phi \in \tilde{P}_\mu(X)$, then X_p will sometimes be denoted by X_ϕ.

Assume that $p\mathcal{T}$ has codimension k. Take a basis $\underline{b} := (b_1, \ldots, b_k)$ relative to $p\mathcal{T}$. As we have seen, $S_{\underline{b},\mu}/S_{\underline{b},\mu,k-1}$ contains a component isomorphic to $N_{p\mathcal{T}}$ whose image inside $S_{X,\mu,k}/S_{X,\mu,k-1}$ we shall denote by $N_{\underline{b},p\mathcal{T}}$. For a given $p\mathcal{T}$ let us set

$$F_{p\mathcal{T}} = \sum_{\underline{b}} N_{\underline{b},p\mathcal{T}} \subset S_{X,\mu,k}/S_{X,\mu,k-1}$$

as $\underline{b}$ runs over the bases relative to $p\mathcal{T}$.

From the previous discussion we deduce the following

Lemma 14.29. *As a $\tilde{W}(\Lambda)$-module $S_{X,\mu,k}/S_{X,\mu,k-1}$ is semisimple, and*

$$S_{X,\mu,k}/S_{X,\mu,k-1} = \oplus_{p\mathcal{T}} F_{p\mathcal{T}},$$

as $p\mathcal{T}$ runs among the elements of the arrangement of codimension k, is the decomposition into its isotypic components.

Proof. The fact that $S_{X,\mu,k}/S_{X,\mu,k-1} = \sum_{p\mathcal{T}} F_{p\mathcal{T}}$ follows from Corollary 14.25 and Proposition 14.26. Since this is a sum of irreducible modules, the module $S_{X,\underline{\mu},k}/S_{X,\underline{\mu},k-1}$ is semisimple. Finally, each $F_{p\mathcal{T}}$ is a sum of irreducibles that are isomorphic, while for two different components $p_1\mathcal{T}, p_2\mathcal{T}$ the corresponding irreducibles are not isomorphic, having different characteristic varieties. Thus the claim follows.

It remains to extract from the family of modules $N_{\underline{b},p\mathcal{T}}$ a subfamily whose sum is direct and equal to $F_{p\mathcal{T}}$. In order to do this, define $V_{p\mathcal{T}}$ to be the subspace of $S_{X,\underline{\mu},k}/S_{X,\underline{\mu},k-1}$ of all solutions of equations (14.15). By Remark 14.15 we know that each nonzero element of $V_{p\mathcal{T}}$ generates a submodule isomorphic to $N_{p\mathcal{T}}$. On the other hand, $N_{\underline{b},p\mathcal{T}} \cap V_{p\mathcal{T}}$ is the one-dimensional space spanned by $\omega_{\underline{b},p\mathcal{T}}$. Thus we deduce that the elements $\omega_{\underline{b},p\mathcal{T}}$ span $V_{p\mathcal{T}}$, and reasoning as in Corollary 4.2, in order to get a family of modules forming a direct sum and adding to $F_{p\mathcal{T}}$, it is equivalent to extract from the elements $\omega_{\underline{b},p\mathcal{T}}$ a basis of $V_{p\mathcal{T}}$. This is done through the theory of unbroken bases.

Take as above a basis $\underline{b} := (b_1, \ldots, b_k)$ relative to $p\mathcal{T}$, so that $p\mathcal{T}$ is a component of the variety of equations $e^{b+\mu_b} = 1$ for each $b \in \underline{b}$. Let $p = e^\phi$.

Assume that $\underline{b}$ is broken in $X_{p\mathcal{T}}$ by an element b. Let $1 \leq e \leq k$ be the largest element such that $b, b_e, \ldots, b_k$ are linearly dependent. Since $p\mathcal{T}$ is contained in the subvariety of equations $e^{b_i+\mu_{b_i}} = e^{b+\mu_b} = 1$, $e \leq i \leq k$, we see that $p\mathcal{T}$ is contained in a unique connected component Z of this subvariety. We consider a point $\psi \in U_{\mathbb{C}}$ with $e^\psi \in Z$ and the associated element $e(\psi)/\prod_{i=e}^{k}(1 - e^{-b_i-\mu_{b_i}})$. We have from (14.20) that

$$\frac{e(\psi)}{\prod_{i=e}^{k}(1 - e^{-b_i-\mu_{b_i}})} = \frac{K}{(1 - e^{-b-\mu_b})}, \tag{14.22}$$

where K is a sum of terms in which the denominator is a product of a proper subset of the elements $(1 - e^{-b_i-\mu_{b_i}})$, $i = e, \ldots, k$.

Lemma 14.30. $e(\phi)e(\psi) = e(\phi) + t$, where $t \in \sum_{i=1}^{k} \mathbb{C}[\Lambda](1 - e^{-b_i-\mu_{b_i}})$.

Proof. Since $p\mathcal{T} \subset Z$, the function $e(\psi)$ equals 1 on $p\mathcal{T}$, hence the function $e(\phi)e(\psi)$ equals 1 on $p\mathcal{T}$ and 0 on the other components of the variety defined by the equations $e^{b_i+\mu_{b_i}} = 1$, $i = 1, \cdots, k$. Since the ideal of definition of this variety is $\sum_{i=1}^{k} \mathbb{C}[\Lambda](1 - e^{-b_i-\mu_{b_i}})$, the claim follows.

Thus substituting

$$\frac{e(\phi)}{\prod_{i=1}^{k}(1 - e^{-b_i-\mu_{b_i}})} = \frac{e(\phi)e(\psi) - t}{\prod_{i=1}^{k}(1 - e^{-b_i-\mu_{b_i}})}, \tag{14.23}$$

the element $-t/\prod_{i=1}^{k}(1 - e^{-b_i-\mu_{b_i}})$ lies in $S_{X,\underline{\mu},k-1}$, while for the term $e(\phi)e(\psi)/\prod_{i=1}^{k}(1 - e^{-b_i-\mu_{b_i}})$, we may apply formula (14.22). As a result, we obtain modulo $S_{X,\underline{\mu},k-1}$ a sum of terms whose denominator is a product $(1 - e^{-b-\mu_b})\prod_{i\neq j}(1 - e^{-b_i-\mu_{b_i}})$ with $e \leq j \leq k$. Once we note that the list $(b_1, \ldots, b_{j-1}, b, b_{j+1}, \ldots b_k)$ is lower than $\underline{b}$ in the lexicographic order, a simple induction proves the following lemma

Lemma 14.31. *$V_{p\mathcal{T}}$ is spanned by the elements $\omega_{\underline{b},p\mathcal{T}}$ as $\underline{b}$ runs over the bases relative to $p\mathcal{T}$ that are unbroken in $X_{p\mathcal{T}}$.*

To finish our analysis we are now going to see that these elements are linearly independent and hence form a basis of $V_{p\mathcal{T}}$.

Let us recall Theorem 10.9. Denote by L the algebra of germs of holomorphic functions around 0 in $\mathbb{C}^s$ and by R the corresponding algebra of polynomials. Let $Z = \{c_1, \ldots, c_m\}$ be a set of homogeneous linear forms (defining a hyperplane arrangement) that span $\mathbb{C}^s$. The algebra $L_Z := L[\prod_{c \in Z} c^{-1}]$ can be filtered in the same way as $R_Z = R[\prod_{c \in Z} c^{-1}]$, and the map

$$i : R_Z/R_{Z,s-1} \to L_Z/L_{Z,s-1}$$

is an isomorphism of $W(s)$-modules.

The next theorem gives the final and precise structure theorem for the composition factors of the $\tilde{W}(\Lambda)$-module $S_{X,\mu,k}/S_{X,\mu,k-1}$.

Theorem 14.32. *$S_{X,\mu,k}/S_{X,\mu,k-1}$, decomposes as a direct sum of the modules $N_{\underline{b},p\mathcal{T}}$ as $p\mathcal{T}$ runs over all components of the arrangement of codimension k and $\underline{b}$ over the unbroken bases on $p\mathcal{T}$.*

For each $p\mathcal{T}$, $F_{p\mathcal{T}}$ is the isotypic component of type $N_{p\mathcal{T}}$.

Proof. By our previous considerations, the only thing we need to show is that the elements $\omega_{\underline{b},p\mathcal{T}}$ as $\underline{b}$ runs over the bases relative to $p\mathcal{T}$ that are unbroken in $X_{p\mathcal{T}}$ are linearly independent.

For this we are going to analyze the algebra $S_{X_{p\mathcal{T}},\mu}$ locally in a neighborhood of a point $e^\phi \in p\mathcal{T}$. As a preliminary step, since from Corollary 14.27 we can assume that $X = X_{p\mathcal{T}}$, translating by p^{-1}, we can also assume that $p = 1$ and all μ_i's are equal to 0.

We want to further reduce to the case in which X spans V, i.e., $\mathcal{T} = \{1\}$. Let M be the sublattice spanned by X and $k = \operatorname{rank}M$.

Set $\tilde{S}_X = \mathbb{C}[\overline{M}][\prod_{a \in X}(1 - e^{-a})^{-1}]$. We have an inclusion of $\tilde{S}_X/\tilde{S}_{X,k-1}$ into $S_X/S_{X_{k-1}}$. It follows that we can work in the space $\tilde{S}_X/\tilde{S}_{X,k-1}$ that contains the elements $\omega_{\underline{b},\mathcal{T}}$ for each basis $\underline{b}$ extracted from X. This clearly gives the desired reduction.

Consider the exponential map $\exp : U_\mathbb{C} \to T_\mathbb{C}$. Notice that if $a \in \Lambda$ and $x \in U_\mathbb{C}$, we have $e^{-a}(\exp(x)) = e^{-\langle x \,|\, a \rangle}$, so by composition, $\exp$ maps the algebra S_X of functions on the complement of the toric arrangement to an algebra of holomorphic functions localized at the elements $1 - e^{-\langle x \,|\, a \rangle}$ as a varies in X, hence holomorphic on the complement of the corresponding periodic hyperplane arrangement. Write $1 - e^{-\langle x \,|\, a \rangle} = \langle x \,|\, a \rangle f(x)$, where f is a holomorphic function with $f(0) = 1$ hence S_X maps to L_X. Furthermore, we have that $\omega_{\underline{b},1} \circ \exp$ equals $\prod_{a \in \underline{b}} a^{-1}g$, with g holomorphic in a neighborhood of 0 and $g(0) = 1$. Therefore, the linear independence of the elements $\omega_{\underline{b},1}$ as $\underline{b}$ runs over the unbroken bases in X reduces to the linear independence of the elements $\prod_{a \in \underline{b}} a^{-1}$, that we have proved in Proposition 8.11.

According to Proposition 14.16, we have that given $p\mathcal{T}$, one can choose co-ordinates (that depend only on the torus of which $p\mathcal{T}$ is a coset) so that $N_{p\mathcal{T},\underline{b}}$ is a free module of rank 1 with generator $\omega_{\underline{b},p\mathcal{T}}$ over the algebra $A_{p\mathcal{T}} := \mathbb{C}[\frac{\partial}{\partial x_1},\dots,\frac{\partial}{\partial x_k},x_{k+1}^{\pm 1},\dots,x_s^{\pm 1}]$. One deduces the following partial fraction expansion

Corollary 14.33.

$$S_{X_{\underline{\mu}}} = \bigoplus_{p\mathcal{T},\underline{b}\in\mathcal{NB}_{X_{p\mathcal{T}}}} A_{p\mathcal{T}}d_{\underline{b},p\mathcal{T}}. \tag{14.24}$$

Remark 14.34. Notice that our coordinates are chosen in such a way that $d_{\underline{b},p\mathcal{T}}$ depends only on $x_1,\dots,x_k$. In particular, $\frac{\partial}{\partial x_j}d_{\underline{b},p\mathcal{T}} = 0, \forall j > k$.

As a special case we can discuss the top part of the filtration relative to the points of the arrangement. We define the *space of polar parts*:

$$\mathcal{SP}_{X,\underline{\mu}} := S_{X,\underline{\mu}}/S_{X,\underline{\mu},s-1}. \tag{14.25}$$

Theorem 14.35. *(1) The $\tilde{W}(\Lambda)$ module $\mathcal{SP}_{X,\underline{\mu}}$ is semisimple of finite length.*
(2) The isotypic components of $\mathcal{SP}_{X,\underline{\mu}}$ are indexed by the points of the arrangement.
(3) Given $e^{\phi} \in P_{\underline{\mu}}(X)$, the corresponding isotypic component $F_{e^{\phi}}$ is the direct sum of the irreducible modules $N_{\underline{b},e^{\phi}}$ generated by the classes $\omega_{\underline{b},e^{\phi}}$ and indexed by the unbroken bases extracted from $X_{e^{\phi}}$.
(4) As a module over the ring $S[U_{\mathbb{C}}]$ of differential operators on $V_{\mathbb{C}}$ with constant coefficients, $N_{\underline{b},e^{\phi}}$ is free of rank 1 with generator the class $\omega_{\underline{b},e^{\phi}}$.

To simplify notation, when referring to a point $e^{\phi} \in P_{\underline{\mu}}(X)$ in subscripts we shall usually write ϕ instead of e^{ϕ}. So $F_{e^{\phi}}$ will be written as F_{ϕ}, $X_{e^{\phi}}$ as X_{ϕ}, and so on.

In what follows we shall often consider the case in which $\mu_i = 0$ for each i. In this case the subscript $\underline{\mu}$ will be omitted.

Thus we have the following canonical decomposition into isotypic components:

$$\mathcal{SP}_{X,\underline{\mu}} = \oplus_{\phi\in\tilde{P}_{\underline{\mu}}(X)}F_{\phi} \quad \text{with} \quad F_{\phi} = \oplus_{\underline{b}\in\mathcal{NB}_{X_{\phi}}}N_{\underline{b},\phi} \tag{14.26}$$

14.3.2 Two Isomorphic Modules

In this section we want to consider the $W^{\#}(\Lambda)$-module $\mathcal{L}_{X,\underline{\mu}}$ generated, in the space of tempered distributions on V, by the element $\mathcal{T}_{X,\underline{\mu}}$.

In analogy with Theorem 8.22, we have the following theorem:

Theorem 14.36. *(1) Under the Laplace transform, $\mathcal{L}_{X,\underline{\mu}}$ is mapped isomorphically onto $S_{X,\underline{\mu}}$. In other words, we get a canonical isomorphism of $\hat{\mathcal{L}}_{X,\underline{\mu}}$ with $S_{X,\underline{\mu}}$ as $\tilde{W}(\Lambda)$-modules.*

(2) $\mathcal{L}_{X,\underline{\mu}}$ *is the space of tempered distributions that are linear combinations of polynomials times* $\xi_{A,\underline{\mu}}$, $A \subset X$ *a linearly independent subset, and their translates under* Λ.

Proof. In view of Proposition 14.4 and Corollary 14.25, everything follows once we observe that as a $\tilde{W}(\Lambda)$-module, $S_{X,\underline{\mu}}$ is generated by the element $\prod_{a \in X}(1 - e^{-a-\mu_a})^{-1}$.

Remark 14.37. Equivalently $\mathcal{L}_{X,\underline{\mu}}$ can be thought of as a space of functions on Λ.

Theorem 14.36 tells us that we can transport our filtration on $S_{X,\underline{\mu}}$ to one on $\mathcal{L}_X$. By formula (14.24), in filtration degree $\leq k$ we have those distributions, or functions, that are supported in a finite number of translates of the sets $\Lambda \cap C(A)$, where A spans a lattice of rank $\leq k$.

Notice that the fact that the Laplace transform of $\mathcal{L}_X$ is an algebra means that $\mathcal{L}_X$ is closed under convolution. This can also be seen directly by analyzing the support of the elements in $\mathcal{L}_X$. Such a support is contained in a finite union of translates of the cone $C(X)$.

14.3.3 A Formula for the Partition Function $\mathcal{T}_X$

In this section we shall assume that $\mu_i = 0$ for each $i = 1, \ldots, m$.

Take the *space of polar parts* (cf. 14.25)

$$SP_X := S_X/S_{X,s-1} = \oplus_{\phi \in \tilde{P}(X)} F_\phi, \quad F_\phi = \oplus_{\underline{b} \in \mathcal{NB}_{X_\phi}} N_{\underline{b},\phi}.$$

Let us consider the element v_X in SP_X, which is the class of the function $\prod_{a \in X}(1 - e^{-a})^{-1}$. Decompose it uniquely as a sum of elements v_{X_ϕ} in F_ϕ. By Theorem 14.35, each one of these elements is uniquely expressed as a sum

$$v_{X_\phi} = \sum_{\underline{b} \in \mathcal{NB}_{X_\phi}} q_{\underline{b},\phi}\omega_{\underline{b},\phi}$$

for suitable polynomials $q_{\underline{b},\phi} \in S[U_\mathbb{C}]$. Thus

$$v_X = \sum_{\phi \in \tilde{P}(X)} \sum_{\underline{b} \in \mathcal{NB}_{X_\phi}} q_{\underline{b},\phi}\omega_{\underline{b},\phi}. \tag{14.27}$$

We are now ready to prove the main formula that one can effectively use for computing the partition function $\mathcal{T}_X$:

Theorem 14.38. *Let* Ω *be a big cell,* $B(X)$ *the box associated to* X. *For every* $x \in \Omega - B(X)$,

$$\mathcal{T}_X(x) = \sum_{\phi \in \tilde{P}(X)} e^{\langle \phi \,|\, x \rangle} \sum_{\underline{b} \in \mathcal{NB}_{X_\phi} \,|\, \Omega \subset C(\underline{b})} |det(\underline{b})|^{-1} q_{\underline{b},\phi}(-x). \tag{14.28}$$

Proof. Formula (14.27), together with the fact that $\omega_{\underline{b},\phi}$ is the class of the element $d_{\underline{b},\phi}$, implies that the generating function $\prod_{a \in X}(1 - e^{-a})^{-1}$ equals the sum $\sum_{\phi \in \tilde{P}(X)} \sum_{\underline{b} \in \mathcal{NB}_{X_\phi}} \mathfrak{q}_{\underline{b},\phi} d_{\underline{b},\phi} + K$ where $K \in S_{X,s-1}$. By Proposition 14.4, $d_{\underline{b},\phi}$ is the Laplace transform of $|\det(\underline{b})|^{-1}\xi_{\underline{b},\phi}$. The function $\xi_{\underline{b},\phi}$ equals $e^{\langle \phi \,|\, x \rangle}$ on $\Lambda \cap C(\underline{b})$ and is 0 elsewhere. Moreover, K is the Laplace transform of a distribution supported on the union of a finite number of translates of lower-dimensional cones. Under the inverse Laplace transform, applying a polynomial differential operator $\mathfrak{q}_{\underline{b},\phi}(\frac{\partial}{\partial x_1}, \ldots, \frac{\partial}{\partial x_s})$ corresponds to multiplying by $\mathfrak{q}_{\underline{b},\phi}(-x_1, \ldots, -x_s)$. We deduce that the right-hand side of ((14.28)), that for fixed Ω is a quasipolynomial, coincides with the partition function on Ω minus possibly a finite number of translates of lower-dimensional cones. Clearly, a quasipolynomial is completely determined by the values that it takes on such a set.

We now apply Theorem 13.52, saying that on $\Omega - B(X)$, the partition function $\mathcal{T}_X$ coincides with a quasipolynomial. Therefore, formula (14.28) expresses the partition function on $\Omega - B(X)$.

For an Euler–Maclaurin sum $\mathcal{T}_{X,\mu}(v)$, defined in Section 14.3.2, one can prove the analogue to Theorem 14.38:

Theorem 14.39. *Let Ω be a big cell. For any $x \in \Omega - B(X)$ we have*

$$\mathcal{T}_{X,\underline{\mu}}(x) = \sum_{\phi \in \tilde{P}_\mu(X)} e^{\langle \phi \,|\, x \rangle} \sum_{\underline{b} \in \mathcal{NB}_{X_\phi} \,|\, \Omega \subset C(\underline{b})} |\det(\underline{b})|^{-1}\mathfrak{q}_{\underline{b},\phi,\mu}(-x). \tag{14.29}$$

Remark 14.40.   (i) As in Theorem 11.35, one can deduce from formula (14.28) that on two different big cells, the function $\mathcal{T}_X$ is given by two different quasipolynomials.

(ii) In the case of the Euler–Maclaurin sums, the reader should observe that the points e^ϕ of the arrangement are not necessarily torsion points in $T_{\mathbb{C}}$, so that $\mathcal{T}_{X,\underline{\mu}}(x)$ is in general not a quasipolynomial.

14.3.4 The Generic Case

In this section we are going to analyze the situation in which the parameters μ are *generic* in the following sense. Consider, as in Section 8.1.4, the set $\mathcal{B}_X$ of subsets of $\{1, \ldots, m\}$ indexing bases extracted from $X = \{a_1, \ldots, a_m\}$. For each $\sigma \in \mathcal{B}_X$, we let $\underline{b}_\sigma$ denote the corresponding basis. Thus $\underline{b}_\sigma$ spans a sublattice $\Lambda_\sigma := \Lambda_{\underline{b}_\sigma}$ and defines a subset $T_\mu(\underline{b}_\sigma)$ of the set $P_{\underline{\mu}}(X)$ of the points of the arrangement, so that by (14.3),

$$P_{\underline{\mu}}(X) := \cup_{\sigma \in \mathcal{B}_X} T_\mu(\underline{b}_\sigma).$$

Definition 14.41. The parameters μ are called *generic* if the sets $T_{\underline{\mu}}(\underline{b}_\sigma)$ are mutually disjoint.

Notice that the parameters are generic if and only if $P_{\underline{\mu}}(X)$ consists of $\delta(X) = \sum_{\sigma \in \mathcal{B}_X} |\det(\underline{b}_\sigma)|$ points.

Recall that linearly independent elements $c_1, \dots, c_h$ of Λ define a surjective homomorphism $T_{\mathbb{C}}^\Lambda \to (\mathbb{C}^*)^h$. Its fibers are cosets of an $(s - h)$-dimensional subgroup, hence finite unions of cosets of an $(s - h)$-dimensional subtorus.

The following easy lemmas are the key facts that allow us to understand the generic case (recall that H_a is defined by the equation $e^{-a} = e^{\mu_a}$).

Lemma 14.42. *Let* $a_1, \dots, a_{r+1} \in \Lambda$ *be linearly independent elements and* μ_i *parameters. If* L *is a component of* $\cap_{i=1}^r H_{a_i}$, *then* $L \cap H_{a_{r+1}} \neq \emptyset$.

Proof. Such a component L is a coset of some subtorus Y of the ambient torus $T_{\mathbb{C}}^\Lambda := T_{\mathbb{C}}$. On Y, the character $e^{a_{r+1}}$ takes as its values all the nonzero complex numbers, so the same is true on L, and the claim follows.

Lemma 14.43. *The parameters* $\underline{\mu}$ *are generic if and only if for any linearly dependent sublist* $(c_1, \dots, c_r) \subset X$, *we have* $\cap_{j=1}^r H_{c_i} = \emptyset$.

Proof. The sufficiency of the condition is clear. To see its necessity, we can clearly assume that both $\{c_2, \dots, c_r\}$ and $\{c_1, \dots, c_{r-1}\}$ are linearly independent. It follows that both $Z := \cap_{j=2}^r H_{c_i}$ and $Z' := \cap_{j=1}^{r-1} H_{c_i}$ are subvarieties of pure codimension $r - 1$ in $T_{\mathbb{C}}$. Since c_r is linearly dependent on $c_1, \dots, c_{r-1}$, both Z and Z' are unions of cosets of the same subtorus S of codimension $r - 1$, the connected component of the identity of the kernel of all the characters e^{c_i}. Since two different cosets of S are disjoint, we immediately get that if $\cap_{j=1}^r H_{c_i}$ is nonempty, then Z and Z' have a coset, and hence a component L, in common. This is an element of the toric arrangement of codimension $r - 1$.

We can choose elements $b_r, \dots, b_s$ in X such that $\{c_1, \dots, c_{r-1}, b_r, \dots, b_s\}$ is a basis.

By the previous lemma there is a point p in $T_{\underline{\mu}}(c_1, \dots, c_{r-1}, b_r, \dots, b_s) \cap L$. Clearly, $\{c_2, \dots, c_{r-1}, c_r, b_r, \dots, b_s\}$ is also a basis and

$$p \in T_{\underline{\mu}}(c_1, \dots, c_r, b_r, \dots, b_s) \cap T_{\underline{\mu}}(c_2, \dots, c_{r-1}, b_r, \dots, b_s)$$

a contradiction.

Corollary 14.44. *Assume* $\underline{\mu}$ *generic. Let* $p\mathcal{T}$ *be an element of the arrangement associated to* X *and* $\underline{\mu}$. *Then* $X_{p\mathcal{T}}$ *is a basis relative to* $p\mathcal{T}$. *In particular, each irreducible module in* $\oplus_{k=1}^s S_{X,\mu,k}/S_{X,\underline{\mu},k-1}$ *appears with multiplicity one.*

Proof. The first part is clear from Lemma 14.43. The second then follows from Theorem 14.32.

Remark 14.45. One can verify that the length (of a composition series) of $S_{X,\underline{\mu}}$ is independent of $\underline{\mu}$. Moreover, $\underline{\mu}$ is generic if and only if all the irreducible factors of a composition series are nonisomorphic.

For each minimally dependent sublist $(a_{i_1}, \ldots, a_{i_r}) \subset X$, there is up to sign a unique primitive relation $R := \sum_{j=1}^{r} k_j a_{i_j} = 0$, $k_i \in \mathbb{Z}$ relatively prime. Take the torus $(\mathbb{C}^*)^m$ of coordinates $(\lambda_1, \ldots, \lambda_m)$. Associated to the relation R we get the character $\prod_{j=1}^{r} \lambda_{i_j}^{k_j}$ whose kernel is an $(m-1)$-dimensional subtorus H_R. The collection of such subtori is a toric arrangement in $(\mathbb{C}^*)^m$, and using Lemma 14.43 we get the following result:

Proposition 14.46. *The elements* $(\lambda_i := e^{\mu_i}) \in (\mathbb{C}^*)^m$ *corresponding to generic parameters* $\underline{\mu}$ *are the complement of the tori* H_R *as* R *varies among the minimal relations.*

Proof. Assume first that the parameters λ_i lie in one of the tori H_R, for a minimal relation $\sum_{j=1} k_j a_{i_j} = 0$, i.e. $\prod_{j=1} \lambda_{i_j}^{k_j} = 1$. We claim that the intersection $\cap_{j=1}^{r} H_{a_{i_j}}$ is not empty. In fact, consider the homomorphism $s : \mathbb{Z}^r \to \Lambda$ defined by $s((h_1, \ldots, h_r)) = \sum_{j=1} h_j a_{i_j}$. This map induces, by duality, a homomorphism $s^* : T_{\mathbb{C}} \to (\mathbb{C}^*)^r$ whose image is the $(r-1)$-dimensional torus, the kernel of the character $(\zeta_1, \ldots, \zeta_r) \mapsto \zeta_1^{k_1} \cdots \zeta_r^{k_r}$. Therefore, the point of coordinates $(\lambda_{i_1}, \ldots, \lambda_{i_r})$ lies in the image of s^*. Its preimage is thus nonempty and equals $\cap_{j=1}^{r} H_{a_{i_j}}$, giving our claim.

The converse is easy.

In the generic case we obtain a particularly simple expression for the function

$$\prod_{a \in X} \frac{1}{(1 - e^{-a - \mu_a})}.$$

Proposition 14.47. *Assume that the parameters* $\underline{\mu}$ *are generic. Then there are elements* $c_\sigma \in \mathbb{C}[\Lambda]$ *for each* $\sigma \in \mathcal{B}_X$ *such that*

$$\prod_{a \in X} \frac{1}{(1 - e^{-a - \mu_a})} = \sum_{\sigma \in \mathcal{B}_X} \frac{c_\sigma}{\prod_{i \in \sigma}(1 - e^{-a_i - \mu_i})}. \tag{14.30}$$

Proof. For each $\sigma \in \mathcal{B}_X$ set

$$D_{\sigma, \underline{\mu}} = D_\sigma := \prod_{i \notin \sigma}(1 - e^{-a_i - \mu_i}).$$

The fact that $\underline{\mu}$ is generic implies that the D_σ generate the unit ideal of $\mathbb{C}[\Lambda]$. Indeed, if $X = \underline{b}_\sigma$, then $D_\sigma = 1$, and there is nothing to prove. So assume $m > s$ and proceed by induction on m. Set $I = \{u \in X | X \setminus \{u\} \text{ spans } V\}$. Take $a \in I$ and set $Y := X \setminus \{a\}$.

By induction, the ideal generated by the D_σ's with $\sigma \subset \mathcal{B}_Y$ is the ideal $(1 - e^{-a - \mu_a})$ defining H_a. Thus it suffices to see that $\cap_{a \in I} H_a = \emptyset$. Let us show that I is a linearly dependent set. If we assume that I is independent, we can complete it to a basis $\underline{b}$. But then $X \setminus \underline{b} \subset I$, giving a contradiction. At this point Lemma 14.43 implies that $\cap_{a \in I} H_a = \emptyset$.

Now write

$$1 = \sum_{\tau \in \mathcal{B}_X} c_\tau D_\tau, \quad c_\tau \in \mathbb{C}[\Lambda]. \tag{14.31}$$

We get

$$\frac{1}{\prod_{a \in X}(1 - e^{-a_i - \mu_i})} = \frac{\sum_{\sigma \in \mathcal{B}_X} c_\sigma D_\sigma}{\prod_{a \in X}(1 - e^{-a_i - \mu_i})} = \sum_{\sigma \in \mathcal{B}_X} \frac{c_\sigma}{\prod_{i \in \sigma}(1 - e^{-a_i - \mu_i})}.$$

Set $\tilde{T}_\mu(\underline{b}_\sigma) \subset \tilde{P}_\mu(X)$, the subset corresponding to $T_\mu(\underline{b}_\sigma)$. We can now expand uniquely $c_\sigma = \sum_{\phi \in \tilde{T}_\mu(\underline{b}_\sigma)} e(\phi) u_\phi$, $u_\phi \in \mathbb{C}[\Lambda_\sigma]$, and

$$\prod_{a \in X}(1 - e^{-a - \mu_a})^{-1} = \sum_{\sigma \in \mathcal{B}_X} \sum_{\phi \in \tilde{T}_\mu(\underline{b}_\sigma)} \frac{e(\phi) u_\phi}{\prod_{i \in \sigma}(1 - e^{-a_i - \mu_{a_i}})}. \tag{14.32}$$

For each $\sigma \in B_X$ and for each $\phi \in \tilde{T}_\mu(\underline{b}_\sigma)$ set $c_\phi := u_\phi(e^\phi) \in \mathbb{C}$.

Lemma 14.48. *In* $S_{X,\mu}/S_{X,\mu,s-1}$,

$$\sum_{\sigma \in \mathcal{B}_X} \sum_{\phi \in \tilde{T}_\mu(\underline{b}_\sigma)} \frac{e(\phi) u_\phi}{\prod_{i \in \sigma}(1 - e^{-a_i - \mu_{a_i}})} \equiv \sum_{\sigma \in \mathcal{B}_X} \sum_{\phi \in \tilde{T}_\mu(\underline{b}_\sigma)} \frac{e(\phi) c_\phi}{\prod_{i \in \sigma}(1 - e^{-a_i - \mu_{a_i}})}.$$

Proof. Since for a given $\phi \in \tilde{T}_\mu(\underline{b}_\sigma)$, $c_\phi - u_\phi$ lies in $\mathbb{C}[\Lambda_\sigma]$ and vanishes in e^ϕ, it lies in the ideal of $\mathbb{C}[\Lambda_\sigma]$ generated by the elements $1 - e^{-a_i - \mu_{a_i}}$, $i \in \sigma$. Thus writing

$$u_\phi = c_\phi + \sum_{i \in \sigma} d_i(1 - e^{-a_i - \mu_{a_i}})$$

and substituting in (14.32) we get our claim.

It remains to compute the numbers c_ϕ. For this we have the following lemma:

Lemma 14.49. *For each* $\sigma \in B_X$ *and for each* ϕ *with* $e^\phi \in T_\mu(\underline{b}_\sigma)$,

$$c_\phi = D_\sigma(e^\phi)^{-1}.$$

Proof. Divide (14.31) by D_σ and apply the definition of D_τ, getting

$$D_\sigma^{-1} = \sum_{\tau \in \mathcal{B}_X} \sum_{\psi \in \tilde{T}_\mu(\underline{b}_\tau)} u_\psi e(\psi) \frac{\prod_{j \in \sigma/\tau}(1 - e^{-a_j - \mu_{a_j}})}{\prod_{i \in \tau/\sigma}(1 - e^{-a_i - \mu_{a_i}})}.$$

Notice that if $\tau \neq \sigma$, it is clear that

$$\frac{\prod_{j \in \sigma/\tau}(1 - e^{-a_j - \mu_{a_j}})}{\prod_{i \in \tau/\sigma}(1 - e^{-a_i - \mu_{a_i}})}$$

vanishes at e^ϕ. On the other hand, if $\tau = \sigma$ but $\psi \neq \phi$, then $e(\psi)$ vanishes at e^ϕ, while $e(\phi)(e^\phi) = 1$. We deduce that

$$D_\sigma(e^\phi)^{-1} = u_\phi(e^\phi) = c_\phi,$$

as desired.

Applying Theorem 14.39 in this case we get that if Ω is a big cell and $x \in \Omega - B(X)$, then

$$\mathcal{T}_{X,\underline{\mu}}(x) = \sum_{\sigma \in \mathcal{B}_X \ | \ \Omega \subset C(\underline{b}_\sigma)} \ \sum_{\phi \in \tilde{T}_{\underline{\mu}}(\underline{b}_\sigma)} \frac{e^{\langle \phi \, | \, x \rangle}}{|\det(\underline{b}_\sigma)| D_\sigma(e^\phi)}. \tag{14.33}$$

Notice that for each $v \in \Lambda$, the value $\mathcal{T}_{X,\underline{\mu}}(v)$ is a polynomial in the $\lambda_i := e^{-\mu_i}$. This suggests that we can use formula (14.33) to get an expression for $\mathcal{T}_X$ by first *clearing denominators* and then specializing the λ_i to 1.

Of course, this procedure is fairly complex, as we can illustrate in the simplest case:

Example 14.50. We take $s = 1$. In this case $\Lambda = \mathbb{Z}$. We take as X the element 1 repeated m times and we consider the parameters $\lambda_j = e^{-\mu_j}$ as variables. A simple computation implies that as a function of $\lambda_1, \ldots, \lambda_m$, $\mathcal{T}_{X,\underline{\mu}}(d)$ is the sum $s_d(\lambda_1, \ldots, \lambda_m)$ of all monomials of degree d. On the other hand, in this case the unique big cell is the positive axis and formula (14.33) gives

$$s_d(\lambda_1, \ldots, \lambda_m) = \sum_{i=1}^m \frac{\lambda_i^{m-1+d}}{\prod_{h \neq i}(\lambda_i - \lambda_h)}.$$

Notice that the right-hand side can be written as a quotient of two antisymmetric polynomials in which the numerator has degree $m - 1 + d + \binom{m-1}{2}$.

In particular, for $-m + 1 \leq d < 0$, the degree is smaller than $\binom{m}{2}$. This implies that the numerator is zero, since every antisymmetric polynomial is a multiple of the Vandermonde determinant, that has degree $\binom{m}{2}$.

More generally, in the case $s = 1$ we can always take the a_i, $i = 1, \ldots, m$ to be positive integers. In this case, set $\lambda_i = e^{-\mu_i}$ and for each i, let $\lambda_{i,1}, \ldots, \lambda_{i,a_i}$ be the set of a_i-th roots of λ_i. The points of the arrangement are the numbers $\lambda_{i,j}$'s, and we are in the generic case if and only if they are distinct. We have

$$\mathcal{T}_{X,\underline{\mu}}(d) = \sum_{i=1}^m \sum_{j=1}^{a_i} \frac{\lambda_{i,j}^{\sum_{h \neq i} a_h + d}}{a_i \prod_{h \neq i}(\lambda_{i,j}^{a_h} - \lambda_h)}.$$

14.3.5 Local Reciprocity Law

We want to discuss a second proof of the reciprocity law (formula (13.25)) based on the theory of modules that we have developed. This proof sheds some light on the basic formula (14.28),

$$\mathcal{T}_X(x) = \sum_{\phi \in \tilde{P}(X)} e^{\langle \phi \mid x \rangle} \sum_{\underline{b} \in \mathcal{NB}X_\phi \mid \mathfrak{c} \subset C(\underline{b})} \mathfrak{q}_{\underline{b},\phi}(-x),$$

proving that in fact, we have a reciprocity law for the summands of this sum.

Let us start by observing that the automorphism $a \mapsto -a$ of Λ induces the automorphism τ of $\mathbb{C}[\Lambda]$ and of $\tilde{W}(\Lambda)$ mapping $e^a \mapsto e^{-a}$, $D_u \mapsto -D_u$. Since $\tau(1 - e^{-a}) = 1 - e^a = -e^a(1 - e^{-a})$, we get that τ extends to an automorphism of S_X.

Clearly τ is compatible with the filtration by polar order and induces a linear isomorphism of $\mathcal{SP}_X$. In fact, this is even semilinear with respect to the module structure and the same automorphism on $\tilde{W}(\Lambda)$.

Geometrically, τ gives a permutation of order two of the elements of the toric arrangement. In particular, it permutes the points of the arrangement mapping a point e^ϕ to the point $e^{-\phi}$. Thus τ permutes the isotypic components of $\mathcal{SP}_X$.

Also,

$$\tau\left(\prod_{a \in X} \frac{1}{1 - e^{-a}} \right) = (-1)^{|X|} e^{-\sum_{a \in X} a} \prod_{a \in X} \frac{1}{1 - e^{-a}}.$$

Apply τ to formula (14.27) to obtain

$$(-1)^{|X|} e^{-\sum_{a \in X} a} v_X = \sum_{\phi \in \tilde{P}(X)} \sum_{\underline{b} \in \mathcal{NB}X_\phi \mid \Omega \subset C(\underline{b})} \tau(\mathfrak{q}_{\underline{b},\phi}) \tau(\omega_{\underline{b},\phi}).$$

The main remark is the following:

Lemma 14.51. *For any point e^ϕ of the arrangement,*

$$\tau(\omega_{\underline{b},\phi}) = (-1)^s \omega_{\underline{b},-\phi}.$$

Proof. Recall that $\omega_{\underline{b},\phi}$ is the class of

$$\frac{\sum_{\lambda \in R_{\underline{b}}} e^{-\langle \phi \mid \lambda \rangle} e^\lambda}{|\det(\underline{b})| \prod_{a \in \underline{b}}(1 - e^{-a})}.$$

Computing, we get

$$\tau\left[\frac{\sum_{\lambda \in R_{\underline{b}}} e^{-\langle \phi \mid \lambda \rangle} e^\lambda}{\prod_{a \in \underline{b}}(1 - e^{-a})} \right] = \frac{(-1)^s e^{-\sum_{a \in \underline{b}} a} \sum_{\lambda \in R_{\underline{b}}} e^{-\langle -\phi \mid -\lambda \rangle} e^{-\lambda}}{\prod_{a \in \underline{b}}(1 - e^{-a})}.$$

When λ runs through a set of coset representatives of $\Lambda/\Lambda_{\underline{b}}$, so does $-\lambda$. It follows that the class of

$$\frac{\sum_{\lambda \in R_{\underline{b}}} e^{-\langle -\phi \mid -\lambda \rangle} e^{-\lambda}}{|\det(\underline{b})| \prod_{a \in \underline{b}}(1 - e^{-a})}$$

equals $\omega_{\underline{b},-\phi}$. This class is an eigenvector of eigenvalue 1 for $e^{-\sum_{a \in \underline{b}} a}$, and the claim follows.

Observe that as polynomials, $\tau(\mathfrak{q}_{\underline{b},\phi})(x) = \mathfrak{q}_{\underline{b},\phi}(-x)$, and that multiplication by $e^{-\sum_{x \in X} x} = e^{-2\rho_X}$ corresponds under the Laplace transform to translation by $-2\rho_X$. Thus we obtain the following theorem:

Theorem 14.52 (Local reciprocity).

$$e^{\phi}\mathfrak{q}_{\underline{b},\phi}(x) = (-1)^{|X|-s}e^{-\phi}\mathfrak{q}_{\underline{b},-\phi}(-x - 2\rho_X). \tag{14.34}$$

It is clear that from this, and (14.28), one obtains formula (13.25).

15

Cohomology of Toric Arrangements

In this chapter we compute the cohomology, with complex coefficients, of the complement of a toric arrangement. A different approach is due to Looijenga [74].

15.1 de Rham Complex

15.1.1 The Decomposition

In this chapter we are going to use the notation of the previous one. Thus we take a real vector space V, a lattice $\Lambda \subset V$ spanning V, a list $X = (a_1, \ldots, a_m)$ of vectors in Λ spanning Λ, and a list $(\mu_1, \ldots, \mu_m) \in \mathbb{C}^m$. In analogy with what we have done in Chapter 10, in this chapter we are going to explain how the results on the structure of $S_{X,\mu}$ as a $\tilde{W}(\Lambda)$-module can be used to compute the cohomology with real coefficients of the complement $\mathcal{A}$ in $T_{\mathbb{C}}$ of the hypersurface $D_{X,\mu}$ of equation $\prod_{a \in X}(1 - e^{-a-\mu_a})$.

The cohomology of $\mathcal{A}$ can be computed as the cohomology of the algebraic de Rham complex $(\Omega^\bullet, d)$, d being the usual de Rham differential.

The description of $\Omega^\bullet$ goes as follows. Consider for each $a \in \Lambda$ the one form $d \log e^a$. Set Γ equal to the graded subalgebra of $\Omega^\bullet$ generated by these forms, for $a \in \Lambda$. If we choose a basis $\{\theta_1, \ldots, \theta_s\}$ of Λ and we set $x_i := e^{\theta_i}$, we can identify Γ with the exterior algebra $\bigwedge(x_1^{-1}dx_1, \ldots, x_s^{-1}dx_s)$. Notice that for each $i = 1, \ldots, s$, we have $x_i^{-1}dx_i = d\theta_i = d \log e^{\theta_i}$.

With this notation, $\Omega^\bullet$ is isomorphic, as a graded algebra, to $S_{X,\mu} \otimes \Gamma$, where $S_{X,\mu}$ is assigned degree zero and Γ is an algebra of closed differential forms on $T_{\mathbb{C}}$ mapping isomorphically onto $H^*(T_{\mathbb{C}}, \mathbb{C})$.

We can filter $\Omega^\bullet$ by setting $\Omega^\bullet_k = S_{X,\mu,k} \otimes \Gamma$, for $k = 0, \ldots, s$.

The fact that $S_{X,\mu,k}$ is a $\tilde{W}(\Lambda)$ module easily implies that the differential d preserves $\Omega^\bullet_k$ so we get a filtration of $(\Omega^\bullet, d)$ by subcomplexes.

Proposition 15.1. *For each $k = 1, \ldots, s$ the exact sequence of complexes*

C. De Concini and C. Procesi, *Topics in Hyperplane Arrangements, Polytopes and Box-Splines*, Universitext, DOI 10.1007/978-0-387-78963-7_15,
© Springer Science+Business Media, LLC 2010

$$0 \to \Omega^\bullet_{k-1} \to \Omega^\bullet_k \to \Omega^\bullet_k/\Omega^\bullet_{k-1} \to 0$$

splits.

Proof. Using the notation of Corollary 14.33, we get that a complement to $S_{X,\underline{\mu},k-1}$ in $S_{X,\underline{\mu},k}$ is given by the direct sum

$$\bigoplus_{p\mathcal{T},\dim\,\mathcal{T}=s-k}\left(\bigoplus_{\underline{b}\in\mathcal{NB}_{X_{p\mathcal{T}}}} A_{p\mathcal{T}}d_{\underline{b},\phi}\right),$$

where $A_{p\mathcal{T}} = \mathbb{C}[\frac{\partial}{\partial x_1},\dots,\frac{\partial}{\partial x_k},x_{k+1}^{\pm 1},\dots,x_s^{\pm 1}]$ for a choice of coordinates x_i with the property that $\mathcal{T}$ has equations $x_i = 1$, $i = 1,\dots,k$. By Remark 14.34, $d_{\underline{b},\phi}$ depends only on $x_1,\dots,x_k$. This immediately implies that $A_{p\mathcal{T}}d_{\underline{b},\phi}\otimes\Gamma$ is a subcomplex of $\Omega^\bullet_k$, proving our claim.

Remark 15.2. Notice that the definition of $A_{p\mathcal{T}}$ depends on the choice of the coordinates x_i, which is quite noncanonical. Hence our splitting is also noncanonical.

We now recall that for a space of the arrangements $p\mathcal{T}$ of codimension k, the image of $\oplus_{\underline{b}\in\mathcal{NB}_{X_{p\mathcal{T}}}}A_{p\mathcal{T}}d_{\underline{b},\phi}$ in $S_{X,\mu,k}/S_{X,\mu,k-1}$ equals the isotypic component $F_{p\mathcal{T}}$. We deduce that the cohomology of the complex $\Omega^\bullet_k/\Omega^\bullet_{k-1}$ is the direct sum of the cohomology of the complexes $F_{p\mathcal{T}}\otimes\Gamma$. Thus the cohomology of $\mathcal{A}$ can be expressed as a direct sum of *local contributions* coming from each of the elements of our arrangement, the cohomologies of the complexes $F_{p\mathcal{T}}\otimes\Gamma$.

The natural Γ-algebra structure on the de Rham cohomology of $\Omega^\bullet$ induces the $H^*(T_\mathbb{C},\mathbb{C})$ algebra structure on $H^*(\mathcal{A},\mathbb{C})$ determined by the inclusion $\mathcal{A}\subset T_\mathbb{C}$.

The Γ-module structure on $F_{p\mathcal{T}}\otimes\Gamma$ clearly induces an $H^*(T_\mathbb{C},\mathbb{C})$-module structure on its cohomology. If $\mathcal{T}$ has equations $x_i = 1$, $i = 1,\dots,k$, the k form $d\log(x_1)\wedge\cdots\wedge d\log(x_k)$ depends only on $\mathcal{T}$ up to sign. With the notation of Section 14.3.1, the vector space $V_{p\mathcal{T}}d\log(x_1)\wedge\cdots\wedge d\log(x_k)$ consists of cocycles in $F_{p\mathcal{T}}\otimes\Gamma$. We denote its image in cohomology by $\Theta_{p\mathcal{T}}$.

Let us observe that if $\underline{b} = \{b_1,\dots,b_k\}$ is a basis relative to $p\mathcal{T}$, the element $\omega_{\underline{b},\phi}d\log(x_1)\wedge\cdots\wedge d\log(x_k)$ is a nonzero constant multiple of the element $\omega_{\underline{b},\phi}db_1\wedge\cdots\wedge db_k$.

We give $H^*(\mathcal{T},\mathbb{C})\otimes\Theta_{p\mathcal{T}}$ the $H^*(T_\mathbb{C},\mathbb{C})$ module structure induced by the inclusion $\mathcal{T}\subset T_\mathbb{C}$.

Proposition 15.3. *(1) There is an isomorphism of graded $H^*(T_\mathbb{C},\mathbb{C})$ modules between $H^*(F_{p\mathcal{T}}\otimes\Gamma)$ and $H^*(\mathcal{T},\mathbb{C})\otimes\Theta_{p\mathcal{T}}$.*
(2) The quotient map from $V_{p\mathcal{T}}d\log(x_1)\wedge\cdots\wedge d\log(x_k)$ to $\Theta_{p\mathcal{T}}$ is an isomorphism.

Proof. Choosing coordinates as in the definition of $A_{p\mathcal{T}}$, we have that the complex $F_{p\mathcal{T}} \otimes \Gamma$ decomposes as the tensor product of two complexes,

$$\mathbb{C}[x_{k+1}^{\pm 1}, \ldots, x_s^{\pm 1}] \bigwedge (d\log(x_{k+1}), \ldots, d\log(x_s)),$$

$$\mathbb{C}[\frac{\partial}{\partial x_1}, \ldots, \frac{\partial}{\partial x_k}] V_{p\mathcal{T}} \bigwedge (d\log(x_1), \ldots, d\log(x_k)).$$

The first complex is the de Rham complex of the torus $\mathcal{T}$. Its cohomology is the exterior algebra $\bigwedge(d\log(x_{k+1}), \ldots, d\log(x_s))$. The other complex is isomorphic to the second type of complex studied in Section 10.1.1 and it has only cohomology in dimension k that is identified isomorphically with the image of $V_{p\mathcal{T}}$. This proves both claims.

Summarizing we have proved the following theorem:

Theorem 15.4. *As an* $H^*(T_{\mathbb{C}}, \mathbb{C})$*-module,* $H^*(\mathcal{A}, \mathbb{C})$ *is isomorphic to*

$$\bigoplus_{p\mathcal{T} \in \mathcal{H}_{X,\mu}} H^*(\mathcal{T}, \mathbb{C}) \Theta_{p\mathcal{T}}.$$

If $p = e^{\phi}$*, then* $\Theta_{p\mathcal{T}}$ *has a basis given by the classes* $\omega_{\underline{b},\phi} db_1 \wedge \cdots \wedge db_k$ *as* $\underline{b} = \{b_1, \ldots b_k\}$ *varies among the unbroken bases relative to* $p\mathcal{T}$*.*

Remark 15.5. Although the decomposition is noncanonical, the filtration of $\Omega^{\bullet}$ induces a canonical filtration in cohomology whose associated graded space is described by the above theorem.

From the previous formula one computes easily the Poincaré polynomial of $\mathcal{A}$.

In particular, the Euler characteristic of $\mathcal{A}$ is given by

$$\chi(\mathcal{A}) = \sum_{p \in P_{\underline{\mu}}(X)} \dim V_p.$$

15.2 The Unimodular Case

Our next task is to compute the algebra structure of cohomology. This we shall do in the unimodular case. In this case we shall also prove formality.

15.2.1 A Basic Identity

We start with a basic formal identity:

$$1 - \prod_{i=1}^{n} x_i = \sum_{I \subsetneq \{1,2,\ldots,n\}} \prod_{i \in I} x_i \prod_{j \notin I} (1 - x_j). \tag{15.1}$$

The proof is by induction on n.

We split the sum into three terms: $I = \{1, \ldots, n-1\}$, $I \subsetneq \{1, \ldots, n-1\}$, and finally $n \in I$. We get

$$\prod_{i=1}^{n-1} x_i(1 - x_n) + \left(1 - \prod_{i=1}^{n-1} x_i\right)(1 - x_n) + x_n\left(1 - \prod_{i=1}^{n-1} x_i\right) = 1 - \prod_{i=1}^{n} x_i.$$

Formula (15.1) implies

$$\sum_{I \subsetneq \{1,2,\ldots,n\}} \frac{1}{(1 - \prod_{i=1}^{n} x_i)} \prod_{i \in I} \frac{x_i}{(1 - x_i)} = \frac{1}{\prod_{i=1}^{n}(1 - x_i)}.$$

We want to interpret this formula as an identity between certain differential forms.

Set, for $i = 1, \ldots, n$, $\omega_i := d\log(1 - x_i)$, $\psi_i := d\log x_i$. Also for each $h = 0, \ldots, n$, set $\theta^{(h)} := d\log(1 - \prod_{i=1}^{h} x_i^{-1} \prod_{j=h+1}^{n} x_j)$. If we take a proper subset $I = \{i_1 < \cdots < i_t\}$ in $\{1, \ldots, n\}$ and let $J = \{j_1 < \cdots < j_{n-t}\}$ be its complement, we can then define the n-forms:

$$\Phi_I^{(h)} = (-1)^{s_I}\omega_{i_1} \wedge \cdots \wedge \omega_{i_t} \wedge \psi_{j_1} \wedge \cdots \wedge \psi_{j_{n-t-1}} \wedge \theta^{(h)}$$

and

$$\Psi_I = (-1)^{s_I}\omega_{i_1} \wedge \cdots \wedge \omega_{i_t} \wedge \psi_{j_1} \wedge \cdots \wedge \psi_{j_{n-t}}$$

with s_I equal to the parity of the permutation $(i_1, \ldots, i_t, j_1, \ldots, j_{n-t})$.

Proposition 15.6. *For each $0 \leq h \leq n$, the n-form $\omega_1 \wedge \cdots \wedge \omega_n$ can be written as a linear combination with integer coefficients of the n-forms $\Phi_I^{(h)}$ and Ψ_I as I varies over the proper subsets of $\{1, \ldots, n\}$.*

Proof. We first deal with the case $h = 0$. In this case, computing, we get

$$\Phi_I^{(0)} = (-1)^{|I|}\prod_{i \in I} \frac{x_i}{(1 - x_i)} \frac{1}{(1 - \prod_{i=1}^{n} x_i)} dx_1 \wedge dx_2 \wedge \cdots \wedge dx_n. \qquad (15.2)$$

Thus our identity (15.1) can be translated into:

$$\sum_{I \subsetneq \{1,2,\ldots,n\}} (-1)^{|I|+n}\Phi_I^{(0)} = \omega_1 \wedge \cdots \wedge \omega_n, \qquad (15.3)$$

proving our claim.

In the general case, let us observe that

$$d\log(1 - x^{-1}) = d\log\frac{x-1}{x} = d\log(1 - x) - d\log x.$$

Therefore, the substitution of x_i with x_i^{-1} for $i = 1, \ldots, s$ corresponds in formula (15.3) to substituting ω_i with $\omega_i - \psi_i$. As a result, we get a formula expressing $\omega_1 \wedge \cdots \wedge \omega_n$ as a linear combination of the forms $\Phi_I^{(h)}$ and Ψ_I with integral coefficients.

15.2.2 Formality

Recall that in Definition 2.51 we have introduced the notion of unimodular list of vectors in Λ.

This notion can in fact be defined, in case of a list of vectors X in a lattice Λ, by either of the following equivalent conditions.

Proposition 15.7. *Given a list of vectors X in Λ, the following are equivalent:*

(1) X is unimodular.

(2) Any sublist Y of X spans a direct summand in Λ.

(3) For any list μ of complex numbers and sublist Y of X, the variety of equation $1 - e^{-a - \mu_a} = 0$, $a \in Y$, is connected.

(4) For any sublist Y of X, the variety of equation $1 - e^{-a} = 0$, $a \in Y$, is connected.

(5) Given linearly independent elements $c_1, \ldots, c_k$ in X and another element $c_0 \in X$ dependent on them, the linear relation $c_0 = \sum_{i=1}^{k} n_i c_i$ has the coefficients $n_i \in \{0, 1, -1\}$.

(6) $d(X) = \delta(X)$.

(7) The two spaces $D(X)$ and $DM(X)$ coincide.

Proof. The proof is left to the reader.

Observe that in the unimodular case, given a linearly independent sublist $\underline{b} = (b_1, \ldots, b_k)$ in X there is a unique element $p\mathcal{T}$ of $\mathcal{H}_{X,\mu}$, of codimension k such that $\underline{b}$ is contained in $X_{p\mathcal{T}}$. We have a corresponding cohomology class represented by the differential form

$$\frac{1}{\prod_{i=1}^{k}(1 - e^{-b_i - \mu_{b_i}})} db_1 \wedge \cdots \wedge db_k. \tag{15.4}$$

Notice that

$$d \log(1 - e^{-a - \mu}) = \frac{e^{-a - \mu}}{1 - e^{-a - \mu}} da = \frac{da}{1 - e^{-a - \mu}} - da.$$

Let us start with a formal construction. Define an exterior algebra in the following generators: a generator λ_a for every $a \in X$ and a generator ξ_χ for every element $\chi \in \Lambda$.

We then impose a set of relations:

(i) $\xi_{\chi_1 + \chi_2} = \xi_{\chi_1} + \xi_{\chi_2}$, for all $\chi_1, \chi_2 \in \Lambda$.

(ii) Take elements $\chi_1, \ldots, \chi_s \in \Lambda$ and $(b_1, \ldots, b_t)$ a sublist of X. If the elements $\chi_1, \ldots, \chi_s, b_1, \ldots, b_t$ are linearly dependent, the product of the elements ξ_{χ_i} and λ_{b_j} is zero.

We come now to the main relations, that generalize the ones for hyperplanes.

(iii) Take a minimal linearly dependent sublist $(b_0, \ldots, b_k)$ of X. By the unimodularity of X, up to reordering the elements, we may assume that $b_0 = \sum_{i=1}^{h} b_i - \sum_{j=h+1}^{k} b_j$. If furthermore

$$e^{\mu_{b_0}} = \prod_{i=1}^{h} e^{\mu_{b_i}} \prod_{j=h+1}^{k} e^{-\mu_{b_j}},$$

we take the formulas obtained in Proposition 15.6, that express the form $\omega_1 \wedge \cdots \wedge \omega_h$ in terms of the forms $\Phi_I^{(h)}$ and Ψ_I and substitute ψ_i with ξ_{χ_i} and ω_i with λ_{a_i}. We get then a formal expression, that we impose as a new relation.

We call $\mathcal{H}$ the algebra defined by the given generators modulo the relations (i), (ii), (iii).

Consider the subalgebra H in $\Omega^\bullet$ generated by the 1-forms $d\log \chi$, $d\log(1 - e^{-a - \mu_a})$ with $\chi \in \Lambda$, $a \in X$. Then H clearly consists of closed forms so that we obtain an algebra homomorphism

$$f : H \to H^*(\mathcal{A}, \mathbb{C}).$$

It is also easy to verify, using Proposition 15.6, that these elements satisfy the previous relations, so that we have an algebra homomorphism $g : \mathcal{H} \to H$ given by

$$g(\lambda_a) := d\log(1 - e^{-a - \mu_a}), \quad g(\xi_\chi) := d\log \chi.$$

Theorem 15.8. *The homomorphisms g, f are isomorphisms.*

Proof. The assumption that X is unimodular implies, by an easy application of formula (15.4) and the description of cohomology of $\mathcal{A}$, that f is surjective.

The fact that g is surjective is clear from its definition. So, in order to prove our claim, it suffices to see that fg is injective.

We apply the theory of no broken circuits. The first set of relations implies that the subalgebra generated by the elements ξ_χ is a homomorphic image of the exterior algebra Γ. So we can consider $\mathcal{H}$ as a Γ-module.

For a fixed component $p\mathcal{T}$ we define the subspace $\mathcal{W}_{p\mathcal{T}} \subset \mathcal{H}$ spanned by the monomials $\lambda_{\underline{b}} = \prod_{a \in \underline{b}} \lambda_a$ as $\underline{b}$ runs among the bases relative to $p\mathcal{T}$.

The second set of relations also implies that $\mathcal{W}_{p\mathcal{T}}$ is annihilated by the elements ξ_χ if the character e^χ is constant on $p\mathcal{T}$.

Given a monomial m in the generators λ_a and ξ_χ, we define its weight as the number of factors of m of the first type λ_a. Notice that the weight of $\lambda_{\underline{b}} \in \mathcal{W}_{p\mathcal{T}}$ equals the codimension of $p\mathcal{T}$. Then, whenever $\underline{b}$ is a broken circuit, the relations of the third type allow us to replace it by a product of elements with lower weight or lower in the lexicographical order. This implies that modulo elements of smaller weight, any element in $\mathcal{W}_{p\mathcal{T}}$ can be written as a linear combination of elements $\lambda_{\underline{b}}$ with $\underline{b}$ unbroken relative to an element of the arrangement less than or equal to $p\mathcal{T}$.

From this our claim follows immediately using Theorem 15.4.

Remark 15.9. Our result shows in particular that in $H^*(\mathcal{A}, \mathbb{C})$, the algebraic relations between the generating forms $d \log(1 - e^{-a-\mu_a})$ and $d \log \xi$ resemble, but are more complicated than, the relations of Orlik–Solomon in the case of hyperplane arrangements (cf. formula (10.3)).

Polar Parts

In this chapter we return to the theory of Chapter 14, using all of its notation, and complete the theory of the partition function.

16.1 From Volumes to Partition Functions

16.1.1 $DM(X)$ as Distributions

In order to study $DM(X)$ as distributions, it is better to work with its complexified form $DM_{\mathbb{C}}(X) = DM(X) \otimes \mathbb{C}$. We apply the methods of Section 5.3 and in particular Theorem 5.29. This implies that

$$DM_{\mathbb{C}}(X) = \oplus_{\phi \in \tilde{P}(X)} e^{\langle \phi \,|\, v \rangle} D_{\phi}, \tag{16.1}$$

where D_{ϕ} is the space of polynomials f such that $e^{\langle \phi \,|\, v \rangle} f \in DM_{\mathbb{C}}(X)$.

Recall that for any list of vectors Y, we have introduced in Definition 11.2 the *differentiable* Dahmen–Micchelli space $D(Y)$, which has also a complexified form $D_{\mathbb{C}}(Y)$. We have the following result, [37]:

Proposition 16.1. *The space D_{ϕ} equals the differentiable Dahmen–Micchelli space $D_{\mathbb{C}}(X_p)$, $p = e^{\phi}$.*

Proof. We have to understand when a polynomial f is such that $e^{\langle \phi \,|\, v \rangle} f$ satisfies the difference equations given by the ideal J_X.

By Corollary 5.30, the function $e^{\langle \phi \,|\, v \rangle} f$ satisfies the difference equation $\nabla_Y e^{\langle \phi \,|\, v \rangle} f = 0$ if and only if f satisfies the twisted difference equation $\nabla_Y^{-\phi} f = \prod_{a \in Y} \nabla_a^{-\phi} f = 0$. On polynomials, the operator $\nabla_a = 1 - \tau_a$ is nilpotent, so that

$$\nabla_a^{-\phi} = 1 - e^{-\langle \phi \,|\, a \rangle} \tau_a = \nabla_a + (1 - e^{-\langle \phi \,|\, a \rangle}) \tau_a$$

is invertible as soon as $e^{-\langle \phi \,|\, a \rangle} \neq 1$.

C. De Concini and C. Procesi, *Topics in Hyperplane Arrangements, Polytopes and Box-Splines*, Universitext, DOI 10.1007/978-0-387-78963-7_16,
© Springer Science+Business Media, LLC 2010

Thus we split X in $X_p \cup (X \setminus X_p)$ and see that for a polynomial f, $\nabla_Y^{-\phi} f = 0$ if and only if $\nabla_{Y \cap X_p} f = 0$. If Y is a cocircuit of X, then $Y \cap X_p$ contains a cocircuit of X_p. Thus $f \in DM_{\mathbb{C}}(X)$ if and only if $\nabla_Z q = 0$ for all cocircuits Z of X_p.

Next observe that if $v \in V$, the difference operator ∇_v and the differential operator ∂_v satisfy $\nabla_v = T_v \partial_v$ with $T_v = \frac{1 - e^{-\partial_v}}{\partial_v}$. The operator T_v is invertible on the space of polynomials and commutes with all the ∇_a. Hence $\nabla_Y = A D_Y$ where A is invertible on polynomials. The equations $\nabla_Y f = 0$ are equivalent to $D_Y f = 0$ for polynomials. Thus we obtain that $e^{\langle \phi \,|\, v \rangle} f \in DM_{\mathbb{C}}(X)$ if and only if $f \in D_{\mathbb{C}}(X_p)$.

In particular, the contribution of the point 1 to $DM_{\mathbb{C}}(X)$ is the space of polynomials $D_{\mathbb{C}}(X)$. In the unimodular case, $DM_{\mathbb{C}}(X) = D_{\mathbb{C}}(X)$. In general, we also have the contribution of the other points, which will be denoted by

$$E(X) := \oplus_{p \in P(X),\ p \neq 1} e^{\langle \phi \,|\, v \rangle} D_{\mathbb{C}}(X_p). \tag{16.2}$$

We want now to explain the nature of $DM_{\mathbb{C}}(X)$ as Fourier coefficients of distributions on the compact torus T (whose character group is Λ), supported on the points of the arrangement. For this we consider the space $D_{\mathbb{C}}(X)$ (or $D_{\mathbb{C}}(X_p)$) as a space of polynomial differential operators on U and hence also on T with constant (complex) coefficients.

Let us normalize the Haar measure on T to be of total mass 1. This allows us to identify generalized functions on T and distributions on T. Under this identification, call $\widehat{DM}(X)$ the space of distributions on T of which $DM_{\mathbb{C}}(X)$ gives the Fourier coefficients. By the previous analysis we are reduced to studying the space $e^{\langle \phi \,|\, v \rangle} D_{\mathbb{C}}(X_p)$.

If $p = e^{\phi} \in T$, the delta distribution δ_p at p has as Fourier coefficients $c_\lambda = e^{-\langle \phi \,|\, \lambda \rangle}$. If $p(v)$ is a polynomial differential operator on U we have that

$$\langle p(v)\delta_p \,|\, e^{-\langle \phi \,|\, \lambda \rangle} \rangle = \langle \delta_p \,|\, p(-v) e^{-\langle \phi \,|\, \lambda \rangle} \rangle = p(\lambda) e^{-\langle \phi \,|\, \lambda \rangle}.$$

We deduce the following statement:

Proposition 16.2. $\widehat{DM}(X)$ *is the direct sum of the spaces of distributions* $D_{\mathbb{C}}(X_p)\delta_p$ *for* $p \in P(X)$.

In particular, we see that $\widehat{DM}(X)$ is supported on the finite set $P(X)$.

16.1.2 Polar Parts

Let us denote by $\tilde{L}_\phi$ the algebra of germs of holomorphic functions around the point $e^{\phi} \in T_{\mathbb{C}}$ and by L_ϕ the algebra of germs of holomorphic functions around the point $\phi \in U_{\mathbb{C}}$. Denote also by $\tilde{\mathcal{W}}_\phi$ the algebra of differential operators (on $\tilde{L}_\phi$) with coefficients in $\tilde{L}_\phi$. Similarly, for $\mathcal{W}_\phi$.

Set $\mathcal{L}_{X,\phi} := \tilde{L}_\phi[\prod_{a \in X}(1 - e^{-a})^{-1}]$ and $\mathcal{R}_{X_\phi} := L_\phi[\prod_{a \in X_\phi}(a - \langle a \,|\, \phi \rangle)^{-1}]$.

The map $x \mapsto e^x$ induces an isomorphism between the algebras $\tilde{L}_\phi$ and L_ϕ and between the algebras $\tilde{\mathcal{W}}_\phi$ and $\mathcal{W}_\phi$.

Under this isomorphism, given $a \in X$, the function $1 - e^{-a}$ is an invertible element of L_ϕ unless $a \in X_\phi$. In this case, $\langle 1 - e^{-a} \,|\, x \rangle = \langle a \,|\, x - \phi \rangle F_a(x)$ with $F_a(\phi) = 1$. Thus we deduce an isomorphism between $\mathcal{L}_{X,\phi}$ and $\mathcal{R}_{X_\phi}$, compatible with the action of the isomorphic algebras $\tilde{\mathcal{W}}_\phi$ and $\mathcal{W}_\phi$ and preserving the filtration by polar order.

We have an induced isomorphism of modules of polar parts:

$$k_\phi : \mathcal{L}_{X,\phi}/\mathcal{L}_{X,\phi,s-1} \to \mathcal{R}_{X_\phi}/\mathcal{R}_{X_\phi,s-1}.$$

For every $a \in X$ set $\mu_a := -\langle a \,|\, \phi \rangle$ and consider the algebra $R_{X_\phi,\underline{\mu}}$. By an obvious extension of Theorem 10.9, the map:

$$R_{X_\phi,\underline{\mu}}/(R_{X_\phi,\underline{\mu}})_{s-1} \to \mathcal{R}_{X_\phi}/(\mathcal{R}_{X_\phi})_{s-1}$$

is also an isomorphism of $W(V)$-modules, where $W(V) \subset \mathcal{W}_\phi$.

Now, restricting to a neighborhood of ϕ, we have the inclusion $S_X \subset \mathcal{L}_{X,\phi}$. So when we compose with the previous isomorphisms, we obtain a map

$$\gamma_\phi : S_X/S_{X,s-1} \to R_{X_\phi,\underline{\mu}}/(R_{X_\phi,\underline{\mu}})_{s-1}.$$

In $R_{X_\phi,\underline{\mu}}/(R_{X_\phi,\underline{\mu}})_{s-1}$ let us take the elements

$$u_{\underline{b}} = \left[\frac{1}{\prod_{a \in \underline{b}} \langle a \,|\, x - \phi \rangle} \right].$$

We have the following result:

Proposition 16.3. *For every $\psi \in \tilde{P}(X)$ and for every basis $\underline{b}$ extracted from X_ψ,*

$$\gamma_\phi(\omega_{\underline{b},e^\psi}) = \begin{cases} u_{\underline{b}} & \text{if } \psi = \phi, \\ 0 & \text{if } \psi \neq \phi. \end{cases}$$

Proof. To prove the proposition we need to study the expansion of $d_{\underline{b},\psi}(e^x)$ around ϕ. We get

$$d_{\underline{b},\psi}(e^x) := \frac{e(\psi)(e^x)}{\prod_{a \in \underline{b}}(1 - e^{-\langle a \,|\, x \rangle + \langle a \,|\, \psi \rangle})} = \frac{1}{\prod_{a \in \underline{b}} \langle a \,|\, x - \psi \rangle} F(x),$$

where $F(x)$ is holomorphic around ϕ.

If $\psi = \phi$ and $\underline{b}$ is an unbroken basis extracted from X_ϕ, we claim that $F(\phi) = 1$. In fact, $e(\phi)(e^x)$ is holomorphic and at $x = \phi$ takes the value 1, while for each factor of the denominator we use the power series expansion $1 - e^{-t} = t(1 - t/2 + \dots)$.

In the case $\psi \neq \phi$ the denominator is holomorphic around ϕ, while by the definition of $e(\psi)$, the numerator vanishes at ϕ.

With the notation of formula (14.26), we get the following corollary:

Corollary 16.4. γ_ϕ *factors through the projection onto the isotypic component* F_ϕ *and induces an isomorphism*

$$j_\phi : F_\phi \to R_{X_\phi,\underline{\mu}}/(R_{X_\phi,\underline{\mu}})_{s-1}$$

of $S[U_\mathbb{C}]$*-modules.*

Proof. It is clear that γ_ϕ is a module map with respect to the algebra $S[U_\mathbb{C}]$ of differential operators with constant coefficients, and the two modules are free over this algebra. By the previous proposition γ_ϕ maps the basis elements in the isotypic components $F_\psi, \psi \neq 0$, to 0 and those of F_ϕ to basis elements of $R_{X_\phi,\underline{\mu}}/(R_{X_\phi,\underline{\mu}})_{s-1}$.

We need now to compare the action of $\mathbb{C}[\Lambda]$ on F_ϕ with that of $S[V]$ on $R_{X_\phi,\underline{\mu}}/(R_{X_\phi,\underline{\mu}})_{s-1}$ under the map j_ϕ. For this, observe that the elements $v - \langle v \,|\, \phi \rangle$ with $v \in V$ are locally nilpotent on $R_{X_\phi,\underline{\mu}}/(R_{X_\phi,\underline{\mu}})_{s-1}$, while by Remark 14.10, the elements $1 - e^{\lambda - \langle \lambda \,|\, \phi \rangle}$ are locally nilpotent on F_ϕ. Thus these two actions extend to the completion of $S[V]$ at the maximal ideal at the point ϕ and the completion of $\mathbb{C}[\Lambda]$ at the maximal ideal at the point p. Furthermore, reasoning as in Proposition 5.23, we get an isomorphism between the two completions that is compatible with the module isomorphism.

16.1.3 A Realization of $\mathbb{C}[\Lambda]/J_X$

Let us take the decomposition given by formula (14.26), $\mathcal{S}P_X = \oplus_{\phi \in \tilde{P}(X)} F_\phi$ into isotypic components of $\mathcal{S}P_X$ considered as a $\tilde{W}(\Lambda)$-module.

The element v_X, which is the class in $\mathcal{S}P_X$ of the generating function $\prod_{a \in X}(1 - e^{-a - \mu_a})^{-1}$, decomposes into a sum of local elements $v_{X_\phi} \in F_\phi$.

We consider next the $\mathbb{C}[\Lambda]$-submodules $\mathcal{Q}_X$ and Q_{X_ϕ} generated in $\mathcal{S}P_X$ by v_X and v_{X_ϕ} respectively. We can now apply Proposition 16.1 and Corollary 16.4 in order to reduce the computation to Theorem 11.20.

Theorem 16.5. *The annihilator of* v_X *is the ideal* J_X. *So we get that the* $\mathbb{C}[\Lambda]$*-modules* $\mathcal{Q}_X$ *and* $\mathbb{C}[\Lambda]/J_X$ *are isomorphic.*

More generally, we obtain a canonical commutative diagram of isomorphisms, compatible with all the identifications made:

$$
\begin{array}{ccc}
\mathbb{C}[\Lambda]/J_X & \xrightarrow{\;\cong\;} & \oplus_{\phi \in \tilde{P}(X)} A_{X_\phi} \\
\cong \downarrow & & \cong \downarrow \\
\mathcal{Q}_X & \xrightarrow{\;\cong\;} & \oplus_{\phi \in \tilde{P}(X)} Q_{X_\phi}
\end{array}
$$

Proof. The map j_ϕ maps $\prod_{a \in X}(1-e^{-a})^{-1}]$ to $\prod_{a \in X_\phi}(a - \langle a \mid \phi \rangle)^{-1} F$, where F is nonzero at ϕ. This implies that the ideal in the completion of $\mathbb{C}[\Lambda]$ annihilating v_{X_ϕ} is identified with the annihilator of the class of $\prod_{a \in X_\phi}(a - \langle a \mid \phi \rangle)^{-1}$ in the completion of $S[V]$. Since all these ideals are of finite codimension, we get the desired isomorphism.

Remark 16.6. This theorem can be considered as a dual version of formula (16.1) and Proposition 16.1.

16.1.4 A Residue Formula

In this section, using the exponential map, we think of the elements of S_X as periodic meromorphic functions on $U_\mathbb{C}$ with poles on the corresponding periodic family of hyperplanes. By formula (14.27) and Corollary 16.4, we can put together the isomorphisms j_ϕ as e^ϕ runs among the points of the arrangement and get an isomorphism of $S[U_\mathbb{C}]$-modules

$$j : SP_X \to \oplus_{\phi \in \tilde{P}(X)} \oplus_{\underline{b} \in \mathcal{NB}X_\phi} S[U_\mathbb{C}]\Big[\prod_{a \in \underline{b}}(a - \langle a \mid \phi \rangle)^{-1}\Big].$$

By Proposition 16.3, the image under j of the class v_X of $\prod_{a \in X}(1 - e^{-a})^{-1}$ decomposes as the sum over ϕ of

$$j_\phi(v_{X_\phi}) = \sum_{\underline{b} \in \mathcal{NB}X_\phi} \mathfrak{q}_{\underline{b},\phi} j(\omega_{\underline{b},\phi}) = \sum_{\underline{b} \in \mathcal{NB}X_\phi} \mathfrak{q}_{\underline{b},X_\phi}\Big(\frac{\partial}{\partial y}\Big) u_{\underline{b}}.$$

We have the *periodic analogue* of (10.10):

Theorem 16.7.

$$\mathfrak{q}_{\underline{b},X_\phi}(-y) = det(\underline{b})\mathrm{res}_{\underline{b},\phi}\Big(\frac{e^{\langle y \mid z \rangle}}{\prod_{a \in X}(1 - e^{-\langle a \mid z \rangle - \langle a \mid \phi \rangle})}\Big). \tag{16.3}$$

Proof. We apply Theorem 10.11, translating ϕ into 0, and setting h equal to the germ of $\prod_{a \in X}(1 - e^{-a})^{-1}$ around ϕ for each ϕ.

Summarizing: formulas (10.10) and (16.3), together with the local computation of residues, provide us with another effective algorithm to compute the functions that we have been studying. Theorem 14.38 becomes:

Theorem 16.8. *Let Ω be a big cell, $B(X)$ the box associated to X. For every $x \in \Omega - B(X)$,*

$$\mathcal{T}_X(x) = \sum_{\phi \in \tilde{P}(X)} e^{\langle \phi \mid x \rangle} \sum_{\underline{b} \in \mathcal{NB}X_\phi \mid \Omega \subset C(\underline{b})} \mathrm{res}_{\underline{b},\phi}\Big(\frac{e^{\langle x \mid z \rangle}}{\prod_{a \in X}(1 - e^{-\langle a \mid z \rangle - \langle a \mid \phi \rangle})}\Big).$$

$$\tag{16.4}$$

It remains to discuss a last algorithmic point. In order to compute the Jeffrey–Kirwan residue, at a given point $v \in C(A)$, it is necessary to determine a big cell Ω for which $v \in \overline{\Omega}$.

In general, the determination of the big cells is a very complex problem, but for our computations much less is needed.

Take a point q internal to $C(A)$ and not lying on any hyperplane generated by $n-1$ vectors of X. This is not difficult to do. Consider the segment qv. This segment intersects these hyperplanes in a finite number of points, and thus we can determine an ϵ sufficiently small for which all the points $tp + (1 - t)q$, with $0 < t < \epsilon$, are regular. If we take one of these points q_0, it lies in a big cell for which p is in the closure.

At this point, for every unbroken basis, we must verify whether q_0 lies in the cone generated by the basis.

16.1.5 The Operator Formula

Here we discuss a formula by Brion–Vergne [27] (see also Guillemin [59]) using our machinery.

We want to analyze the action of multiplication by $1 - e^{-a}$ on F_ϕ. If $a \notin X_\phi$, then $e^{\langle a \,|\, \phi \rangle} \neq 1$, and $(1 - e^{-a})$ is invertible in the algebra of germs, so it gives an invertible operator on F_ϕ.

If $a \in X_\phi$, we have that the function

$$\frac{1 - e^{-a}}{a - \langle a \,|\, \phi \rangle}$$

takes the value 1 at ϕ. Thus it is invertible in $\mathcal{L}_\phi$. We set

$$Q_\phi = \prod_{a \notin X_\phi} \frac{1}{1 - e^{-a}} \prod_{a \in X_\phi} \frac{a - \langle a \,|\, \phi \rangle}{1 - e^{-a}}.$$

This gives by multiplication an invertible operator on F_ϕ; hence via j_ϕ, it also gives an invertible operator on Υ_ϕ. We deduce

Proposition 16.9. *Let*

$$j : \oplus_{\phi \in \tilde{P}(X)} F_\phi \to \oplus_{\phi \in \tilde{P}(X)} \Upsilon_\phi$$

denote the isomorphism j of coordinates j_ϕ. Then

$$j\left(\prod_{a \in X} (1 - e^{-a})^{-1} \right) = \sum_{\phi \in P(X)} Q_\phi \prod_{a \in X_\phi} (a - \langle \phi \,|\, a \rangle)^{-1}. \qquad (16.5)$$

Proof. In $\mathcal{R}_{X,\phi}$ we have the obvious identity

$$\prod_{a \in X} \frac{1}{1 - e^{-a}} = \prod_{a \notin X_\phi} \frac{1}{1 - e^{-a}} \prod_{a \in X_\phi} \frac{a - \langle a \,|\, \phi \rangle}{1 - e^{-a}} \frac{1}{\prod_{a \in X_\phi} (a - \langle a \,|\, \phi \rangle)},$$

where each factor of the function

$$\prod_{a \notin X_\phi} \frac{1}{1 - e^{-a}} \prod_{a \in X_\phi} \frac{a - \langle a \mid \phi \rangle}{1 - e^{-a}}$$

takes a nonzero value at ϕ (so it is a germ of an invertible holomorphic function at ϕ). So everything follows.

Using Theorem 9.5 we deduce that the class of $1/\prod_{a \in X_\phi}(a - \langle a \mid \phi \rangle)$ equals $\sum_{\underline{b} \in \mathcal{NB}_{X_\phi}} \mathfrak{p}_{\underline{b},X_\phi} u_{\underline{b}}$. Equating terms we have

$$\mathfrak{q}_{\underline{b},X_\phi} u_{\underline{b}} = Q_\phi \mathfrak{p}_{\underline{b},X_\phi} u_{\underline{b}}. \tag{16.6}$$

We want to transform the previous identity into a formula expressing $\mathfrak{q}_{\underline{b},X_\phi}$.

Lemma 16.10. *Let $\widehat{S[V]}$ be the completion of $S[V]$ at the point ϕ. We have that $\widehat{S[V]}$ acts as differential operators on the space $e^\phi S[U_\mathbb{C}]$.*

Proof. Clearly, $S[V]$ acts on this space by differential operators. The elements in the maximal ideal of ϕ are locally nilpotent, so this action extends to an action on $\widehat{S[V]}$.

Notice that since for every $v \in V$ we have $v u_{\underline{b}} = \langle v \mid \phi \rangle u_{\underline{b}}$, the module generated by $u_{\underline{b}}$ is isomorphic to the module N_ϕ over an algebra of differential operators on $U_\mathbb{C}$ mapping $u_{\underline{b}}$ to the delta function at ϕ.

Its Fourier transform is isomorphic to the module of functions generated by e^ϕ. Applying this to (16.6) we have

$$\hat{\mathfrak{q}}_{\underline{b},X_\phi} e^\phi = \hat{Q}_\phi \hat{\mathfrak{p}}_{\underline{b},X_\phi} e^\phi = \hat{Q}_\phi(\mathfrak{p}_{\underline{b},X_\phi}(-x)e^\phi). \tag{16.7}$$

Where recall that $\hat{\mathfrak{p}}_{\underline{b},X_\phi} = \mathfrak{p}_{\underline{b},X_\phi}(-x)$ and $\hat{Q}_\phi$ is a differential operator applied to the function $\mathfrak{p}_{\underline{b},X_\phi}(-x)e^\phi$. Now on the module of functions pe^ϕ with p a polynomial, the element $\hat{Q}_\phi$ acts as a differential operator of finite order on each function.

Substituting in formula (14.27), we deduce the final general operator formula (16.8) for $x \in \Lambda \cap \Omega - B(X)$:

Theorem 16.11. *Given a point $x \in \Lambda \cap \Omega - B(X)$, we have*

$$\mathcal{T}_X(x) = \sum_{\phi \in \tilde{P}(X)} \hat{Q}_\phi\Big(e^{\langle \phi \mid x \rangle} \sum_{\underline{b} \in \mathcal{NB}_{X_\phi} \mid \Omega \subset C(\underline{b})} |\det(\underline{b})|^{-1} \mathfrak{p}_{\underline{b},X_\phi}(-x)\Big). \tag{16.8}$$

Proof. Identifying functions on Λ with distributions supported on Λ, we know that $\mathcal{T}_X$ is the inverse Laplace transform of $\prod_{a \in X} \frac{1}{1-e^{-a}}$. From formula (14.27) we have that

$$\prod_{a \in X} \frac{1}{1 - e^{-a}} = \sum_{\phi \in \tilde{P}(X)} \sum_{\underline{b} \in \mathcal{NB}_{X_\phi}} \mathfrak{q}_{\underline{b},\phi} d_{\underline{b},\phi} + L$$

where L is in lower filtration and its inverse Laplace transform is supported in a finite union of cones of dimension $< s$ (Theorem 14.36). Thus let us first concentrate on the first part of the sum. We know by Proposition 14.4 that $|\det(\underline{b})|d_{\underline{b},\phi}$ is the Laplace transform of $\xi_{\underline{b},\phi}$, which coincides with the function e^{ϕ} on the points of $\Lambda \cup C(\underline{b})$ and it is 0 elsewhere. Therefore, possibly outside some set of dimension $< s$, we have that on $\Omega \cap \Lambda$

$$\mathcal{T}_X(x) = \sum_{\phi \in \tilde{P}(X)} \sum_{\underline{b} \in \mathcal{NB}_{X_\phi}} \hat{\mathfrak{q}}_{\underline{b},\phi} e^{\langle \phi \mid x \rangle} = \sum_{\phi \in \tilde{P}(X)} \sum_{\underline{b} \in \mathcal{NB}_{X_\phi}} \mathfrak{q}_{\underline{b},\phi}(-x) e^{\langle \phi \mid x \rangle}.$$

We now substitute for $\hat{\mathfrak{q}}_{\underline{b},\phi}(x)e^{\langle \phi \mid x \rangle}$ its expression given by Formula 16.7 and finally obtain our desired formula up to the terms given by L. Now we use the fact that we know already that on $\Omega - B(X)$ the function $\mathcal{T}_X(x)$ is a quasipolynomial. The right hand side of Formula (16.8) is already a quasipolynomial and hence we must have that Formula (16.8) holds on $\Omega - B(X)$ unconditionally.

Observe that this expresses the partition function as a sum distributed over the points $\phi \in \tilde{P}(X)$. The contribution of each ϕ is obtained by applying a differential operator to the local multivariate splines for the corresponding lists X_ϕ. Let $C^{\text{reg}}(X)$ be the set of strongly regular points of $C(X)$.

Theorem 16.12. *On the intersection of Λ with the open set of strongly regular points we have*

$$\mathcal{T}_X = \sum_{\phi \in P(X)} \widehat{Q}_\phi \mathcal{T}_{X_\phi,\underline{\phi}}. \tag{16.9}$$

Proof. The explicit formula is a consequence of formula (16.5) plus the previous discussion.

16.1.6 Local Reciprocity

Finally, let us give a third proof of the reciprocity (13.25) and (14.34) based on formula (16.8), that is, the connection between volumes and the partition function. Take one of the terms $q_{\underline{b},\phi}(-x) := \widehat{Q}_\phi\big(e^{\langle \phi \mid x \rangle}\mathfrak{p}_{\underline{b},X_\phi}(-x)\big)$ with

$$\widehat{Q}_\phi = \prod_{a \notin X_\phi} \frac{1}{1 - e^{-D_a}} \prod_{a \in X_\phi} \frac{D_a - \langle \phi \mid a \rangle}{1 - e^{-D_a}}$$

and compute $\mathfrak{q}_{\underline{b},\phi}(x)$, that is (with τ as in (14.34))

$$\mathfrak{q}_{\underline{b},\phi}(x) = \tau(\widehat{Q}_\phi)\big(e^{-\phi}\mathfrak{p}_{\underline{b},X_\phi}(x)\big).$$

We have, since $X_\phi = X_{-\phi}$ and $\mathfrak{p}_{\underline{b},X_\phi}(x)$ is homogeneous of degree $|X_\phi| - s$, that $\mathfrak{p}_{\underline{b},X_\phi}(x) = (-1)^{|X_\phi|-s}\mathfrak{p}_{\underline{b},X_{-\phi}}(-x)$. Now

$$\tau(\widehat{Q}_\phi) = \prod_{a \notin X_\phi} \frac{1}{1 - e^{D_a}} \prod_{a \in X_\phi} \frac{-D_a + \langle \phi \mid a \rangle}{1 - e^{D_a}}$$

$$= (-1)^{|X|+|X_\phi|} e^{-\sum_{a \in X} D_a} \prod_{a \notin X_\phi} \frac{1}{1 - e^{-D_a}} \prod_{a \in X_\phi} \frac{D_a - \langle -\phi \mid a \rangle}{1 - e^{-D_a}}$$

$$= (-1)^{|X|+|X_\phi|} e^{-\sum_{a \in X} D_a} \widehat{Q}_{-\phi}.$$

Since for any function f

$$e^{-\sum_{a \in X} D_a} f(x) = f\Big(x - \sum_{a \in X} a\Big) = f(x - 2\rho_X),$$

we deduce that

$$\begin{aligned}
\mathfrak{q}_{\underline{b},\phi}(x) &= (-1)^{|X_\phi|-s} \tau(\widehat{Q}_\phi) e^{-\phi} \mathfrak{p}_{\underline{b}, X_{-\phi}}(-x) \\
&= (-1)^{|X|-s} e^{-\sum_{a \in X} D_a} \widehat{Q}_{-\phi} e^{-\phi} \mathfrak{p}_{\underline{b}, X_{-\phi}}(-x) \\
&= (-1)^{|X|-s} e^{-\sum_{a \in X} D_a} e^{-\phi} \mathfrak{q}_{\underline{b},-\phi}(-x) \\
&= (-1)^{|X|-s} e^{-\phi} \mathfrak{q}_{\underline{b},-\phi}(-x - 2\rho_X)
\end{aligned}$$

as desired.

16.2 Roots of Unity

16.2.1 Dedekind Sums

We now make a further remark about the computations. Clearly the partition function is integer-valued, while the expressions we found involve roots of 1. This has to be dealt with as follows. The points of finite order of T correspond to the homomorphisms of Λ to $\mathbb{C}^*$ whose image is in the group of roots of 1. Thus we get an action of the Galois group Γ of the maximal cyclotomic extension of $\mathbb{Q}$ on these points. This action preserves $P(X)$, and clearly X_ϕ is constant on an orbit of the Galois group. Thus we decompose $P(X)$ into Γ-orbits, and for each such orbit $\mathcal{O}$ we set $X_\mathcal{O} := X_\phi$, $e^\phi \in \mathcal{O}$. The characters e^ϕ, $\phi \in \mathcal{O}$ have all the same kernel that we denote by $\Lambda_\mathcal{O}$. $\Lambda/\Lambda_\mathcal{O}$ is a cyclic group of order n, the order of e^ϕ. Then $\mathcal{O}$ consists of the $\phi(n)$ (Euler's ϕ function) injective maps of $\Lambda/\Lambda_\mathcal{O}$ into the group of roots of 1.

Each orbit contributes to (16.8) with the term

$$\sum_{e^\phi \in \mathcal{O}} \widehat{Q}_\phi \Big(e^{\langle \phi \mid x \rangle} \sum_{\underline{b} \in \mathcal{NB}_{X_\mathcal{O}} \mid \Omega \subset C(\underline{b})} |\det(\underline{b})|^{-1} \mathfrak{p}_{\underline{b}, X_\mathcal{O}}(-x) \Big). \tag{16.10}$$

We can compute using the fact that

$$(D_a - \langle \phi \mid a \rangle) e^{\langle \phi \mid x \rangle} = e^{\langle \phi \mid x \rangle} D_a,$$

so that

$$\sum_{e^\phi \in \mathcal{O}} \widehat{Q}_\phi e^{\langle \phi \mid x \rangle} = \left[\sum_{e^\phi \in \mathcal{O}} e^{\langle \phi \mid x \rangle} \prod_{a \notin X_\mathcal{O}} \frac{1}{1 - e^{-D_a - \langle \phi \mid a \rangle}} \right] \prod_{a \in X_\mathcal{O}} \frac{D_a}{1 - e^{-D_a}}. \tag{16.11}$$

The operator

$$\tilde{Q}_\mathcal{O} := \sum_{\phi \in \mathcal{O}} e^{\langle \phi \mid x \rangle} \prod_{a \notin X_\mathcal{O}} \frac{1}{1 - e^{-D_a - \langle \phi \mid a \rangle}} \tag{16.12}$$

can be viewed as a *trace*. For each $e^\phi \in \mathcal{O}$, $e^{\langle \phi \mid x \rangle}$ is constant on a fixed coset of $\Lambda_\mathcal{O}$ in Λ. Thus on each such coset the expression (16.12) gives a differential operator with constant coefficients, which, being invariant under the Galois group, are rational numbers. Let us give an algorithm to compute $\tilde{Q}_\mathcal{O}$. Set

$$E(t) := e^{-t} - 1 = \sum_{k=1}^{\infty} \frac{(-t)^k}{k!}.$$

A simple manipulation gives

$$\frac{1}{1 - e^{-D_a - \langle \phi \mid a \rangle}} = \left(1 - e^{-\langle \phi \mid a \rangle}\right)^{-1} \sum_{j=0}^{\infty} \left(\frac{e^{-\langle \phi \mid a \rangle}}{1 - e^{-\langle \phi \mid a \rangle}}\right)^j E(D_a)^j.$$

Denote by $b_1, \ldots, b_r$ the elements of $X \setminus X_\mathcal{O}$. When we develop the product, in a term of (16.12), we get terms of type

$$E(D_{b_1})^{j_1} E(D_{b_2})^{j_2} \cdots E(D_{b_r})^{j_r}$$

multiplied by the coefficient

$$C_{j_1, \ldots, j_r}(\phi) := \prod_{h=1}^{r} (1 - e^{-\langle \phi \mid b_h \rangle})^{-1} \left(\frac{e^{-\langle \phi \mid b_h \rangle}}{1 - e^{-\langle \phi \mid b_h \rangle}}\right)^{j_h}.$$

Then (16.11) is a sum of terms

$$\sum_{j_1, \ldots, j_r} s_\mathcal{O}(j_1, \ldots, j_r)[x] E(D_{b_1})^{j_1} E(D_{b_2})^{j_2} \ldots E(D_{b_r})^{j_r} \prod_{a \in X_\mathcal{O}} \frac{D_a}{1 - e^{-D_a}},$$

where $s_\mathcal{O}(j_1, \ldots, j_r)[x]$ equals

$$\sum_{e^\phi \in \mathcal{O}} e^{\langle \phi \mid x \rangle} C_{j_1, \ldots, j_r}(\phi) = \sum_{\phi \in \mathcal{O}} \frac{e^{\langle \phi \mid x - \sum_i j_i b_i \rangle}}{\prod_{k=1}^{r} \left(1 - e^{-\langle \phi \mid b_k \rangle}\right)^{j_k + 1}}. \tag{16.13}$$

Pick once and for all $e^{\phi_0} \in \mathcal{O}$ and assume that $\mathcal{O}$ consists of elements of order n. Let $\zeta_n := e^{\frac{2\pi i}{n}}$. We have, for each $x \in \Lambda$, there is h and n_k with

$$e^{\langle \phi_0 \mid x \rangle} = \zeta_n^h, \qquad e^{-\langle \phi_0 \mid b_k \rangle} = \zeta_n^{n_k}, \qquad 0 \le h < n, \ 0 < n_k < n, \ \forall k,$$

so that

$$\sum_{e^\phi \in \mathcal{O}} e^{\langle \phi \mid x \rangle} C_{j_1,\ldots,j_r}(\phi) = \sum_{0<i<n,\ (i,n)=1} \zeta_n^{ih} \prod_{k=1}^{r} \frac{\zeta_n^{in_k j_k}}{(1-\zeta_n^{in_k})^{j_k+1}}. \tag{16.14}$$

This expression, which a function of x, through the number h, is the trace from $\mathbb{Q}(\zeta_n)$ to $\mathbb{Q}$ of the element $\zeta_n^h \prod_{k=1}^{r} \frac{\zeta_n^{n_k j_k}}{(1-\zeta_n^{n_k})^{j_k+1}}$. It thus takes rational values that are constant on each coset of $\Lambda_\mathcal{O}$ in Λ. It can be computed by suitable computations in the cyclotomic field $\mathbb{Q}(\zeta_n)$. We will give some examples of these computations in the next section, when we treat in detail the 1-dimensional case.

Expressions of this type are known as *generalized Dedekind sums.*

We can go one step further, calling a sublist $A \subset X$ *saturated* if it is of the form $A = X_\mathcal{O}$ for some orbit $\mathcal{O}$.

Then for such A define $\tilde{P}_A := \prod_{a \in A} D_a/(1 - e^{-D_a})$ and $\tilde{Q}_A := \sum_\mathcal{O} \tilde{Q}_\mathcal{O}$ as $\mathcal{O}$ varies over the orbits such that $X_\mathcal{O} = A$. Then we see that formula (16.9) can be rewritten as a formula computing the partition function from the multivariate splines:

$$\mathcal{T}_X = \sum_{\substack{A \subset X, \\ A \text{ saturated}}} \tilde{Q}_A \tilde{P}_A(T_A). \tag{16.15}$$

16.3 Back to Decomposing Integers

16.3.1 Universal Formulas

In order to use formula (16.15) for computations, we proceed in various steps.

  (i) First of all, we determine the saturated sets A and the multivariate splines T_A.
  (ii) Secondly, if A is a saturated set, we determine the Galois orbits $\mathcal{O}$ in $P(X)$ such that $X_\mathcal{O} = A$.
  (iii) If A has k elements, T_A is a polynomial of degree $k - s$. So we then need to compute the differential operator of infinite order $\tilde{Q}_A \tilde{P}_A$ up to degree $k - s$. This is done by the method explained in the previous section.

We want to analyze this strategy in the case of numbers, that is, when $s = 1$ and $X = (a_1, \ldots, a_m)$ is a sequence of m positive integers.

In this case the points of the arrangement are all the primitive d-th roots of 1, as d varies among the set $\mathcal{D}_X$ of divisors of the numbers $a_1, \ldots, a_m$.

Let us also denote by $\tilde{\mathcal{D}}_X$ the elements of $\mathcal{D}_X$ that are greatest common divisors of sublists of X.

Orbits under the Galois group are indexed by $\mathcal{D}_X$. Given $k \in \mathcal{D}_X$, the corresponding orbit consists of the k-th primitive roots of 1. A sublist A of X is saturated if and only if it is of the form

$$A_d := \{a_i \mid d \text{ divides } a_i, \ d \in \tilde{\mathcal{D}}_X\}.$$

In other words, given a sublist A of X, set $d(A)$ equal to the greatest common divisor of the elements of A. Then A is saturated if and only if every element of $X \setminus A$ is not a multiple of $d(A)$.

Given a saturated list A and $a := d(A)$, let us then write the elements in A in the form $ae_i, i = 1, \ldots, p$. Also, for any $0 < h < a$ let c_h be the number of elements in $X \setminus A$ that are congruent to h modulo a.

We have already seen how to write the coefficients of the expansion

$$\tilde{Q}_A = \sum_{j=0}^{\infty} C_j \frac{d^j}{dt^j}.$$

To show the type of computations one has to perform, let us compute C_0. Set $X \setminus A = \{b_1, \ldots, b_r\}$. Let us use (16.14) and the definition of $\tilde{Q}_A$. A root of 1, ζ is such that $X_\zeta = A$ if and only if it belongs to the set

$$\Theta_a^X := \{\zeta \mid \zeta^a = 1, \ \zeta^{b_i} \neq 1, \forall i = 1, \ldots, r\},$$

and therefore

$$C_0 := \sigma_x(b_1, \ldots, b_r; a) = \sum_{\zeta \in \Theta_a^X} \frac{\zeta^x}{\prod_{i=1}^r (1 - \zeta^{-b_i})} \tag{16.16}$$

where x varies among the cosets modulo A so that C_0 is a degree zero quasipolynomial.

The expression $\sigma_x(b_1, \ldots, b_r; a)$ is called a generalized Dedekind sum.[1] If $x = 0$, we shall write $\sigma(b_1, \ldots, b_r; a)$ instead of $\sigma_0(b_1, \ldots, b_r; a)$. If $a = 1$ then by definition, $\sigma_x(b_1, \ldots, b_r; 1) = 0$.

Turning to $\tilde{P}_A$, we see that

$$\tilde{P}_A := \prod_{a \in A} \frac{D_a}{1 - e^{-D_a}} = \prod_{i=1}^p \frac{e_i a \frac{d}{dt}}{1 - e^{-e_i a \frac{d}{dt}}} = \sum_{j=0}^{\infty} B_j(e_1, \ldots, e_p) a^j \frac{d^j}{dt^j}. \tag{16.17}$$

The elements $B_j(e_1, \ldots, e_p)$ are universal symmetric polynomials in the e_i's involving Bernoulli numbers. As an example, set $s_i := s_i(e_1, \ldots, e_p)$ equal to the i-th elementary symmetric function in the e_j's. That is

$$\prod_{i=1}^p (1 + e_i t) = \sum_{j=0}^p s_j(e_1, \ldots, e_p) t^j \tag{16.18}$$

Then

$$B_0 = 1, \quad B_1 = \frac{s_1}{2}, \quad B_2 = \frac{s_1^2 + s_2}{12}, \quad B_3 = \frac{s_1 s_2}{24}.$$

We deduce the existence of *universal formulas* for the quasipolynomials describing the solution to decomposing integers. In the next two sections we shall work out a number of examples in special situations.

---

[1] We find it useful to sum over primitive roots when defining Dedekind sums rather than follow the classical definition [14].

16.3.2 Some Explicit Formulas

Let us develop a special case, the one in which the numbers a_i are pairwise coprime.

Given $k \in \mathcal{D}_X$, the corresponding orbit consists of all primitive k-th roots of 1. As for $X_{\mathcal{O}}$, we have that if $k = 1$, then $X_{\mathcal{O}} = X$. If $k > 1$ and a_i is the only element in X divisible by k, then $X_{\mathcal{O}} = \{a_i\}$. So the saturated sets are X and the $\{a_i\}$, $i = 1, \ldots, m$. Since $\tilde{Q}_X = 1$ let us now compute

$$\tilde{Q}_{\{a_i\}} \tilde{P}_{\{a_i\}} T_{\{a_i\}}.$$

Here $T_{\{a_i\}}$ is the constant a_i^{-1}. It follows that we only need to compute the constant term of $\tilde{Q}_{\{a_i\}} \tilde{P}_{\{a_i\}}$. We note that $\tilde{P}_{\{a_i\}}$ has constant term 1. Finally, by (16.16), one immediately computes the constant term of $\tilde{Q}_{\{a_i\}}$, getting

$$\sum_{h=1}^{a_i-1} \zeta_{a_i}^{hx} \prod_{j \neq i} \frac{1}{1 - \zeta_{a_i}^{-ha_j}} = \sigma_x(a_1, \ldots, \overset{\vee}{a_i}, \ldots a_m; a_i). \qquad (16.19)$$

It remains to discuss the contribution coming from 1. In this case (cf. (9.1)), $T_X(t) = t^{m-1}/(m-1)! \prod_i a_i$. Thus we obtain, by formula (16.17), the polynomial with rational coefficients

$$\sum_{j=0}^{m-1} \frac{B_j(a_1, \ldots, a_m) t^{m-j-1}}{(m-j-1)! \prod_i a_i}. \qquad (16.20)$$

Now recall that $\mathcal{T}_X$ is a quasipolynomial relative to the subgroup of multiples of $M := \prod_{i=1}^{m} a_i$. Applying our formulas, we get

$$\mathcal{T}_X(t) = \sum_{j=0}^{m-1} \frac{B_j(a_1, \ldots, a_m) t^{m-j-1}}{(m-j-1)! \prod_i a_i} + \sum_{i=1}^{m} \frac{\sigma_t(a_1, \ldots, \overset{\vee}{a_i}, \ldots, a_m; a_i)}{a_i}. \qquad (16.21)$$

As an application we get a reciprocity law for generalized Dedekind sums as follows. The equation $\sum_{i=1}^{n} a_i x_i = 0$, $x_i \geq 0$, has as unique solution $(0, \ldots, 0)$; hence $\mathcal{T}_X(0) = 1$. Computing, with the above method, we get

$$\frac{B_{m-1}(a_1, \ldots, a_m)}{a_1 \cdots a_m} + \sum_{i=1}^{m} \frac{\sigma(a_1, \ldots, \overset{\vee}{a_i}, \ldots, a_m; a_i)}{a_i} = 1. \qquad (16.22)$$

Using the fact that $\mathcal{T}_X(t) = 0, -\sum_{i=1}^{m} a_i < t < 0$ we have also other reciprocity laws, which determine the partition function on these cosets.

A related problem that has been extensively studied is the following. Consider in $\mathbb{R}^n$ the simplex with vertices $(0, \ldots, 0, \alpha_i, 0, \ldots, 0)$ with α_i a positive integer for each $i = 1, \ldots, n$ and the origin 0. We want to count the number of integer points in this simplex.

A point with coordinates x_i is in this simplex if and only if

$$\sum_{i=1}^{n} \frac{x_i}{\alpha_i} \le 1, \quad x_i \ge 0, \ \forall i.$$

Multiply by the least common multiple m of the α_i and set $a_i := m/\alpha_i$. We then have that the integral points in this simplex are in one-to-one correspondence with the nonnegative integer solutions $(x_1, \ldots, x_n, w)$ of

$$\sum_{i=1}^{n} a_i x_i + w = m, \quad a_i > 0, \ \gcd(a_1, \ldots, a_n) = 1.$$

Thus we have reduced this question to a computation involving the partition function $\mathcal{T}_X(t)$ with $s = 1$, the list X being $(a_1, \ldots, a_n, 1)$ and the a_i coprime. In fact we need only the polynomial coinciding with $\mathcal{T}_X(t)$ on the multiples of all the a_i.

In this case, in principle, the formulas we have developed allow us to compute the quasipolynomial $\mathcal{T}_X(t)$ for all choices of the a_i's. However, explicit computations have been performed only for $n \le 3$.

For $n = 2$ let us denote by a, b the two elements a_1, a_2. Since a, b, and 1 are mutually coprime, we can apply formula(16.21):

$$\mathcal{T}_X(t) = \frac{t^2}{2ab} + \frac{(a + b + 1)}{2ab} t + \frac{a^2 + b^2 + 1}{12ab} + \frac{ab + a + b}{4ab} + \frac{\sigma_t(b, 1; a)}{a} + \frac{\sigma_t(a, 1; b)}{b}.$$

As for the reciprocity law, we get setting $t = 0$

$$1 = \frac{a^2 + b^2 + 1}{12ab} + \frac{ab + a + b}{4ab} + \frac{\sigma(b, 1; a)}{a} + \frac{\sigma(a, 1; b)}{b},$$

which is a form of the classical *reciprocity for Dedekind sums*.

Finally, using Theorem 13.54, we have similar formulas for any t with $-a - b - 1 < t < 0$. In fact $\mathcal{T}_X(t)$ is zero at these integers, and we get the identity

$$0 = \frac{t^2}{2ab} - \frac{(a + b + 1)}{2ab} t + \frac{a^2 + b^2 + 1}{12ab} + \frac{ab + a + b}{4ab} + \frac{\sigma_t(b, 1; a)}{a} + \frac{\sigma_t(a, 1; b)}{b}$$

at any such integer.

This allows us to compute simply the term $\sigma_t(b, 1; a)/a + \sigma_t(a, 1; b)/b$, and thus determine the polynomial expressing $\mathcal{T}_X$ on any residue class modulo ab having a representative t with $-a - b \le t \le 0$. To get the terms $\sigma_t(b, 1; a)/a + \sigma_t(a, 1; b)/b$, in the general case, we can use formula (12.3) that tells us that the quasipolynomial F expressing $\mathcal{T}_X$ for $t > -1 - a - b$ satisfies the Euler relation

$$\nabla_X F(x) = F(x) - F(x - a) - F(x - b) + F(x - a - b) = 0.$$

Let us now pass to the case $n = 3$. Denote by a, b, c the three coprime elements a_1, a_2, a_3. Let M be the least common multiple of a, b, c and $p(t)$ the polynomial coinciding with $\mathcal{T}_X$ on the nonnegative multiples of M. We concentrate our analysis on computing $p(t)$. We have that $p(0) = 1$ and the leading term of $p(t)$ is easily seen to be $t^3/6abc$. So, we only need to compute the two coefficients of t, t^2. We thus write explicitly only the parts that contribute to these coefficients.

The points of the arrangement are the divisors of the three numbers a, b, c. Since a, b, c are not necessarily pairwise coprime, the methods we have already developed have to be slightly modified.

For the point 1 we can apply directly the expansion of $\tilde{P}_X$. We get, with $s_i = s_i(a, b, c, 1)$,

$$(6s_4)^{-1}\{t^3 + \frac{3s_1}{2}t^2 + \frac{s_1^2 + s_2}{2}t + \frac{s_1 s_2}{24}\}.$$

Since any three elements among $a, b, c, 1$ have greatest common divisor equal to one, we see that saturated sets different from X have cardinality either 1 or 2. In particular, they do not contribute the coefficient of t^2 in $p(t)$. We now have to compute the contributions to the coefficient of t. A saturated subset can give a contribution if it consists of two not coprime elements. So we have the three possibilities

$$\{b, c\}, \ \{a, c\}, \ \{a, b\}.$$

Let us analyze $\{b, c\}$ so $T_{\{b,c\}} = t/bc$. Set $A = \gcd(b, c)$ and assume $A \neq 1$. Applying formula (16.16), we get that the contribution of $\{b, c\}$ to the coefficient of t is given by $\sigma(a, 1; A)/bc$. Similarly, if $B = \gcd(a, c)$ and $C = \gcd(a, b)$, the corresponding contributions are given by $\sigma(b, 1; B)/ac$ and $\sigma(c, 1; C)/ab$. We deduce that

$$p(t) = \frac{1}{6s_4}\left\{t^3 + \frac{3s_1}{2}t^2 + \frac{s_1^2 + s_2}{2}t\right\} + \left[\frac{\sigma(a, 1; A)}{bc} + \frac{\sigma(b, 1; B)}{ac} + \frac{\sigma(c, 1; C)}{ab}\right]t + 1.$$

As for $n = 2$, one could perform a direct computation of the constant term of $p(t)$. This computation would involve several Dedekind sums. The very complicated expression adding to 1, could be considered as a generalized *reciprocity formula* for Dedekind sums.

16.3.3 Computing Dedekind Sums

In this section we want to make a few remarks on how one could actually compute the sum of terms

$$\zeta^x \prod_j \frac{1}{1 - \zeta^{-a_j}}, \tag{16.23}$$

where ζ runs over the primitive k-th roots of 1. This is a computation inside the cyclotomic field $\mathbb{Q}(\zeta_k)$, with $\zeta_k := e^{2\pi i/k}$ giving rise to a rational number. In fact, such a sum is the trace $\text{tr}_{\mathbb{Q}}^{\mathbb{Q}(\zeta_k)}$ of each of its terms. These are standard facts of arithmetic, that we recall for completeness.

Short of having enough reciprocity laws allowing us to simplify the expression, the standard way to proceed is in two steps. First one has to clear the denominator and obtain an expression of type $b = \sum_{i=0}^{\phi(k)-1} c_i \zeta_k^i$, and then one has to compute $\text{tr}_{\mathbb{Q}}^{\mathbb{Q}(\zeta_k)}(b) = \sum_{i=0}^{\phi(k)-1} c_i \text{tr}(\zeta_k^i)$.

The element ζ_k^i is a primitive root of 1 of order $t = k/(k,i)$. The value of $\text{tr}(\zeta_k^i)$ equals $\text{tr}(\zeta_t)$. We have

$$\text{tr}_{\mathbb{Q}}^{\mathbb{Q}(\zeta_k)}(\zeta_t) = [\mathbb{Q}(\zeta_k) : \mathbb{Q}(\zeta_t)]\text{tr}_{\mathbb{Q}}^{\mathbb{Q}(\zeta_t)}(\zeta_t) = \frac{\phi(k)}{\phi(t)}\text{tr}_{\mathbb{Q}}^{\mathbb{Q}(\zeta_t)}(\zeta_t).$$

Where $\phi(k) = [\mathbb{Q}(\zeta_k) : \mathbb{Q}]$ is Euler's function. As for $\text{tr}_{\mathbb{Q}}^{\mathbb{Q}(\zeta_t)}(\zeta_t)$, it equals $-a(t)$, where $a(t)$ is the second coefficient of the cyclotomic polynomial $\phi_t(x)$ of degree t.

In order to compute $a(t)$, consider the identity $\prod_{d\,|\,n} \phi_d(x) = x^n - 1$. We deduce that if $n > 1$, we have $\sum_{d\,|\,n} a(d) = 0$.

Proposition 16.13. $-a(t)$ *coincides with the classical Möbius function* $\mu(t)$:

$$\begin{cases} \mu(t) = 0 & \text{if } t \quad \text{is divisible by a square of a prime,} \\ \mu(t) = (-1)^h & \text{if } t \quad \text{is a product of } h \text{ distinct primes,} \\ \mu(1) = 1. \end{cases}$$

Proof. $\phi_1(x) = x - 1$ so $a(1) = -1$.

First, for a prime p we have $a(p) + a(1) = 0$, hence $a(p) = 1$. Assume that $n = \prod_{i=1}^{h} p_i$ is a product of distinct primes. For a divisor d that is a product of $k < h$ primes, we have by induction $-a(d) = (-1)^k$; thus we have

$$a(n) - \sum_{j=0}^{k-1} \binom{k}{j}(-1)^j = 0.$$

Since $(1 + (-1))^k = \sum_{j=0}^{k} \binom{k}{j}(-1)^j = 0$, we have that $-a(n) = (-1)^k$.

Next assume that $n = p^k$ is a prime power. By induction $a(p^i) = 0$ if $1 < i < k$, so $a(p^k) + a(p) + a(1) = 0$ implies $a(p^k) = 0$.

In the case $n = \prod_i p_i^{n_i}$, where at least one n_i is grater than 1, by induction, for all proper divisors d that contain a square of a prime we have $a(d) = 0$. The remaining divisors are divisors of $\prod_i p_i$, so that $\sum_{d\,|\,\prod p_i} a(d) = 0$. Then $a(n) + \sum_{d\,|\,\prod p_i} a(d) = 0$ implies $a(n) = 0$.

As for clearing a denominator $\prod_j (1 - \zeta^{-a_j})$, observe that the elements ζ^{-a_j} are primitive h-th roots of 1 for numbers $h\,|\,k$, $h > 1$.

Let us denote by $G_h := \{\eta_1, \ldots, \eta_{\phi(h)}\}$ the primitive h-th roots of 1. We have for some positive integers $0 \leq n_{h,i}$, $1 \leq i \leq \phi(h)$ determined by the list a_i, that

$$\prod_j (1 - \zeta^{-a_j}) = \prod_{1 < h \mid k} \prod_{i=1}^{\phi(h)} (1 - \eta_i)^{n_{h,i}}.$$

In other words, each primitive h-th root of 1 may appear with some multiplicity depending on the list of numbers a_i. Let $n_h := \max(n_{h,i})$, for $i = 1, \ldots, \phi(h)$. We can write

$$\prod_j \frac{1}{1 - \zeta^{-a_j}} = \frac{\prod_{1 < h \mid k} \prod_{i=1}^{\phi(h)} (1 - \eta_i)^{n_h - n_{h,i}}}{\prod_{1 < h \mid k} \prod_{i=1}^{\phi(h)} (1 - \eta_i)^{n_h}}. \tag{16.24}$$

Once this is done, we easily compute the denominator of formula (16.24), and we obtain a positive integer.

In fact, $\prod_{i=1}^{\phi(h)} (1 - \eta_i)$ is the *norm* of each of its factors in the cyclotomic field of h-th roots of 1. We thus need to recall the following formula:

Proposition 16.14.

$$M(k) := \prod_{\eta \in G_k} (1 - \eta) = \begin{cases} 1 & \text{if } k \text{ is not a prime power,} \\ p & \text{if } k \text{ is a prime power } p^h. \end{cases}$$

Proof. Let $\phi_k(x)$ denote the k-th cyclotomic polynomial, so that $M(k) = \phi_k(1)$. We have $\prod_{i \mid k, \, i \neq 1} \phi_k(x) = 1 + x + x^2 + \cdots + x^{k-1}$; hence

$$\prod_{i \mid k, \, i \neq 1} \phi_k(1) = k.$$

So for a prime p we have $M(p) = p$.

Next, if $k = p^h$, we have the divisors p^i, $i = 1, \ldots, h$, and hence by induction, $p^h = M(p^h) \prod_{i=1}^{h-1} M(p^i) = M(p^h) p^{h-1}$ implies $M(p^h) = p$.

Now consider $k = \prod_i^h p_i$, a product of $h > 1$ distinct primes. Suppose we know by induction the result for products of fewer primes:

$$k = \prod_{A \subset \{1, \ldots, h\}} M \prod_{i \in A} p_i = M(k) \prod_i M(p_i) = M(k)k$$

implies $M(k) = 1$.

Finally, in general, take $k = \prod_i p_i^{h_i}$. We have again by induction that $k = \prod_{d \mid k} M(d) = M(k) \prod_i \prod_{j=1}^{h_i} M(p_i^j) = M(k)k$ so $M(k) = 1$.

To finish we have to expand (set $\zeta_k := e^{\frac{2\pi i}{k}}$)

$$\prod_{1 < h \mid k} \prod_{i=1}^{\phi(h)} (1 - \eta_i)^{n_h - n_{h,i}} = \sum_{j=0}^{k-1} c_j \zeta_k^j,$$

and using the theory already developed, compute the trace of

$$\sum_{h=0}^{k-1} c_h \sum_{\zeta \in \Gamma_k} \zeta^{x+h}. \tag{16.25}$$

This is a computation that cannot be further reduced.

Other algorithms involve an opposite point of view, that is, computing Dedekind sums using partition functions. For a recent survey and further references see [13].

16.4 Algorithms

By Theorem 13.19, given a chamber $\mathfrak{c}$, $DM(X)$ is identified with the space of integral valued functions on $\delta(\mathfrak{c}\,|\,X)$. The value of a function on a given other point is obtained by recursion using the defining equations $\nabla_Y f = 0$, where Y runs over the cocircuits in X.

Of course, the space of functions on $\delta(\mathfrak{c}\,|\,X)$ has a natural basis given by the characteristic functions of its $\delta(X)$ points. This basis depends on $\mathfrak{c}$, and it does not describe in any sense the elements of $DM(X)$ as functions, but only as initial values of a recursive algorithm.

We also have the basis given by Theorem 13.69, but this does not seem the best for actual computations, since it is somewhat tautological, as Example 13.70 shows. It does not show in any explicit way the quasipolynomial nature of the functions.

16.4.1 Rational Space $DM_{\mathbb{Q}}(X)$.

For any abelian group A we have by the same recursion equations a space $DM_A(X)$ of A valued functions given by initial conditions an A-valued function on $\delta(\mathfrak{c}\,|\,X)$.

In this section we want to describe a linear basis for the space $DM_{\mathbb{Q}}(X)$ that exhibits in a clear way the quasipolynomial nature of these functions and at the same time discuss a possible algorithm that expresses the partition function in this basis.

We have the description given by Proposition 16.1, which indeed presents the elements of $DM_{\mathbb{C}}(X)$ as quasipolynomials. Nevertheless, this description will present a rational-valued function as a sum over terms involving roots of 1. Thus we want to start from this description and eliminate the roots of 1.

We assume $\Lambda = \mathbb{Z}^s$, so that X is a list of integral vectors. Each point $p \in P(X)$ of the arrangement has as coordinates some roots of 1. Given a positive integer m, we have set $\zeta_m := e^{2\pi i/m}$. Let $R_m \subset \mathbb{C}^*$ be the group of m-th roots of 1 and $\mathbb{Q}(\zeta_m)$ the cyclotomic field generated by R_m. We have that $\mathbb{Q}(\zeta_m)$ is a Galois extension of $\mathbb{Q}$ of degree $\phi(m)$, where $\phi(m)$ is Euler's function.

Now take a point $p \in P(X)$, $p = (\zeta_1, \ldots, \zeta_s)$, where ζ_i is a root of 1 of some order k_i. Let m be the least common multiple of the k_i, which is also the order of p in the group T. The coordinates of p generate over $\mathbb{Q}$ the field $\mathbb{Q}(\zeta_m)$, and the point p belongs to an orbit $\mathcal{O}$ under the Galois group that is contained in $P(X)$ and is formed of $\phi(m)$ different elements.

Let $\mathbb{Q}(\mathcal{O})$ denote the functions on $\mathcal{O}$ obtained by evaluating the elements of $\mathbb{Q}[\Lambda]$, and denote by $\rho_{\mathcal{O}} : \Lambda \to \mathbb{Q}(\mathcal{O})$ the evaluation. Let $I_{\mathcal{O}}$ be the ideal of $\mathbb{Q}[\Lambda]$ vanishing at $\mathcal{O}$. The following facts are easy to verify:

- The evaluation of the characters Λ at a point $p \in \mathcal{O}$ gives homomorphisms $\rho_p : \Lambda \to \mathbb{C}^*$, $\rho_p : \mathbb{Q}[\Lambda] \to \mathbb{C}$.
- The image of Λ is the group R_m of m-th roots of 1, while the image of $\mathbb{Q}[\Lambda]$ is $\mathbb{Q}(\zeta_m)$.
- $\mathbb{Q}(\mathcal{O}) := \mathbb{Q}[\Lambda]/I_{\mathcal{O}}$ is a field.
- In this way we have the $\phi(m)$ different embeddings of $\mathbb{Q}(\mathcal{O})$ into $\mathbb{C}$ all isomorphisms with $\mathbb{Q}(\zeta_m)$.
- $\mathbb{Q}(\mathcal{O}) \otimes \mathbb{C}$ is the space of complex-valued functions on $\mathcal{O}$.

The set X_p of characters in X that are 1 at p equals $X \cap \ker \rho_{\mathcal{O}}$; it depends only on the orbit, and we can denote it by $X_{\mathcal{O}}$.

We want to make explicit $\mathbb{Q}(\mathcal{O})$ as a space of functions on Λ as follows. Consider the usual trace map $\mathrm{tr}_{\mathbb{Q}}^{\mathbb{Q}(\mathcal{O})} : \mathbb{Q}(\mathcal{O}) \to \mathbb{Q}$. Given $a \in \mathbb{Q}(\mathcal{O})$, we have the function $T_a : \Lambda \to \mathbb{Q}$ defined by

$$T_a(\gamma) := \mathrm{tr}_{\mathbb{Q}}^{\mathbb{Q}(\mathcal{O})}(\gamma a).$$

Let $T(\mathcal{O})$ be the space of functions T_a on Λ thus induced.

In particular, $T_1(\gamma) := \mathrm{tr}_{\mathbb{Q}}^{\mathbb{Q}(\mathcal{O})}(\gamma) = \mathrm{tr}_{\mathbb{Q}}^{\mathbb{Q}(\zeta_m)}(\rho_p(\gamma))$ is independent of p in the orbit and we can denote it by $\psi_{\mathcal{O}}(\gamma)$. A translate $\psi_{\mathcal{O}}(\gamma + \lambda)$ depends only on the class of λ modulo $\ker \rho_{\mathcal{O}}$. We have m different translates corresponding to the m-th roots of 1.

These translates are of course not linearly independent, since a basis of $\mathbb{Q}(\zeta_m)$ is given by the elements ζ_m^i, $0 \le i < \phi(m)$. So a basis for $T(\mathcal{O})$ can be obtained as follows. If $\lambda \in \Lambda$ is such that $\rho_p(\lambda) = \zeta_m$, we have that

$$\mathrm{tr}_{\mathbb{Q}}^{\mathbb{Q}(\zeta_m)}(\rho_p(\gamma)\zeta_m^i) = \mathrm{tr}_{\mathbb{Q}}^{\mathbb{Q}(\zeta_m)}(\rho_p(\gamma + i\lambda)) := \psi_m(\gamma + i\lambda).$$

Thus a basis for $T(\mathcal{O})$ is given by these $\phi(m)$ translates of the function $\psi_{\mathcal{O}}$, and the main point is to understand ψ_m. In Section 16.3.3 we have computed $\mathrm{tr}_{\mathbb{Q}}^{\mathbb{Q}(\zeta_m)}(\xi)$ when ξ is a root of 1 of some order k dividing m, getting

$$\mathrm{tr}_{\mathbb{Q}}^{\mathbb{Q}(\zeta_m)}(\xi) = \frac{\phi(m)}{\phi(k)}\mu(k).$$

Lemma 16.15. *Given $f \in DM_{\mathbb{Q}}(X_{\mathcal{O}})$ and $a \in \mathbb{Q}(\mathcal{O})$, the product $T_a f$ of these two functions lies in $DM_{\mathbb{Q}}(X)$.*

Proof. In fact, if $b \in X_{\mathcal{O}}$, we have that $\nabla_b T_a = 0$, and we see immediately that if $f \in DM_{\mathbb{Q}}(X_{\mathcal{O}})$ and $Y \subset X_{\mathcal{O}}$, we have $\nabla_Y(T_a f) = T_a \nabla_Y(f)$.

If Y is a cocircuit in X, we have that $Y \cap X_{\mathcal{O}}$ is a cocircuit in $X_{\mathcal{O}}$ and this gives the claim.

Associated to $X_{\mathcal{O}}$ we have $D_{\mathbb{Q}}(X_{\mathcal{O}})$ considered as a space of polynomials on Λ and $D_{\mathbb{Q}}(X_{\mathcal{O}}) \subset DM_{\mathbb{Q}}(X_{\mathcal{O}})$. On the other hand, $DM_{\mathbb{Q}}(X_{\mathcal{O}}) \subset DM_{\mathbb{Q}}(X)$, and thus $T(\mathcal{O})D_{\mathbb{Q}}(X_{\mathcal{O}}) \subset DM_{\mathbb{Q}}(X)$.

Proposition 16.16.

$$DM_{\mathbb{Q}}(X) := \oplus_{\mathcal{O} \subset P(X)} T(\mathcal{O}) D_{\mathbb{Q}}(X_{\mathcal{O}}).$$

Proof. We prove this equality by complexifying and then remarking that

$$T(\mathcal{O})D_{\mathbb{Q}}(X_{\mathcal{O}}) \otimes_{\mathbb{Q}} \mathbb{C} = (T(\mathcal{O}) \otimes_{\mathbb{Q}} \mathbb{C}) \otimes D_{\mathbb{Q}}(X_{\mathcal{O}}) = \oplus_{p=e^{\phi} \in \mathcal{O}} e^{\phi} D(X_p).$$

So the claim follows from formula (16.1).

At this point one can easily describe an explicit basis for $DM_{\mathbb{Q}}(X)$ using the basis we developed for $T(\mathcal{O})$ and for instance the dual of the one discussed in Theorem 11.20, for $D_{\mathbb{Q}}(X_{\mathcal{O}})$. We then ask the following question. Given a big cell Ω, express in this basis the element $f \in DM_{\mathbb{Q}}(X)$ coinciding with $\mathcal{T}_X$ on $(\Omega - B(X)) \cap \Lambda$. For this we can apply Theorem 13.54, which tells us that this element is given by its initial conditions $f(0) = 1$ and $f(a) = 0$, $\forall a \in \delta(\mathfrak{c} \,|\, X)$, $a \neq 0$.

This finally gives the coefficients of f in the given basis as the solution of an explicit system of linear equations by computing the $\delta(X)$ elements of the basis on the $\delta(X)$ points of $\delta(\mathfrak{c} \,|\, X)$.

Let us discuss the example of dimension $s = 1$. In this case X is a sequence of positive integers a_i and there is a unique big cell, the positive line, so the geometry is trivial.

- The set $P(X)$ consists of all roots ζ of 1 such that $\zeta^{a_i} = 1$ for some i.
- If $\zeta_k \in P(X)$, we have the set $X_{\zeta_k} = \{a_i \in X \,|\, k \text{ divides } a_i\}$, and denote by $m(k)$ its cardinality.
- Set G_k to be the set of all primitive k-th roots of 1 and let $\mathcal{D}_X$ be the set of divisors of the numbers in X.
- The decomposition into Galois orbits is $P(X) = \cup_{k \in \mathcal{D}_X} G_k$.
- The space $D_{\mathbb{Q}}(X_{\zeta})$ equals the space of polynomials in x with rational coefficients and of degree $< m(k)$. We denote it by $P(m(k))$.
- Set $\psi_k(x) := \mathrm{tr}_{\mathbb{Q}}^{\mathbb{Q}(k)}(\zeta_k^x)$, $x \in \mathbb{Z}$.

We have thus, by the previous discussions

$$DM_{\mathbb{Q}}(X) := \oplus_{k \in \mathcal{D}_X} T(G_k) P(m(k));$$

in other words, the space $DM_{\mathbb{Q}}(X)$ has as basis the quasipolynomial functions $\psi_k(x+j)x^h$, $h < m(k)$, $0 \le j < \phi(k)$, where $k \in \mathcal{D}_X$.

At this point set $A := \sum_{a \in X} a$. The partition function agrees for positive integers with the unique element $\sum_{k \in \mathcal{D}_X} \sum_{j=0}^{m(k)-1} c_{k,j} \psi_k(x+j)x^h$ with the property that

$$\sum_{k \in \mathcal{D}_X} \sum_{j=0}^{m(k)-1} c_{k,j} \psi_k(i+j)i^h = \begin{cases} 0 & \text{if } -A < i < 0, \\ 1 & \text{if } i = 0. \end{cases}$$

The $A \times A$ matrix with entries $\psi_k(i+j)i^h$, $-A < i \le 0$, is invertible, and so the coefficients $c_{k,j}$ can be solved by resolving the system of linear equations.

The reader will notice that most of the complexity of this algorithm consists in computing the set $\mathcal{D}_X$ and the Euler function, since this requires factoring several numbers into primes.

Example 16.17. $X = 2, 3, 4, 4, 6, 9, 10, 40$, write $\psi(k,x) := \psi_k(x)$.

The quasipolynomial associated to the partition function has been computed with Mathematica, and it is

$$\frac{1262093963}{2985984000} + \frac{252100403\,x}{2508226560} + \frac{81263\,x^2}{2457600} + \frac{3303691\,x^3}{995328000} + \frac{27443\,x^4}{199065600} + \frac{1639\,x^5}{597196800} + \frac{13\,x^6}{497664000} +$$

$$\frac{x^7}{10450944000} + \frac{1173757\,\psi(2,x)}{4096000} + \frac{483129\,x\,\psi(2,x)}{4096000} + \frac{26273\,x^2\,\psi(2,x)}{2457600} + \frac{1621\,x^3\,\psi(2,x)}{4423680} +$$

$$\frac{13\,x^4\,\psi(2,x)}{2457600} + \frac{x^5\,\psi(2,x)}{36864000} + \frac{1009\,\psi(3,x)}{5832} - \frac{11\,x\,\psi(3,x)}{4374} - \frac{x^2\,\psi(3,x)}{8748} - \frac{149\,\psi(3,1+x)}{972} -$$

$$\frac{13\,x\,\psi(3,1+x)}{729} - \frac{x^2\,\psi(3,1+x)}{4374} + \frac{81\,\psi(4,x)}{1280} + \frac{39\,x\,\psi(4,x)}{10240} + \frac{x^2\,\psi(4,x)}{20480} + \frac{117\,\psi(4,1+x)}{10240} +$$

$$\frac{3\,x\,\psi(4,1+x)}{10240} + \frac{89\,\psi(5,x)}{10000} + \frac{x\,\psi(5,x)}{5000} + \frac{123\,\psi(5,1+x)}{10000} + \frac{3\,x\,\psi(5,1+x)}{10000} + \frac{111\,\psi(5,2+x)}{10000} +$$

$$\frac{3\,x\,\psi(5,2+x)}{10000} + \frac{67\,\psi(5,3+x)}{10000} + \frac{x\,\psi(5,3+x)}{5000} - \frac{\psi(6,x)}{648} + \frac{\psi(6,1+x)}{324} - \frac{\psi(8,1+x)}{640} - \frac{2\,\psi(9,x)}{81} +$$

$$\frac{\psi(9,2+x)}{81} - \frac{\psi(9,3+x)}{81} - \frac{\psi(9,4+x)}{81} + \frac{13\psi(10,x)}{10000} + \frac{169\psi(10,1+x)}{10000} + \frac{x\psi(10,1+x)}{2000} - \frac{221\psi(10,2+x)}{10000}$$

$$- \frac{x\,\psi(10,2+x)}{2000} + \frac{13\,\psi(10,3+x)}{10000} + \frac{\psi(20,1+x)}{200} - \frac{\psi(20,2+x)}{200} + \frac{\psi(20,3+x)}{400} + \frac{\psi(20,4+x)}{400} +$$

$$\frac{\psi(20,5+x)}{400} - \frac{\psi(20,6+x)}{200} + \frac{\psi(20,7+x)}{200} - \frac{3\,\psi(40,3+x)}{80} - \frac{\psi(40,4+x)}{80} - \frac{\psi(40,5+x)}{80} - \frac{\psi(40,6+x)}{20} -$$

$$\frac{\psi(40,7+x)}{40} - \frac{\psi(40,9+x)}{20} - \frac{\psi(40,11+x)}{40} - \frac{\psi(40,12+x)}{20} - \frac{\psi(40,13+x)}{80} - \frac{\psi(40,14+x)}{80} -$$

$$\frac{3\,\psi(40,15+x)}{80}$$

Question: how to describe $DM_{\mathbb{Z}}(X)$ as quasipolynomials? The computation of the partition function, as in the previous example, seems to imply that describing the elements of the basis given by Theorem 13.69 as quasipolynomials, may not be computationally very explicit.

Part IV

Approximation Theory

Convolution by $B(X)$

In this chapter we fix a lattice $\Lambda \subset V$ and a list X of vectors in Λ spanning V. We would like to give a streamlined presentation of some of the applications in this case. Most results are taken from the papers of Dahmen and Micchelli, or from [40], with minor variations of the proofs.

For further details and more information the reader should look at the original literature.

17.1 Some Applications

17.1.1 Partition of 1

Fix a lattice $\Lambda \subset V$ spanning V such that each vector in the list X lies in Λ. We assume that the Lebesgue measure on V is normalized in such a way that a fundamental domain for Λ has volume 1. In more concrete terms, we may assume $V = \mathbb{R}^s, \Lambda = \mathbb{Z}^s$ and the standard measure.

Let $C^{\mathrm{sing}}(X)$ denote the set of strongly singular points (see Definition 1.50) of the cone $C(X)$. We have seen in Proposition 1.55 that the set of all translates $\cup_{\lambda \in \Lambda} C^{\mathrm{sing}}(X)$ equals the *cut locus*, that is the periodic hyperplane arrangement generated under translation by Λ by the hyperplanes generated by subsets of X. Thus the complement of the cut locus is the set of regular points of this periodic hyperplane arrangement.

Proposition 17.1. *(1) The complement of the cut locus is the union of all translates of a finite number of chambers, each an interior of a (compact) polytope.*
(2) Over each such chamber the functions $B_X(x - \lambda)$, $\lambda \in \Lambda$, are polynomials in the space $D(X)$ (introduced in Definition 11.2).

Proof. (1) Since it is clear that the hyperplanes generated by subsets of X intersect in 0, the first statement is a simple consequence of parts 3 and 4 of Theorem 2.7.

C. De Concini and C. Procesi, *Topics in Hyperplane Arrangements, Polytopes and Box-Splines*, Universitext, DOI 10.1007/978-0-387-78963-7_17,
© Springer Science+Business Media, LLC 2010

(2) This follows from formula (7.10), Remark 9.11, and the fact that $\tilde{D}(X)$ is stable under translation.

Next one easily proves the following fundamental fact:

Theorem 17.2. *If X spans V, the translates of B_X form a partition of unity:*

$$1 = \sum_{\lambda \in \Lambda} B_X(x - \lambda). \tag{17.1}$$

Proof. If $X = (a_1, \ldots, a_s)$ is a basis, we have chosen B_X to be the characteristic function of the half-open parallelepiped $B(X) = \{\sum_{i=1}^{s} t_i a_i,\ 0 \le t_i < 1\}$ with basis X divided by its volume d. In fact, d equals the number of points in $\Lambda \cap B(X)$, and the claim easily follows. In general, one can use the iterative description of the box spline (formula (7.17)) and get, if $X = \{Y, v\}$, with Y spanning V,

$$1 = \int_0^1 dt = \int_0^1 \sum_{\lambda \in \Lambda} B_Y(x - tv - \lambda) dt = \sum_{\lambda \in \Lambda} B_X(x - \lambda).$$

17.1.2 Semidiscrete Convolution

Consider the box spline $B_X(x)$, that we know is supported in the zonotope $B(X) = \{\sum_{a \in X} t_a a,\ 0 \le a \le 1\}$.

Recall that in Definition 6.3 we have introduced the cardinal spline space $\mathcal{S}_X := \mathcal{S}_{B_X}$ as the image of the functions on Λ (*mesh functions*) by the operation of discrete convolution

$$B_X * a(x) = \sum_{\lambda \in \Lambda} B_X(x - \lambda) a(\lambda)$$

of mesh functions with B_X. When a is the restriction to Λ of a function f on V, we have defined $B_X * a = B_X *' f(x)$ and called it *semidiscrete convolution* (cf. (6.6)).

Using the polynomiality of the multivariate splines on each big cell, formula (7.11) and Proposition 1.55 give a first qualitative description of the functions in $\mathcal{S}_X$:

Proposition 17.3. *Each translate of B_X, and hence each element of the cardinal spline space, is a polynomial on each chamber.*

We now pass to semidiscrete convolution and prove one of the main results of the theory:

Theorem 17.4. *When $p \in D(X)$, also $B_X *' p$ lies in $D(X)$.*

This defines a linear isomorphism F of $D(X)$ to itself, given explicitly by the invertible differential operator $F_X := \prod_{a \in X} \frac{1 - e^{-D_a}}{D_a}$.

Proof. We can immediately reduce, using Proposition 7.14, to the case of X nondegenerate. If $X = a$ is a number and $s = 1$, we have that $D(X)$ reduces to the constants, and then the statement reduces to the fact that the translates of B_a sum to the constant function 1. In the other cases B_X is a continuous function on $\mathbb{R}^s$, and a way to understand this convolution is by applying the Poisson summation formula to the function $B_X(x+y)p(-y)$ of y that is continuous and with compact support.

Its Laplace transform is obtained from the function $e^x \prod_{a \in X}(1 - e^{-a})/a$, Laplace transform of $B_X(x+y)$ by applying the polynomial $\hat{p}$ as a differential operator.

In our definition of the Laplace transform we have

$$Lf(\xi) = (2\pi)^{n/2}\hat{f}(i\xi),$$

where $\hat{f}$ denotes the usual Fourier transform. We want to apply the classical Poisson summation formula (cf. [121]), which gives, for a function ϕ with suitable conditions, and in particular if ϕ is continuous with compact support,

$$\sum_{\mu \in \Lambda^*} L\phi(\mu) = \sum_{\lambda \in \Lambda} \phi(\lambda),$$

where μ runs over the *dual lattice* Λ^* of elements for which $\langle \mu \,|\, \lambda \rangle \in 2\pi i\mathbb{Z}$, $\forall \lambda \in \Lambda$.

If we are in the situation that $L\phi(\mu) = 0, \forall \mu \neq 0,\ \mu \in \Lambda^*$, we have

$$L\phi(0) = \sum_{\lambda \in \Lambda} \phi(\lambda).$$

The fact that this holds for $\phi(y) = B_X(x+y)p(-y)$ is a key result of Dahmen and Micchelli. This will imply our theorem.

Before proving this key result, let us write in a suitable form the action of a differential operator given by a polynomial $p(\frac{\partial}{\partial x_1}, \ldots, \frac{\partial}{\partial x_s})$ of some degree k on a power series $F(x_1, \ldots, x_s)$.

Lemma 17.5. *Introducing auxiliary variables $t_1, \ldots, t_s$, we have*

$$p\Big(\frac{\partial}{\partial x_1}, \ldots, \frac{\partial}{\partial x_s}\Big)[F(x_1, \ldots, x_s)] \tag{17.2}$$

$$= F\Big(x_1 + \frac{\partial}{\partial t_1}, \ldots, x_s + \frac{\partial}{\partial t_s}\Big)[p(t_1, \ldots, t_s)]_{t_1 = \cdots = t_s = 0}.$$

Proof. Start from the obvious identity:

$$p\Big(\frac{\partial}{\partial x_1}, \ldots, \frac{\partial}{\partial x_s}\Big)[F(x_1 + t_1, \ldots, x_s + t_s)]$$

$$= p\Big(\frac{\partial}{\partial t_1}, \ldots, \frac{\partial}{\partial t_s}\Big)[F(x_1 + t_1, \ldots, x_s + t_s)],$$

from which we have

$$p\Big(\frac{\partial}{\partial x_1},\ldots,\frac{\partial}{\partial x_s}\Big)[F(x_1,\ldots,x_s)]$$

$$= p\Big(\frac{\partial}{\partial t_1},\ldots,\frac{\partial}{\partial t_s}\Big)[F(x_1+t_1,\ldots,x_s+t_s)]_{t_1=\cdots=t_s=0}.$$

Now use the fact that, if p,q are two polynomials in the variables t_i we have

$$p\Big(\frac{\partial}{\partial t_1},\ldots,\frac{\partial}{\partial t_s}\Big)[q(t_1,\ldots,t_s)]_{t_1=\cdots=t_s=0}$$

$$=q\Big(\frac{\partial}{\partial t_1},\ldots,\frac{\partial}{\partial t_s}\Big)[p(t_1,\ldots,t_s)]_{t_1=\cdots=t_s=0}.$$

The main observation of Dahmen and Micchelli is the following:

Lemma 17.6. *If $p(x) \in D(X)$, the Laplace transform of $B_X(x + y)p(-y)$ (thought of as a function of y) vanishes at all points $\mu \neq 0$, $\mu \in \Lambda^*$.*

Proof. In order to avoid confusion, let us use ξ as the symbol denoting the variables in the Laplace transform. Thus if $a = (a_1,\ldots,a_s)$, we are using the notation

$$a = \langle a \,|\, \xi\rangle = \sum_i a_i\xi_i,\quad e^a = e^{\sum_i a_i\xi_i},\quad e^x = e^{\sum_i x_i\xi_i}.$$

We may assume that $p(y)$ is homogeneous of some degree k. The Laplace transform of $B_X(x+y)p(-y)$ is computed by recalling that the Laplace transform of $B_X(y)$ is $\prod_{a\in X}(1 - e^{-a})/a$. So the Laplace transform of $B_X(x + y)$ is $L_X(\xi) := e^x \prod_{a\in X}(1 - e^{-a})/a$ and that of $B_X(x + y)p(-y)$ finally is

$$p\Big(\frac{\partial}{\partial \xi_1},\ldots,\frac{\partial}{\partial \xi_s}\Big)\Big[e^x \prod_{a\in X} \frac{1 - e^{-a}}{a}\Big].$$

Let us develop a simple identity:

$$\frac{1 - e^{-x-y}}{x + y} = -\sum_{k=0}^{\infty}(-x - y)^k/(k + 1)! = \frac{1 - e^{-x}}{x} + yH(x,y),$$

where $H(x,y) = \sum_{k=1}^{\infty}\sum_{j=1}^{k} \binom{k}{j}/(k + 1)!\,y^{j-1}x^{k-j}$.

Now we can apply formula (17.2). Substitute ξ_i with $\xi_i + \frac{\partial}{\partial t_i}$. We get

$$\frac{1 - e^{-\langle a \,|\, \xi+\partial_t\rangle}}{\langle a \,|\, \xi + \partial_t\rangle} = \frac{1 - e^{-\langle a \,|\, \xi\rangle}}{\langle a \,|\, \xi\rangle} + H_a(a, D_a)D_a,$$

where we set $\langle a \,|\, \partial_t\rangle = \sum_i a_i\frac{\partial}{\partial t_i} = D_a$ and $H_a(a, D_a)$ is a convergent power series in a, D_a. Notice that since they involve disjoint variables, $a = \sum_i a_i\xi_i$ and $D_b = \sum_i b_i\frac{\partial}{\partial t_i}$ commute for all $a, b \in A$.

$$L_X(\xi + \partial_t) = e^{\langle x \mid \xi + \partial_t \rangle} \prod_{a \in X} \left(\frac{1 - e^{-\langle a \mid \xi \rangle}}{\langle a \mid \xi \rangle} + H_a(a, D_a)D_a \right)$$

$$= \sum_{B \subset X} e^{\langle x \mid \xi + \partial_t \rangle} \prod_{a \notin B} \left(\frac{1 - e^{-\langle a \mid \xi \rangle}}{\langle a \mid \xi \rangle} \right) \prod_{a \in B} H_a(a, D_a)D_a.$$

Take a summand relative to B and the corresponding function

$$e^{\langle x \mid \xi + \partial_t \rangle} \prod_{a \notin B} \left(\frac{1 - e^{-\langle a \mid \xi \rangle}}{\langle a \mid \xi \rangle} \right) \prod_{a \in B} H_a(a, D_a)D_a[p(t_1, \ldots, t_s)]_{t_1 = t_2 = \cdots = t_s = 0}.$$

We have that either B is a cocircuit or $X \setminus B$ contains a basis. In the first case, since $p(t) \in D(X)$, we have $\prod_{a \in B} D_a p(t) = 0$. These terms are identically zero.

In the second case, since the elements of $X \setminus B$ span the vector space, if $\mu \neq 0$, at least one $a_0 \in A \setminus B$ does not vanish at μ.

- If μ is in the lattice Λ^* we have $\langle \mu \mid b \rangle \in 2\pi i \mathbb{Z}$,
- All elements $1 - e^{-b}$, $b \in \Lambda^*$, vanish at μ.
- We deduce that $(1 - e^{-a_0})/a_0$ vanishes at μ and thus the entire product vanishes.

We return to the proof of Theorem 17.4. We have shown that in our case, Poisson summation degenerates to the computation at 0.

Recall that $e^x = e^{\langle x \mid \xi \rangle} = e^{\sum_{i=1}^s x_i \xi_i}$. Consider the duality $\langle p \mid f \rangle$ defined as follows. We take a polynomial $p(\xi_1, \ldots, \xi_s)$, compute it in the derivatives $\frac{\partial}{\partial \xi_i}$, apply it as differential operator to the function f, and then evaluate the resulting function at 0. We have for each i that

$$\langle p \mid \xi_i f \rangle = \left\langle \frac{\partial}{\partial \xi_i}(p) \mid f \right\rangle.$$

Thus, setting $f_X := \prod_{a \in X} \frac{1 - e^{-a}}{a}$ and $F_X := \prod_{a \in X} \frac{1 - e^{-D_a}}{D_a}$, we have

$$\hat{p}\left(\frac{\partial}{\partial \xi_i} \right)\left(e^x \prod_{a \in X} \frac{1 - e^{-a}}{a} \right)(0) = \langle p \mid e^x f_X \rangle = \langle F_X p \mid e^x \rangle = F_X p(x),$$

since for any polynomial q we have

$$\left\langle q\left(\frac{\partial}{\partial \xi} \right) \mid e^{\langle x \mid \xi \rangle} \right\rangle = q(x).$$

Since $D(X)$ is stable under derivatives and F_X is clearly invertible, both our claims follow.

Let us deduce from this a corollary that clarifies the relationship between the polynomials expressing the multivariate spline on the big cells and the top-degree part of $D(X)$.

Corollary 17.7. *(1) The polynomials expressing locally the box spline and their translates by elements of Λ linearly span $D(X)$.*
(2) The polynomials expressing locally the multivariate spline on the big cells generate the top-degree part of $D(X)$.

Proof. (1) From the definitions and Theorem 11.6 we know that all these polynomials lie in $D(X)$. For the converse, take a polynomial $f \in D(X)$. We know by the previous theorem that it lies in the cardinal spline space. But on each chamber a function in the cardinal spline space is a linear combination of translates of the polynomials describing B_X locally. The claim follows.

(2) By construction, the box spline on a chamber has as top homogeneous part a linear combination of the polynomials expressing the multivariate spline. By the previous theorem, on a given chamber a polynomial in $D(X)$ is a linear combination of translates of the polynomials building the box spline; hence the top-degree part is a linear combination of the polynomials expressing the multivariate spline.

Theorem 17.4 tells us that $D(X) \subset \mathcal{S}_X$. On the other hand, we have the following result:

Corollary 17.8. *If p is a polynomial in $\mathcal{S}_X$, then $p \in D(X)$. Thus $D(X)$ coincides with the space of all polynomials in $\mathcal{S}_X$.*

Proof. By Proposition 17.3 we know that each function f in the space $\mathcal{S}_X$, once restricted to a chamber $\mathfrak{c}$ of the strongly regular points coincides with a polynomial $f_{\mathfrak{c}}$ in $D(X)$. If f is itself a polynomial, then it must coincide with $f_{\mathfrak{c}}$ everywhere.

Corollary 17.9. *On $D(X)$, the inverse of F_X is given by the differential operator of infinite order*

$$Q := \prod_{x \in X} \frac{D_x}{1 - e^{-D_x}}.$$

Notice that Q is like a *Todd operator*; its factors can be expanded using the Bernoulli numbers B_n by the defining formula:

$$\frac{D_x}{1 - e^{-D_x}} = \sum_{k=0}^{\infty} \frac{B_n}{n!}(-D_x)^n,$$

and it acts on $D(X)$ as an honest differential operator.

Let us recall that in Proposition 7.14 we have seen that if $z \in X$, then $D_z B_X = \nabla_z B_{X \setminus z}$. Moreover, for any function a on Λ, and any B we also have $(\nabla_z B) * a = B * \nabla_z a$. Using this, we get, for any subset $Y \subset X$,

$$D_Y(B_X * a) = B_{X \setminus Y} * \nabla_Y a. \tag{17.3}$$

This gives another insight into Corollary 17.8. Assume that a polynomial p is of the form $p = B_X * a$ for some function a on Λ. Let Y be a cocircuit.

Then $D_Y p = D_Y(B_X * a) = B_{X \setminus Y} * \nabla_Y a$ is a distribution supported on the subspace $\langle X \setminus Y \rangle$. Since it is also a polynomial, it must be equal to 0, showing again that $p \in D(X)$.

17.1.3 Linear Independence

We now assume that we are in the unimodular case (cf. Section 15.2.2). In this case $\delta(X) = d(X)$. Choose a strongly regular point x_0 in the interior of a chamber $\mathfrak{c}$. According to Proposition 2.50, we have the $d(X)$ points

$$\delta(x_0|X) := (x_0 - B(X)) \cap \Lambda = \{p_1, \ldots, p_{d(X)}\}.$$

By Remark 13.2 we have $\delta(x_0|X) = \delta(\mathfrak{c}|X)$.

Moreover, for any function a on Λ and $x \in \mathfrak{c}$ we have

$$B_X * a(x) = \sum_{\lambda \in \Lambda} B_X(x - \lambda)a(\lambda) = \sum_{\lambda \in b(\mathfrak{c}|X)} B_X(x - \lambda)a(\lambda). \qquad (17.4)$$

Proposition 17.10. *Evaluation of polynomials at the points in $\delta(\mathfrak{c}|X)$ establishes a linear isomorphism between $D(X)$ and the space $\mathbb{C}^{d(X)}$ (or $\mathbb{R}^{d(X)}$ if we restrict to real polynomials).*

Proof. Since $\dim(D(X)) = d(X)$, it suffices to prove that the only polynomial $p(x) \in D(X)$ vanishing on these points is $p(x) = 0$. Consider $F_X p(x) = B_X *' p$. If p vanishes on $\delta(\mathfrak{c}|X)$, formula (17.4) implies that $F_X p(x) = 0$ on $\mathfrak{c}$ and hence everywhere. Since F_X is invertible on $D(X)$, this proves the proposition.

We can prove the *linear independence of the translates of the box spline*:

Theorem 17.11. *For X unimodular and any function $f(\lambda)$ on Λ not identically 0, we have*
$$B_X * f \neq 0.$$

Proof. Assume $f(a_0) \neq 0$ for some $a_0 \in \Lambda$. We can choose a chamber $\mathfrak{c}$ in such a way that $a_0 \in \delta(\mathfrak{c}|X)$. Thus there is a nonzero polynomial $p(x) \in D(X)$ coinciding on $\delta(\mathfrak{c}|X)$ with f. We deduce $\sum_{a \in \Lambda} B_X(x - a)f(a) = F_X p(x) \neq 0$ on the set $\mathfrak{c}$.

Remark 17.12. Notice that given a chamber $\mathfrak{c}$ and a function a on Λ, the restriction of $B_X * a$ to $\mathfrak{c}$ coincides with the polynomial $F_X p$, where p is the unique polynomial in $D(X)$ coinciding with a on $\delta(\mathfrak{c}|X)$.

In fact, we have a stronger statement, that is also useful in interpolation theorems.

Proposition 17.13. *Consider an open set Ω and the set $\delta(\Omega|X)$. If $B * a$ restricted to Ω equals zero, then $a(j) = 0$, $\forall j \in \delta(\Omega|X)$.*

Proof. By definition, given $j \in \delta(\Omega \,|\, X)$, there is a chamber $\mathfrak{g} \subset j + B(X)$ with nonempty intersection with Ω. Thus from Remark 13.2, $j \in \delta(\mathfrak{g} \,|\, X)$. Since $B * a$ restricted to $\mathfrak{g}$ is a polynomial, our assumption implies that this polynomial is identically zero. By Remark 17.12, we have thus that $a(j) = 0$, as desired.

Unimodularity is a necessary condition

Proposition 17.14. *Semidiscrete convolution maps $DM_{\mathbb{C}}(X)$ onto $D(X)$ with kernel $E(X)$ (see formula (16.2)).*

Proof. The first statement follows immediately from identity (17.3).

Let $\lambda \in \Lambda$, then $(\tau_\lambda f)(a) = f(a - \lambda)$ so

$$B_X *' (\tau_\lambda f) = \sum_{a \in \Lambda} B(x - a) f(a - \lambda) = \sum_{a \in \Lambda} B(x - \lambda - a) f(a) = \tau_\lambda (B *' f).$$

Therefore semidiscrete convolution is a map of $\mathbb{C}[\Lambda]$–modules.

Now for each $p \in P(X)$, $e^{\langle \phi \,|\, v \rangle} D_{\mathbb{C}}(X_p)$ is the generalized eigenspace in $DM_{\mathbb{C}}(X)$ for the character $\lambda \mapsto e^{\langle \phi \,|\, \lambda \rangle}$.

In particular $D(X)$, which we know to be equal to the image of $DM_{\mathbb{C}}$ by semidiscrete convolution with B_X, is the generalized eigenspace relative to the trivial character. It follows that for $p \neq 1$ $e^{\langle \phi \,|\, v \rangle} D_{\mathbb{C}}(X_p)$ lies in the kernel of semidiscrete convolution with B_X. Thus $E(X)$ does, proving our claim.

We have in fact a general

Theorem 17.15. $B_X * a = 0$ *if and only if $a \in E(X)$.*

Proof. By the previous proposition, it is enough to show that $a \in DM_{\mathbb{C}}(X)$. We may assume that X is irreducible. In the 1-dimensional case $X = \{k\}$, we see this directly as Fourier analysis over $\mathbb{Z}/(k)$. In fact, B_X is the characteristic function of $[0, k)$, and $B_X * a = 0$ implies that for $x \in \mathbb{Z}$ we have

$$\sum_{x - i \in [0, k)} a(i) = \sum_{i \in x - [0, k)} a(i) = \sum_{i=0}^{k-1} a(x - i) = 0.$$

This implies that $0 = \sum_{i=0}^{k-1} a(x - i) - \sum_{i=0}^{k-1} a(x + 1 - i) = a(x) - a(x + k)$, so $a(x)$ is constant on cosets modulo $\mathbb{Z}k$, and thus it is in $DM_{\mathbb{C}}(X)$.

In general, let $X = \{Z, y\}$, so that (Proposition 7.14) $D_y B_X = \nabla_y B_Z$. If $B_X * a = 0$ we have also $D_y B_X * a = \nabla_y B_Z * a = B_Z * \nabla_y a = 0$, so by induction, $\nabla_y a \in DM_{\mathbb{C}}(Z)$. Now given any cocircuit Y in X let $y \in Y$ and $Z := X \setminus y$. Clearly, $Y \setminus y$ is a cocircuit in Z, so that $\nabla_Y a = \nabla_{Y \setminus y} \nabla_y a = 0$, as desired.

We leave to the reader to verify that the previous theorem implies a new characterization of the space $DM_{\mathbb{C}}(X)$.

Corollary 17.16. *A function a on Λ belongs to $DM_{\mathbb{C}}(X)$ if and only if $B_X * a$ is a polynomial.*

We apply the previous theory to the partition function $\mathcal{T}_X(a)$ and obtain another interesting formula of Dahmen–Micchelli.

Proposition 17.17.

$$T_X(x) = B_X * \mathcal{T}_X = \sum_{a \in \Lambda} B_X(x - a)\mathcal{T}_X(a).$$

Proof. Compute the Laplace transform

$$L\Big(\sum_{a \in \Lambda} B_X(x - a)\mathcal{T}_X(a)\Big) = \sum_{a \in \Lambda} e^{-a}L(B_X)\mathcal{T}_X(a)$$

$$= \prod_{a \in X} \frac{1}{1 - e^{-a}} \prod_{a \in X} \frac{1 - e^{-a}}{a} = \prod_{a \in X} \frac{1}{a} = LT_X.$$

This proves the identity at least in a weak sense. In order to prove it as functions we can reduce to the irreducible case, and we have two cases. If X is nondegenerate, both sides are continuous and so identical. Otherwise, in the case of a single vector $X = \{a\}$, the identity is trivial by the convention that in this case, the box spline is the characteristic function of $[0, a)$.

In the unimodular case, where we have the linear independence of translates of B_X, we recover the results of Section 16.1.

We have already remarked that if

$$Q := \prod_{x \in X} \frac{D_x}{1 - e^{-D_x}},$$

we have $QFp = p = FQp$ on $D(X)$. Take a big cell Ω for which T_X coincides with some polynomial $p_\Omega \in D(X)$. Set $q_\Omega := Qp_\Omega$. We have on Ω,

$$T_X(x) = FQp_\Omega = \sum_{a \in \Lambda} Qp_\Omega(a)B(x - a).$$

Since $T_X(x) = \sum_{a \in \Lambda} \mathcal{T}_X(a)B_X(x - a)$, we have by linear independence the following formula, that is nothing else than Theorem 16.11 in the special case in which X is unimodular:

$$Qp_\Omega(a) = \mathcal{T}_X(a) \tag{17.5}$$

for any $a \in \overline{\Omega}$.

18

Approximation by Splines

In this chapter we want to give a taste to the reader of the wide area of approximation theory. This is a very large subject, ranging from analytical to even engineering-oriented topics. We merely point out a few facts more closely related to our main treatment. We refer to [70] for a review of these topics.

We start by resuming and expanding the ideas and definitions already given in Chapter 6.

18.1 Approximation Theory

As usual, we take an s-dimensional real vector space V in which we fix a Euclidean structure, and denote by dx the corresponding Lebesgue measure. We also fix a lattice $\Lambda \subset V$ and a list X of vectors in Λ spanning V as a vector space and generating a pointed cone.

18.1.1 Scaling

We use the notation of Section 17.1.2. Corollary 17.8 tells us that the space $D(X)$ coincides with the space of polynomials in the cardinal spline space $\mathcal{S}_X$. This has a useful application for approximation theory. In order to state the results, we need to introduce some notation.

For every positive real number h we have the *scale operator* (see (6.8))

$$(\sigma_h f)(x) := f(x/h).$$

In particular, we shall apply this when $h = n^{-1}$, $n \in \mathbb{N}^+$, so that $h\Lambda \supset \Lambda$ is a *refinement* of Λ.

Recall that in Proposition 6.5 we have seen that the space $\sigma_h(\mathcal{S}_X)$ equals the cardinal space $\mathcal{S}_{hX}$ with respect to the lattice $h\Lambda$.

C. De Concini and C. Procesi, *Topics in Hyperplane Arrangements, Polytopes and Box-Splines*, Universitext, DOI 10.1007/978-0-387-78963-7_18,
© Springer Science+Business Media, LLC 2010

Remark 18.1. If f is supported in a set C, then $\sigma_h f$ is supported in hC. If U is a domain, we have

$$\int_U f\,dx = h^{-s} \int_{hU} \sigma_h f\,dx.$$

Corollary 18.2. *For box splines we have*

$$\sigma_h B_X = h^s B_{hX}$$

Proof. By definition,

$$\int_V B_X(x)f(x)dx = \int_0^1 \cdots \int_0^1 f\Big(\sum_{j=1}^m t_j a_j\Big)dt_1 \cdots dt_m.$$

Thus

$$\int_V \sigma_h(B_X(x)f(x))dx = h^s \int_0^1 \cdots \int_0^1 f\Big(\sum_{j=1}^m t_j a_j\Big)dt_1 \cdots dt_m$$

$$= h^s \int_0^1 \cdots \int_0^1 (\sigma_h f)\Big(\sum_{j=1}^m t_j h a_j\Big)dt_1 \cdots dt_m$$

$$= \int_V B_{hX}(x)(\sigma_h f)(x)dx.$$

So the claim follows from the definition.

We define the scaling operator on distributions by duality. So on a test function f,

$$\langle \sigma_h(T) \,|\, f \rangle := \langle T \,|\, \sigma_h^{-1}(f) \rangle = \langle T \,|\, \sigma_{1/h}(f) \rangle. \tag{18.1}$$

In particular, $\sigma_h \delta_a = \langle \delta_a \,|\, f(hx) \rangle = f(ha)$, so that

$$\sigma_h \delta_a = \delta_{ha}, \quad \sigma_h \sum_a f(a)\delta_a = \sum_a f(a/h)\delta_a.$$

Observe that σ_h acts as an automorphism with respect to convolution.

Notice that if T is represented by a function g, i.e., $\langle T \,|\, f \rangle = \int_V g(x)f(x)dx$, we have

$$\langle \sigma_h(T) \,|\, f \rangle = \int_V g(x)f(hx)dx = h^{-s} \int_V g(h^{-1}x)f(x)dx,$$

and thus $\sigma_h(T)$ is represented by the function $h^{-s}\sigma_h(g)$ and not $\sigma_h(g)$.

The relation between scaling and the Laplace transform is

$$L(\sigma_h f) = h^s \sigma_{1/h} L(f). \tag{18.2}$$

In fact,

$$L(\sigma_h f)(y) = \int_V e^{-\langle x \mid y \rangle} f(x/h)\,dx = h^s \int_V e^{-\langle u \mid hy \rangle} f(u)\,du = h^s \sigma_{1/h} L(f)(y).$$

Notice the commutation relations between σ_h and a derivative D_v, a difference operator ∇_a, or a translation τ_a:

$$D_v \sigma_h = h^{-1}\sigma_h D_v \qquad \sigma_h \nabla_a = \nabla_{ha}\sigma_h \qquad \tau_a \sigma_h = \sigma_h \tau_{a/h}. \tag{18.3}$$

18.1.2 Mesh Functions and Convolution

Consider the group algebra $\mathbb{C}[\Lambda]$ for the lattice Λ. If we identify an element $\lambda \in \Lambda$, with the distribution given by $\langle \delta_\lambda \mid f \rangle = f(\lambda)$, we see that multiplication in the group algebra becomes convolution of distributions.

The space $\mathcal{C}[\Lambda] := \mathbb{C}[\Lambda]^*$ of mesh-functions, i.e., functions on Λ, is a module over $\mathbb{C}[\Lambda]$ by duality. This module structure is also described in the language of distributions by convolution, since $\delta_\lambda * a = \tau_\lambda(a)$. The mesh functions of finite support are clearly a free module of rank 1 generated by the characteristic function of $\{0\}$.

The cardinal spline space $\mathcal{S}_X$ is also a $\mathbb{C}[\Lambda]$ module by the action of translation, and the basic map $B_X * - : \mathcal{C}[\Lambda] \to \mathcal{S}_X$ is a morphism of $\mathbb{C}[\Lambda]$-modules, an isomorphism in the unimodular case by Theorem 17.11. In other words, given $T \in \mathbb{C}[\Lambda]$ and a function a on Λ, $TB_X * a = B_X * Ta$ is well-defined and in $\mathcal{S}_X$.

We shall use the relation $D_Y(B_X) = \nabla_Y B_{X \setminus Y}$ proved in Proposition 7.14. The commutative diagram of $\mathbb{C}[\Lambda]$-modules

$$
\begin{array}{ccc}
\mathcal{C}[\Lambda] & \xrightarrow{\ B_X * \ } & \mathcal{S}_X \\
\downarrow{\scriptstyle \nabla_Y} & & \downarrow{\scriptstyle D_Y} \\
\mathcal{C}[\Lambda] & \xrightarrow{\ B_{X \setminus Y} * \ } & \mathcal{S}_{X \setminus Y}
\end{array}
$$

gives the following result:

Proposition 18.3. *The map $D_Y : \mathcal{S}_X \to \mathcal{S}_{X \setminus Y}$ is surjective.*

Proof. It is enough to prove that ∇_Y is surjective. By induction, one reduces to proving it for ∇_a with $a \in \Lambda \setminus \{0\}$. One can find a basis of Λ such that $a = (k, 0, \dots, 0)$ for some $k > 0$, and one reduces simply to the case $\Lambda = \mathbb{Z}$ and $a = k$. Furthermore, decomposing $\mathbb{Z}$ into the cosets modulo k, one finally reduces to $k = 1$. Thus we need to see that given a sequence $(a_i)_{i \in \mathbb{Z}}$, there is a sequence $(c_i)_{i \in \mathbb{Z}}$ with $a_i = c_i - c_{i-1}$. Setting $c_0 = 0$, we see that for $i > 0$ we define recursively $c_i = c_{i-1} + c_i$, while $c_{i-1} = c_i - a_i$ for $i < 0$.

As an application we can characterize the cardinal spline spaces $\mathcal{S}_m$ generated by the box spline b_m discussed in Example 6.2.

Proposition 18.4. *For all $m \in \mathbb{N}$ we have that $\mathcal{S}_m$ coincides with the space A_m of all the C^{m-1} functions that, restricted to each interval $[i, i+1)$, $i \in \mathbb{Z}$, are polynomials of degree $\leq m$.*

Proof. By Example 6.2 we have $\mathcal{S}_m \subset A_m$ and want to prove the reverse inclusion. We proceed by induction on m. For $m = 0$ we have that $b_0(x)$ is the characteristic function of $[0, 1)$, so that the cardinal space is the space of functions that are constant on the intervals $[i, i + 1)$. Notice next that if $f \in A_m$ and $m \geq 1$, its derivative f' lies in A_{m-1} (as a function if $m \geq 2$ or as a distribution when $m = 1$) and $f(x) = f(0) + \int_0^x f'(t)dt$. For $m = 0$ we have proved the required equality. In general, we obtain it by applying the operator $\frac{d}{dx}$, which, by the previous proposition, maps $\mathcal{S}_m$ surjectively onto $\mathcal{S}_{m-1}$ with kernel the constant functions. The same is true for the same operator from A_m to A_{m-1}. Since by induction $\mathcal{S}_{m-1} = A_{m-1}$, the claim follows.

It is also clear that if the function a on Λ has finite support, the function $f := B_X * a \in \mathcal{S}_X$ has compact support. We have the following converse in the unimodular case.

Proposition 18.5. *If X is unimodular and a is a function on Λ, then $B_X * a$ has compact support if and only if a has finite support.*

Proof. Assume by contradiction that $f = B_X * a$ has compact support while a has infinite support. We can then choose a chamber $\mathfrak{c}$ such that there is a point $i_0 \in \delta(\mathfrak{c} \,|\, X)$ with $a(i_0) \neq 0$, and also $\mathfrak{c}$ is disjoint from the support of f, that is, $f = 0$ on $\mathfrak{c}$. Let $p(x) \in D(X)$ be the nonzero polynomial coinciding on $\delta(\mathfrak{c} \,|\, X)$ with a. By Remark 17.12, we deduce $f(x) = F_X p(x) \neq 0$ on $\mathfrak{c}$, a contradiction.

In the general case we can nevertheless prove the following theorem:

Theorem 18.6. *If $F \in \mathcal{S}_X$ has compact support, there is a unique a with finite support such that $F = B_X * a$.*

We need first a lemma:

Lemma 18.7. *Assume that a is a mesh function such that for every cocircuit Y in X we have that $\nabla_Y a$ has finite support. Then there is a unique function $b \in DM_{\mathbb{C}}(X)$ such that $a - b$ has finite support.*

Proof. Let A contain all the supports of $\nabla_Y a$, choose a chamber $\mathfrak{c}$ sufficiently far from A, and let $b \in DM_{\mathbb{C}}(X)$ be defined by the property that $b = a$ on $\delta(\mathfrak{c} \,|\, X)$. The equations $\nabla_Y b = 0$ allow us to compute b recursively, and as long as also $\nabla_Y a = 0$ in the recursion steps we see that a and b agree. Clearly, this happens if all the terms appearing in the recursion are evaluations of a outside A. This easily implies that $a = b$ outside of some finite set. The uniqueness is clear, since no nonzero element of $DM_{\mathbb{C}}(X)$ has finite support.

Let us now prove Theorem 18.6. Let $F = B_X * a$ have compact support. Arguing as in Theorem 17.15, we see that $\nabla_Y a$ has finite support for every cocircuit Y. Hence by the previous lemma there is a $b \in DM_{\mathbb{C}}(X)$ with $a - b$ of finite support. Now by hypothesis it follows that $B_X * b$ also has finite support, but by Proposition 17.14, $B_X * b \in D(X)$ is a polynomial, and thus it must vanish. Uniqueness follows from Theorem 17.15.

Let us then denote by

$$\mathcal{S}_X^f := \mathbb{C}[\Lambda]B_X = B_X * \mathbb{C}[\Lambda]\delta_0 \tag{18.4}$$

the functions $F \in \mathcal{S}_X$ with compact support.

Notice that the functions F in $\mathcal{S}_{X \setminus Y}^f$ in the image under D_Y of $\mathcal{S}_X^f$ have the property that $F *' q = 0$ for all polynomials q with $\nabla_Y q = 0$.

Further algebraic constructions are useful. There is also a *tensor product construction* associated to convolution. If $X = \{X_1, X_2\}$, we also have two convolution products:

$$\mathcal{S}_{X_1}^f \otimes \mathcal{S}_{X_2} \xrightarrow{\ F*G\ } \mathcal{S}_X,$$

$$\mathcal{S}_{X_1} \otimes \mathcal{S}_{X_2}^f \xrightarrow{\ F*G\ } \mathcal{S}_X,$$

and of course

$$\mathcal{S}_{X_1}^f \otimes \mathcal{S}_{X_2}^f \xrightarrow{\ F*G\ } \mathcal{S}_X^f.$$

When either X_1 or X_2 does not span V this has to be understood as convolution of distributions.

Assume that $X = X_1 \cup X_2$ is a decomposition such that $V = V_1 \oplus V_2$, where V_i is the span of X_i. If $\Lambda = \Lambda_1 \oplus \Lambda_2$ where $\Lambda_i = V_i \cap \Lambda$, we can extend the tensor product to an isomorphism:

$$\mathcal{S}_{X_1} \otimes \mathcal{S}_{X_2} \xrightarrow{\ F*G\ } \mathcal{S}_X.$$

18.1.3 A Complement on Polynomials

We start by interpreting the Taylor series of a function as the formal identity: $\nabla_x = 1 - e^{-D_x}$. The meaning of this identity in general resides in the fact that the vector field D_x is the infinitesimal generator of the 1-parameter group of translations $v \mapsto v + tx$.

On the space of polynomials this is an identity as operators, since both D_x and ∇_x are locally nilpotent.

This identity gives

$$D_x = -\log(1 - \nabla_x) = \sum_{k=1}^{\infty} \frac{\nabla_x^i}{i}. \tag{18.5}$$

Let us denote by P_k the space of polynomials of degree $< k$.

Corollary 18.8. *On P_k, the differential operator D_x and the difference operator ∇_x induce the same algebra of operators.*

Definition 18.9. We set A_k to be the commutative algebra of linear operators on P_k generated by all the operators D_a, $a \in V$, or equivalently ∇_a, $a \in \Lambda$.

The first remark on A_k is that every nonzero A_k submodule of P_k contains the constant polynomials.

In other words, we see that the dual P_k^*, thought of as a module under A_k by transposition, is cyclic and generated by the linear form $\phi_0 : q \mapsto q(0)$.

We deduce the following result:

Proposition 18.10. *A linear map T on P_k commutes with the algebra A_k of difference operators on P_k if and only if $T \in A_k$.*

Proof. The map T commutes with A_k if and only if its transpose T^* commutes with A_k in the dual module. Since this module is cyclic, this happens if and only if $T \in A_k$. Indeed, $T^*\phi_0 = a^*\phi_0$ for some $a \in A_k$. Given $\psi \in P_k^*$, write $\psi = b^*\phi_0$, $b \in A_k$. Then

$$T^*\psi = T^*b^*\phi_0 = b^*T^*\phi_0 = b^*a^*\phi_0 = a^*b^*\phi_0 = a^*\psi,$$

so $T = a$.

18.1.4 Superfunctions

The goal of this and the following sections is to approximate a function by a sequence of functions obtained by rescaling semidiscrete convolutions (cf. Definition 6.3) with B_X.

We have defined the approximation power of a function B in Definition 6.6 and introduced the Strang–Fix conditions in Theorem 6.7. A remarkable class of functions that have approximation power k for simple reasons is given by the following:

Definition 18.11. We say that $S(x)$ is a *superfunction* of power k if $S(x)$ is continuous with compact support and $S *' q = q$ for all polynomials of degree $< k$.[1]

For a list X and $S \in \mathcal{S}_X$ we say that S is a superfunction if $S *' q = q$ for all $q \in D(X)$.

Remark 18.12. A superfunction $S \in \mathcal{S}_X$ is also a superfunction of power $m(X)$.

---

[1] For $k = 0$ we may also want to allow the characteristic function of $[0, 1)$ as a superfunction.

Let us now introduce the approximation algorithm we seek. For a function g and a scale h, define $A_X^h(g) := g_h$ by

$$g_h := \sigma_h(S *' (\sigma_{1/h}g)) = \sum_{b \in h\Lambda} S((x-b)/h)g(b) = \sum_{a \in \Lambda} S(x/h - a)g(ha). \quad (18.6)$$

This is the semidiscrete convolution with respect to the scaled lattice $h\Lambda$. For a given point x, the sum $g_h(x) = \sum_a S(x/h - a)g(ha)$ restricts to those a such that $x/h - a$ lies in the support of S, that we can assume to be contained in a disk Δ_R of radius R centered at 0. Let A be the maximum number of points in Λ contained in a disk $\Delta_R + y$, $y \in V$. Let $M = \max|S(x)|$ be the ∞ norm of S.

Let f be a C^k function such that the sum of the absolute values of all the k derivatives is bounded by some number N.

Theorem 18.13. *If $S(x)$ is a superfunction of power $k \geq 1$, then*

$$|f - \sigma_h(S *' (\sigma_{1/h}f))| = |f - f_h| \leq \frac{(sR)^k}{k!} AMNh^k.$$

In particular, $S(x)$ has approximation power k.

Proof. Take any point x_0 and let q be the polynomial expressing the Taylor series of f around x_0 at order $< k$. Let $g := f - q$. By definition of superfunction we have $q - \sigma_h(S *' (\sigma_{1/h}q)) = 0$. Thus

$$|f(x) - \sigma_h(S *' (\sigma_{1/h}f))(x)| = |g(x) - \sigma_h(S *' (\sigma_{1/h}g))(x)|.$$

Since $|x_0/h - a| < R$ means that $|x_0 - ha| < hR$ we have

$$\sigma_h(S *' (\sigma_{1/h}g))(x_0) = \sum_{a \in \Lambda} S(x_0/h - a)g(ha) = \sum_{\substack{a \in \Lambda, \\ |x_0 - ha| < hR}} S(x_0/h - a)g(ha).$$

Moreover, since $k \geq 1$, we have $g(x_0) = 0$ hence

$$|f(x_0) - \sigma_h(S *' (\sigma_{1/h}f))(x_0)| = \left| \sum_{a \in \Lambda, |x_0 - ha| < hR} S(x_0/h - a)g(ha) \right|$$

and by Theorem 6.1, for such a we have $|g(ha)| \leq \frac{(sR)^k}{k!} Nh^k$, so that

$$|f(x_0) - f_h(x_0)| = \left| \sum_{a \in \Lambda, |x_0 - ha| < hR} S(x_0/h - a)g(ha) \right| \leq \frac{(sR)^k}{k!} AMNh^k.$$

Remark 18.14. Given a function $M(x)$ with compact support, if there is a finite difference operator Q such that $S(x) := QM(x)$ is a superfunction of power k, then also $M(x)$ has approximation power k. In fact, we have that

$$(QM) *' f = M *' Qf.$$

Lemma 18.15. *Assume that semidiscrete convolution by $M(x)$ maps polynomials of degree $\leq h$ into themselves for all $h < k$ and the translates of $M(x)$ form a partition of 1.*

Then there is a finite difference operator Q such that $S(x) := QM(x)$ is a superfunction of power k.

Proof. Denote by P_k the space of polynomials of degree $< k$ and by T the operator $q \mapsto M(X) *' q$ on P_k. By the previous remark it is thus sufficient to see that T is invertible and its inverse Q is a difference operator.

Clearly, T commutes with the algebra A_k of difference operators on P_k, thus by Proposition 18.10 it lies in A_k. Thus we only need to see that T is invertible. If T is not invertible, its kernel is a proper submodule and thus contains 1. This is excluded by the hypothesis that $M *' 1 = 1$.

By Theorem 17.4, when $p \in D(X)$ we have that also $B_X *' p \in D(X)$. We thus have that the approximation power of B_X equals at least $m(X)$. In the next section we see that it is equal to $m(X)$ and make explicit a superfunction of power $m(X)$ associated to $B(X)$.

18.1.5 Approximation Power

Theorem 18.16. *The approximation power of B_X equals $m(X)$.*

Proof. We ave already seen in the previous paragraph that B_X has approximation power at least $m(X)$. Thus we need to show that B_X does not have approximation power larger than $m(X)$.

We start with a simple fact.

Lemma 18.17. *Let us consider a cube C in V with sides of some length r. Let z be a vector in the direction of one of the sides. Assume that f is a function such that $D_z f = 1$ and g is a function such that $D_z g = 0$. Then*

$$\int_C |f - g|\, dx \geq |z|^{-1} \frac{r^{s+1}}{4}.$$

Proof. With respect to a suitable orthonormal basis, C is the standard cube $[0, r]^s$ and $D_z = |z|\frac{\partial}{\partial x_1}$. We use the simple estimate

$$\int_0^r |ax + c|\, dx \geq |a|\frac{r^2}{4}.$$

By assumption, that $f - g = |z|^{-1} x_1 + h(x_2, \ldots, x_s)$, we thus have

$$\int_0^r \cdots \int_0^r |f - g|\, dx_1 \cdots dx_s \geq |z|^{-1} \frac{r^2}{4} \int_0^r \cdots \int_0^r dx_2 \cdots dx_s = |z|^{-1} \frac{r^{s+1}}{4}.$$

Let us now go back to the proof of Theorem 18.16 . Consider a minimal cocircuit Y in X and take a polynomial f of degree $|Y|$ with $D_Y f = 1$. We want to show that f cannot be approximated to order $m(X)+1 = |Y|+1$ by elements of $\mathcal{S}_X$.

Let us fix p with $1 \le p \le \infty$. Suppose by contradiction that for the bounded domain G, there exists a sequence $f_h \in \sigma_h(\mathcal{S}_X)$, $h = 1/n$, $n \in \mathbb{N}$, such that we have

$$\|f_h - f\|_{L^p(G)} = o(h^{|Y|}).$$

Let $m = m(X)$. Given any list of vectors T in V, restrict the linear operator $D_T = \prod_{u \in T} D_u$ to the finite-dimensional space P^m of polynomials of degree $\le m$. We then have a constant $c_T = c > 0$ with

$$\|D_T g\|_{L^1(G)} < c\|g\|_{L^1(G)}, \ \forall g \in P^m.$$

Since the space P^m is invariant under translations, we also have, for every vector v:

$$\|D_T g\|_{L^1(G+v)} < c\|g\|_{L^1(G+v)}, \ \forall g \in P^m,$$

with the same constant c.

We want to apply this inequality to the infinitely many bounded connected components of the complement of the cut region. Since, up to translations, there are only finitely many of these components, there is a constant d such that for each such component U,

$$\|D_T g\|_{L^1(U)} < d\|g\|_{L^1(U)}, \ \forall g \in P^m.$$

Under the change of scale one has

$$\|D_T g\|_{L^1(hU)} < dh^{-|T|}\|g\|_{L^1(hU)}, \ \forall g \in P^m. \tag{18.7}$$

Let us consider $Y = \{Z, z\}$, so that $D_z(D_Z f) = 1$.

We estimate $\|f_h - f\|_{L^1(G)}$ as follows. Consider all the components U such that $hU \subset G$.

Lemma 18.18. *Given any domain G, there exists a positive constant K such that for $n \in \mathbb{N}$ large enough and $h = n^{-1}$, there are at least $n^s K$ components U with $hU \subset G$.*

Proof. Choosing a basis $\underline{b}$ out of X, the components of the cut locus for X are contained in those for $\underline{b}$. It is thus enough to prove the same statement when X is a basis. With a coordinate change we may even assume that X is the canonical basis. Then the components $n^{-1}U$ are just the hypercubes of the refined standard lattice. Moreover, we may also restrict G and assume that it is a hypercube parallel to the standard one, and the claim is an easy exercise that we leave to the reader.

Set now $G_h = \cup_{hU \subset G} hU$. We have

$$\|D_Z f_h - D_Z f\|_{L^1(G_h)} = \sum_{hU \subset G} \|D_Z f_h - D_Z f\|_{L^1(hU)}.$$

We can now use the fact that f_h, on each set hU, coincides with a polynomial in $D(X)$, in particular of degree $\leq m$. Hence we can apply inequality (18.7) and get

$$\|D_Z f_h - D_Z f\|_{L^1(hU)} < dh^{-|Y|+1}\|f_h - f\|_{L^1(hU)}.$$

Summing over all the components $hU \subset G$, we have

$$\|D_Z f_h - D_Z f\|_{L^1(G_h)} < dh^{-|Y|+1}\|f_h - f\|_{L^1(G)}. \tag{18.8}$$

Choose for each component U a cube $C_U \subset U$ with sides of some size r and with a side parallel to z. Since by construction $D_z(D_Z f) = D_Y f = 1$ and $D_z(D_Z f_h) = D_Y f_h = 0$ on each U (since $f_h \in D(X)$ on hU), we can apply Lemma 18.17 to the functions $D_Z f_h, D_Z f$ and the cubes hC_U. Setting $C := K|z|^{-1}r^{s+1}/4$, we estimate

$$\|D_Z f_h - D_Z f\|_{L^1(G_h)} \geq \sum_{hU \subset G} \|D_Z f_h - D_Z f\|_{L^1(hC_U)}$$

$$> \frac{K|z|^{-1}h^{-s}(hr)^{s+1}}{4} = Ch.$$

Combining this inequality with (18.8), we deduce

$$Ch < dh^{-|Y|+1}\|f_h - f\|_{L^1(G)},$$

or

$$\|f_h - f\|_{L^1(G)} > \frac{C}{d}h^{|Y|}.$$

By Hölder's inequality there is a constant H with

$$\frac{C}{d}h^{|Y|} < \|g_h - f\|_{L^1(G)} \leq H\|f_h - f\|_{L^p(G)} = o(h^{|Y|}),$$

yielding a contradiction.

18.1.6 An Explicit Superfunction

We want to make explicit the superfunction associated to B_X. Recall (Corollary 17.9) that the operator

$$Q := \prod_{a \in X} \frac{D_a}{1 - e^{-D_a}}$$

has the property that

$$Q(B_X *' q) = q$$

for every polynomial $q \in D(X)$.

We write Q formally as a difference operator as follows using equation (18.5):

$$D_x = -\log(1 - \nabla_x) = \sum_{k=1}^{\infty} \frac{\nabla_x^i}{i},$$

from which we deduce that

$$\frac{D_x}{1 - e^{-D_x}} = \sum_{i=0}^{\infty} \frac{\nabla_x^i}{i+1}.$$

Thus

$$Q := \prod_{a \in X} \left(\sum_{i=0}^{\infty} \frac{\nabla_a^i}{i+1} \right). \tag{18.9}$$

Remark 18.19. One has that

$$\frac{D_x}{1 - e^{-D_x}} = \sum_{i \geq 0} \frac{B_i}{i!} (-D_x)^i,$$

where B_i is the i-th Bernoulli number. Thus for each $n \geq 0$,

$$B_n = (-1)^n \left[\sum_{i \geq 0} \frac{B_i}{i!} (-D_x)^i \right] (x^n)_{|x=0} = \left[\sum_{i=0}^{n} \frac{\nabla_1^i}{i+1} \right] (x^n)_{|x=0}.$$

Indeed, $\sum_{i=0}^{n} (\nabla_1^i/(i+1)) x^n$ is a variant of the n-th Bernoulli polynomial.

In fact, the difference between the n-th Bernoulli polynomial and the n-th previously defined polynomial equals $nx^{n-1} = D_x(x^n)$.

Define Q_X to be the difference operator expressed by Q truncated at order $m(X) - 1$ (or maybe higher). We have that Q_X acts as the inverse of $q \mapsto B_X *' q$ on the polynomials of degree $\leq m(X) - 1$.

Theorem 18.20. *The function $Q_X B_X$ is a superfunction of power $m(X)$.*

Discrete convolution $Q_X B_X *' f = B_X *' Q_X f$ generates under scaling an approximation algorithm of power $m(X)$.

Example 18.21. The hat function $b_1(x)$:

$$b_1(x) = \begin{cases} x & \forall\, 0 \leq x \leq 1, \\ 2 - x & \forall\, 1 \leq x \leq 2, \\ 0 & \text{otherwise.} \end{cases}$$

In this case $X = \{1,1\}$, $m(X) = 2$. We have seen in Proposition 18.4 that the cardinal spline space coincides with the space of all continuous functions on $\mathbb{R}$ that are linear on the intervals $[i, i+1]$, $i \in \mathbb{Z}$. Such a function is completely determined by the values $f(i)$ on the integers. Moreover, given any function f on $\mathbb{Z}$, we see that $b_1 * f$ at i equals $f(i-1) = \tau_1 f(i)$. As for the operator Q_X, we have that it equals $1 + \nabla_1 = 2 - \tau_1$. On the other hand, on the space of linear polynomials we see that $2 - \tau_1 = \tau_{-1}$ so in this particular case the most convenient choice of Q_X is τ_{-1}. In general, it is a difficult question to describe, in the simplest possible way, the operator Q_X as a linear combination of translations.

Remark 18.22. Our algorithm (18.6) is not too sensitive to the way we truncate Q. Suppose we take two truncations that differ by a difference operator T with terms only of order $N \geq m(X)$. The difference of the resulting approximating functions is

$$\sigma_h[B_X *' T\sigma_{1/h}(g(x))].$$

The following lemma tells us that these terms contribute to order $O(h^N)$ and hence do not change the statement of the theorem.

Let $A = (a_1, \ldots, a_q)$ be a list of q nonzero vectors in $\mathbb{R}^s$. We use the notation $|f|_\infty$ for the L^∞ norm on the whole space.

Lemma 18.23. *Let f be a function of class C^q with bounded derivatives on the space V. We have, for any positive h,*

$$|\nabla_A \sigma_{1/h} f|_\infty \leq h^q |D_A f|_\infty. \tag{18.10}$$

Proof. By elementary calculus, for any vector a we have the simple estimate $|\nabla_a \sigma_{1/h} f|_\infty \leq h|D_a f|_\infty$.

By induction, set $A := \{B, a\}$. We have

$$|\nabla_B \nabla_a \sigma_{1/h} f| = |\nabla_B \sigma_{1/h} \nabla_{ha} f| \leq h^{q-1} |D_B \nabla_{ha} f|_\infty$$
$$= h^{q-1} |\sigma_{1/h} D_B \nabla_{ha} f|_\infty = h^{q-1} |\nabla_a \sigma_{1/h} D_B f|_\infty \leq h^q |D_A f|_\infty.$$

Example 18.24. Consider $s = 1$, $X = 1^m$, so $B_X = b_{m-1}$, $m(X) = m$, and $B(X) = [0, m]$. If $\nabla = \nabla_1 = 1 - \tau_1$, we have that Q_X is the truncation, at order $m - 1$ in ∇, of

$$\left(\sum_{i=0}^{m-1} \frac{\nabla^i}{i+1} \right)^m = \left(\sum_{i=0}^{m-1} \frac{(1-\tau_1)^i}{i+1} \right)^m.$$

Thus Q_X has the form $\sum_{j=0}^{m-1} c_j \tau_j$. The corresponding superfunction $Q_X B_X$ is thus supported in $[0, 2m - 1]$. The chambers are the open intervals $(n, n+1)$, $n \in \mathbb{Z}$.

18.1.7 Derivatives in the Algorithm

We want to see next how our algorithm behaves with respect to derivatives. In fact, the theory of Strang–Fix ensures that when we have approximation power m, we can approximate a function f with a spline to order $O(h^m)$ and at the same time its derivatives of order k as $k < m$ to order $O(h^{m-k})$. We want to show that our proposed algorithm, under a suitable choice of the truncation for Q_X, indeed satisfies this property. Since we shall work also with subsets Y of X, let us use the notation $A_X^h(f) := f_h$, the approximant to f in the algorithm associated to X at the step $n = 1/h$.

Again, the algorithm being local, we may assume that f and all of its derivatives up to order $m(X)$ are continuous and bounded everywhere by some constant C.[2]

By Remark 18.22, the choice of truncation is not essential for the final Theorem 18.26. For our next theorem we make a suitable choice of the truncation. Let us establish some notation and normalize the truncation, defining

$$T_a^{(m)} := \sum_{i=0}^{m-1} \frac{\nabla_a^i}{i+1}, \quad \nabla_a^{[m]} := \sum_{i=1}^{m} \frac{\nabla_a^i}{i} = \nabla_a T_a^{(m)},$$

$$Q_X := \prod_{a \in X} \left(\sum_{i=0}^{m(X)-1} \frac{\nabla_a^i}{i+1} \right) = \prod_{a \in X} T_a^{(m(X))}. \tag{18.11}$$

Observe that this truncation is not the optimal one; it is convenient only in order to carry out the proof.

Lemma 18.25. *Given a nonempty list $A = (a_1, \ldots, a_k)$ of nonzero vectors in V and $m \geq k$, consider the two operators*

$$D_A := D_{a_1} \cdots D_{a_k}, \qquad \nabla_A^{[m]} := \nabla_{a_1}^{[m]} \cdots \nabla_{a_k}^{[m]}.$$

Then for any function g of class C^{m+1}, $h > 0$, and $x_0 \in V$,

$$|([\nabla_A^{[m]} - D_A]\sigma_{1/h}g)(x_0/h)| \leq h^{m+1} K \sum_{|\alpha|=m+1} \|D^\alpha g\|_{L^\infty(x_0+\Delta_h r)}.$$

Here K, r are constants (dependent on A and m) independent of g and x_0.

Proof. On polynomials of degree less than or equal to m, every difference operator consisting of terms of order $> m$ is zero. Hence on such polynomials $\nabla_A^{[m]}$ acts as $\prod_{i=1}^{k} \sum_{i=1}^{\infty} \nabla_{a_i}^i / i = D_A$.

The operator $\nabla_A^{[m]}$ is a finite difference operator. It follows that

---

[2]If we assume only that f has bounded derivatives of order $\leq t$, we shall get the same results but only for these derivatives.

$$(\nabla_A^{[m]})f = \sum_{i=1}^{k} c_i f(x_0 - u_i),$$

for some $u_i \in \Lambda$ and for every function f. Therefore, one has the uniform estimate

$$|(\nabla_A^{[m]} f)(x_0)| \leq c\|f\|_{L^\infty_{x_0 + \Delta_r}}, \tag{18.12}$$

where $c = \sum_{i=1}^{k} |c_i|$, $r = \max(|u_i|)$, and as usual, Δ_r denotes the disk of radius r centered at 0.

Given f, let q be the Taylor series of f at x_0 truncated at degree $\leq m$. We have then $[\nabla_A^{[m]} - D_A](f) = [\nabla_A^{[m]} - D_A](f - q)$. Since $m \geq k$, we have also $D_A(f - q)(x_0) = 0$; hence

$$[\nabla_A^{[m]} - D_A](f)(x_0) = (\nabla_A^{[m]}(f - q))(x_0). \tag{18.13}$$

We now apply this to the function $f = \sigma_{1/h} g$ at the point x_0/h. Denote by q the Taylor series of g at x_0 truncated at degree $\leq m$. We obtain the estimate, using (18.12), (18.13) and (6.1):

$$\begin{aligned}
|[\nabla_A^{[m]} - D_A](\sigma_{1/h} g)(x_0/h)| &= (\nabla_A^{[m]} \sigma_{1/h}(g - q))(x_0/h) \\
&\leq c\|\sigma_{1/h}(g - q)\|_{L^\infty(x_0/h + \Delta_r)} = c\|g - q\|_{L^\infty(x_0 + \Delta_{hr})} \\
&\leq h^{m+1} K \sum_{|\alpha| = m+1} \|D^\alpha g\|_{L^\infty(x_0 + \Delta_{hr})}.
\end{aligned}$$

The constant $K = c\,Cr^{m+1}$ is an absolute constant independent of g, x_0 (but depends on m and A).

Theorem 18.26. *Under the algorithm $f_h = A_X^h(f)$, for any domain G and for every multi-index $\alpha \in \mathbb{N}^s$ with $|\alpha| \leq m(X) - 1$, we have*

$$\|\partial^\alpha f_h - \partial^\alpha f\|_{L^\infty(G)} = O(h^{m(X) - |\alpha|}).$$

Proof. Given any domain H whose closure lies in G, we may extend the restriction of f to H to a function defined on V having uniformly bounded derivatives up to order $m(X) - 1$. If we prove the estimate for such functions, we are easily done.

We prove first the estimate

$$\|D_Y f_h - D_Y f\|_\infty = O(h^{m(X) - |Y|}) \tag{18.14}$$

for the differential operators D_Y, where Y is a sublist in X that is not a cocircuit and then show how the required estimate follows easily from this. Recall that (cf. Proposition 7.14), we have $D_Y(B_X) = \nabla_Y B_{X \setminus Y}$ as distributions. Under our assumptions, $D_Y(B_X)$ as well as $D_Y g$ for any $g \in \mathcal{S}_X$ is in fact a function, although when $|Y| = m(X)$ it need not be continuous.

Let $(D_Y f)_h = A^h_{X\setminus Y}(D_Y f)$. Since we have $m(X\setminus Y) \geq m(X) - |Y|$, we deduce from Theorem 18.20 that $\|(D_Y f)_h - D_Y f\|_\infty = O(h^{m(X)-|Y|})$, hence the theorem is proved once we establish the estimate

$$\|D_Y f_h - (D_Y f)_h\|_\infty = O(h^{m(X)-|Y|}).$$

Using the commutation rules (18.3) and the choice (18.11) of Q_X, we get

$$D_Y f_h = h^{-|Y|}\sigma_h(D_Y B_X *' Q_X \sigma_{1/h} f) = h^{-|Y|}\sigma_h(\nabla_Y B_{X\setminus Y} *' Q_X \sigma_{1/h} f).$$

Set $\tilde{Q}_{X\setminus Y} = \prod_{z\in X\setminus Y} T_z^{(m(X))}$ and observe that $\nabla_Y \prod_{z\in Y} T_z^{(m(X))} = \nabla_Y^{[m(X)]}$ (as in Lemma 18.25). We get finally

$$D_Y f_h = \sigma_h(B_{X\setminus Y} *' \tilde{Q}_{X\setminus Y} h^{-|Y|} \nabla_Y^{[m(X)]} \sigma_{1/h} f). \tag{18.15}$$

Compare (18.15) with the approximants to $D_Y f$:

$$(D_Y f)_h = \sigma_h(B_{X\setminus Y} *' Q_{X\setminus Y} \sigma_{1/h} D_Y f).$$

By construction, $\tilde{Q}_{X\setminus Y} - Q_{X\setminus Y}$ is a sum of terms of order $\geq m(X\setminus Y)$ hence these approximants differ (by Lemma 18.23) from the approximants

$$(\widetilde{D_Y f})_h := \sigma_h(B_{X\setminus Y} *' \tilde{Q}_{X\setminus Y} \sigma_{1/h} D_Y f)$$

by order $O(h^{m(X\setminus Y)})$.

From the estimate

$$|D_Y f_h - (D_Y f)_h| \leq |D_Y f_h - (\widetilde{D_Y f})_h| + |(\widetilde{D_Y f})_h - (D_Y f)_h|$$

we are reduced to showing that $|D_Y f_h - (\widetilde{D_Y f})_h| = O(h^{m(X)-|Y|+1})$.

We have

$$D_Y f_h - (\widetilde{D_Y f})_h = \sigma_h\left(B_{X\setminus Y} *' \tilde{Q}_{X\setminus Y}\left[h^{-|Y|}\nabla_Y^{[m(X)]}\sigma_{1/h} f - \sigma_{1/h} D_Y f\right]\right).$$

Clearly

$$|\sigma_h(B_{X\setminus Y} *' \tilde{Q}_{X\setminus Y}[h^{-|Y|}\nabla_Y^{[m(X)]}\sigma_{1/h} f - \sigma_{1/h} D_Y f])|_\infty$$

$$= |\tilde{Q}_{X\setminus Y} B_{X\setminus Y} *' [h^{-|Y|}(\nabla_Y^{[m(X)]} - D_Y)\sigma_{1/h} f]|_\infty.$$

The norm $|f * a|_\infty$ of the convolution $f * a$ of a function with compact support f by a sequence a is clearly bounded by $C|a|_\infty$, where $C = |f|_\infty d$ and d is the maximum number of points in Λ that lie in the support of a translate $\tau_a f$ of f as a varies in Λ. We can now apply the estimate to the sequence $(\nabla_Y^{[m(X)]} - D_Y)\sigma_{1/h} f$ (computed on points of Λ), given by Lemma 18.25. We obtain an estimate $O(h^{m(X)+1})$ proving formula (18.14).

Now let $m \leq m(X) - 1$. Since for any sublist Y of X with m elements we have that $X\setminus Y$ still generates V, we deduce by a simple induction that the operators D_Y linearly span the same space spanned by the partial derivatives ∂^α, $\alpha \in \mathbb{N}^s$, with $|\alpha| = m$. Thus the claim of the theorem follows.

Remark 18.27. Theorem 18.26 is also independent of the way in which we perform the truncation, or in other words, on which superfunction we choose in $\mathcal{S}_X$. In fact, we may apply Lemma 18.23 to any derivative $\partial^\alpha f$ and see that the algorithm, applied to a function f of class $C^n, n \geq m(X)$, changes by a function that goes to zero as fast as $h^k C$, where $k = \min(m(X), n - |\alpha|)$.

18.2 Super and Nil Functions

18.2.1 Nil Functions

Definition 18.28. Let F be a function with compact support.

We say that F is *nilpotent* of order r if for every polynomial q of degree strictly smaller than r, we have $F *' q = 0$.

For a list X and $F \in \mathcal{S}_X$ we say that F is nilpotent if $F *' q = 0$ for all $q \in D(X)$.

Remark 18.29. A nilpotent $F \in \mathcal{S}_X$ is also nilpotent of order $m(X)$.

We clearly have the following

Proposition 18.30. *If F_1 is a superfunction and $F_2 \in \mathcal{S}_X$ has compact support, F_2 is a superfunction if and only if $F_1 - F_2$ is nilpotent.*

Take an element $F \in \mathcal{S}_X$ of the form $B_X * a$, where a is a sequence with finite support. Thus F has compact support, and by formula (6.7), we have, for every f,

$$F *' f = (B_X * a) *' f = B_X *' (f * a).$$

If $f \in D(X)$, we also have $f * a \in D(X)$ and $B_X *' (f * a) = 0$ if and only if $f * a = 0$. Thus $B_X * a$ is nilpotent if and only if $f * a = 0$ for every $f \in D(X)$. We interpret the convolution with a sequence with finite support as the action of the corresponding difference operator. Thus using the ideal J_X of difference equations satisfied by $D(X)$ (Theorem 16.5), we get the following result:

Theorem 18.31. *A function $F = B_X * a$, where a is a sequence with finite support, is nilpotent if and only if $a \in J_X$, that is, $F \in J_X B_X$.*

We have proved that a function $B_X * a$ with compact support can also be expressed as $B_X * b$ with b of finite support (Theorem 18.6). We deduce the following corollary:

Corollary 18.32. *The space of nilpotent functions in the cardinal spline space equals $J_X B_X$.*

Moreover, let J be the ideal of $\mathbb{C}[\Lambda]$ generated by the difference operators ∇_a, $a \in \Lambda$. For any $T \in \mathbb{C}[\Lambda]$, the function $F = T B_X$ is nilpotent of order $\geq r$ if and only if $T \in J^r$.

We want to relate the notion of nilpotent functions with the algorithm (18.6).

Consider $F = TB_X$ and the family of functions $\sigma_h(F *' \sigma_{1/h}g)$ when g is bounded with its derivatives up to order $m(X)$

Proposition 18.33. *Let $r \leq m(X)$ and $F = TB_X$. Then F is nilpotent of order $\geq r$ if and only if (for any such g) the sequence $\sigma_h(F *' \sigma_{1/h}g)$ converges to zero as $O(h^r)$.*

Proof. In one direction this is an immediate consequence of Lemma 18.23. In the other we argue as follows. Assume that $\sigma_h(F *' \sigma_{1/h}g)$ converges to zero as $O(h^r)$. Let j be the minimum such that $T \in J^j$. Since $T \notin J^{j-1}$, there is a homogeneous polynomial q of degree j with $Tq \neq 0$, and since $T \in J^j$, we have that Tq is of degree 0 and we may assume $Tq = 1$. Choose g coinciding with q on some open set H. On H we have

$$\sigma_h(TB_X *' \sigma_{1/h}g) = h^j \sigma_h(B_X *' Tq) = h^j.$$

Thus $j \geq r$.

As a consequence, we get our next theorem:

Theorem 18.34. *Given $F = TB_X \in \mathcal{S}_X^f$ with T a finite difference operator for which the algorithm $g - \sigma_h(F *' \sigma_{1/h}g)$ converges to zero as $O(h^{m(X)})$, then F is a superfunction of power $m(X)$.*

Proof. By Proposition 18.30 we need to show that $F = Q_X B_X + N$ with N nilpotent of order $\geq m(X)$. We make the computations locally on some large disk, so we can take among our input functions also polynomials. Let p be a homogeneous polynomial of degree $t < m(X)$. We have

$$\sigma_h(F *' \sigma_{1/h}p) = \sigma_h(TB_X *' h^t p) = h^t \sigma_h(TQ_X^{-1}(p)),$$

where Q_X^{-1} is the inverse of Q_X modulo $J^{m(X)}$. If $T \in J^{m(X)}$, the polynomial $TQ_X^{-1}(p)$ is zero. Otherwise, develop the polynomial $TQ_X^{-1}(p)$ as $TQ_X^{-1}(p) = \sum_{j=0}^{t} q_j(x)$, where $q_j(x)$ is the homogeneous component of degree j. We have $h^t \sigma_h(TQ_X^{-1}(p)) = \sum_{j=0}^{t-1} q_j(x)h^{t-j}$. Therefore, $p - \sigma_h(F *' \sigma_{1/h}p) = p - \sum_{j=0}^{t} q_j(x)h^{t-j}$ is a polynomial in h of degree at most t; hence it is of order $= O(h^{m(X)})$ if and only if $p - \sum_{j=0}^{t-1} q_j(x)h^{t-j} = 0$, equivalently $(1 - TQ_X^{-1})p = 0$. If we assume that this is valid for all p of degree smaller than $m(X)$, we need to have that $1 - TQ_X^{-1} \in J^{m(X)}$. Hence $Q_X - T \in J^{m(X)}$, and the claim follows.

It may be of some interest to choose among all superfunctions some that are particularly well behaved, in particular with support as small as possible, in order to minimize the cost of the algorithm. For this we have to choose a possibly different normalization for Q_X, that is uniquely determined only modulo J_X. One normalization is the following. We have proven in Theorem 13.19 that given a chamber $\mathfrak{c}$, the classes of elements e^a as $a \in \delta(\mathfrak{c}|X)$ form a basis of $\mathbb{C}[\Lambda]/J_X$. Thus we can write, modulo J_X, any difference operator uniquely in the form $\sum_{a \in \delta(\mathfrak{c}|X)} c_a \tau_a$. We deduce that the superfunction can be chosen with support contained in $\cup_{a \in \delta(\mathfrak{c}|X)} a + B(X)$. A different choice of $\mathfrak{c}$ may yield a different normalization.

Remark 18.35. In the 1-dimensional case, set $\tau = \tau_1$, and $\mathbb{C}[\Lambda] = \mathbb{C}[\tau, \tau^{-1}]$. If $X = 1^m$, we have that $J_X = ((1 - \tau)^m)$, and $D(X)$ is the space of polynomials in x of degree $< m$. Thus the usual argument using the Vandermonde determinant shows that given any distinct integers $a_1, \ldots, a_m$, the classes of the elements $\tau_{a_i} = \tau^{a_i}$ form a basis of $\mathbb{C}[\tau, \tau^{-1}]/((1 - \tau)^m)$. In particular, we want to keep these numbers a_i close to 0, so for $m = 2k+1$ odd we may choose $-k, \ldots, 0, \ldots, k$, while for $m = 2k$ even we may choose $-k, \ldots, 0, \ldots, k - 1$.

Examples. Let us use a notation as in Example 7.9. We denote by $s_m(x) = (\sum_{i=0}^{m} \frac{\nabla^i}{i+1})^{m+1} b_m(x)$ a superfunction associated to $b_m(x)$. Given $a > 0$, we define $n_{a,m} = \nabla^a b_m$. This is a nilpotent function of degree a.

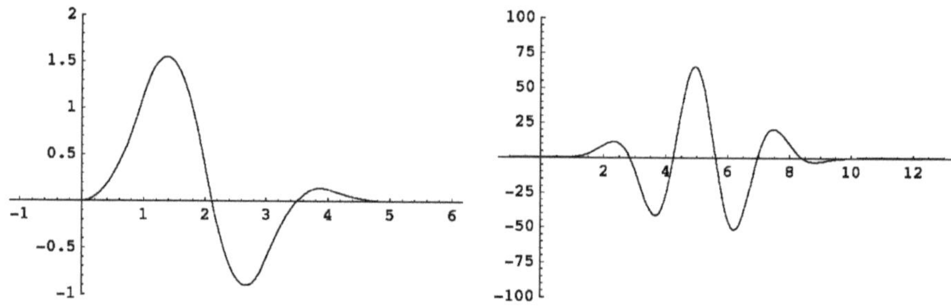

Fig. 18.1. $s_2(x)$ and $s_5(x)$

We add some nilpotent functions.

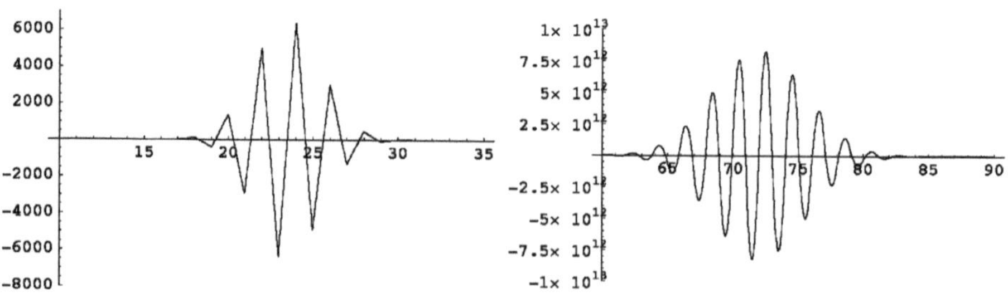

Fig. 18.2. $n_{2,14}(x)$ and $n_{3,46}(x)$

Remark 18.36. For any integrable function f we have $\int_V \nabla_a f \, d\mu = 0$, where $d\mu$ is a translation-invariant measure.

Thus given any polynomial $p(x_1, \ldots, x_i)$ and elements $a_i \in X$, we have

$$\int_V p(\nabla_{a_1}, \ldots, \nabla_{a_s}) B_X \, d\mu = p(0).$$

In particular, by the previous remarks and the expression (18.9) for Q_X we deduce that $\int_V F \, d\mu = 1$ for any superfunction F associated to B_X.

18.3 Quasi-Interpolants

18.3.1 Projection Algorithms

In this section we analyze some algorithms of interpolation different from the one proposed in Section 18.1.6. We follow closely the two papers [75], [76].

The starting point of any such algorithm is a linear functional L mapping functions (with some regularity restrictions) to splines in the cardinal space $\mathcal{S}_X$. Once such a functional is given, one can repeat the approximation scheme associating to a function f and a parameter h the approximant $f_h := \sigma_h L \sigma_{1/h} f$. Of course, in order for such an algorithm to be a real approximation algorithm, one requires suitable hypotheses on L. Let us review some of the principal requirements that we may ask of L.

(1) *Translation invariance.* By this we mean that L commutes with the translations τ_a, $\forall a \in \Lambda$.
(2) *Locality.* In this case one cannot impose a strong form of locality, but one may ask that there be a bounded open set Ω such that if $f = 0$ on Ω, then Lf also equals 0 on Ω.
   If we assume translation invariance, we then have also locality for all the open sets $a + \Omega, a \in \Lambda$.
(3) *Projection property.* One may ask that L be the identity on suitable subspaces of $\mathcal{S}_X$. For instance, the algorithm treated in Section 18.1.6 is the identity on the space $D(X)$ of polynomials contained in $\mathcal{S}_X$.

Let us start to analyze the previous conditions. We shall do this in the *unimodular case,* that is, when all the bases extracted from $X \subset \Lambda$ are in fact integral bases of the lattice Λ. In this case one knows the linear independence of translates of the box spline (Theorem 17.11).

In this case, the functional L can be expressed through its *coordinates*:

$$Lf = \sum_{a \in \Lambda} L_a(f) B_X(x - a).$$

If we assume translation invariance $L\tau_b = \tau_b L$, we have

$$\tau_b L f = \sum_{a \in \Lambda} L_a(f) B_X(x - a - b) = \sum_{a \in \Lambda} L_{a+b}(f) B_X(x - a).$$

On the other hand, $L\tau_b f = \sum_{a \in \Lambda} L_a(\tau_b f) B_X(x - a)$, so that by linear independence, we get the identities $L_{a+b}(f) = L_a \tau_b f$, $\forall a, b \in \Lambda$. In particular, L is determined by the unique linear functional L_0 by setting $L_a f := L_0 \tau_a f$.

A question studied in [75], [76] is whether we can construct L_0 and hence L so that L is the identity on the entire cardinal space $\mathcal{S}_X$. From the previous discussion, assuming X unimodular, we only need to construct a linear functional L_0 such that on a spline $g = \sum_{a \in \Lambda} c_a B_X(x - a)$, we have $L_0 g = c_0$ and set $Lf = \sum_{a \in \Lambda} L_0(f(x - a)) B_X(x - a)$.

A way to construct such a linear functional L_0 is the following. Consider a bounded open set Ω with $B(X) \cap \Omega \neq \emptyset$. The set

$$A = \{\alpha_1, \ldots, \alpha_k\} := \{\alpha \in \Lambda \mid (\alpha + B(X)) \cap \Omega \neq \emptyset\}$$

is finite and $0 \in A$, say $0 = \alpha_1$. Since B_X is supported on $B(X)$, only the translates $B_X(x - \alpha)$, $\alpha \in A$, do not vanish on Ω. We have shown in Proposition 17.13 that the restrictions of the functions $B_X(x - \alpha)$, $\alpha \in A$, to Ω are still linearly independent. Let us denote by $\mathcal{S}_X(A)$ the k-dimensional space with basis the elements $B_X(x - \alpha)$, $\alpha \in A$.

Notice that, if we have such an L_0 that is local on Ω, we must have

$$L_0\left(\sum_{a \in \Lambda} c_a B_X(x - a)\right) = c_0 = L_0\left(\sum_{\alpha \in A} c_\alpha B_X(x - \alpha)\right) = \sum_{\alpha \in A} c_\alpha L_0(B_X(x - \alpha)).$$

In order to determine an L_0 that satisfies the previous identity, we proceed as follows.

Suppose that we show the existence of k linear functionals T_i (on the given space of functions that we want to approximate), $i = 1, \ldots, k$, local on Ω and such that the $k \times k$ matrix with entries $a_{i,j} := T_i B_X(x - \alpha_j)$ is invertible. We have then on the space $\mathcal{S}_X(A)$ that $T_j = \sum_{i=1}^{k} c_{\alpha_i} a_{j,i}$. If $b_{h,k}$ are the entries of the inverse of the matrix $(a_{i,j})$, we have

$$c_0 = c_{\alpha_1} = \sum_{h=1}^{k} b_{1,h} T_h \tag{18.16}$$

as a linear functional on $\mathcal{S}_X(A)$. So we set $L_0 := \sum_{h=1}^{k} b_{1,h} T_h$.

18.3.2 Explicit Projections

There are several possible approaches to the construction of k functionals T_i with the required properties. One is to consider the Hilbert space structure and define for $\alpha_i \in A$ the functional $T_i f := \int_\Omega f(x) B_X(x - \alpha_i) dx$. By the local linear independence, these functionals are clearly linearly independent on the space $\mathcal{S}_X(A)$.

A second method consists in observing that by the linear independence, one can find k points $p_1, \ldots, p_k \in \Omega$ such that the evaluation of functions in $\mathcal{S}_X(A)$ at these points establishes a linear isomorphism between $\mathcal{S}_X(A)$ and $\mathbb{R}^k$. In other words, the $k \times k$ matrix with entries $B_X(p_i - a)$, $i = 1, \ldots, k$, $a \in A$ is invertible, and we can define $T_i f := f(p_i)$ as the evaluation of f at the point p_i.

In general, it seems to be difficult to exhibit explicit points with the required property (although most k-tuples of points satisfy the property).

Let $X = \{a_1, \ldots, a_m\}$. We make some remarks in the case in which we choose $\Omega := \{\sum_{i=1}^m t_i a_i, \ 0 < t_i < 1\} = \mathring{B}(X)$, the interior of the zonotope.

Let $\sigma_X := \sum_{i=1}^m a_i$ and observe the symmetry condition $a \in \Omega \iff \sigma_X - a \in \Omega$.

Let $A := \Omega \cap 2^{-1}\Lambda$ be the set of half-integral points contained in Ω and $B := \{j \in \Lambda \,|\, [\Omega - j] \cap \Omega \neq \emptyset\}$.

Proposition 18.37. *The mapping $\phi : a \mapsto 2a - \sigma_X$, is a bijiection between A and B with inverse $\phi^{-1} : b \mapsto \frac{1}{2}(b + \sigma_X)$.*

Proof. First let us show that if $a \in A$, then $2a - \sigma_X \in B$. In fact, we have that
$$\sigma_X - a = a - (2a - \sigma_X) \in \Omega \cap (\Omega - (2a - \sigma_X)).$$

Conversely, assume $b \in B$ and let $a := \frac{1}{2}(b + \sigma_X)$. Clearly $a \in 2^{-1}\Lambda$. We claim that also $a \in \Omega$. By assumption, there is an element $c \in \Omega$ with $c - b \in \Omega$, and by symmetry $\sigma_X - c + b \in \Omega$. Since Ω is convex, we finally have that $\frac{1}{2}(\sigma_X - c + b + c) = a \in \Omega$.

One may thus conjecture that the half-integral points in Ω give rise to an invertible matrix C with entries $c_{a,b} := B_X(a - 2b + \sigma_X)$. This can be verified in simple examples.

If $c^{a,b}$ are the entries of the matrix C^{-1}, we deduce from (18.16) that
$$L_0 f = \sum_{b \in A} c^{0,b} f(b)$$

and the projection operator L is given by
$$L f = \sum_{a \in \Lambda} \left[\sum_{b \in A} c^{0,b} f(b - a) \right] B_X(x - a).$$

Of course, a further analysis of this formula, from the algorithmic point of view, is desirable, but we shall not try to go into this.

Stationary Subdivisions

In this chapter we want to sketch a theory that might be viewed as an inverse or dual to the spline approximations developed from the Strang-Fix conditions. Here the main issue is to approximate or fit discrete data through continuous or even smooth data. In this setting, an element of the cardinal spline space $\sum_{\alpha \in \Lambda} B_X(x - \alpha)g(\alpha)$ is obtained as a limit of discrete approximations. The main reference is the memoir of Cavaretta, Dahmen, Micchelli [31], where the reader can find proofs of the statements we make; see also Dyn and Levin [52].

19.1 Refinable Functions and Box Splines

19.1.1 Refinable Functions

The relevance of box splines for the type of algorithms that we shall discuss in this section comes from a special property that they possess.

For the box splines, the values at half-integral points are related to the values at integral points by a linear relation:

Definition 19.1. Let $d > 1$, $d \in \mathbb{N}$. We say that a continuous function F is d-**refinable**[1] if there is a finite difference operator $T = \sum_i c_i \tau_{a_i}$, $a_i \in \mathbb{Z}^s$, with constant coefficients such that

$$F(x) = \sigma_{1/d} T F(x) = \sum_i c_i F(dx - a_i). \tag{19.1}$$

T is called the *mask* of the refinement.

Remark 19.2. We can write the refining equation also as $F(x) = T'\sigma_{1/d}F(x)$, where $T' = \sum_i c_i \tau_{a_i/d}$.

---

[1]This is not the most general definition of refinable, cf. [31].

C. De Concini and C. Procesi, *Topics in Hyperplane Arrangements, Polytopes and Box-Splines*, Universitext, DOI 10.1007/978-0-387-78963-7_19,
© Springer Science+Business Media, LLC 2010

Thus the condition is that $F(x)$ lies in the space spanned by the translates of $F(dx)$ under $d^{-1}\mathbb{Z}^s$. A 2-refinable function will usually be called refinable.

This condition, plus the condition that $F(x)$ has compact support, are a starting point (usually for $d = 2$) for building a theory of *wavelets*, a basic tool in signal-processing analysis (cf. [38], [78]).

Remark 19.3. (i) Let $d \in \mathbb{N}, d > 1$, and take $x = a/d^k$, with $a \in \mathbb{Z}^s$. The identity (19.1) shows that $F(a/d^k)$ is computed through the values of F at the points b/d^{k-1}, $b \in \mathbb{Z}^s$, and so recursively at the integer points. Since we assume F continuous and the points $x = a/d^k$, $a \in \mathbb{Z}^s$ are dense, this determines F completely once we know its values at integer points.

(ii) The same equation puts a constraint also on integer points. As one easily sees by the first examples, these constraints are insufficient to determine the integer values. Nevertheless, we shall see that up to a multiplicative constant, F is uniquely determined by T (Section 19.1.4).

Clearly, if F is d-refinable it is also d^k-refinable for all $k \in \mathbb{N}$. In fact, in most applications we take $d = 2$ and insist that the vectors a_i be integral vectors. If F has support in a compact set A, then $F(2^k x)$ has support in $2^{-k} A$, and refining F expresses it as a linear combination of functions of smaller and smaller support.

Proposition 19.4. *The box spline B_X is refinable:*

$$B_X(x) = \frac{1}{2^{|X|-s}} \prod_{a \in X} (1 + \tau_{a/2}) \sigma_{\frac{1}{2}} B_X(x) \tag{19.2}$$

$$= \frac{1}{2^{|X|-s}} \sigma_{\frac{1}{2}} \prod_{a \in X} (1 + \tau_a) B_X(x) = \frac{1}{2^{|X|-s}} \sum_{S \subset X} B_X(2x - a_S).$$

Proof. The Laplace transform of B_X is $\prod_{a \in X} (1 - e^{-a})/a$, so that by (18.2), the Laplace transform of $B_X(2x)$, is $2^{-s} \prod_{a \in X} (1 - e^{-a/2})/(a/2)$. We have

$$\prod_{a \in X} \frac{1 - e^{-a}}{a} = 2^{-|X|} \prod_{a \in X} (1 + e^{-a/2}) \prod_{a \in X} \frac{1 - e^{-a/2}}{a/2}.$$

Therefore, by the properties of the Laplace transform (cf. (3.4)), we have in the nondegenerate case

$$B_X = 2^{s-|X|} \sigma_{\frac{1}{2}} \prod_{a \in X} (1 + \tau_a) B_X.$$

In the one-dimensional case the property is straightforward hence the identity follows unconditionally.

In general, given a refinable function F with mask T, it is a difficult question to decide whether a function $G := AF$ obtained from F by applying a difference operator $A = p(\tau_a)$ is still refinable.

A function G is refinable if there is a difference operator S with $G = \sigma_{\frac{1}{2}} S G$. Thus substituting, yields $SAF = \sigma_{\frac{1}{2}}^{-1} A \sigma_{\frac{1}{2}} T F$. A sufficient condition, that is often also necessary, is that $SA = \sigma_{\frac{1}{2}}^{-1} A \sigma_{\frac{1}{2}} T = p(\tau_a^2) T$.

Let us see this in the one-variable case. We have that A, T, S are Laurent polynomials and we need to study the equation (in the unknown S)

$$S(x)A(x) = A(x^2)T(x). \tag{19.3}$$

Assume that A has no multiple roots. The compatibility condition for equation (19.3) is that if α is a root of A, then either α is a root of T or α^2 is a root of A. For instance, if we start from the spline $F = b_m$, where $T = 2^{-m}(1+x)^{m+1}$, we see that the roots of A are a subset of the set I of roots of 1 such that if $\alpha \in I$, then either $\alpha^2 \in I$ or $\alpha = -1$. In this way one generates various new polynomials and corresponding functions. We shall return to this problem in the multivariable case later on.

19.1.2 Cutting Corners

The starting idea in this theory, used in computer graphics and numerical analysis, goes back to de Rham (cf. [49]), who showed how to construct a sufficiently smooth curve fitting a given polygonal curve by an iterative procedure of *cutting corners*. Of course, a polygonal is given by just giving its sequence of endpoints P_i, $i = 1, \dots, m$, that have to be joined by segments. This can be thought of as a function $f(i) := P_i$ with $f(i + m) = f(i)$. In general, we can start with a vector-valued function a defined on the integer lattice $\mathbb{Z}^s$ as input. The algorithm proceeds then by steps:

- Construct successive potential smoothings that are functions on the refined integer lattice $2^{-k}\mathbb{Z}^s$.
- Pass to the limit to determine a continuous or even differentiable (up to some degree) function that represents the *smooth* version of the discrete initial datum.

Since the work of de Rham, it has appeared clear that unless one chooses extremely special procedures, one tends to define objects with a very fractal nature. As we shall see, one of the efficient procedures is based on the box splines, and this justifies including this topic here.

Let us start now in a more formal way. We are going to consider a mesh function on a lattice Λ as a distribution $D = 2^{-s}\sum_{\lambda \in \Lambda} a(\lambda)\delta_\lambda$. In what follows we are going to choose D with finite support.

Choosing coordinates, i.e., an integral basis e_i of Λ, and $\lambda = \sum_i m_i e_i$, we identify $e^\lambda = e^{\sum_i m_i e_i}$ with the monomial $\prod_{i=1}^s x_i^{m_i}$ and the Fourier-Laplace transform of D with the Laurent polynomial $2^{-s}\sum_\lambda a(\lambda)x^\lambda$.

We now consider $T := \sigma_{\frac{1}{2}}(D) = 2^{-s}\sum_{\lambda\in\mathbb{Z}^s} a(\lambda)\delta_{\lambda/2}$ and the following recursive algorithm, associated to T. Given $g = \sum_\lambda g(\lambda)\delta_\lambda$ a discrete distribution supported on Λ, that represents the discrete data that we want to fit with a continuous function, we recursively define

$$g_0 := g, \quad g_{k+1} := \sigma_{\frac{1}{2^k}}(T) * g_k = \sigma_{\frac{1}{2^k}}(T) * \cdots * \sigma_{\frac{1}{2}}(T) * T * g. \qquad (19.4)$$

The notation we use is

$$T = 2^{-s}\sum_{\lambda\in\Lambda} a(\lambda)\delta_{\lambda/2}, \quad g_k := 2^{-sk}\sum_{\lambda\in\Lambda} g_\lambda^{(k)}\delta_{\lambda/2^k}.$$

Then we have

$$g_{k+1} = \sigma_{\frac{1}{2^k}}(T) * g_k = 2^{-s(k+1)}\sum_\mu\sum_\lambda a(\lambda)g_\mu^{(k)}\delta_{\lambda/2^{k+1}+\mu/2^k},$$

or

$$g_\lambda^{(k+1)} = \sum_\mu a(\lambda - 2\mu)g_\mu^{(k)}.$$

By induction, g_k is a distribution supported on the lattice $2^{-k}\mathbb{Z}^s$.

One of the main goals of the theory is to understand under which conditions this sequence of distributions converges, in a suitable way and for all reasonable input functions g, to a limit function $\tilde{g}$ that is continuous. In this case we call $\tilde{g}$ a *continuous or smooth fitting* of the discrete data g. Of course, in computer graphics the purpose is to plot the smooth function $\tilde{g}$ by plotting one of its discrete approximants g_k.

The fundamental case to which the general case can be reduced is that in which we start with $g = \delta_0$. Let us call the resulting sequence T_k. We have

$$T_0 = \delta_0, \quad T_1 = T, \ldots, T_{k+1} = \sigma_{\frac{1}{2^k}}(T) * T_k,$$

and for a general g, $g_k = T_k * g$. Notice that if X_0 is the support of the distribution T, then for each k, we deduce that the distribution T_k is supported in $X_0 + \frac{1}{2}X_0 + \cdots + \frac{1}{2^k}X_0$. Thus when this sequence converges to a distribution, it converges to one with compact support contained in the convex envelope of $2X_0$.

The general form of the iteration given by (19.4) and the possible limit is now

$$\tilde{g} = \lim_k g_k, \quad \lim_k T_k = \tilde{T} \implies \tilde{g} = \tilde{T} * g.$$

Thus assuming that $\tilde{T}$ is a function, the function $\tilde{g}(x) = \sum_k \tilde{T}(x-k)g(k)$ lies in the cardinal space generated by $\tilde{T}$.

Since for every $k \geq 0$ we have $T_{k+1} = \sigma_{\frac{1}{2}}(T_k) * T$ and $D = 2^{-s}\sum_\alpha a(\alpha)\delta_\alpha$, if we can pass to the limit under this identity we have the following proposition:

Proposition 19.5. $\tilde{T} = \sigma_{\frac{1}{2}}(\tilde{T}) * T = \sigma_{\frac{1}{2}}(D * \tilde{T}).$

The important remark made in [31] is that the previous proposition, in which the limit distribution $\tilde{T}$ exists and coincides with a continuous function ϕ, implies the following *functional equation* for ϕ:

$$2^s T * \sigma_{\frac{1}{2}}(\phi) = \phi, \quad \sum_{\alpha \in \Lambda} a(\alpha)\phi(2x - \alpha) = \phi(x). \qquad (19.5)$$

This functional equation expresses the fact that ϕ is refinable, in the sense of Definition 19.1.

Conversely, let us begin with a continuous refinable function $\phi(x)$ satisfying $\sum_{\alpha \in \Lambda} a(\alpha)\phi(2x-\alpha) = \phi(x)$. One can then take the corresponding distribution $T = 2^{-s} \sum_{\alpha \in \Lambda} a(\alpha)\delta_{\alpha/2}$ and ask whether the sequence T_k converges to ϕ.

We refer to [31] for a thorough discussion of the convergence problem in general. In the next paragraph we shall treat in some detail the case that ϕ is a box spline.

Remark 19.6. We still use the word *mask* for T to mean either the sequence of coefficients a_α or the Laurent polynomial $\sum_{\alpha \in \Lambda} a(\alpha)x^\alpha$.

19.1.3 The Box Spline

We want to show how the previous discussion applies to box splines. Let us start with a basic example.

Example 19.7. Let $s = 1$ and take the box spline $b_0(x)$, that is, the characteristic function of the half-open interval $[0, 1)$. Then $b_0(x)$ satisfies the refinement equation $b_0(x) = (1 + \tau_{\frac{1}{2}})b_0(2x) = \sigma_{\frac{1}{2}}((1 + \tau)b_0)$. Thus the associated distribution is $T = \frac{1}{2}(\delta_0 + \delta_{\frac{1}{2}})$, and

$$T_k = \frac{1}{2^k} \sum_{i=0}^{2^k - 1} \delta_{i/2^k}.$$

For any function f, $T_k * f$ is the Riemann sum $2^{-k} \sum_{i=0}^{2^k - 1} f(i/2^k)$. Thus the distributions T_k are measures and have as weak limit the characteristic function of the interval $[0, 1)$.

A way to analyze the previous theory for box splines is to reduce to the basic example using a convolution argument. If we start with two distributions T, S with finite support, then also $T * S$ has finite support, and we see by an easy induction that the sequences of distributions $T_k, S_k, (T * S)_k$ are related by:

$$T_k * S_k = \sigma_{\frac{1}{2^k}}(T) * T_{k-1} * \sigma_{\frac{1}{2^k}}(S) * (S_{k-1}) = \sigma_{\frac{1}{2^k}}(T * S) * (T_{k-1} * S_{k-1})$$

$$= (T * S)_k.$$

Given a list of vectors $X = (a_1, \ldots, a_N)$, consider the measures $\frac{1}{2}(\delta_0 + \delta_{a_i/2})$ and their convolution product

$$T^X := 2^{-N} \prod_{i=1}^{N} (\delta_0 + \delta_{a_i/2}) = 2^{-N} \sum_{S \subset X} \delta_{a_S/2}.$$

The following lemma is immediate from the definitions.

Lemma 19.8. *The measure T_k^X is the push-forward measure, under the map $(t_1, \ldots, t_N) \mapsto \sum_i t_i a_i$, of the measure $2^{-kN} \sum_{p \in 2^{-k}\mathbb{Z}^N \cap [0,1)^N} \delta_p$, supported in the lattice $2^{-k}\mathbb{Z}^N$ restricted to the cube $[0,1)^N$ with value 2^{-kN} at these points.*

In other words, the value of T_k^X against any function f is the Riemann sum associated to $\int_0^1 \cdots \int_0^1 f(\sum_{i=1}^N t_i a_i) dt_1 \cdots dt_N$ for the lattice $2^{-k}\mathbb{Z}^N$ restricted to the cube $[0,1)^N$.

Theorem 19.9. *For $T^X := 2^{-|X|} \prod_{a \in X}^{*} (\delta_0 + \delta_{a/2})$, the sequence T_k^X converges weakly to B_X.*

For a discrete datum g, the corresponding sequence $g_k = a_k^X * g$ converges weakly to $B_X *' g$.

Proof. Everything follows from the convergence of the basic example by applying convolution. We leave the details to the reader.

Example 19.10. One of the first algorithms used to *round corners* is that based on the quadratic spline b_2 of class C^1. It is known as the Chaikin algorithm (see [32]). The mask is $2D = \frac{1}{4}(\delta_0 + 3\delta_1 + 3\delta_2 + \delta_3)$ and the recursive rule is

$$g_{2i}^{k+1} = \frac{3}{4}g_i^k + \frac{1}{4}g_{i+1}^k, \qquad g_{2i+1}^{k+1} = \frac{1}{4}g_i^k + \frac{3}{4}g_{i+1}^k.$$

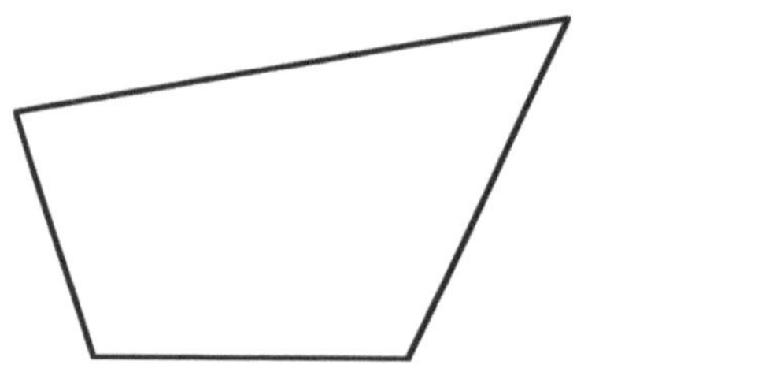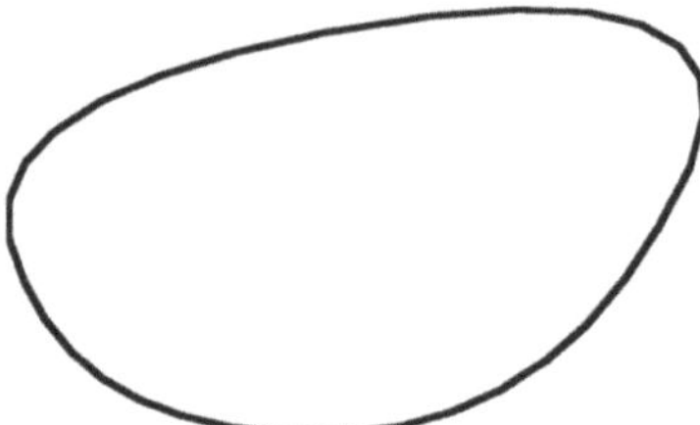

Fig. 19.1. The curve on the right is obtained applying the algorithm three times to the polygon on the left.

Summarizing, we have that to a discrete input data given by a function $g(\alpha)$ one associates as limiting function the element of the cardinal spline space given by $B_X *' g$. Such functions appear as approximations of smooth functions in the Strang–Fix approach and also as continuous interpolations of discrete data in the subdivision algorithm.

Putting together these two things, we see that if we have a function g and we know its values at the points of the lattice $2^{-k}\mathbb{Z}^s$, we have two possible approximations by splines, one continuous using the superfunction and the algorithm (18.6), the other discrete using the box spline and the cutting corners algorithm.

19.1.4 The Functional Equation

There is nothing special about the number 2 in formula (19.5), except perhaps the fact that computers use binary codes. In general, given a distribution T with finite support and a positive integer h, one can study the functional equation

$$T * \sigma_{1/h}(\phi) = \phi, \tag{19.6}$$

for an unknown distribution ϕ.

For this one may apply the Paley–Wiener theorem ([121] Section VI.3):

Theorem 19.11. *An entire function F on $\mathbb{C}^n$ is the Fourier–Laplace transform of a distribution v of compact support if and only if for all $z \in \mathbb{C}^n$, one has*

$$|F(z)| \le C(1 + |z|)^N e^{B|\mathrm{Im}(z)|} \tag{19.7}$$

for some constants C, N, B. The distribution v in fact will be supported in the closed ball of center 0 and radius B.

Additional growth conditions on the entire function F impose regularity properties on the distribution v.

We can thus try to study the functional equation (19.6) in terms of the Laplace transform (let us point out that since we use Laplace transforms, we have to replace the imaginary part of z with the real part). Recall the identities

$$L(\sigma_{1/h}(\phi)) = h^{-s}\sigma_h L(\phi), \quad L(T) = \sum_{\alpha \in \Lambda} a(\alpha)e^{\alpha}.$$

We know that Laplace transforms changes convolution into product. Setting $F(z) := L(\phi)$, the functional equation (19.6) is transformed into

$$h^{-s}\left(\sum_{\alpha \in \Lambda} a(\alpha)e^{\langle \alpha \,|\, z \rangle}\right)F(z/h) = F(z). \tag{19.8}$$

We shall call this equation the refining equation for $F(z)$. Now we have the following easy result:

Proposition 19.12. *Let $A(z)$ be a formal power series in the s variables $z = (z_1, \ldots, z_s)$. Consider the functional equation in the algebra of formal power series*

$$A(z)F(z/h) = F(z); \quad assume \quad h^k \neq 0, 1, \forall k \in \mathbb{N}. \tag{19.9}$$

(1) There exists a nonzero solution to (19.9) if and only if $A(0) = h^t$ for some integer $t \geq 0$. In this case the solution has the form $F(z) = \sum_{j \geq t} b_j$, b_j a homogeneous polynomial of degree j with $b_t \neq 0$, and it is uniquely determined by b_t.

(2) Assume $A(z) = \sum_k a_k(z)$, where $a_k(z)$ denotes the homogeneous part of degree k of $A(z)$. If $|a_k(z)| \leq C\frac{(E|z|)^k}{k!}$, for two positive constants C, E and $2 \leq h \in \mathbb{R}$, then $F(z)$ is an entire function.

Proof. (1) Assume that there is a solution of the form $F(z) = \sum_{j \geq t} b_j$ with $b_t \neq 0$. Equating the terms of degree t, we have $b_t(z) = A(0)b_t(z/h)$; hence $1 = A(0)h^{-t}$.

Assume now $A(0) = h^t$ and choose arbitrarily $b_t \neq 0$ homogeneous of degree t. Write $A(z) = \sum_{k=0}^{\infty} a_k$ with a_k the part of degree k. We want to determine a solution $F(z) = \sum_{k=t}^{\infty} b_k$ to (19.9). We have $F(z/h) = \sum_{k=t}^{\infty} h^{-k}b_k$, and hence (19.9) we is equivalent to the identities

$$\sum_{j=t}^{k-1} h^{-j}b_j a_{k-j} = b_k(1 - h^{t-k}).$$

These allow us to compute b_k recursively, starting from b_t.

(2) Under the assumption $|a_k| \leq C\frac{(E|z|)^k}{k!}$ we will find a constant D with $|b_j(z)| \leq (hD)^j(E|z|)^j/j!$ for all j. Set $K := 2C \geq (1 - h^{t-k})^{-1}C$, $\forall k > t$. Assume that D has been found satisfying the previous inequality for all $j < k$.

$$|b_k(z)| \leq (1 - h^{t-k})^{-1} \sum_{j=t}^{k-1} h^{-j}|b_j||a_{k-j}|$$

$$\leq C(1 - h^{t-k})^{-1} \sum_{j=t}^{k-1} h^{-j}|b_j|\frac{(E|z|)^{k-j}}{(k-j)!} \leq K\left(\sum_{j=t}^{k-1} \frac{h^{-j}(hD)^j}{j!(k-j)!}\right)(E|z|)^k$$

$$\leq K\left(\sum_{j=0}^{k-1}\binom{k}{j}D^j\right)\frac{(E|z|)^k}{k!} \leq K(1+D)^{k-1}\frac{(E|z|)^k}{k!}.$$

As soon as $D \geq \max((h-1)^{-1}, Kh^{-1})$, we have $K(1+D)^{k-1} \leq (hD)^k$. Thus this choice of D satisfies, by recursive induction, all the inequalities.

Remark 19.13. The proof is rather formal, and it applies also when $A(z)$ is a matrix of formal power series and F a vector. In this case we assume that $A(0)$ is an invertible matrix.

Consider the special case in which $A(z)$ is of the form $\sum_{i=1}^{m} a_i e^{\langle \alpha_i \mid z \rangle}$. Setting $C = \sum_{i=1}^{m} |a_i|$ and $E = \max(|\alpha_i|)$, we have the estimate $|a_k| \leq C\frac{(E|z|)^k}{k!}$. Therefore, if $\sum_i a_i = 1$, the refining equation (19.8) has a solution, unique up to a multiplicative constant, that is an entire function.

We want to apply this to find a solution to the functional equation (19.6), that is a distribution with compact support. We need to verify the conditions of the Paley–Wiener theorem.

Recursively, assuming $1 < h \in \mathbb{R}^+$, we have

$$F(z) = A(z)A(z/h)\ldots A(z/h^k)F(z/h^{k+1}).$$

Let us now make an estimate on $F(z)$, under the assumption that $A(z)$ is of the form $\sum_\alpha a_\alpha e^{\langle \alpha \mid z \rangle}$ with all the α real. We first estimate $|A(z)| \le C e^{B\,|\mathrm{Re}(z)|}$ if all the α are in a ball of radius B. We have thus

$$|A(z)A(z/h)\cdots A(z/h^k)| \le C^k e^{B\,|\mathrm{Re}(z)|(\sum_{i=0}^k h^{-i})},$$

$$|F(z)| \le C^k e^{B(1-h^{-k-1})(1-h^{-1})^{-1}|\mathrm{Re}(z)|}|F(z/h^{k+1})|.$$

So if $h^k \le 1 + |z| \le h^{k+1}$, we can estimate, setting $C_0 = \max_{|z|\le 1}|F(z)|$,

$$|F(z)| \le C^k e^{B(1-h^{-k-1})(1-h^{-1})^{-1}|\mathrm{Re}(z)|}C_0 \le C^k e^{B'\,|\mathrm{Re}(z)|}C_0$$

$$\le C_0(1+|z|)^N e^{B'|\mathrm{Re}(z)|},$$

where $B' = hB/(h-1)$ and N is chosen such that $C \le h^N$.

Theorem 19.14. *If $T = \sum_{\alpha \in \Lambda} a(\alpha)\tau_\alpha$ with $\sum_{\alpha \in \Lambda} a(\alpha) = h^s$, the refinement equation has a solution, unique up to a multiplicative constant, that is a distribution with compact support.*

19.1.5 The Role of Box Splines

We want to finish this chapter by showing that box splines can be characterized as those distributions satisfying a refining identity and for which the Fourier–Laplace transform is of the form

$$\frac{\sum_{j=1}^m c_j e^{\ell_j}}{\prod_{i=1}^r a_i}, \tag{19.10}$$

where $\ell_j,\ j = 1,\ldots,m$, $a_i,\ i = 1,\ldots,r$ are linear forms and the ℓ_j's lie in the lattice $\mathbb{Z}^s$.

We can now state and prove the main result of this section.[2]

Theorem 19.15. *Let $h > 1$ be a positive integer. Take an algebraic hypersurface S in a torus T stable under the map $x \mapsto x^h$. Then there are finitely many codimension-one tori U_i and for each i a finite set A_i of torsion points of T/U_i stable under $x \mapsto x^h$ such that if $\pi_i : T \to T/U_i$ is the projection, we have $S = \cup_i \pi_i^{-1}(A_i)$.*

---

[2]We thank David Masser for explaining to us the context of this problem [79] and Umberto Zannier for pointing out this simple proof.

Proof. Let us first remark that if on a subvariety V of a torus T, a character χ has the property that $|\chi(x)| = 1$, $\forall x \in V$, we must have that χ is constant on each connected component of V. This follows from the theorem of Chevalley that the values of an algebraic map on an affine algebraic variety form a constructible set, so a function on a connected variety is either constant or takes all values except possibly finitely many.

Let us write the equation of the hypersurface S stable under $x \mapsto x^h$ as $\sum_{r=1}^{t} a_r \chi_r = 0$ with the χ_r characters and $a_r \neq 0$. Take an irreducible component S_0 of S. We claim that S_0 is a torsion coset. i.e., that there is a primitive character χ^3 taking value on S_0 a fixed root of unity. For this it is enough to show that there are two distinct indices $i, j \leq t$ such that for $x \in S_0$, we have $|\chi_i(x)| = |\chi_j(x)|$. Indeed, if this is the case, by the previous remark the character $\chi = \chi_i \chi_j^{-1}$ takes a constant value ζ on S_0. On the other hand, the set of transforms of S_0 by the map $x \mapsto x^h$ is finite, so that also the set ζ^{h^k}, $k \in \mathbb{N}$, is finite, showing that ζ is a root of unity.

If $\chi = m\psi$ with ψ primitive we have that ψ also takes as values in a finite set of roots of 1. Since S_0 is irreducible it takes as value on S_0 a single root of 1.

If for each $x \in S_0$, there are two distinct indices i, j with $|\chi_i(x)| = |\chi_j(x)|$, we claim that for one such pair the equality holds for every $x \in S_0$. Indeed, there is a pair i, j such that the closed set $A_{i,j} \subset S_0$ for which the equality holds has nonempty interior. But a holomorphic function that has constant absolute value on an open set is necessarily constant, proving our claim.

Assume by contradiction that there is a point x for which all the numbers $|\chi_i(x)|$ are distinct. In particular, there is a unique i_0 for which $|\chi_{i_0}(x)|$ is maximum. Dividing the equation $\sum_i a_i \chi_i = 0$ by χ_{i_0}, we get a new equation that produces the identities of the form $a_{i_0} + \sum_{i \neq i_0} a_i \chi_{i_0}^{-1} \chi_i(x^{h^k}) = 0$ for all k. Notice now that $|\chi_{i_0}^{-1} \chi_i(x^{h^k})| = (|\chi_i(x)|/|\chi_{i_0}(x)|)^{h^k}$ tends to zero, implying $a_{i_0} = 0$, a contradiction.

At this point we have shown that S is a union of torsion cosets. The remaining properties are left to the reader.

The equation of such a hypersurface is given as follows, each U_i is the kernel of a primitive character χ_i and T/U_i is identified to $\mathbb{C}^*$. In $\mathbb{C}^*$ we choose a finite set of roots of 1: $\zeta_1, \cdots, \zeta_{m_i}$ closed under the map $\zeta \mapsto \zeta^h$ and then the desired equation is of the form $\prod_i \prod_{j=1}^{m_i} (\chi_i - \zeta_j)$.

In general we may ask what are the polynomials $p(x_1, \ldots, x_s)$ that have the property of dividing $p(x_1^h, \ldots, x_s^h)$. Such a polynomial defines a hypersurface of the previous type hence it is of the form

$$\prod_i \prod_{j=1}^{m_i} (\chi_i - \zeta_j)^{t_j}, \ t_j > 0.$$

---

[3]Recall that a character is *primitive* if it is part of an integral basis of the character group, or if it is not a multimple $\chi = m\psi$ with $m > 1$.

As a consequence, if $p(x_1, \ldots, x_s)$ has no multiple factors, it must be equal (up to a monomial factor) to a product of elements $\prod_{i=1}^{s} x_i^{n_i} \zeta - 1$ with ζ running over a finite set of roots of 1 closed under $\zeta \mapsto \zeta^h$. In the general case of multiple factors the situation is combinatorially more complicated, and we will not analyze it.

Let us then consider the following question. Assume that ϕ is a refinable function with compact support such that its Laplace transform is of the form (19.10) that is,

$$F = L\phi = \frac{\sum_{j=1}^{m} c_j e^{\langle \ell_j \,|\, z \rangle}}{\prod_{i=1}^{r} \langle a_i \,|\, z \rangle}.$$

We want to prove the following result:

Theorem 19.16. *A function ϕ as described above is obtained from a box spline by applying a suitable translation operator. Thus ϕ is in a cardinal spline space $\mathcal{S}_X$, where X is a list of integral vectors.*

Proof. Write the refining equation in the form $A(hz)F(z) = F(hz)$. First remark that $\prod_{i=1}^{r} \langle a_i \,|\, hz \rangle = h^r \prod_{i=1}^{r} \langle a_i \,|\, z \rangle$, hence $\sum_{j=1}^{m} c_j e^{\ell_j}$ also satisfies a similar refining equation. Then use the coordinates in the torus $e^{\langle \ell \,|\, z \rangle} = x^\ell$ so the refining equation implies that $\sum_{j=1}^{m} c_j x^{\ell_j}$ divides $\sum_{j=1}^{m} c_j x^{h\ell_j}$. In other words, the refining equation implies that the hypersuface $\sum_{j=1}^{m} c_j x^{\ell_j} = 0$ satisfies the conditions of the previous theorem. Therefore, the Laurent polynomial $\sum_{j=1}^{m} c_j x^{\ell_j}$ is a product of factors of type $\prod_{i=1}^{s} x_i^{n_i} \zeta - 1 = e^{\sum_{i=1}^{s} n_i z_i} \zeta - 1$. We observe that $e^{\sum_{i=1}^{s} n_i z_i} \zeta - 1$ vanishes at the origin if and only if $\zeta = 1$. In this case $(e^{\sum_{i=1}^{s} n_i z_i} - 1)/(\sum_{i=1}^{s} n_i z_i)$ is still holomorphic and nonvanishing at 0.

It follows that, in order for the function F to be holomorphic, each linear forms $a_j = \sum_{i=1}^{s} n_{i,j} z_i$ appearing in the denominator, must cancel a zero appearing in a factor of the numerator. This implies that $a_j = c_j \sum_{i=1}^{s} n_{i,j} z_i$, with $n_{i,j} \in \mathbb{Z}$ and c_j a nonzero constant, and that the corresponding factor is of the form $e^{\sum_{i=1}^{s} n_{i,j} z_i} - 1$.

In particular, each a_j gives a factor $(e^{a_j} - 1)/a_j$ of F. We see that in this way we obtain only the Laplace transforms of box splines associated to lists of integer vectors multiplied by a product of factors $e^{\sum_{i=1}^{s} n_i z_i} \zeta - 1$. These factors have to satisfy further combinatorial restrictions in order for the refining identity to hold. In particular, we see that a function TB_X in the cardinal space is refinable only if the mask of T is one of these special polynomials we have described.

Part V

The Wonderful Model

Minimal Models

One of the purposes of this chapter is to explain the geometric meaning of the notion of residue introduced in Section 10.2.1.

This chapter is quite independent of the rest of the book and can be used as an introduction to the theory developed in [42] and [46].

The main point to be understood is that the nonlinear coordinates u_i used in Section 10.2.1, represent local coordinates around a point at infinity of a suitable geometric model of a completion of the variety $\mathcal{A}_X$. In fact, we are thinking of models, proper over the space $U \supset \mathcal{A}_X$, in which the complement of $\mathcal{A}_X$ is a divisor with normal crossings. In this respect the local computation done in Section 10.2.1, corresponds to a model in which all the subspaces of the arrangement have been blown up, but there is a subtler model that gives rise to a more intricate combinatorics but possibly to more efficient computational algorithms, due to its minimality.

In order to be able to define our models, we need to introduce a number of notions of a combinatorial nature.

20.1 Irreducibles and Nested Sets

20.1.1 Irreducibles and Decompositions

As usual, let us consider a list $X := (a_1, \ldots, a_m)$ of nonzero vectors in V, that in this section we assume to be a complex vector space. We are going to be interested only in the hyperplanes defined by the equation $\langle a_i \,|\, x \rangle = 0$, and therefore we shall assume that the a_i's are pairwise nonproportional and we may think of X as a set of vectors.

Definition 20.1. Given a subset $A \subset X$, the list $\overline{A} := X \cap \langle A \rangle$ will be called the **completion** of A. In particular, A is called **complete** if $A = \overline{A}$.

The space of vectors $\phi \in U$ such that $\langle a|\phi \rangle = 0$ for every $a \in A$ will be denoted by $A^{\perp}$. Notice that clearly $\overline{A}$ equals the list of vectors $a \in X$ that vanish on $A^{\perp}$.

C. De Concini and C. Procesi, *Topics in Hyperplane Arrangements, Polytopes and Box-Splines*, Universitext, DOI 10.1007/978-0-387-78963-7_20,
© Springer Science+Business Media, LLC 2010

From this we see that we get an order-reversing bijection between the complete subsets of X and subspaces of the arrangement defined by X.

A central notion in what follows is given in the following definition:

Definition 20.2. Given a complete set $A \subset X$, a **decomposition** of A is a partition $A = A_1 \cup A_2 \cup \cdots \cup A_k$ into nonempty sets such that

$$\langle A \rangle = \langle A_1 \rangle \oplus \langle A_2 \rangle \oplus \cdots \oplus \langle A_k \rangle.$$

Clearly, the sets $A_1, A_2, \ldots, A_k$ are necessarily complete.

We shall say that a complete set A is **irreducible** if it does not have a nontrivial decomposition.

If $A = A_1 \cup A_2$ is a decomposition of a complete set and $B \subset A$ is complete, we have $B = B_1 \cup B_2$, where $B_1 = A_1 \cap B$, $B_2 = A_2 \cap B$. Also $\langle B \rangle = \langle B_1 \rangle \oplus \langle B_2 \rangle$, and we have the following:

Lemma 20.3. $B = B_1 \cup B_2$ *is a decomposition unless one of the two sets is empty.*

We deduce immediately the following result:

Proposition 20.4. *If $A = A_1 \cup A_2$ is a decomposition and $B \subset A$ is irreducible, then $B \subset A_1$ or $B \subset A_2$.*

From this get our next theorem:

Theorem 20.5. *Every set A can be decomposed as $A = A_1 \cup A_2 \cup \cdots \cup A_k$ with each A_i irreducible and*

$$\langle A \rangle = \langle A_1 \rangle \oplus \langle A_2 \rangle \oplus \cdots \oplus \langle A_k \rangle.$$

This decomposition is unique up to order.

Proof. The existence of such a decomposition follows by a simple induction.

Let $A = B_1 \cup B_2 \cup \cdots \cup B_h$ be a second decomposition with the same properties. Proposition 20.4 implies that every B_i is contained in an A_j and vice versa.

Thus the A_j's and the B_i's are the same up to order.

We call $A = A_1 \cup A_2 \cup \cdots \cup A_k$ the *decomposition into irreducibles* of A.

20.1.2 Nested Sets

We say that two sets A, B are *comparable* if one is contained in the other.

Definition 20.6. A family $\mathcal{S}$ of irreducibles A_i is called *nested* if given elements $A_{i_1}, \ldots, A_{i_h} \in \mathcal{S}$ mutually incomparable, we have that their union $C := A_{i_1} \cup A_{i_2} \cup \cdots \cup A_{i_h}$ is a complete set with its decomposition into irreducibles.

Remark 20.7. If $A_1, \ldots, A_k$ is nested, we have that $\cup_i A_i$ is complete. In fact, this union can be obtained by taking the maximal elements, which are necessarily noncomparable, and then applying the definition of nested.

We are in particular interested in *maximal nested sets*, that we denote by MNS.

Nested sets can be inductively constructed by combining the following two procedures, that are easily justified.

Proposition 20.8. *(1) Suppose we are given a nested set $\mathcal{S}$, a minimal element $A \in \mathcal{S}$, and a nested set $\mathcal{P}$ whose elements are contained in A. Then we have that $\mathcal{S} \cup \mathcal{P}$ is nested.*

(2) Suppose we are given a nested set $\mathcal{S}$ and a complete set A containing each element of $\mathcal{S}$. Then if $A = A_1 \cup \cdots \cup A_k$ is the decomposition of A into irreducibles, $\mathcal{S} \cup \{A_1, \ldots, A_k\}$ is nested.

Theorem 20.9. *Assume that $\langle X \rangle = V$. Let $\mathcal{S} := \{A_1, \ldots, A_k\}$ be an MNS in X.*

Given $A \in \mathcal{S}$, let $B_1, \ldots, B_r$ be the elements of $\mathcal{S}$ contained properly in A, and maximal with this property.

(1) $C := B_1 \cup \cdots \cup B_r$ is complete and decomposed by the B_i.
(2) $\dim\langle A \rangle = \dim\langle C \rangle + 1$.
(3) $k = \dim(V)$.

Proof. (1) is the definition of nested set, since the B_i, being maximal, are necessarily noncomparable.

(2) Let us consider $\langle C \rangle = \oplus_{i=1}^r \langle B_i \rangle \subset \langle A \rangle$.
We have $\langle C \rangle \neq \langle A \rangle$. Otherwise, since C is complete, by the definition of nested set we must have $A = C$. Since A is irreducible and the B_i's are properly contained in A, this is a contradiction.

Therefore, there exists an element $a \in A$ such that $a \notin \langle C \rangle$. Let us define $A' := X \cap \langle C, a \rangle$. We have $C \subsetneq A' \subset A$. We claim that $A = A'$.

Otherwise, by Proposition 20.8, adding all the irreducibles that decompose A' to $\mathcal{S}$, we obtain a nested family that properly contains $\mathcal{S}$. This contradicts the maximality of $\mathcal{S}$. Clearly, $A = A'$ implies that $\langle A \rangle = \langle C \rangle \oplus \mathbb{C}a$ and thus $\dim\langle A \rangle = \dim\langle C \rangle + 1$.

(3) We proceed by induction on $s = \dim(V)$.
If $s = 1$ there is nothing to prove, there is a unique set complete and irreducible, namely X.

Let $s > 1$. Decompose $X = \cup_{i=1}^h X_h$ into irreducibles.
We have that an MNS in X is the union of MNS's in each X_i. Then $s = \dim(\langle X \rangle) = \sum_{i=1}^h \dim(\langle X_i \rangle)$.

Thus we can assume that X is irreducible. In this case we have that $X \in \mathcal{S}$ for every MNS $\mathcal{S}$.

Let $B_1, \ldots, B_s$ be the elements of $\mathcal{S}$ properly contained in X and maximal with this property.

The set $\mathcal{S}$ consists of X and the subsets $\mathcal{S}_i := \{A \in \mathcal{S} \mid A \subseteq B_i\}$.

Clearly, $\mathcal{S}_i$ is an MNS relative to the set B_i (otherwise, we could add an element to $\mathcal{S}_i$ and to $\mathcal{S}$, contradicting the maximality of $\mathcal{S}$).

By induction, $\mathcal{S}_i$ has $\dim\langle B_i \rangle$ elements, and thus by (2) the claim follows.

Given an MNS $\mathcal{S}$, let us define a map

$$p_{\mathcal{S}} : X \to \mathcal{S}$$

as follows. Since $\cup_{A \in \mathcal{S}} A = X$, every element $a \in X$ lies in at least one $A \in \mathcal{S}$. Also, if an element $a \in X$ appears in two elements $A, B \in \mathcal{S}$, the two elements must be necessarily comparable, and thus there exists a minimum among the two. It follows that for any element $a \in X$ there is a minimum element $p_{\mathcal{S}}(a) \in \mathcal{S}$ containing a.

Now a new definition:

Definition 20.10. We shall say that a basis $\underline{b} := (b_1, \ldots, b_s) \subset X$ of V is *adapted* to the MNS $\mathcal{S}$ if the map $b_i \mapsto p_{\mathcal{S}}(b_i)$ is a bijection.

Such a basis always exists. It suffices to take, as in the proof of Theorem 20.9, for every $A \in \mathcal{S}$ an element $a \in A - \cup_i B_i$, where the B_i are the elements of $\mathcal{S}$ properly contained in A.

Given any basis $\underline{b} := (b_1, \ldots, b_s) \subset X$, we shall build an MNS $\mathcal{S}_{\underline{b}}$ to which it is adapted, in the following way. Consider for any $1 \leq i \leq s$ the complete set $A_i := X \cap \langle \{b_i, \ldots, b_s\} \rangle = \overline{\{b_i, \ldots, b_s\}}$. Clearly, $A_1 = X \supset A_2 \supset \cdots \supset A_s$.

For each i consider all the irreducibles in the decomposition of A_i. Clearly, for different i we can obtain the same irreducible several times. In any case we have the following statement:

Theorem 20.11. *The family $\mathcal{S}_{\underline{b}}$ of all the (distinct) irreducibles that appear in the decompositions of the sets A_i form an MNS to which the basis $\underline{b}$ is adapted.*

Proof. By induction. Decompose $X = A_1 = B_1 \cup B_2 \cup \cdots \cup B_k$ into irreducibles. By construction,

$$s = \dim\langle A_1 \rangle = \sum_{i=1}^{k} \dim\langle B_i \rangle.$$

We have that $A_2 = (A_2 \cap B_1) \cup (A_2 \cap B_2) \cup \cdots \cup (A_2 \cap B_k)$ is a decomposition of A_2, not necessarily into irreducibles.

Since $\dim\langle A_2 \rangle = s - 1$, we have

$$s - 1 = \dim\langle A_2 \rangle = \sum_{i=1}^{k} \dim\langle A_2 \cap B_i \rangle.$$

Therefore, $\dim\langle A_2 \cap B_i\rangle < \dim\langle B_i\rangle$ for exactly one index i_0. In other words, we must have that $A_2 \cap B_i = B_i$ for all the $i \neq i_0$. For such an index, necessarily $b_1 \in B_{i_0}$.

By induction, the family of all the (distinct) irreducibles that appear in the decompositions of the sets A_i, $i \geq 2$, form an MNS for $\langle A_2\rangle$, with adapted basis $\{b_2, \ldots, b_s\}$. To this set we must thus only add B_{i_0} in order to obtain $\mathcal{S}_{\underline{b}}$. Thus $\mathcal{S}_{\underline{b}}$ is a nested set with s elements, hence maximal, and the basis $\underline{b}$ is adapted.

Remark 20.12. One can easily verify that conversely, every MNS $\mathcal{S}$ is of the form $\mathcal{S}_{\underline{b}}$ for each of its adapted bases.

20.1.3 Non Linear Coordinates

Now we introduce a nonlinear change of coordinates associated to a maximal nested set.

So, choose an MNS $\mathcal{S}$ and a basis $\underline{b} := (b_1, \ldots, b_s)$ adapted to $\mathcal{S}$. Let us consider the b_i's as a *system of linear coordinates on U*.

If $p_{\mathcal{S}}(b_i) = A$, b_i will be denoted by b_A. We now build new coordinates z_A, $A \in \mathcal{S}$, using the monomial expressions

$$b_A := \prod_{B \in \mathcal{S}, \ A \subseteq B} z_B. \tag{20.1}$$

Given $A \in \mathcal{S}$, let $\mathcal{S}_A := \{B \subseteq A, B \in \mathcal{S}\}$. Clearly, $\mathcal{S}_A$ is an MNS for A (in place of X), and the elements b_B with $B \in \mathcal{S}_A$ form a basis of A, adapted to $\mathcal{S}_A$.

Take $a \in X$ with $p_{\mathcal{S}}(a) = A$ and write $a = \sum_{B \in \mathcal{S}_A} c_B b_B$, $c_B \in \mathbb{C}$. Let us now substitute the expressions (20.1), getting

$$a := \sum_{B \in \mathcal{S}_A} c_B \prod_{C \in \mathcal{S}, \ B \subseteq C} z_C$$
$$= \prod_{B \in \mathcal{S}, \ A \subset B} z_B \left(c_A + \sum_{B \in \mathcal{S}_A, B \neq A} c_B \prod_{C \in \mathcal{S}, \ B \subseteq C \subsetneq A} z_C \right). \tag{20.2}$$

Since A is the minimum set of $\mathcal{S}$ containing a, we must have $c_A \neq 0$.

At this point we can proceed as in Section 10.2.1 and define an embedding of R_X into the ring of Laurent series in the variables z_A, $A \in \mathcal{S}$.

Thus for a top form ψ we can define a *local residue* $\mathrm{res}_{\mathcal{S},\underline{b}}\psi$ (or $\mathrm{res}_{\mathcal{S}}\psi$ if $\underline{b}$ is clear from the context).

20.1.4 Proper Nested Sets

We are going to tie the concepts of the previous section to that of unbroken bases. Thus we start by totally ordering X. Our goal is to define a bijection between $\mathcal{NB}(X)$ and a suitable family of MNS's, that we shall call *proper nested sets*. We need some preliminary steps.

Assume thus that the basis $\underline{b} := (b_1, \ldots, b_s) \subset X$ is unbroken.

Lemma 20.13. *b_i is the minimum element of $p_{\mathcal{S}_{\underline{b}}}(b_i)$ for every i.*

Proof. Let $A := p_{\mathcal{S}_{\underline{b}}}(b_i)$. By the definition of $\mathcal{S}_{\underline{b}}$, we must have that A appears in the decomposition into irreducibles of one of the sets $A_k = \langle b_k, \ldots, b_s \rangle \cap X$. Since $b_i \in A$, necessarily it must be $k \leq i$.

On the other hand, b_i belongs to one of the irreducibles of A_i, which therefore, for each $h \leq i$, is contained in the irreducible of A_h that contains b_i.

It follows that A must be one of the irreducibles decomposing A_i. By definition, of unbroken basis, b_i is the minimum element of $A_i = \langle b_i, \ldots, b_s \rangle \cap X$, hence also the minimum element of A.

This property suggests the following definition:

Definition 20.14. Let $\mathcal{S}$ be a MNS. We say that $\mathcal{S}$ is *proper* if the elements $a_A := \min_{a \in A} a$, as A runs over $\mathcal{S}$, form a basis.

Lemma 20.15. *If $\mathcal{S}$ is proper, $p_{\mathcal{S}}(a_A) = A$. Thus the basis $\{a_A\}_{A \in \mathcal{S}}$ is adapted to $\mathcal{S}$.*

Proof. Let $B \in \mathcal{S}$ with $a_A \in B$. Since $a_A \in A$, we must have that A and B are comparable. We must verify that $A \subset B$. If $B \subset A$, then $a_A = \min a \in B$, and thus $a_A = a_B$.

Since the elements a_A are a basis, this implies that $A = B$.

If $\mathcal{S}$ is proper, we order its elements $A_1, \ldots, A_s$ using the increasing order of the elements a_A. We then set $b_i := a_{A_i}$, and we have the following theorem:

Theorem 20.16. *(1) The basis $\underline{b}_{\mathcal{S}} := (b_1, \ldots, b_s)$ is unbroken.*
(2) The map $\mathcal{S} \to \underline{b}_{\mathcal{S}}$ is a one-to-one correspondence between proper MNS's and unbroken bases.

Proof. (1) From Remark 20.7, setting $S_i := \cup_{j \geq i} A_j$, we have that S_i is complete and decomposed by those A_j that are maximal.

Clearly, by definition, b_i is the minimum of S_i. It suffices to prove that $S_i = \langle b_i, \ldots, b_s \rangle \cap X$, hence that $\langle S_i \rangle = \langle b_i, \ldots, b_s \rangle$, since S_i is complete.

We prove it by induction. If $i = 1$, the maximality of $\mathcal{S}$ implies that $S_1 = X$, so b_1 is the minimum element in X. Now $b_1 \notin S_2$, so $S_2 \neq X$, and since S_2 is complete, $\dim(\langle S_2 \rangle) < s$.

Clearly, $\langle S_2 \rangle \supset \langle b_2, \ldots, b_s \rangle$, so $\langle S_2 \rangle = \langle b_2, \ldots, b_s \rangle$.

At this point it follows that by induction, $\langle b_2, \ldots, b_s \rangle$ is an unbroken basis in S_2, and thus since b_1 is minimum in X, $\langle b_1, b_2, \ldots, b_s \rangle$ is an unbroken basis.

(2) From the proof, it follows that the two constructions, of the MNS associated to a unbroken basis and of the unbroken basis associated to a proper MNS, are inverses of each other, and thus the one-to-one correspondence is established.

20.2 Residues and Cycles

We can now complete our analysis performing a residue computation. Recall that if $\underline{b} := (b_1, \ldots, b_s) \subset X$ is a basis, we set

$$\omega_{\underline{b}} := d\log(b_1) \wedge \cdots \wedge d\log(b_s).$$

Theorem 20.17. *Given two unbroken bases* $\underline{b}, \underline{c}$, *we have*

$$\mathrm{res}_{\mathcal{S}_{\underline{b}}} \omega_{\underline{c}} = \begin{cases} 1 & \text{if} \quad \underline{b} = \underline{c} \\ 0 & \text{if} \quad \underline{b} \neq \underline{c}. \end{cases}$$

Proof. We prove first that $\mathrm{res}_{\mathcal{S}_{\underline{b}}} \omega_{\underline{b}} = 1$.

By definition, $b_i = \prod_{A_i \subset B} z_B$; hence $d\log(b_i) = \sum_{A_i \subset B} d\log(z_B)$.

When we substitute and expand the product we get a sum of products of type $d\log(z_{B_1}) \wedge d\log(z_{B_2}) \wedge \cdots \wedge d\log(z_{B_n})$ with $A_i \subset B_i$.

Now, in a finite poset $\mathcal{P}$ there is only one nondecreasing and injective map of $\mathcal{P} \to \mathcal{P}$ into itself, which is the identity.

Therefore, the map $A_i \mapsto B_i$ is injective only when it is the identity. Therefore, setting $z_i := z_{A_i}$ all the monomials of the sum vanish except for the monomial $d\log(z_1) \wedge d\log(z_2) \wedge \cdots \wedge d\log(z_n)$, which has residue 1.

The second case $\underline{b} \neq \underline{c}$ follows immediately from the next lemma.

Lemma 20.18. *Let* $\underline{b}, \underline{c}$ *be two unbroken bases.*

(1) If $\underline{b} \neq \underline{c}$, *the basis* $\underline{c}$ *is not adapted to* $\mathcal{S}_{\underline{b}}$.
(2) If a basis $\underline{c}$ *is not adapted to an MNS* $\mathcal{S}$, *we have* $\mathrm{res}_{\mathcal{S}} \omega_{\underline{c}} = 0$.

Proof. (1) Let $\underline{c} = \{c_1, \ldots, c_n\}$ be adapted to $\mathcal{S}_{\underline{b}} = \{S_1, \ldots, S_n\}$. We want to prove that we have $\underline{b} = \underline{c}$. We know that $c_1 = b_1$ is the minimum element of X.

Since $\mathcal{S}_{\underline{b}}$ is proper, we have that $S_1 = p_{\mathcal{S}_{\underline{b}}}(c_1)$ must be a maximal element that is the irreducible component of X containing $b_1 = c_1$.

Set $X' := S_2 \cup \cdots \cup S_n$. Then X' is complete. The set $S_2, \ldots, S_n$ coincides with the proper MNS $\mathcal{S}_{\underline{b}'}$ associated to the unbroken basis $\underline{b}' := \{b_2, \ldots, b_n\}$ of $\langle X' \rangle = \langle \underline{b}' \rangle$. Clearly, $\underline{c}' = \{c_2, \ldots, c_n\}$ is adapted to $\mathcal{S}_{\underline{b}'}$. Therefore, $\underline{b}' = \underline{c}'$ by induction, and so $\underline{b} = \underline{c}$.

Using the first part, the proof of (2) follows the same lines as the proof of Lemma 8.14, and we leave it to the reader.

Remark 20.19. We end this section pointing out that by what we have proved, it follows that for any unbroken basis $\underline{b}$ we have

$$\mathrm{res}_{\underline{b}} = \mathrm{res}_{\mathcal{S}_{\underline{b}}},$$

with $\mathrm{res}_{\underline{b}}$ as defined in Section 10.2.1.

The advantage of this new definition is that one uses monomial transformations of smaller degree, and this could provide more efficient algorithms.

20.3 A SpecialCase: Graphs and Networks

We shall now investigate the notions we have just introduced in the special case of the graph arrangements introduced in Section 2.1.3.

Let us take a graph Γ with vertex set V_Γ and edge set L_Γ. We are going to assume that Γ has no labels, that is, $m(a) = 1$ for all $a \in L_\Gamma$.

Fix an orientation on the edges, so that each edge a has an initial vertex $i(a)$ and a final vertex $f(a)$, and the corresponding list X_Γ, consisting of the vectors $x_a = e_{f(a)} - e_{i(a)}$, $a \in L_\Gamma$. Given $A \subset X_\Gamma$ we denote by Γ_A the graph spanned by the edges $a \in L_\Gamma$ such that $x_a \in A$.

In Corollary 2.16 we have seen that graph arrangements are unimodular; in particular, we deduce that the corresponding partition function is a polynomial on each big cell (instead of just a quasipolynomial). In particular, this applies to the example of magic-squares that we shall treat presently (cf. Stanley [104], Dahmen–Micchelli [37] and [3]).

20.3.1 Complete Sets

Now recall that a subgraph is called complete if whenever it contains two vertices it also contains all edges between them. On the other hand a subset $A \subset X_\Gamma$ is complete in the sense of arrangements if and only if $\langle A \rangle \cap X_\Gamma = A$. If A is complete in this sense, we shall say that the corresponding subgraph Γ_A is X-complete.

Proposition 20.20. *A subgraph of Γ is X-complete if and only if all its connected components are complete.*

Proof. The fact that a complete subgraph is also X-complete is clear, and from this the statement in one direction follows.

Assume that Λ is X-complete. We have to show that each connected component is complete. Otherwise, there is an edge $a \notin \Lambda$ with vertices in a connected component Π of Λ. By connectedness, a is thus in a cycle in which the other edges are in Π and so is dependent on X_Π and so on X_Λ. By the X-completeness of Λ we have a contradiction.

Corollary 20.21. *Given a connected graph Γ, a proper subgraph Λ is maximal X-complete if and only if:*

(1) either Λ is a connected subgraph obtained from Γ, by deleting one vertex and all the edges from it;

(2) or it is a graph with two connected components obtained from Γ, by deleting a set of edges each of which joins the two components.

Proof. If we remove one vertex and all the edges from it and the resulting graph Λ with edges C is still connected, it follows that the corresponding subspace $\langle C \rangle$ has codimension-one. Since Λ is clearly complete, it is also maximal.

In the second case the resulting graph Λ has two connected components, each of which is complete. So it is X-complete. Since the number of vertices is unchanged, X_Λ spans a codimension-one subspace; hence it is maximal.

Conversely, if Λ is maximal X-complete with w vertices and c connected components, we must have $v - 2 = w - c$, so either $c = 1$ and $w = v - 1$, or $w = v$ and $c = 2$. It is now easy to see that we must be in one of the two preceding cases (from the description of X-complete subgraphs and Lemma 2.12).

A set of edges such that the remaining graph has two connected components and all the deleted edges join the two components will be called a *simple disconnecting set*. Finding such a set is equivalent to decomposing the set of vertices into two disjoint subsets V_1, V_2 such that each of the two complete subgraphs of vertices V_1, V_2 is connected.

20.3.2 Irreducible Graphs

Definition 20.22. Given two graphs Γ_1, Γ_2 with a preferred vertex w_1, w_2 in each, the **wedge** $\Gamma_1 \vee \Gamma_2$ of the two graphs is given by forming their disjoint union and then identifying the two vertices.

Clearly, if v_1, v_2, v (resp. c_1, c_2, c) denote the number of vertices (resp. of connected components) of $\Gamma_1, \Gamma_2, \Gamma_1 \vee \Gamma_2$, we have $v = v_1 + v_2 - 1$, $c = c_1 + c_2 - 1$, and hence $v - c = v_1 - c_1 + v_2 - c_2$.

We shall now say that an X-complete subgraph is irreducible if the corresponding subset of X_Γ is irreducible. The previous formulas show that the decomposition of a graph Γ, as wedge or into connected components, implies a decomposition of the corresponding list of vectors X_Γ, so that in order to be irreducible, a graph must be connected and cannot be expressed as the wedge of two smaller subgraphs. The following proposition shows that these conditions are also sufficient.

Proposition 20.23. *A connected graph Γ is irreducible if and only if it is not a wedge of two graphs. This also means that there is no vertex that disconnects the graph.*

Proof. We have already remarked that the decomposition of a graph as wedge implies a decomposition (in the sense of Section 20.1.1) of X_Γ.

On the other hand, let Γ be connected and suppose that X_Γ has a nontrivial decomposition $X_\Gamma = A \cup B$ (in the sense of Definition 20.2). If A' is the set of vectors associated to the edges of a connected component of Γ_A, then $A = A' \cup (A \setminus A')$ is a decomposition of A, $X_\Gamma = A' \cup (A \setminus A') \cup B$ is also a decomposition, and we may thus reduce to the case that Γ_A is connected.

Denote by V_A (resp. V_B) the set of vertices of Γ_A (resp. Γ_B). We must have that $V_A \cap V_B$ is not empty, since Γ is connected.

The fact that $X_\Gamma = A \cup B$ is a decomposition implies $\langle X_\Gamma \rangle = \langle A \rangle \oplus \langle B \rangle$ so from Lemma 2.12 we deduce $v - 1 = |V_A| - b_A + |V_B| - b_B$ (with b_A, b_B the number of connected components of the two graphs with edges A, B respectively). Since $v = |V_A| + |V_B| - |V_A \cap V_B|$, we have $1 + |V_A \cap V_B| = b_A + b_B$.

We are assuming that $b_A = 1$, so we get $|V_A \cap V_B| = b_B$. Since Γ is connected, each connected component of Γ_B must contain at least one of the vertices in $V_A \cap V_B$. The equality $|V_A \cap V_B| = b_B$ implies then that each connected component of Γ_B contains exactly one of the vertices in $V_A \cap V_B$. Thus Γ is obtained by attaching, via a wedge operation, each connected component of Γ_B to a different vertex of Γ_A.

Remark 20.24. (i) Notice that the first case of Corollary 20.21 can be considered as a degenerate case of the second, with Γ_A reduced to a single vertex.

(ii) In general, a complete decomposition presents a connected graph as an iterated wedge of irreducible graphs.

20.3.3 Proper Maximal Nested Sets

With the previous analysis it is easy to give an algorithm that allows us to describe all proper maximal nested sets in X_Γ.

The algorithm is recursive and based on the idea of building a proper maximal flag of complete sets.

Step 1. We choose a total order of the edges.

Step 2. We decompose the graph into irreducibles.

Step 3. We proceed by recursion separately on each irreducible where we have the induced total order.

We assume thus that we have chosen one irreducible.

Step 4. We build all the proper maximal complete sets that do not contain the minimal edge a. These are of two types:

(i) For each of the two vertices v of a we have a maximal complete set by removing the edges out of v. By Proposition 20.23, this operation produces in both cases a connected graph.

(ii) Remove the simple disconnecting sets containing the minimal edge.

Step 5. Keeping the induced order, go back to Step 2 for each of the proper maximal complete sets constructed.

From the residue formulas it is clear that in the previous algorithm, given a vector u, in order to compute the volume or the number of integer points of the polytope Π_u, it is necessary only to compute the proper nested sets $\mathcal{S}$ that satisfy the further condition $u \in C(\mathcal{S})$, that we will express by the phrase $\mathcal{S}$ *is adapted to* u. The previous algorithm explains also how to take into account this condition.

In fact, let $\Lambda \subset \Gamma$ be a proper maximal complete set. We can see as follows whether this can be the first step in constructing an $\mathcal{S}$ adapted to u.

In fact, the basis $\phi(\mathcal{S})$ is composed of the minimal element x_a and the basis of the span $\langle \Lambda \rangle$ corresponding to the part $\mathcal{S}'$ of the nested set contained in $\langle \Lambda \rangle$. Thus u can be written uniquely in the form $\lambda x_a + w$ with $w \in \langle \Lambda \rangle$.

We observe that $\mathcal{S}$ is adapted to u if and only if $\lambda \geq 0$ and $\mathcal{S}'$ is adapted to w.

This gives a recursive way of proceeding if we can compute λ. Let us do this in the second case (since the first is a degenerate case). Denote the decomposition of the maximal complete subset by $A \cup B$; thus $\langle \Lambda \rangle$ can be identified with the space of functions on the vertices whose sum on A and on B equals 0. Let us assume that the orientation of the arrow of the minimum edge a points toward A, so that x_a as a function on the vertices takes the value 0 on all vertices except its final point in A, where it takes the value 1, and its initial point in B, where it takes the value -1. The vector u is just a function on the vertices with the sum 0. Let λ equal the sum of the values of u on the vertices of A; thus $-\lambda$ equals the sum of the values of u on the vertices of B.

We then have that $u - \lambda x_a = w \in \langle \Lambda \rangle$, giving the following result: see that:

Proposition 20.25. *In the decomposition of the maximal complete subset Λ as $A \cup B$ let us assume that the orientation of the arrow of the minimum edge a points toward A. Then if Λ is the first step of a proper flag adapted to u, we must have that the sum of the values of u on the vertices of A is nonnegative.*

Remark 20.26. (i) We have seen in Theorem 9.16 that the decomposition into big cells can be detected by any choice of ordering and the corresponding unbroken bases. Nevertheless, the number of unbroken bases adapted to a given cell depends strongly on the order. For a given cell it would thus be useful to minimize this number in order to optimize the algorithms computing formulas (5), (6).

(ii) Let us point out that in [56] and [88] some polytopes associated to graphs have been constructed and studied.

20.3.4 Two Examples: A_n and Magic Arrangements

The first interesting example we want to discuss is that of the complete graph Γ_s with s vertices, that we label with the indices $\{1, 2, \ldots, s\}$. In this case $X_{\Gamma_s} = \{e_i - e_j\}$ with $i < j$, so that it consists of the positive roots of the root system of type A_{s-1}. Thus the hyperplanes H_α are the corresponding root hyperplanes.

In this case a connected subgraph is complete if and only if it is complete as an abstract graph, that is, it contains all edges joining two of its vertices. Clearly, it is also irreducible. Given a complete subgraph C, the irreducible components of C are nothing else than its connected components.

It follows that irreducible sets are in bijection with subsets of $\{1, \ldots, s\}$ with at least two elements. If S is such a subset, the corresponding irreducible is $I_S = \{e_j - e_i | \{i, j\} \subset S\}$. Using this correspondence we identify the set of irreducibles with the set $\mathcal{I}_s$ of subsets in $\{1, \ldots, s\}$ with at least two elements. In particular, the elements of X_{Γ_s} are identified with the subsets with two elements and displayed as pairs (i, j) with $i < j$. It also follows that a family $\mathcal{S}$ of elements $\mathcal{I}_s$ is nested if given $A, B \in \mathcal{S}$, either A and B are comparable or they are disjoint.

This immediately implies that if $\mathcal{S}$ is a maximal nested set and $A \in \mathcal{S}$ has more that two elements, then A contains either a unique maximal element $B \in \mathcal{S}$ with necessarily $a - 1$ elements or exactly two maximal elements $B_1, B_2 \in \mathcal{S}$ with $A = B_1 \cup B_2$.

We can present such an MNS in a convenient way as a planar binary rooted tree with s leaves labeled by $\{1, \ldots, s\}$. Every internal vertex of this tree corresponds to the set of numbers that appear on its leaves.

For example, the tree

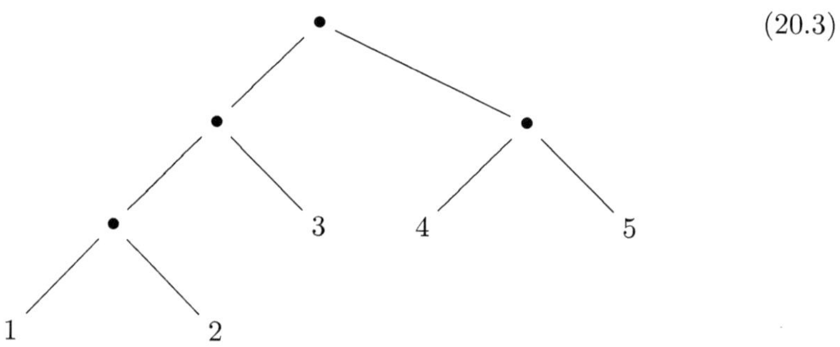

$$(20.3)$$

represents the MNS $\{\{1, 2\}, \{1, 2, 3\}, \{4, 5\}, \{1, 2, 3, 4, 5\}\}$.

We now fix a total ordering on X_{Γ_s} by defining $(i, j) \leq (h, k)$ if either $k - h < j - i$ or $k - h = j - i$ and $i \leq h$.

Define for any subset $A \in \mathcal{I}_s$ the pair $c(A) := (i, j)$, $i := \min(A)$, $j := \max(A)$. Notice that with respect to the chosen ordering, $c(A)$ is the minimum element in the set of pairs extracted from A.

Lemma 20.27. *An MNS $\mathcal{S}$ of subsets of $\{1, 2, \ldots, s\}$ is proper if and only if the pairs $c(A)$ for $A \in \mathcal{S}$ are distinct.*

Proof. In one direction the statement is clear.

We only have to show that if the elements $c(A)$ for $A \in \mathcal{S}$, are distinct, then $\mathcal{S}$ is proper. Let $x_A := e_i - e_j$ with $(i, j) = c(A)$. To say that $\mathcal{S}$ is proper means that the $n - 1$ elements x_A are linearly independent. In other words, it means that for any $k = 1, \ldots, s$, the vectors e_k and x_A as A varies in $\mathcal{S}$, span the entire s-dimensional space.

We argue by induction, starting from $\{1, 2, \ldots, s\} \in \mathcal{S}$, to which we associate $e_1 - e_s$. As we have seen, for the maximal subsets of $\mathcal{S}$ properly contained in $\{1, 2, \ldots, s\}$, we have two possibilities. They reduce either to a unique element B, with $s - 1$ elements, or to two disjoint subsets B_1, B_2, whose union is $\{1, 2, \ldots, s\}$.

In the first case we must have (since $\mathcal{S}$ is proper) either $B = \{2, \ldots, s\}$ or $B = \{1, 2, \ldots, s - 1\}$. By induction, adding e_s in the first case, and e_1 in the second, we easily deduce our claim.

If instead $\{1, 2, \ldots, s\} = B_1 \cup B_2$, always using the fact that $\mathcal{S}$ is proper, we can assume that $1 \in B_1$, $s \in B_2$. By induction, the space generated by x_C, $C \subset B$, and e_1 coincides with the space generated by the e_i, $i \in B_1$, and similarly for B_2 and e_s. If we add e_1 to all the x_A, $A \in \mathcal{S}$, we have also $e_s = e_1 - (e_1 - e_s)$, and thus we obtain all the basis elements e_i.

To finish our first example, we establish a one-to-one correspondence between such proper MNS's and the permutations of s elements that fix s (proving in particular that their number is $(s - 1)!$).

It is better to formulate this claim in a more abstract way, starting with any totally ordered set S with s elements instead of $\{1, 2, \ldots, s\}$. Of course, using its canonical bijection with $\{1, 2, \ldots, s\}$, we can transfer to S all the notions of MNS and proper MNS.

Proposition 20.28. *There is a bijection p between the set of proper MNS's associated to S and that of the permutations of S fixing the maximal element.*

Proof. Let a be the maximum element of S. We think of these permutations as words (without repetitions) of length s in the elements of S ending with a.

Start by identifying S with $\{1, \ldots, s\}$ using the total order. Take a proper MNS $\mathcal{S}$ given by a sequence $\{S_1, \ldots, S_{s-1}\}$ of subsets of $\{1, \ldots, s\}$.

We may assume that $S_1 = \{1, 2, \ldots, s\}$. We know that $\mathcal{S}' := \mathcal{S} - \{S_1\}$ has either one or two maximal elements.

We treat first the case of a unique maximal element. We can assume that it is S_2. Since $\mathcal{S}$ is proper, we have seen that either $S_2 = \{2, 3, \ldots, s\}$ or $S_2 = \{1, 2, \ldots, s - 1\}$.

In the first case we take by induction the word $p(\mathcal{S}')$, formed with the elements $2, \ldots, s$ and terminating with s. We then define $p(\mathcal{S}) := 1p(\mathcal{S}')$.

We obtain in this way all the words in $1, 2, \ldots, s - 1, s$ that start with 1 and end with s.

In the second case we use the opposite ordering $(s - 1, \ldots, 2, 1)$ on S_2. Still by induction we have a word $p(\mathcal{S}')$ in $1, \ldots, s - 1$ that ends with 1. We set $p(\mathcal{S}) := p(\mathcal{S}')s$.

We obtain all the words in $1, 2, \ldots, s - 1, s$ that end with $1, s$.

It remains to get the words ending with s and in which 1 appears in the interior of the part of the word preceding s.

We now assume that there are two maximal elements, denoted by S_2 and S_3 (each with at least 2 elements), in $\mathcal{S}'$ whose disjoint union is $\{1, \ldots, s\}$. Let us denote by $\mathcal{S}_i$, $i = 1, 2$, the corresponding induced MNS.

As we have seen, since $\mathcal{S}$ is proper, we can assume that $1 \in S_2$, $s \in S_3$.

By induction, we have a word $p(\mathcal{S}_2)$ for S_2, relative to the opposite order terminating with 1. Similarly, we have a word $p(\mathcal{S}_3)$ for S_3 relative to the usual ordering terminating with s. We set $p(\mathcal{S}) := p(\mathcal{S}_2)p(\mathcal{S}_3)$, precisely a word of the desired type.

It is clear that this construction can be inverted and that in this way, we have established a one-to-one correspondence.

We illustrate this bijection with a few special cases:
1. If the MNS $\mathcal{S}$ is formed by the subsets

$$\{i, i+1, \ldots, s\}, \quad i = 1, \ldots, s-1,$$

its associated permutation is the identity.
2. If the MNS $\mathcal{S}$ is formed by the subsets

$$\{1, 2, \ldots, i\}, \quad i = 2, \ldots, s,$$

its associated permutation is $(s-1, s-2, \ldots, 1, s)$.
3. The example (20.3) is proper. Its associated permutation is $(3, 2, 1, 4, 5)$.

The second example we want to analyze is the following:

Given two positive integers m, n, we define the arrangement $M(m, n)$ as follows. We start from the vector space $\mathbb{R}^{m+n}$ with basis elements e_i, $i = 1, \ldots, m$, f_j, $j = 1, \ldots, n$, and let V be the hyperplane where the sum of the coordinates is 0. The arrangement is given by the nm vectors $\Delta(m, n) := \{(i|j) := e_i - f_j, \ i = 1, \ldots, m, \ j = 1, \ldots, n\}$. It is the graph arrangement associated to the full bipartite graph formed of all oriented edges from a set X with n elements to a set Y of m elements.

The polytopes associated to this arrangement are related to *magic squares*, or rather to a weak formulation of magic squares.

The classical definition of a magic square of size m is a filling of the square of size m (with m^2 cases) with the numbers $1, 2, \ldots, m^2$ in such a way that the sum on each column, on each row, and on the two diagonals is always the same number, that necessarily must be $m(m^2 + 1)/2$.

A weaker condition is to assume equality only on rows and columns, and an even weaker is to fix the value of the sum without insisting that the summands be distinct and coincide with the first m^2 integers.

One may call these squares *semi-magic*.

We can view a filling of a square, or more generally a rectangle, with positive integers as a vector $\sum_{i,j} a_{i,j}(i|j) = \sum_h c_h e_h - \sum_k d_k f_k$, where $c_h = \sum_j a_{h,j}$ is the sum on the h-th row and d_k the sum on the k-th column. Thus the number of semi-magic squares, adding to some number b, is an instance of the counting problem on polytopes, in this case the polytope associated to the vector $b(\sum_h e_h - \sum_k f_k)$. Notice that for a filling of the square of size m such that the sum on each column and on each row is always $m(m^2 + 1)/2$ the condition that we have all the numbers $1, 2, \ldots, m^2$ is equivalent to the condition that the filling consists of distinct numbers. This number can in theory be computed using the formula developed in Theorem 13.68. Still, the explicit computations involved are quite formidable. In fact, the number of magic squares has been computed only for $m \le 5$.

Even the simpler computation of the volume of the corresponding polytope has been carried out only up to dimension 10, a general formula is not available. Let us in any case discuss the general theory in this example.

We discuss the notions of irreducible, nested, and proper nested in the example $M(m, n)$. We need some definitions. Given two nonempty subsets $A \subset \{1, 2, \ldots, m\}, B \subset \{1, 2, \ldots, n\}$, we denote by $A \times B$ the set of vectors $(i|j), i \in A, \ j \in B$, and call it a *rectangle*. We say that the rectangle is degenerate if either A or B consists of just one element (and we will speak of a row or a column respectively).

In particular, when A and B both have two elements, we have a *little square*. We define a *triangle* as a subset with three elements of a little square.

Lemma 20.29. *(1) The four elements of a little square $\{i, j\} \times \{h, k\}$ form a complete set. They span a 3-dimensional space and satisfy the relation $(i|h) + (j|k) = (i|k) + (j|h)$.*
(2) The completion of a triangle is the unique little square in which it is contained.
(3) Any rectangle is complete.

The proof is clear.

Theorem 20.30. *For a subset $S \subset \Delta(m, n)$, the following conditions are equivalent:*

(1) Γ_S is X-complete.
(2) If a triangle T is contained in S, then its associated little square is also contained in S.
(3) $S = \cup_{i=1}^{h} A_i \times B_i$, where the A_i are mutually disjoint and the B_j are mutually disjoint.

Proof. Clearly, (1) implies (2). Let us now show that (2) implies (3). For this it is enough, by induction, to consider a maximal rectangle $A \times B$ contained in S and prove that $S \subset A \times B \cup C(A) \times C(B)$. Suppose this is not the case. Then there is an element $(i|k) \in S$ where either $i \in A, k \notin B$ or $i \notin A, k \in B$. Let us treat the first case; the second is similar. If $A = \{i\}$, then $A \times (B \cup \{k\})$ is a larger rectangle contained in S, a contradiction. Otherwise, take $j \in A$, $j \ne i, h \in B$. We have that $(i|h), (j|h), (i|k)$ are in S and form a triangle, so by assumption, also $(j|k) \in S$. This means that again $A \times (B \cup \{k\})$ is a larger rectangle contained in S, a contradiction. Now we can observe that $S \cap C(A) \times C(B)$ is also complete, and we proceed by induction.

(3) implies (1) follows from Proposition 20.20.

Theorem 20.30 now gives the decomposition of a complete set into irreducibles.

Corollary 20.31. *A nondegenerate rectangle is irreducible. Given a complete set of the form $S = \cup_{i=1}^{h} A_i \times B_i$, where the A_i are mutually disjoint and the B_j are mutually disjoint, its irreducible components are the nondegenerate rectangles $A_i \times B_i$ and the single elements of the degenerate rectangles $A_i \times B_i$.*

Theorem 20.30 also implies the structure of the maximal proper complete subsets of $\Delta(h,k)$.

Corollary 20.32. *A proper subset S of $\Delta(h,k)$ is maximal complete if and only if it is of one of the following types:*

$A \times B \cup C(A) \times C(B)$ *with A, B proper subsets.*

$A \times \{1, \ldots, n\}$, *where A has $m - 1$ elements.*

$\{1, \ldots, m\} \times B$, *where B has $n - 1$ elements.*

All these considerations allow us to find all proper flags in the case of the magic arrangement. Of course, in order even to speak about proper flags, we have to fix a total ordering among the pairs (i, j). Let us use as order the lexicographic order, so that $(1|1)$ is the minimum element. It follows that if S is a proper maximal complete subset, in order to be the beginning of a proper flag one needs that $(1|1) \notin S$.

It is then clear that once we have started with such a proper maximal complete subset, we can complete the flag to a proper flag by taking a proper flag in S for the induced order. This gives a recursive formula for the number $b(m, n)$ of proper flags, which is also the number of unbroken bases and thus, by Theorem 10.3, the top Betti number of the complement of the corresponding hyperplane arrangement. We have from our discussion the following recursive formula for $b(m, n)$:

$$b(m - 1, n) + b(m, n - 1) + \sum_{a,c} \binom{m - 1}{a - 1} \binom{n - 1}{c} b(a, c) b(m - a, n - c).$$

Remark 20.33. In [1] a number of interesting quantitative results on the enumeration of magic squares have been obtained. It could be interesting to use our results above to see whether one can push those computations a little bit further.

20.4 Wonderful Models

20.4.1 A Minimal Model

Although we do not use it explicitly, it may useful, and we present it in this section, to understand the origin of the nonlinear coordinates we have been using and of the entire theory of irreducibles and nested sets that we have built in [42].

Consider the hyperplane arrangement $\mathcal{H}_X$ and the open set

$$\mathcal{A}_X = U \setminus (\cup_{H \in \mathcal{H}_X} H),$$

the complement of the union of the given hyperplanes.

Let us denote by $\mathcal{I}$ the family of irreducible subsets in X. In [42] we construct a minimal smooth variety Z_X containing $\mathcal{A}_X$ as an open set with complement a normal crossings divisor, plus a proper map $\pi : Z_X \to U$ extending the identity of $\mathcal{A}_X$. The construction is the following.

For any irreducible subset $A \in \mathcal{I}$ take the vector space $V/A^{\perp}$ and the projective space $\mathbb{P}(V/A^{\perp})$. Notice that since $A^{\perp} \cap \mathcal{A}_X = \emptyset$, we have a natural projection $\pi_A : \mathcal{A}_X \to \mathbb{P}(V/A^{\perp})$. If we denote by $j : \mathcal{A}_X \to U$ the inclusion, we get a map

$$i := j \times (\times_{a \in \mathcal{I}} \pi_A) : \mathcal{A}_X \to U \times (\times_{a \in \mathcal{I}} \mathbb{P}(U/A^{\perp})). \tag{20.4}$$

Definition 20.34. The wonderful model Z_X is the closure of the image $i(\mathcal{A}_X)$ in $U \times (\times_{a \in \mathcal{I}} \mathbb{P}(U/A^{\perp}))$.

We describe some of the geometric features of Z_X. First of all, notice that $\times_{a \in \mathcal{I}} \mathbb{P}(U/A^{\perp})$ is a projective variety. Thus the restriction $\pi : Z_X \to U$ of the projection of $U \times (\times_{a \in \mathcal{I}} \mathbb{P}(U/A^{\perp}))$ to the first factor U is a projective and hence proper map.

Secondly, notice that i is injective. We identify $\mathcal{A}_X$ with its image under i. This image is closed inside $\mathcal{A}_X \times (\times_{a \in \mathcal{I}} \mathbb{P}(U/A^{\perp}))$, that is open in $U \times (\times_{a \in \mathcal{I}} \mathbb{P}(U/A^{\perp}))$. Thus Z_X contains $\mathcal{A}_X$ as a dense open subset, and the restriction of π to $\mathcal{A}_X$ coincides with the inclusion j.

In order to have a better understanding of the geometry of Z_X we are now going to describe some dense affine open sets in Z_X.

Take an MNS $\mathcal{S}$. Choose among all the bases adapted to $\mathcal{S}$ the basis $\underline{b} = \{b_1, \ldots, b_s\}$ that is lexicographically minimum (this is not necessary, but it allows us to choose exactly one adapted basis for each MNS). As usual, if $p_{\mathcal{S}}(b_i) = A$, denote b_i by b_A. The b_i form a system of linear coordinates in U.

Consider now $\mathbb{C}^s$ with coordinates $\underline{z} = \{z_A\}_{A \in \mathcal{S}}$, and a map $f_{\mathcal{S}} : \mathbb{C}^s \to U$ defined in the given coordinates, by

$$b_A = \prod_{B \supseteq A} z_B.$$

We have an inverse map $i_{\mathcal{S}}$, defined on the open set where the $b_A \neq 0$ given by $z_B := b_B$ if B is maximal and $z_B = b_B/b_C$ when B is not maximal and C is the minimal element of $\mathcal{S}$ properly containing B.

Now take $a \in X \setminus \underline{b}$ and let $p_{\mathcal{S}}(a) = A$. Thus $a \in A$, and by the nested property it does not lie in the space spanned by the elements $B \in \mathcal{S}$ properly contained in A. Therefore, we can write $a = \sum_{B \subseteq A} \gamma_B b_B$, $\gamma_B \in \mathbb{C}$, with $\gamma_A \neq 0$. Making the nonlinear change of coordinates, we get

$$a = \sum_{B \subseteq A} \gamma_B \prod_{C \supseteq B} z_C = \left(\gamma_A \prod_{C \supseteq A} z_B \right) p_a(\underline{z}), \tag{20.5}$$

where $p_a(\underline{z}) = 1 + q_a(\underline{z})$ with $q_a(\underline{z})$ a polynomial vanishing at the origin.

Set now $\mathcal{V}_\mathcal{S}$ equal to the open set in $\mathbb{C}^s$ that is the complement of the divisor of equation $\prod_{a \in X \setminus \underline{b}} p_a(\underline{z}) = 0$.

Notice that the previously defined map $i_\mathcal{S}$ lifts the open embedding $i : \mathcal{A}_X \to U$ to an open embedding

$$i_\mathcal{S} : \mathcal{A}_X \to \mathcal{V}_\mathcal{S}.$$

By its very construction, the complement in $\mathcal{V}_\mathcal{S}$ of the union of the divisors $z_A = 0$ is mapped isomorphically to $\mathcal{A}_X$ by the morphism $f_\mathcal{S}$ that is the inverse of $i_\mathcal{S}$.

Lemma 20.35. *Let $A \in \mathcal{I}$ and $\mathcal{S}$ be an MNS. Then there exists a map $\pi_{A,\mathcal{S}} : \mathcal{V}_\mathcal{S} \to \mathbb{P}(U/A^\perp)$ such that the diagram*

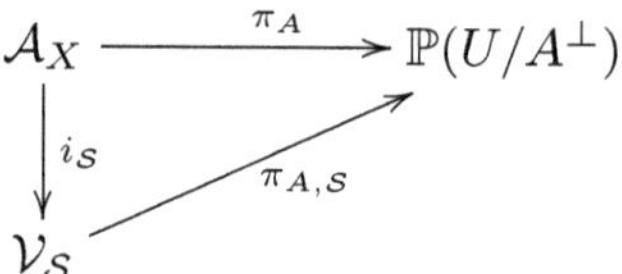

is commutative.

Proof. Since A is irreducible, there exists a minimum $B \in \mathcal{S}$ such that $A \subset B$.

We can choose projective coordinates in $\mathbb{P}(U/A^\perp)$ by choosing a basis of $\langle A \rangle$ contained in X. Call it $c_1, \ldots, c_k$. It is now clear that for each i, $p_\mathcal{S}(c_i) \subseteq B$. From the irreducibility of A, we have that for at least one i, which we can assume to be 1, we have $p_\mathcal{S}(c_i) = B$. Indeed, A would be contained in the union of the irreducible subsets in $\mathcal{S}$ properly contained in B, and thus by its irreducibility in one of them.

Substituting our nonlinear coordinate, we get that each c_i is divisible by $\prod_{B \subseteq C} z_C$, and moreover, $c_1 / \prod_{B \subseteq C} z_C$ is invertible on Z_X. Since we are working in projective space, our claim follows.

The previous lemma clearly implies that there exists a map $j_\mathcal{S} : \mathcal{V}_\mathcal{S} \to Z_X$ such that the diagram

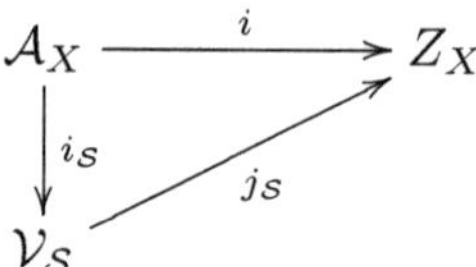

commutes.

Notice that if $B \in \mathcal{S}$, we can choose as basis of $\langle B \rangle$ the set of elements b_C such that $C \subset B$ and then $b_C / b_B = \prod_{C \subseteq D \subsetneq B} z_D$. In particular, if C is maximal in B, then $b_C / b_B = z_C$. If, on the other hand, C is maximal in $\mathcal{S}$, we have that $b_C = z_C$. From the proof of the previous lemma, $\mathcal{V}_\mathcal{S}$ maps into the affine subset of $\mathbb{P}(U/B^\perp)$, where $b_B \neq 0$. Thus the coordinates z_B of $\mathcal{V}_\mathcal{S}$ are restrictions of functions on the Cartesian product of U with all these affine sets. This proves that $\mathcal{V}_\mathcal{S}$ embeds as a closed subset into an affine open set in $U \times (\times_{B \in \mathcal{S}} \mathbb{P}(U/B^\perp))$ and implies our next result.

Proposition 20.36. *For each MNS $\mathcal{S}$, the map*

$$j_{\mathcal{S}} : \mathcal{V}_{\mathcal{S}} \longrightarrow Z_X$$

is a embedding onto an affine open set.

Remark 20.37. As we have already stated, the special choice of the basis $\underline{b}$ is not necessary. Indeed, choose another basis $\underline{b}'$ adapted to $\mathcal{S}$. Repeating our construction, we can construct an open set $\mathcal{V}_{\mathcal{S},\underline{b}'}$ in $\mathbb{C}^s$ with coordinates z'_A, $A \in \mathcal{S}$, and an embedding $j_{\mathcal{S},\underline{b}'} : \mathcal{V}_{\mathcal{S},\underline{b}'} \to Z_X$.

Since both open sets naturally contain the open set $\mathcal{A}_X$, we can think the z_A's (resp. z'_A's) as rational functions on $\mathcal{V}_{\mathcal{S}}$ (resp. on $\mathcal{V}_{\mathcal{S},\underline{b}'}$).

If $A \in \mathcal{S}$ and A is not maximal, we know by the nested property that there is a minimal element $B \in \mathcal{S}$ such that $A \subsetneq B$. We denote such a B by $c(A)$ (the *consecutive* to A).

It follows that $z_A = b_A / b_{c(A)}$ and $z'_A = b'_A / b'_{c(A)}$. Using this and (20.5), we easily see that for each $A \in \mathcal{S}$,

$$z'_A = z_A G_A(\underline{z}),$$

where the function $G(\underline{z})$ is invertible on $\mathcal{V}_{\mathcal{S}}$. We deduce that the open set $\mathcal{V}_{\mathcal{S},\underline{b}'}$ is naturally isomorphic to $\mathcal{V}_{\mathcal{S}}$ by an isomorphism ψ taking the locus of equation $z'_A = 0$ to that of equation $z_A = 0$ for each $A \in \mathcal{S}$ and that the diagram

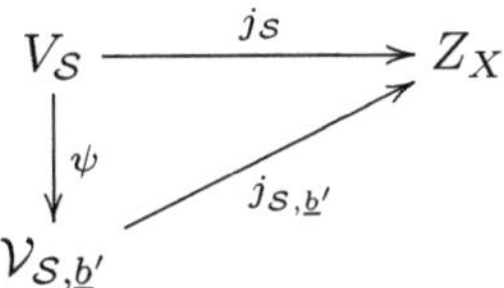

commutes. Since $\underline{b}, \underline{b}'$ play symmetric roles, ψ is an isomorphism.

Let us now identify $\mathcal{V}_{\mathcal{S}}$ with its image $j_{\mathcal{S}}(\mathcal{V}_{\mathcal{S}})$.

Theorem 20.38. $\cup_{\mathcal{S}} \mathcal{V}_{\mathcal{S}} = Z_X$. *In particular, Z_X is a smooth variety.*

Proof. Let $Y_X := \cup_{\mathcal{S}} \mathcal{V}_{\mathcal{S}}$. It is clearly enough to show that Y_X is closed in $U \times (\times_{A \in \mathcal{I}} \mathbb{P}(U/A^\perp))$. We claim that for this it is enough to see that the projection map of Y_X to U is proper. In fact, assume that this is true; the projection of $U \times (\times_{A \in \mathcal{I}} \mathbb{P}(U/A^\perp))$ to U is proper and so the embedding of Y_X into $U \times (\times_{A \in \mathcal{I}} \mathbb{P}(U/A^\perp))$ is also proper, and finally, its image is closed.

Now in order to prove the required properness we shall use a standard tool from algebraic geometry, the properness criterion given by curves, also known as the *valuative criterion of properness* (cf. [61]). Over the complex numbers this amounts to showing that for any curve $f : D_\varepsilon \to U$, $D_\varepsilon = \{t \in \mathbb{C} | \, |t| < \varepsilon\}$ such that $f(D_\varepsilon \setminus \{0\}) \subset \mathcal{A}_X$, there exists a curve $\tilde{f} : D_\varepsilon \to Y_X$ such that the diagram

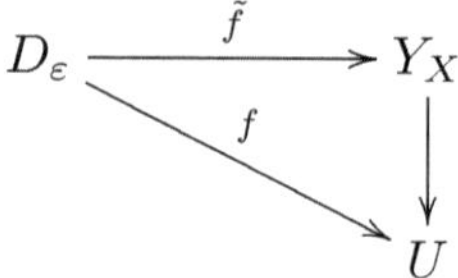

commutes.

Take such an f. For any $a \in X$, we can write $\langle a \,|\, f(t) \rangle = t^{n_a} \kappa_a(t)$ near $t = 0$, where $n_a \geq 0$ and $\kappa_a(0) \neq 0$.

For each $h \geq 0$, we define $A_h = \{a \in X \,|\, n_a \geq h\}$. Clearly $A_0 = X$, and for each h, $A_{h+1} \subset A_h$ and each A_h is complete.

If we now decompose each A_h into irreducibles, we get a collection of elements in $\mathcal{I}$ that, due to the fact that $A_{h+1} \subset A_h$, is clearly a nested set. Let us complete this nested set to an MNS $\mathcal{S}$.

We claim that the map $f : D_\varepsilon \setminus \{0\} \to \mathcal{A}_X$ extends to a map $f : D_\varepsilon \to \mathcal{V}_{\mathcal{S}}$. To see this, we need to show that for each $C \in \mathcal{S}$, $z_C(f(t))$ is defined at 0 and that for each $a \in X \setminus \underline{b}$, $p_a(f(t)) \neq 0$ at 0.

For the first claim notice that if C is maximal in X, then $z_C = b_C$, and so $z_C(f(t)) = \langle b_C \,|\, f(t) \rangle$ is defined at 0. If C is not maximal, let B be the minimum element in $\mathcal{S}$ properly containing C. Then by the definition of $\mathcal{S}$, $n_{b_C} \geq n_{b_B}$. Also

$$z_C(f(t)) = \frac{\langle b_C \,|\, f(t) \rangle}{\langle b_B \,|\, f(t) \rangle} = t^{n_{b_C} - n_{b_B}} \frac{\kappa_{b_B}(t)}{\kappa_{b_B}(t)},$$

so that $z_C(f(t))$ is clearly defined at 0.

For the second claim, consider $p_{\mathcal{S}}(a) = B$. Then by the definition of $\mathcal{S}$, $n_a = n_{b_B}$. Thus

$$p_a(f(0)) = \frac{\langle a \,|\, f(t) \rangle}{\langle b_B \,|\, f(t) \rangle} = \frac{\kappa_a(0)}{\kappa_{b_B}(0)} \neq 0.$$

20.4.2 The Divisors

We need a technical fact.

Lemma 20.39. *Let $\mathcal{S}$ and $\mathcal{T}$ be two MNS's in X. Let $\underline{b}^{\mathcal{S}}$ (resp. $\underline{b}^{\mathcal{T}}$) denote the associated adapted basis and let $z_A^{\mathcal{S}}$, $A \in \mathcal{S}$ (resp. $z_A^{\mathcal{T}}$, $A \in \mathcal{T}$) be the corresponding set of coordinates on $\mathcal{V}_{\mathcal{S}}$ (resp. $\mathcal{V}_{\mathcal{T}}$). Then*

(1) If $A \in \mathcal{S} \setminus \mathcal{T}$, $z_A^{\mathcal{S}}$ is invertible as a function on $\mathcal{V}_{\mathcal{S}} \cap \mathcal{V}_{\mathcal{T}}$.
(2) If $A \in \mathcal{T} \setminus \mathcal{S}$, $z_A^{\mathcal{T}}$ is invertible as a function on $\mathcal{V}_{\mathcal{S}} \cap \mathcal{V}_{\mathcal{T}}$.
(3) If $A \in \mathcal{S} \cap \mathcal{T}$, $z_A^{\mathcal{S}}/z_A^{\mathcal{T}}$ is regular and invertible as a function on $\mathcal{V}_{\mathcal{S}} \cap \mathcal{V}_{\mathcal{T}}$.

Proof. Assume that A is a maximal element in $\mathcal{I}$. This implies $A \in \mathcal{S} \cap \mathcal{T}$, so we have to prove (3). We have $z_A^{\mathcal{S}} = b_A^{\mathcal{S}}$. Let $B = p_{\mathcal{T}}(b_A^{\mathcal{S}}) \subseteq A$. Then by (20.5),

$$z_A^{\mathcal{S}} = \prod_{C \in \mathcal{T}, B \subseteq C} z_C^{\mathcal{T}} g = z_A^{\mathcal{T}} \prod_{C \in \mathcal{T}, B \subseteq C, C \neq A} z_C^{\mathcal{T}} g,$$

where g is invertible.

Similarly, $z_A^{\mathcal{T}} = b_A^{\mathcal{T}}$ and if $B' = p_{\mathcal{S}}(b_A^{\mathcal{T}}) \subseteq A$, then

$$z_A^{\mathcal{T}} = z_A^{\mathcal{S}} \prod_{D \in \mathcal{S}, B \subseteq D, D \neq A} z_D^{\mathcal{S}} \tilde{g},$$

where $\tilde{g}$ is invertible.

Substituting and using the fact that $A \supset B$ and $A \supset B'$, we get

$$z_A^{\mathcal{S}} = z_A^{\mathcal{S}} \prod_{C \in \mathcal{T}, B \subseteq C, C \neq A} z_C^{\mathcal{T}} g \prod_{D \in \mathcal{S}, B \subseteq D, D \neq A} z_D^{\mathcal{S}} \tilde{g}.$$

Thus

$$\prod_{C \in \mathcal{T}, B \subseteq C, C \neq A} z_C^{\mathcal{T}} g \prod_{D \in \mathcal{S}, B \subseteq D, D \neq A} z_D^{\mathcal{S}} \tilde{g} = 1.$$

Since

$$\frac{z_A^{\mathcal{S}}}{z_A^{\mathcal{T}}} = \prod_{C \in \mathcal{T}, B \subseteq C, C \neq A} z_C^{\mathcal{T}} g,$$

our claim follows in this case.

In the general case take $A \in \mathcal{S}$. By induction, we can assume that our claim holds for each $B \supsetneq A$ in $\mathcal{S} \cup \mathcal{T}$.

Since A is irreducible, there is a minimum element $B \in \mathcal{T}$ containing A. We know that there exists $a \in A$ such that $p_{\mathcal{T}}(a) = B$.

Set $E = p_{\mathcal{S}}(a) \subset A$. Notice that if C is irreducible and $B \supsetneq C \supseteq E$, then necessarily $C \notin \mathcal{T}$.

Using (20.5) let us write

$$a = \prod_{C \in \mathcal{T}, B \subseteq C} z_C^{\mathcal{T}} g = \prod_{D \in \mathcal{S}, E \subseteq D} z_D^{\mathcal{S}} \tilde{g},$$

where g and $\tilde{g}$ are invertible. We deduce

$$1 = \prod_{D \in \mathcal{S} \setminus \mathcal{T}, E \subseteq D} z_D^{\mathcal{S}} \prod_{C \in \mathcal{T} \setminus \mathcal{S}, B \subseteq C} \frac{1}{z_C^{\mathcal{T}}} \prod_{D \in \mathcal{S} \cap \mathcal{T}, B \subseteq D} \frac{z_D^{\mathcal{S}}}{z_D^{\mathcal{T}}} \tilde{g} g^{-1}. \tag{20.6}$$

If $A \notin \mathcal{T}$, then $B \supsetneq A$. We claim that all the factors in (20.6) are regular on $\mathcal{V}_{\mathcal{S}} \cap \mathcal{V}_{\mathcal{T}}$. Indeed, the factors of the form $z_D^{\mathcal{S}}$ with $D \in \mathcal{S} \setminus \mathcal{T}, E \subseteq D$ are obviously regular. The factors $1/z_C^{\mathcal{T}}$ for $C \in \mathcal{T} \setminus \mathcal{S}, B \subseteq C$ and $z_D^{\mathcal{S}}/z_D^{\mathcal{T}}$ for $D \in \mathcal{S} \cap \mathcal{T}, B \subseteq D$ are regular by the inductive assumption, since they involve elements properly containing A. Since $z_A^{\mathcal{S}}$ appears as one of the factors in (20.6), its invertibility follows.

If, on the other hand, $A \in \mathcal{T}$, so that $B = A$, reasoning as above, all the factors in (20.6) are regular on $\mathcal{V}_{\mathcal{S}} \cap \mathcal{V}_{\mathcal{T}}$ with the possible exception of $z_A^{\mathcal{S}}/z_A^{\mathcal{T}}$. But then this implies that $z_A^{\mathcal{T}}/z_A^{\mathcal{S}}$ is regular and invertible.

If we now exchange the roles of $\mathcal{S}$ and $\mathcal{T}$, all our claims follow.

We can now use the previous lemma to define for each $A \in \mathcal{I}$ the divisor $D_A \subset Z_X$ as follows. Let us take an MNS $\mathcal{S}$ with $A \in \mathcal{S}$. In the open set $\mathcal{V}_\mathcal{S}$ let us take the divisor of equation $z_A = 0$. Then D_A is the closure of this divisor in Z_X. Our previous lemma clearly implies that the definition of D_A is independent of the choice of $\mathcal{S}$, and furthermore if $\mathcal{T}$ is any MNS, then $Z_X \setminus \mathcal{V}_\mathcal{S} = \cup_{A \notin \mathcal{S}} D_A$.

From these considerations we deduce the following:

Theorem 20.40. *Let $D = \cup_{A \in \mathcal{I}} D_A$. Then:*

(1) $\mathcal{A}_X = Z_X \setminus D$.
(2) D is a normal crossing divisor whose irreducible components are the divisors D_A, $A \in \mathcal{I}$.
(3) Let $\mathcal{H} \subset \mathcal{I}$. Set $D_\mathcal{H} = \cap_{A \in \mathcal{H}} D_A$. Then $D_\mathcal{H} \neq \emptyset$ if and only if $\mathcal{H}$ is nested.
(4) If $\mathcal{H} \subset \mathcal{I}$ is nested, $D_\mathcal{H}$ is smooth and irreducible.

Proof. Let us fix an MNS $\mathcal{S}$. Then (1) follows from the fact that

$$Z_X \setminus D = \mathcal{V}_\mathcal{S} \setminus (\cup_{A \in \mathcal{S}}(D_A \cap \mathcal{V}_\mathcal{S})) = \mathcal{A}_X.$$

All the other statements are immediate consequence of Lemma 20.39.

20.4.3 Geometric Side of Residues

Theorem 20.40 singles out for each MNS $\mathcal{S}$ the point $D_\mathcal{S}$, $D_\mathcal{S} \in \mathcal{V}_\mathcal{S}$.

If we now take a basis $\underline{b}$ in X, we can construct as in Theorem 20.11 a MNS $\mathcal{S}$ to which $\underline{b}$ is adapted.

We can define a small parametrized real s-dimensional torus $T_\mathcal{S}$ (*centered at $D_\mathcal{S}$*) as the set of points in $\mathcal{V}_\mathcal{S}$ such that $|z_A| = \varepsilon$ for each $A \in \mathcal{S}$ (ε a small enough positive constant).

Notice that $T_\mathcal{S} \subset \mathcal{A}_X$ and that $T_\mathcal{S}$ is naturally oriented using the total orientation induced on $\mathcal{S}$ by that on the adapted basis $\underline{b}$.

We can consider the map $H_s(T_\mathcal{S}, \mathbb{Z}) \to H_s(\mathcal{A}_X, \mathbb{Z})$ induced by the inclusion and taking the image of the fundamental class in $H_s(T_\mathcal{S}, \mathbb{Z})$ to get a class $\alpha_{\underline{b}} \in H_s(\mathcal{A}_X, \mathbb{Z})$. The very definition of $\mathrm{res}_{\underline{b}}$ then implies the following result:

Proposition 20.41. *For any basis $\underline{b}$ and any s-form ω on $\mathcal{A}_X$, $\mathrm{res}_{\underline{b}}\omega$ equals the evaluation of the cohomology class $[\omega]$ on the homology class $\alpha_{\underline{b}}$.*

In particular, the classes $\alpha_{\underline{b}}$ as $\underline{b}$ varies among the proper unbroken bases form a basis of $H_s(\mathcal{A}_X, \mathbb{Z})$.

20.4.4 Building Sets

The choice of the irreducible subsets in the construction of the embedding of $\mathcal{A}_X$ in $U \times (\times_{A \in \mathcal{I}} \mathbb{P}(U/A^\perp))$ is not essential for obtaining the properties of smoothness and normal crossings previously described. In fact, one can easily see that also taking the collection $\mathcal{C}$ of *all* complete subsets, one has the same properties for the embedding $\mathcal{A}_X \to U \times (\times_{A \in \mathcal{C}} \mathbb{P}(U/A^\perp))$. Now, though, we have many more divisors at infinity indexed by all complete subsets, and the nested property means that the family $A_1, \ldots, A_s$ of complete sets is a flag.

In fact, there are several families of complete sets that provide smooth models in which the complement is a divisor with normal crossings. They are given by the notion of *building set*:

Definition 20.42. A family $\mathcal{G}$ of complete sets in X is said to be **building** if given any complete set A, the maximal elements $B_1, \ldots, B_k$ of $\mathcal{G}$ contained in A have the property that $A = \cup_i B_i$ is a decomposition.

We leave to the reader to verify that $\mathcal{G}$ always contains $\mathcal{I}$. The closure $Z_X^{\mathcal{G}}$ of the image of $\mathcal{A}_X$ in $U \times (\times_{A \in \mathcal{G}} \mathbb{P}(U/A^\perp))$ is a smooth variety, the complement of $\mathcal{A}_X$ is a divisor with normal crossings, and its irreducible components are indexed by the elements of $\mathcal{G}$. One has a similar definition of nested sets that correspond to divisors with nonempty intersection. Finally, one has clearly that if $\mathcal{G}_1 \subset \mathcal{G}_2$ are both building families, there is a commutative diagram

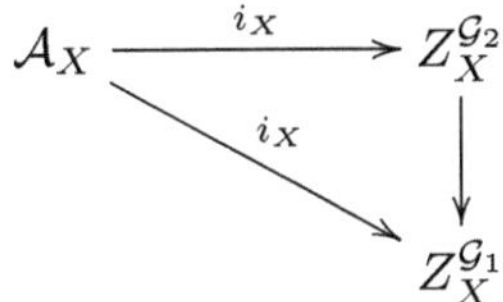

For further details the reader should see [42], [77].

20.4.5 A Projective Model

In this section we are going to assume that X itself is irreducible. If this is the case, we want to note two facts. The first is that since $X^\perp = \{0\}$, we have the projection $\pi_V : Z_X \to \mathbb{P}(U)$. The second is that we have the divisor D_X in Z_X.

Lemma 20.43. *Consider the projection $\pi : Z_X \to U$. Then $\pi(D_X) = \{0\}$.*

Proof. To prove our claim, it clearly suffices to see that for any point $x \in D_X$, $a \in V$, $\langle a \,|\, \pi(x) \rangle = 0$. Take an MNS $\mathcal{S}$, and consider the open set $\mathcal{V}_{\mathcal{S}}$ and the corresponding adapted basis $\underline{b}$. Since X is the unique maximal element in $\mathcal{I}$, we have $X \in \mathcal{S}$. Also on $\mathcal{V}_{\mathcal{S}}$, $b_A = \prod_{B \supset A} z_B$ is divisible by z_X. It follows that $\langle b_A \,|\, \pi(x) \rangle = 0$ for each $x \in \mathcal{V}_{\mathcal{S}} \cap D_X$ and for each $A \in \mathcal{S}$. Since $\underline{b}$ is a basis and $\mathcal{V}_{\mathcal{S}} \cap D_X$ is dense in D_X, the claim follows.

Remark 20.44. In a completely analogous fashion one can show that for each $A \in \mathcal{I}$, $\pi(D_A) = A^\perp$.

Using this lemma we immediately get that D_X is smooth and projective and that if we consider the map $g := \times_{a \in \mathcal{I}} \pi_A : Z_X \to \prod_{A \in \mathcal{I}} \mathbb{P}(U/A^\perp)$, we have the following:

Lemma 20.45. *The restriction of g to D_X is injective and $g(Z_X) = g(D_X)$.*

Proof. The first statement follows immediately from the previous lemma. As for the second, notice that if $v \in \mathcal{A}_X$, then $g(v) = g(tv)$ for any nonzero $t \in \mathbb{C}$. It follows that $g(Z_X)$ is an irreducible subvariety of dimension $s - 1$. Since $g(D_X)$ is closed and of the same dimension, the claim follows.

Let us now identify D_X with its image $g(D_X)$. We deduce a projection, that we can again denote by g, of Z_X onto D_X. Recall that on $\mathbb{P}(U)$ we have the tautological line bundle $\mathcal{L} = \{(u, \ell) \in U \times \mathbb{P}(U) | u \in \ell\}$. We can thus consider the line bundle $L := \pi_U^*(\mathcal{L})$ on D_X with projection $p : L \to D_X$.

Let us introduce an action of the group $\mathbb{C}^*$ on Z_X as follows: $\mathbb{C}^*$ acts on $U \times \prod_{A \in \mathcal{I}} \mathbb{P}(U/A^\perp)$ by acting by scalar multiplication on the first factor U. It is then clear the $i(\mathcal{A}_X)$ is stable under this action, so that Z_X, being its closure, is also stable. Notice that $\mathbb{C}^*$ acts also on L by multiplication on each fiber.

Theorem 20.46. *There is a $\mathbb{C}^*$ isomorphism $\delta : Z_X \to L$ such that the diagram*

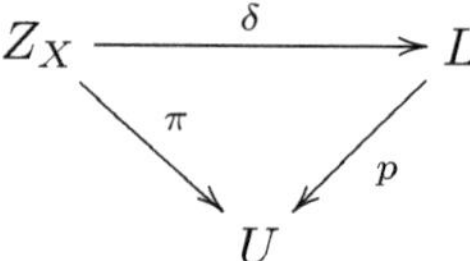

commutes.

Proof. Notice that by definition, $L = \{(u, x) \in U \times D_X | u \in \pi_U(x)\}$. The definition of Z_X then implies that we have a well-defined $\mathbb{C}^*$-equivariant map $\delta : Z_X \to L$ whose composition with the inclusion of L in $U \times D_X$ is the inclusion of Z_X. In particular, δ is an inclusion. Since L and Z_X are irreducible of the same dimension, δ is necessarily an isomorphism.

Not only is then Z_X identified with the total space of a line bundle over a projective variety D_X, we also have that D_X is a compactification of the image $\tilde{\mathcal{A}}_X := \mathcal{A}_X/\mathbb{C}^* \subset \mathbb{P}(U)$. Indeed, we have the following commutative diagram:

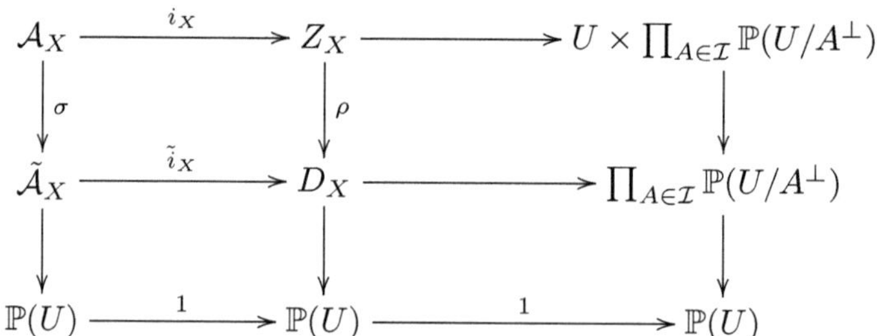

Here σ is the projection to projective space and ρ the projection of the total space of the line bundle to its basis.

Now again the boundary of $\tilde{\mathcal{A}}_X$ in D_X is a divisor with normal crossings whose irreducible components are indexed by the proper irreducible subsets of X, and so on.

References

[1] Maya Ahmed, Jesús De Loera, and Raymond Hemmecke. Polyhedral cones of magic cubes and squares. In *Discrete and Computational Geometry*, volume 25 of *Algorithms Combin.*, pages 25–41. Springer, Berlin, 2003.

[2] A. A. Akopyan and A. A. Saakyan. A system of differential equations that is related to the polynomial class of translates of a box spline. *Mat. Zametki*, 44(6):705–724, 861, 1988.

[3] Harsh Anand, Vishwa Chander Dumir, and Hansraj Gupta. A combinatorial distribution problem. *Duke Math. J.*, 33:757–769, 1966.

[4] Federico Ardila. Computing the Tutte polynomial of a hyperplane arrangement. *Pacific J. Math.*, 230(1):1–26, 2007.

[5] Paul Bachmann. *Niedere Zahlentheorie*. Zweiter Teil. B. G. Teubner's Sammlung von Lehrbüchern auf dem Gebiete der mathematischen Wissenschaften mit Einschluss ihrer Anwendungen, Band X, 2., Leipzig, 1902.

[6] W. Baldoni-Silva, J. A. De Loera, and M. Vergne. Counting integer flows in networks. *Found. Comput. Math.*, 4(3):277–314, 2004.

[7] A. I. Barvinok. A polynomial time algorithm for counting integral points in polyhedra when the dimension is fixed. In *34th Annual Symposium on Foundations of Computer Science (Palo Alto, CA, 1993)*, pages 566–572. IEEE Comput. Soc. Press, Los Alamitos, CA, 1993.

[8] A. I. Barvinok, A. M. Vershik, and N. E. Mněv. Topology of configuration spaces, of convex polyhedra and of representations of lattices. *Trudy Mat. Inst. Steklov.*, 193:37–41, 1992.

[9] Alexander Barvinok. Computing the Ehrhart quasi-polynomial of a rational simplex. *Math. Comp.*, 75(255):1449–1466 (electronic), 2006.

[10] Alexander Barvinok, Matthias Beck, Christian Haase, Bruce Reznick, and Volkmar Welker, editors. *Integer points in polyhedra—geometry, number theory, algebra, optimization*, volume 374 of *Contemporary Mathematics*, Providence, RI, 2005. American Mathematical Society.

[11] Alexander Barvinok and James E. Pommersheim. An algorithmic theory of lattice points in polyhedra. In *New perspectives in algebraic combinatorics (Berkeley, CA, 1996–97)*, volume 38 of *Math. Sci. Res. Inst. Publ.*, pages 91–147. Cambridge Univ. Press, Cambridge, 1999.

[12] Alexander I. Barvinok. A polynomial time algorithm for counting integral points in polyhedra when the dimension is fixed. *Math. Oper. Res.*, 19(4):769–779, 1994.

[13] M. Beck and S. Robins. Dedekind sums: a combinatorial-geometric viewpoint. arXiv:math.NT/0112076.

[14] Matthias Beck and Sinai Robins. *Computing the continuous discretely.* Undergraduate Texts in Mathematics. Springer, New York, 2007. Integer-point enumeration in polyhedra.

[15] E. T. Bell. Interpolated denumerants and Lambert series. *Amer. J. Math.*, 65:382–386, 1943.

[16] Anders Björner. The homology and shellability of matroids and geometric lattices. In *Matroid applications*, volume 40 of *Encyclopedia Math. Appl.*, pages 226–283. Cambridge Univ. Press, Cambridge, 1992.

[17] Armand Borel. *Linear algebraic groups*, volume 126 of *Graduate Texts in Mathematics*. Springer-Verlag, New York, second edition, 1991.

[18] Raoul Bott and Loring W. Tu. *Differential forms in algebraic topology*, volume 82 of *Graduate Texts in Mathematics*. Springer-Verlag, New York, 1982.

[19] N. Bourbaki. *Éléments de mathématique. XIV. Première partie: Les structures fondamentales de l'analyse. Livre II: Algèbre. Chapitre VI: Groupes et corps ordonnés. Chapitre VII: Modules sur les anneaux principaux.* Actualités Sci. Ind., no. 1179. Hermann et Cie, Paris, 1952.

[20] N. Bourbaki. *Éléments de mathématique. Fasc. XXXIV. Groupes et algèbres de Lie. Chapitre IV: Groupes de Coxeter et systèmes de Tits. Chapitre V: Groupes engendrés par des réflexions. Chapitre VI: systèmes de racines.* Actualités Scientifiques et Industrielles, No. 1337. Hermann, Paris, 1968.

[21] Egbert Brieskorn. Sur les groupes de tresses [d'après V. I. Arnol'd]. In *Séminaire Bourbaki, 24ème année (1971/1972), Exp. No. 401*, pages 21–44. Lecture Notes in Math., Vol. 317. Springer, Berlin, 1973.

[22] Michel Brion. Points entiers dans les polyèdres convexes. *Ann. Sci. École Norm. Sup. (4)*, 21(4):653–663, 1988.

[23] Michel Brion. Polyèdres et réseaux. *Enseign. Math. (2)*, 38(1-2):71–88, 1992.

[24] Michel Brion and Michèle Vergne. Une formule d'Euler-Maclaurin pour les fonctions de partition. *C. R. Acad. Sci. Paris Sér. I Math.*, 322(3):217–220, 1996.

[25] Michel Brion and Michèle Vergne. Une formule d'Euler-Maclaurin pour les polytopes convexes rationnels. *C. R. Acad. Sci. Paris Sér. I Math.*, 322(4):317–320, 1996.

[26] Michel Brion and Michèle Vergne. An equivariant Riemann-Roch theorem for complete, simplicial toric varieties. *J. Reine Angew. Math.*, 482:67–92, 1997.

[27] Michel Brion and Michèle Vergne. Residue formulae, vector partition functions and lattice points in rational polytopes. *J. Amer. Math. Soc.*, 10(4):797–833, 1997.

[28] Michel Brion and Michèle Vergne. Arrangement of hyperplanes. I. Rational functions and Jeffrey-Kirwan residue. *Ann. Sci. École Norm. Sup. (4)*, 32(5):715–741, 1999.

[29] Michel Brion and Michèle Vergne. Arrangement of hyperplanes. II. The Szenes formula and Eisenstein series. *Duke Math. J.*, 103(2):279–302, 2000.

[30] Tom Brylawski. The broken-circuit complex. *Trans. Amer. Math. Soc.*, 234(2):417–433, 1977.

[31] Alfred S. Cavaretta, Wolfgang Dahmen, and Charles A. Micchelli. Stationary subdivision. *Mem. Amer. Math. Soc.*, 93(453):vi+186, 1991.

[32] G.M. Chaikin. An algorithm for high speed curve generation. *Computer Graphics and Image Processing*, 3(12):346–349, 1974.

[33] Charles K. Chui and Harvey Diamond. A natural formulation of quasi-interpolation by multivariate splines. *Proc. Amer. Math. Soc.*, 99(4):643–646, 1987.

[34] C. Cochet. *Réduction des graphes de Goresky–Kottwitz–MacPherson: nombres de Kostka et coefficients de Littlewodd-Richardson*. 2003. Thèse de Doctorat: Mathématiques. Universit Paris 7, Paris, France.

[35] S. C. Coutinho. *A primer of algebraic D-modules*, volume 33 of *London Mathematical Society Student Texts*. Cambridge University Press, Cambridge, 1995.

[36] Henry H. Crapo. The Tutte polynomial. *Aequationes Math.*, 3:211–229, 1969.

[37] Wolfgang Dahmen and Charles A. Micchelli. The number of solutions to linear Diophantine equations and multivariate splines. *Trans. Amer. Math. Soc.*, 308(2):509–532, 1988.

[38] Ingrid Daubechies. *Ten lectures on wavelets*, volume 61 of *CBMS-NSF Regional Conference Series in Applied Mathematics*. Society for Industrial and Applied Mathematics (SIAM), Philadelphia, PA, 1992.

[39] C. de Boor and R. DeVore. Approximation by smooth multivariate splines. *Trans. Amer. Math. Soc.*, 276(2):775–788, 1983.

[40] C. de Boor, K. Höllig, and S. Riemenschneider. *Box splines*, volume 98 of *Applied Mathematical Sciences*. Springer-Verlag, New York, 1993.

[41] Carl de Boor, Nira Dyn, and Amos Ron. On two polynomial spaces associated with a box spline. *Pacific J. Math.*, 147(2):249–267, 1991.

[42] C. De Concini and C. Procesi. Wonderful models of subspace arrangements. *Selecta Math. (N.S.)*, 1(3):459–494, 1995.

[43] C. De Concini and C. Procesi. Hyperplane arrangements and box splines. With an appendix by A. Björner. *Michigan Math. J.*, 57:201–225, 2008.

[44] C. De Concini, C. Procesi, and M. Vergne. Partition function and generalized Dahmen-Micchelli spaces. arXiv:math 0805.2907.

[45] C. De Concini, C. Procesi, and M. Vergne. Vector partition functions and index of transversally elliptic operators. arXiv:0808.2545v1.

[46] Corrado De Concini and Claudio Procesi. Nested sets and Jeffrey-Kirwan residues. In *Geometric methods in algebra and number theory*, volume 235 of *Progr. Math.*, pages 139–149. Birkhäuser Boston, Boston, MA, 2005.

[47] Jesús A. De Loera. The many aspects of counting lattice points in polytopes. *Math. Semesterber.*, 52(2):175–195, 2005.

[48] Jesús A. De Loera, Raymond Hemmecke, Jeremiah Tauzer, and Ruriko Yoshida. Effective lattice point counting in rational convex polytopes. *J. Symbolic Comput.*, 38(4):1273–1302, 2004.

[49] Georges de Rham. Sur une courbe plane. *J. Math. Pures Appl. (9)*, 35:25–42, 1956.

[50] Leonard Eugene Dickson. *History of the Theory of Numbers. Vol. II: Diophantine Analysis*. Chelsea Publishing Co., New York, 1966.

[51] N. Dyn and A. Ron. Local approximation by certain spaces of exponential polynomials, approximation order of exponential box splines, and related interpolation problems. *Trans. Amer. Math. Soc.*, 319(1):381–403, 1990.

[52] Nira Dyn and David Levin. Subdivision schemes in geometric modelling. *Acta Numer.*, 11:73–144, 2002.

[53] E. Ehrhart. Sur un problème de géométrie diophantienne linéaire. I. Polyèdres et réseaux. *J. Reine Angew. Math.*, 226:1–29, 1967.

[54] E. Ehrhart. *Polynômes arithmétiques et méthode des polyèdres en combinatoire.* Birkhäuser Verlag, Basel, 1977. International Series of Numerical Mathematics, Vol. 35.

[55] David Eisenbud. *Commutative Algebra*, volume 150 of *Graduate Texts in Mathematics.* Springer-Verlag, New York, 1995. With a view toward algebraic geometry.

[56] Eva Maria Feichtner and Bernd Sturmfels. Matroid polytopes, nested sets and Bergman fans. *Port. Math. (N.S.)*, 62(4):437–468, 2005.

[57] A. Grothendieck. On the de Rham cohomology of algebraic varieties. *Inst. Hautes Études Sci. Publ. Math.*, (29):95–103, 1966.

[58] Branko Grünbaum. *Convex Polytopes*, volume 221 of *Graduate Texts in Mathematics.* Springer-Verlag, New York, second edition, 2003. Prepared and with a preface by Volker Kaibel, Victor Klee, and Günter M. Ziegler.

[59] Victor Guillemin. Riemann-Roch for toric orbifolds. *J. Differential Geom.*, 45(1):53–73, 1997.

[60] Mark Haiman. Hilbert schemes, polygraphs and the Macdonald positivity conjecture. *J. Amer. Math. Soc.*, 14(4):941–1006 (electronic), 2001.

[61] Robin Hartshorne. *Algebraic Geometry.* Springer-Verlag, New York, 1977. Graduate Texts in Mathematics, No. 52.

[62] O. Holtz and A. Ron. Zonotopal algebra. arXiv:math. AC/0708.2632.

[63] Hiroki Horiuchi and Hiroaki Terao. The Poincaré series of the algebra of rational functions which are regular outside hyperplanes. *J. Algebra*, 266(1):169–179, 2003.

[64] Lars Hörmander. *Linear Partial Differential Operators.* Third revised printing. Die Grundlehren der mathematischen Wissenschaften, Band 116. Springer-Verlag New York Inc., New York, 1969.

[65] T. Huettemann. On a theorem of Brion. arXiv:math. CO/0607297.

[66] James E. Humphreys. *Linear Algebraic Groups.* Springer-Verlag, New York, 1975. Graduate Texts in Mathematics, No. 21.

[67] James E. Humphreys. *Reflection groups and Coxeter groups*, volume 29 of *Cambridge Studies in Advanced Mathematics.* Cambridge University Press, Cambridge, 1990.

[68] Lisa C. Jeffrey and Frances C. Kirwan. Localization for nonabelian group actions. *Topology*, 34(2):291–327, 1995.

[69] Rong Qing Jia. Translation-invariant subspaces and dual bases for box splines. *Chinese Ann. Math. Ser. A*, 11(6):733–743, 1990.

[70] Rong-Qing Jia and Qing-Tang Jiang. Approximation power of refinable vectors of functions. In *Wavelet analysis and applications (Guangzhou, 1999)*, volume 25 of *AMS/IP Stud. Adv. Math.*, pages 155–178. Amer. Math. Soc., Providence, RI, 2002.

[71] V. P. Khavin. *Commutative Harmonic Analysis. I*, volume 15 of *Encyclopaedia of Mathematical Sciences.* Springer-Verlag, Berlin, 1991. Translation by D. Khavinson and S. V. Kislyakov, Translation edited by V. P. Khavin and N. K. Nikol′skiĭ.

[72] Bertram Kostant. Lie algebra cohomology and the generalized Borel-Weil theorem. *Ann. of Math. (2)*, 74:329–387, 1961.

[73] G. I. Lehrer and Louis Solomon. On the action of the symmetric group on the cohomology of the complement of its reflecting hyperplanes. *J. Algebra*, 104(2):410–424, 1986.

[74] Eduard Looijenga. Cohomology of $\mathcal{M}_3$ and $\mathcal{M}_3^1$. In *Mapping Class Groups and Moduli Spaces of Riemann Surfaces (Göttingen, 1991/Seattle, WA, 1991)*, volume 150 of *Contemp. Math.*, pages 205–228. Amer. Math. Soc., Providence, RI, 1993.

[75] Tom Lyche, Carla Manni, and Paul Sablonnière. Quasi-interpolation projectors for box splines. *J. Comput. Appl. Math.*, 221(2):416–429, 2008.

[76] Tom Lyche and Larry L. Schumaker. Local spline approximation methods. *J. Approximation Theory*, 15(4):294–325, 1975.

[77] R. MacPherson and C. Procesi. Making conical compactifications wonderful. *Selecta Math. (N.S.)*, 4(1):125–139, 1998.

[78] Stéphane Mallat. A theory for multiresolution signal decomposition: The wavelet representation. *IEEE Trans. Pattern Anal. Mach. Intell.*, 11(7):674–693, 1989.

[79] D. W. Masser. A vanishing theorem for power series. *Invent. Math.*, 67(2):275–296, 1982.

[80] P. McMullen. Weights on polytopes. *Discrete Comput. Geom.*, 15(4):363–388, 1996.

[81] Peter McMullen. The polytope algebra. *Adv. Math.*, 78(1):76–130, 1989.

[82] Theodore S. Motzkin. *Selected papers*. Contemporary Mathematicians. Birkhäuser Boston, Mass., 1983. Edited and with an introduction by David Cantor, Basil Gordon, and Bruce Rothschild.

[83] Peter Orlik and Louis Solomon. Combinatorics and topology of complements of hyperplanes. *Invent. Math.*, 56(2):167–189, 1980.

[84] Peter Orlik and Louis Solomon. Coxeter arrangements. In *Singularities, Part 2 (Arcata, Calif., 1981)*, volume 40 of *Proc. Sympos. Pure Math.*, pages 269–291. Amer. Math. Soc., Providence, RI, 1983.

[85] Paolo Papi. A characterization of a special ordering in a root system. *Proc. Amer. Math. Soc.*, 120(3):661–665, 1994.

[86] Paul-Émile Paradan. Jump formulas in Hamiltonian Geometry. arXiv:math/0411306.

[87] Paul-Émile Paradan. Localization of the Riemann–Roch character. *J. Funct. Anal.*, 187:442–509, 2001.

[88] A. Postnikov. Permutaedra, associaedra and beyond. arXiv:math CO/0507163.

[89] Claudio Procesi. *Lie Ggroups*. Universitext. Springer, New York, 2007. An approach through invariants and representations.

[90] A. V. Pukhlikov and A. G. Khovanskiĭ. The Riemann-Roch theorem for integrals and sums of quasipolynomials on virtual polytopes. *Algebra i Analiz*, 4(4):188–216, 1992.

[91] R. Tyrrell Rockafellar. *Convex Analysis*. Princeton Mathematical Series, No. 28. Princeton University Press, Princeton, N.J., 1970.

[92] Amos Ron. Exponential box splines. *Constr. Approx.*, 4(4):357–378, 1988.

[93] Amos Ron. *Lecture Notes on Hyperplane Arrangements and Zonotopes*. University of Wisconsin, Madison, Wi., 2007.

[94] Gian-Carlo Rota. On the foundations of combinatorial theory. I. Theory of Möbius functions. *Z. Wahrscheinlichkeitstheorie und Verw. Gebiete*, 2:340–368 (1964), 1964.

[95] I. J. Schoenberg. *Cardinal Spline Interpolation.* Society for Industrial and Applied Mathematics, Philadelphia, Pa., 1973. Conference Board of the Mathematical Sciences Regional Conference Series in Applied Mathematics, No. 12.

[96] Larry L. Schumaker. *Spline Functions: Basic Theory.* John Wiley & Sons Inc., New York, 1981. Pure and Applied Mathematics, A Wiley-Interscience Publication.

[97] L. Schwartz. *Théorie des distributions. Tome I.* Actualités Sci. Ind., no. 1091 = Publ. Inst. Math. Univ. Strasbourg 9. Hermann & Cie., Paris, 1950.

[98] Laurent Schwartz. *Théorie des distributions. Tome II.* Actualités Sci. Ind., no. 1122 = Publ. Inst. Math. Univ. Strasbourg **10**. Hermann & Cie., Paris, 1951.

[99] Laurent Schwartz. *Théorie des distributions.* Publications de l'Institut de Mathématique de l'Université de Strasbourg, No. IX-X. Nouvelle édition, entiérement corrigée, refondue et augmentée. Hermann, Paris, 1966.

[100] P. D. Seymour. Decomposition of regular matroids. *J. Combin. Theory Ser. B*, 28(3):305–359, 1980.

[101] G. C. Shephard. Combinatorial properties of associated zonotopes. *Canad. J. Math.*, 26:302–321, 1974.

[102] T. A. Springer. *Invariant Theory.* Springer-Verlag, Berlin, 1977. Lecture Notes in Mathematics, Vol. 585.

[103] T. A. Springer. *Linear Algebraic Groups*, volume 9 of *Progress in Mathematics*. Birkhäuser Boston, Mass., 1981.

[104] Richard P. Stanley. *Combinatorics and commutative algebra*, volume 41 of *Progress in Mathematics*. Birkhäuser Boston Inc., Boston, MA, 1983.

[105] Gilbert Strang and George J. Fix. *An Analysis of the Finite Element Method.* Prentice-Hall Inc., Englewood Cliffs, N. J., 1973. Prentice-Hall Series in Automatic Computation.

[106] Gilbert Strang and George J. Fix. A Fourier analysis of the finite element variational method. In *Constructive Aspects of Functional Analysis*, pages 793–840. 1973.

[107] J. J. Sylvester. *The Collected Mathematical Papers of James Joseph Sylvester. Volume II (1854-1873). With two plates.* Cambridge: At the University Press. XVI u. 731 S. gr. Lex.-8°, 1908.

[108] András Szenes. Iterated residues and multiple Bernoulli polynomials. *Internat. Math. Res. Notices*, (18):937–956, 1998.

[109] András Szenes. Residue theorem for rational trigonometric sums and Verlinde's formula. *Duke Math. J.*, 118(2):189–227, 2003.

[110] András Szenes and Michèle Vergne. Residue formulae for vector partitions and Euler-MacLaurin sums. *Adv. in Appl. Math.*, 30(1-2):295–342, 2003. Formal power series and algebraic combinatorics (Scottsdale, AZ, 2001).

[111] Hiroaki Terao. Free arrangements of hyperplanes and unitary reflection groups. *Proc. Japan Acad. Ser. A Math. Sci.*, 56(8):389–392, 1980.

[112] Hiroaki Terao. Generalized exponents of a free arrangement of hyperplanes and Shepherd-Todd-Brieskorn formula. *Invent. Math.*, 63(1):159–179, 1981.

[113] Hiroaki Terao. Algebras generated by reciprocals of linear forms. *J. Algebra*, 250(2):549–558, 2002.

[114] W. T. Tutte. A contribution to the theory of chromatic polynomials. *Canadian J. Math.*, 6:80–91, 1954.

[115] W. T. Tutte. Lectures on matroids. *J. Res. Nat. Bur. Standards Sect. B*, 69B:1–47, 1965.

[116] Michèle Vergne. Residue formulae for Verlinde sums, and for number of integral points in convex rational polytopes. In *European Women in Mathematics (Malta, 2001)*, pages 225–285. World Sci. Publ., River Edge, NJ, 2003.

[117] Hassler Whitney. A logical expansion in mathematics. *Bull. AMS*, 38(8):572–579, 1932.

[118] Hassler Whitney. On the Abstract Properties of Linear Dependence. *Amer. J. Math.*, 57(3):509–533, 1935.

[119] Herbert S. Wilf. Which polynomials are chromatic? In *Colloquio Internazionale sulle Teorie Combinatorie (Roma, 1973), Tomo I*, pages 247–256. Atti dei Convegni Lincei, No. 17. Accad. Naz. Lincei, Rome, 1976.

[120] Edward Witten. Two-dimensional gauge theories revisited. *J. Geom. Phys.*, 9(4):303–368, 1992.

[121] Kôsaku Yosida. *Functional analysis*, volume 123 of *Grundlehren der Mathematischen Wissenschaften [Fundamental Principles of Mathematical Sciences]*. Springer-Verlag, Berlin, sixth edition, 1980.

[122] Thomas Zaslavsky. Facing up to arrangements: face-count formulas for partitions of space by hyperplanes. *Mem. Amer. Math. Soc.*, 1(154):vii+102, 1975.

[123] Günter M. Ziegler. *Lectures on Polytopes*, volume 152 of *Graduate Texts in Mathematics*. Springer-Verlag, New York, 1995.

Index